The Vital Vastness
—Volume One

The Living Cosmos Society Series—Volume One

The Vital Vastness —Volume One

Our Living Earth

Richard Michael Pasichnyk

The Vital Vastness—Volume One
Our Living Earth

iUniverse books may be ordered through booksellers or by contacting:

iUniverse
1663 Liberty Drive
Bloomington, IN 47403
www.iuniverse.com
1-800-Authors (1-800-288-4677)

ISBN: 978-0-5952-1078-7 (sc)
ISBN: 978-0-5958-2181-5 (e)

Print information available on the last page.

iUniverse rev. date: 07/17/2020

To all the living things in our living cosmos.
And to all those people who nurture, preserve and care for—love—life.

"Our task must be to free ourselves—by widening
our circle of compassion to embrace all living creatures and the whole
of nature and its beauty."

"The search for truth and knowledge is one of the highest
qualities of man."

"The life of the individual has meaning only insofar as it aids
in making the life of every living thing nobler and more beautiful.
Life is sacred, that is to say, it is the supreme value, to which all other
values are subordinate."

Albert Einstein

LIST OF ILLUSTRATIONS OF VOLUMES ONE AND TWO

Note: The first number of these illustrations is the Tome number, and the second is the number of that illustration in that tome. For example, Figure 1.2 is the second figure of Tome One.

Figures

Volume One

Plates

PREFACE

A recent surge of public interest in new ideas in the sciences is a refreshing new social phenomena. What lies at the heart of this is the extraordinary power to explain the world, along with a sort of mysticism of where we are, where we're going, and how we can make more sense of the world. The entire cosmos, from the minute subatomic particles to the grand scale of the Cosmos, can be consolidated into a single theoretical framework. Such a concept, a complete theory of the Cosmos and our place in it, cannot fail to be deeply inspiring.

Science has always been an ever-changing discipline and old ideas give way to the new. Yet, this path has always been a difficult one. Changing concepts that are long held to be true involves letting go of theories that have been defended for a long time, often over an entire lifetime for individuals and longer for a given disicpline as a whole. This is not an easy path, as new ideas are most often resisted. This is now a hot topic in the philosophy of science, where science is seen as a social problem because of its elitism and resistance to ideas outside of the conventionally recognized arenas.

Many famous scientists, such as Copernicus, Albert Einstein, Alfred Wegener, Charles Darwin, James Lovelock, Lynn Margolis, Halton Arp, and many more, have gone this route, been admonished and ridiculed, only to end up changing the face of science. The problem is even more acute, as ideas and trends in science can change overnight and new discoveries occur every day.

Many of the ideas presented in this work are established, others will, no doubt, remain controversial, some would even say speculative.

Notwithstanding, science thrives on speculation, controversy and conjecture. Scientific controversy tends to involve confrontation between theories, rather than between theory, and facts and observations. Scientific progress can still be made by keeping with the rules of logic and reasoning. I believe that most of the essential ideas embodied in this work are correct, and will become more widely accepted as time passes.

What follows, then, is a new look at the Universe and our—your—place in it.

ACKNOWLEDGMENTS

Some of the most inspiring and provoking thoughts have come from pioneering scientists. There is James Lovelock, who was brave enough to suggest that the Earth's biosphere was a single super organism he called Gaia. Lynn Margulis contributed much to the Gaia theory, and biology in general; especially noteworthy is her Endosymbiotic Theory, revealing the fact that larger organisms are communities of smaller organisms. Paul Davies is a physicist who wrote about new and complex theories in physics in a way that was easier to assimilate, and often embodied a touch of mysticism. Halton Arp brought forth his bold theories on quasars as ejection phenomena leading to the birth of new galaxies, and redshifts as intrinsic and evolving characteristics. Supporting his work, and bold enough to suggest theories in cosmology that are not mainstream Big Bang theories, but Steady State, is Geoffery Burbidge. Fred Hoyle has, it seems, always looked at the Universe as more life oriented.

Some of the great, free-thinking scientists are no longer with us, but their work was and is also inspiring. One of the great minds of the 20th Century, was Albert Einstein, whose theories will forever continue to change the face of science. Especially noteworthy is Hannes Alfven, a Nobel laureate for his contributions in plasma physics, whose theories will continue to make contributions in physics and cosmology. An "outsider," Alfred Wegener daringly brought forth the theory that the continents drifted on the Earth's surface, but was not himself a geologist. Nicolas Copernicus was even jailed to dare say that it was the Sun, not the Earth, that is at the center of our solar system.

There are also the great philosophers of science. Thomas S. Kuhn, an American historian of science, is noted for *The Structure of Scientific Revolutions* (1962), one of the most influential works of history and philosophy written in the 20th Century. Karl Popper, a philosopher of natural and social science who subscribed to anti-determinist metaphysics, and believed that knowledge evolves from the experience of the mind. David Hume, a philosopher, among other things, was known especially for his philosophical empiricism and skepticism.

All of these scientists and philosophers were a great inspiration, arousing a thirst for knowledge, much of which is presented in *The Vital Vastness*. They all taught me that new ideas will most often be negatively reacted to by people in the profession until the evidence is really looked at, and that determination is an absolutely essential ingredient for bringing forth new ideas.

I would also like to acknowledge Arizona State University for its great libraries that offered a wealth of knowledge. Also, I thank the Interlibrary Loan Department at ASU, who were able to locate hard-to-find books and articles, especially those old, out-of-print historical works of *In Defense of Nature—The History Nobody Told You About*.

CONTENTS

Tome Two
How Small is Small If One Principle
Connects it All?
The Microcosm and Reality

Chapter 15
Intimacy of a Different Sort: The Earth and the Life Upon it

Chapter 16
You Reap What You Sow

Tome Three
Prehistoric Recollections
Geological and Paleontological Investigations of Earth History with Special Reference to the Cretaceous/Tertiary Boundary

Selected References

INTRODUCTION

Did You Know That There's Another World Out There...

...and that the key to it is in *your* hands?

At some point in time each of us has kicked back in our favorite chair and contemplated the world around us. As we turned on the television or radio, picked up a newspaper or looked out the window, we have often seen a world that, in many ways, is growing in disorder and chaos. As hopeless, or helplessly victimized, any of us may have felt at that point, it was only because we had not envisioned just how much control was possible, even from that easy chair.

Science has brought about great understandings, and all things that claim to be true should be tested scientifically. After all, we are all susceptible to self-deception and bias. As a result, all conclusions must be critically examined if we are to arrive at the truth.

Once a hypothesis has been made, it is necessary to examine its premise by observing predictions based on that theory. Due to the fact that incomplete understandings can still lead to predictions, fantasy and illusion do not only exist in fiction writing, poetry and scripts. Some of the scientific theories we hold to be true at one time or another, even today, are just as imaginative. However, this is the basis of knowledge itself, which through the accumulation of experimental evidence, becomes more complete with time. Recent discoveries have disclosed that some so-called "truths" are not only unsupported in numerous ways by many of our present-day observations, but are even contradicted.

Throughout human history we humans have often originated theories and then looked to support them. In the process, facts were gathered and commented upon almost always in order to defend those preconceived ideas. Wherever we looked, our perception was tainted by those assumptions, and thereby, information was reported, observed and written in the perspective of what was expected to be.

In spite of it all, facts and observations were discovered that reached outside of expectations and theories. As a by-product, and rarely, if ever discussed, there were the accumulated facts that did not vindicate conventional theories, but do say something about how those assumptions need to be revised. An unfortunate product of the information explosion has been the backlash of overlooked facts that could not be applied to those theories. Meanwhile, ignored or as yet unexamined facts have amassed into waves which have washed upon the "shores of reality," bringing new insights to light.

Progress in science or scientific discovery depends on one's selection of what is considered pertinent data, and also one's instruction during learning. Notwithstanding, it is essential to probe into the possible weaknesses of theories and attempt to refute them. Most of the time, studies are selected to fit already existing beliefs, which gives short-term success, but according to the history of science, long-term failure. The problem is that when we are trained to see something in a particular way, our mindset often merely recovers what has been learned, and nothing fundamentally new will be uncovered. I have made a point not to let this happen by being as interdisciplinary as is humanly possible, probing into the weaknesses of present theories, searching for new interpretations of the empirical evidence, and have only two basic questions in mind: "Do the facts tell a different story than we are being told?" and "How important is life itself on Earth and in the Universe" This amounted to a two-decade journey that culminated with this book.

Moreover, scientific progress is not uniform, except in minor matters. Changing assumptions or theories, known as paradigms, involves so much trauma that it is resisted until the accepted paradigm has been shown to be unmistakably wrong. Then there comes a new paradigm with a swift transition in what could be called a "scientific revolution."[1] In many ways science is admittedly in trouble today. A scientific revolution is literally "at the door," so to speak.

One of the worst problems inhibiting scientific progress has been our mind-set that humans are merely observers of—detached from—the physical world. At times we have even claimed, as we often do today, that we were and are: "Victims of forces out of our control." Yet, this is assuredly not so. As we will discover, and have scientifically uncovered, life is an all-powerful stabilizing force, particularly if we envision ourselves as participators in the altruistic care of other living things (love), who in turn stabilize the physical world.

The whole experience has been as though we stood before a pond and tossed stone after stone into it. Then in making our observations we concluded that all ponds have ripples and stony bottoms. Through the entire process we made no real attempt to acknowledge our involvement in creating those observations to begin with. It is in this way, particularly, that we have led ourselves astray on the road to the discovery of genuine scientific truth, free from illusion. For we *are* participators in the physical world, as quantum mechanics have unquestionably revealed.

With the unfolding of this book you are about to embark on a journey into ancient and present mysteries, a wide scope of human experiences, and the far reaches of the Cosmos. What awaits your curiosity are not predetermined opinion, theory or expectations molding concepts, but factual insights that have lead to the discovery of a long-overlooked reality. Our story is the result of millions of observers, both past and

---

[1]   Kuhn, T.S. (1970) *The Structure of Scientific Revolutions*. Ill, Univ Chicago Press.

present, not a handful, not one discipline, but all the natural and physical sciences (Volumes One and Two), and world history and prehistory (*In Defense of Nature—The History Nobody Told You About*). Present theories and facts will not be discarded, but offered in a revised state so that our understanding encompasses most, if not all, of the observed facts. Due to the understanding that this knowledge is as far from opinion as is humanly possible, it is best not considered the work of one author. In fact, it is not even merely a book written by all people, but, instead, one composed by all of life itself.

"Speaking" in their own "language," every human and living thing has communicated something of that truth. Whether written in the remnants of long ago—fossils, artifacts, and ancient writings—or in the gathering of information from living beings and the physical world in the present, the story revealed demonstrates the existence of a new world, a new Universe. You have, or for that matter, each human being and living thing has, lent a contribution to comprehending our place— *your* place—in the "Grand Scheme" of things.

Nearly five decades ago one of the greatest scientists left this world and an array of theories behind him. Every day scientists perform experiments to examine that elite logic, though, like Einstein himself, they have yet to prove or disprove the existence of a unifying principle in the physical world (i.e., the Unified Field). This unifying principle, embodied within the general theory of relativity, holds that all things in the physical world, regardless of their differences in time, space, matter and energy, converge at certain points so that all are somehow connected. It involves a flowing of forces from the smallest and simplest of things to the largest and most complex. The adventure you embark upon with the parting of these pages is a long journey into another dimension that includes a new Earth and a new universe, and your awesome power to shape, control and perfect that world.

Whether you are a scientist, or a non-scientific person, this book is understandable and addresses the issue of what that unifying principle entails. Scientific information will be set forth in such a way that it is separated by parenthesis and in italics (*like this; only in scientific wording*), while along side is the more commonly understood descriptions. As a result, there is more written in this book than the non-scientific reader would need to read in order to complete the entire work. This book appears larger than it actually is, and can also be made even more brief if one so chooses to ignore these words in italics.

The arrangement just described was produced so that scientists can examine the understandings presented with their own language. Often scientific words do not have any truly definitive common synonym or phrase. Meanwhile, those who have no scientific background, or the scientist in another discipline, can read what is written on any subject. Facts from all the natural and physical sciences, and world prehistory and history have been compiled in this single work, yielding one of the most interdisciplinary works ever produced.

At a glance many of the topics may appear as though they do not concern the average person's life. However, they are relevant to our daily existence. Due to the fact that the unifying principle (i.e., the Unified Field) links all things, it is essential to know how each of us interacts with the Universe. As a result, completely reading this work will bring the best results into the world and your personal life. If you consider yourself someone who loves—hopefully all of us—then the entire book will enable you to better interact with the Cosmos, and make the world a better place.

Due to the fact that innovative insights are set forth here, some people may wonder whether a given assertion is factual. Well, rest assured, because it is substantiated by a supportive scientific and historically accurate bibliography (though, merely a portion of the support). Consequently, if any doubt exists then one should consult the references in the back of the book for confirmation. References are presented as numbers in superscript at the end of sentences or phrases. A person can then obtain the article(s) and/or book(s), and read about what might seem questionable.

Those statements that appear without references are based on previous discussions in the text, and if anything is questioned about them, then the index should be consulted to find where the topic is discussed. This work involves facts accumulated from more than 7,500 references (nearly 5,000 selected) and more than two decades of research, ranking it as one of the largest bibliographies ever produced in support of a single work. This approach was taken in order to eliminate, as far as is humanly possible, opinion and bias.

Each and every person's life will change as they come to know some of the hidden truth in the world, and the awesome power of what it is to be human in a physical world in which we *are* participators. This is not only a book that deals with an objective and subjective look at the world around us. Think of *The Vital Vastness* as a vehicle in which you are, so to speak, in the "driver's seat" and heading for a new world that literally can be best described as "Paradise." Have a magnificent ride!

TOME ONE

The Earth: Is It Alive?
An Analogy Between Geophysical Properties and Life: The Gaia Theory Extended

"It may be useful for practical purposes to consider
the Earth as if it were a living organism."
James Lovelock ("Geophysiology, The Science of Gaia"
Reviews in Geophysics 27:215, 1989)

CHAPTER 1

Views from the Ether
(Satellite Observations of Geodesy)

Throughout history the Earth has had many images in the minds of men and women. Surely, in one way or another, each of us has come across the fact that at one time it was believed the Earth was flat. The voyage of Christopher Columbus (1446-1506) began to change that misconception for many when he "discovered" America. Afterwards, the Earth was viewed as a perfect sphere blemished only by mountains and was thought to be the center of not only the Solar System, but of the entire Cosmos. Nicholas Copernicus (1473-1543) helped to correct that perception by demonstrating the Sun was really at the center, with the Earth as one of the orbiting planets. Outside of the Old World philosophers, from one corner of the globe to the other, the Earth has been conceived of as anything from a revered living entity and Mother to the "third stone from the Sun" whose purpose bears nothing but exploitation.

Clear perception has probably been the quality most lacking in humanity on its long arduous journey through time. Although we now know why the Sun appears to rise each day, much about the Earth and our relationship with it is still in an immature stage. Science, though, goes about its business of cataloging bits of information in the hope of resolving the abounding mysteries that still prevail. Some of the biggest riddles lie in the "whys" and "hows" about the Earth. This uncertainty has lead science to send space probes to the Moon and planets in an attempt to strike a few questions from the list.

With all the data-gathering drudgery that science demands, some aspects of the Earth's "whyness" have come into focus. Geologists and geophysicists can now explain most of the questions they once asked less than a decade ago. However, for each answerable question, a myriad of new ones (as yet unanswerable) has taken its place. And one cannot help but wonder just how absolute or permanent today's theoretical visions are.

Many of the so-called "truths" of the past have been superceded. The image of the Earth as a sphere or globe has given way to a more or less pear-shaped object (actually a tetrahedron). It wasn't until the latter half of the Twentieth Century with its physics, mathematics and artificial satellites that the true shape became evident. Likewise, and more recently, the web of life on this planet has been reevaluated. The physical and chemical conditions on the Earth have been and are actively made fit and comfortable by life's presence. This is in contrast to conventional "wisdom," which states that life adapted to planetary conditions as it and they evolved their separate ways. Now, some scientists consider the planet's surface and the life on it—the biosphere—as connected and mutually, even symbiotically, reinforcing to the point of being considered a single super-organism. Contemplating these two understandings immediately brings to mind a major question that seems to be the next logical step. It may insult the senses of those who have already come to the conclusions about this planet: Is the Earth itself alive?

This may sound like a strange question to some of us, especially those who have been educated to think of the Earth as a huge lifeless mass of metal and stone. However, it is a valid question and has recently been answered, in part, beginning with one scientist, who coined the name Gaia Hypothesis.[258] Gaia is an ancient Greek goddess of the Earth.

Furthermore, there are many observations to support an investigation into such an inquiry. Much of what brings an affirmative answer has been unexpected, and therefore, not remarked upon by many scientists. However, comments in support of the idea have been noted in scientific observations, and written in books and journals as something "interesting," "unusual" or "unknown." At the heart of the dilemma, which involves this restricted scientific mind-set, is the fact that those who do examine the Earth scientifically have not been trained to even begin to consider it as a living thing. As a result, any such consideration has been ignored, but it has not gone unobserved.

Meanwhile, analogy—the method of comparing similar functions in known living things—has always been the most basic way in which to identify new life forms.[337] Yet, never has this been attempted in our examination of the Earth until recently, and then only in part. This tome is merely a fundamental understanding of the Earth that is different from conventional theory, though not in conflict with scientific observations. Scientists have also recently realized that today's theories about the Earth are flawed with numerous observations contradicting the theoretical status quo.[6]

Later chapters, Volume Two, and *In Defense of Nature—The History Nobody Told You About* though quite different in subject matter, help to confirm this new insight about the Earth. As a result, new vistas open when other disciplines, such as prehistory, history, plate tectonics, weather, planetary phenomena, life's origin and evolution, and a host of other subjects are examined, lending unshakable support to a new Earth model. Even if you have already decided that the Earth cannot be alive, you will need to face how it is extremely life-like, how it is stabilized by and supports life, and how it is far more dynamic than we could have ever anticipated from our concept of a planet.

Therefore, proceed with this tome knowing that analogy is only a portion of this penetrating look at the planet we live on. Also to be kept in mind is the fact that scientific and historical observations bring further confirmation of that understanding, which also leads to the most fundamental unifying principle of the entire Universe.

It was physics, mathematics, and artificial satellites that revealed, in more precise detail, our current view of the Earth's shape. The rate of backward movement (regression) towards the west by an eastward moving satellite made it evident that a bulge exists at the equator. A comparison of circumferences show that the equator (40,075 km.; 24,902 mi.) is over 67 kilometers (42 miles) greater than from pole to pole (40,008 km.; 24,860 mi.).

This observed flattening makes a few characteristics about the Earth definite. One is that it has an elastic, deformable body that allows it to bulge at the equator. Another is that when calculating the ratio of flattening (1/298.25), it demonstrates that the Earth is not uniform in composition, and in fact, at least some portions of its interior are more dense than the crust. The centrifugal force of the Earth's rotation on its axis is not sufficient to cause the bulge, and it cannot be due to an earlier faster rotation, because the bulge shifts along with shifts in the Earth's magnetic field. With all the observations and calculations integrated, the Earth is then known to be as the labels indicate, a tetrahedron, "a upside-down, rounded pyramid" or "pear-shaped" object.

On these understandings rest some imposing questions and observations. We might pick up a rock or climb a mountain, thinking all the time that it's hard and lifeless. To the 5,876 quintillion (10^{18}) metric-ton Earth, it's as soft and elastic as our skin is to us. If we were reduced to a small enough size and handled a particle of our skin that was not reduced, it would also seem as hard and lifeless as a rock. It is merely a matter of perception, colored by perspective. Just looking at the Earth from outer space brings to mind that harmonious whole, perfect in structure and balance, the ultimate in order in contrast to what might be considered "chaos"; that understanding of what life should possess handed down to us from Hippocrates (467-357 BC).

CHAPTER 2

Enormous Cells
(Are the Organelles of the Cell an Analogue of the Habitat's Components?)

The Earth spins upon its axis with the grace and timing of a ballerina. To the East, the shining Sun shows itself as it seems to ascend a lake's rippling surface. The soil is cool and moist, but the first few warm rays begin to heat a cool clod where newly sprouted roots spread. There, drinking nutrient-rich water, is a youthful herb unfolding itself as it sheds its seed cover and breaks ground to feed on the life-giving light. Meanwhile, scurrying can be heard as a trail of bending plants in the low dense cover can be seen heading towards the sprout. Beginning to open its leaves now, the newcomer is face to face with a ground squirrel. A few sniffs and the shocked youth is cut clean by the sharp piercing teeth of the trekker.

Belly filled, the squirrel moves on, unobservant of what is flying over head. Piercing the cumulus, wings tucked in, the hawk begins diving at a rapid pace, closing in quickly. Strong claws and the force of descent stop the squirrel's journey, making him the hawk's next meal. Blades of grass bow before gusts of wind as powerful wings raise a contented hawk skyward. What remains is now visited by insects and bacteria, eventually breaking it down into soil nutrients. Beneath is a seed, patiently awaiting the next rain to begin its life when it will replace its lost relative. It seems to be something of justice that a new herb grows from its relative's slayer. Still, a mother squirrel and her young also await the sprouts that come after. And the hawk's eyes become clear and sharp as hunger again sets in.

2.1 Habitats

From one viewpoint, the contributing of one level of life to another might seem simple. Soil, water, and sunshine make plant life possible. Plant life gives life to animals, who in turn give life to other animals, and so on. All things eventually make their contribution to the soil. Remaining at this level, it is clear that plants are the beginning link in the chain of life. Taking this understanding a step back, it is realistic to declare the fundamental importance of those factors which govern the various types of plant growth. Looking at the interrelationships from this point, the idea can no longer be made simple, except by labeling it a habitat and assigning it a name accordingly.

A habitat might be considered the smallest unit on the Earth's surface which has the capability of maintaining a distinct type of life system within it. It consists of physical properties and characteristics that govern the growth of certain vegetation types, and thereby, the range of animals that depend on those plants for sustenance. Habitats always comprise a unique combination of climate, the physical features and altitude of land forms (*topography*), soil type, adjacent weathering rock (*parent rock*), underlying rock (*bed rock*), available water, and the type and frequency of waterways.

Habitats always have a system of life that is analytical from the perspective of certain classes or types of individuals, as well. These labels reflect an organism's role in energy transfer within the food chain. A food chain consists of those various members of a habitat that rely on each other as food sources. Minerals are the basic substances responsible for the growth of primary producers. Plants are primary producers (also thermosynthetic and photosynthetic bacteria) and are designated as such to indicate their relationship to other life; that is, they produce the first and most basic food source.

Those animals which consume plants are called primary consumers and are mainly herbivores or plant-eaters. Individuals called secondary consumers use primary consumers for their food. In this category we find mostly carnivores or meat-eaters. Omnivores, those that eat a variety of plants and animals, fall into both the primary and secondary consumer categories. For example, humans are omnivores, if observed collectively and the question of proper diet is not discussed. When corn (primary producer) is eaten by a man, he is then a primary consumer. But, if a chicken (primary consumer) were to eat the corn (primary producer), and then that man eats the chicken, that man is then categorized as a secondary consumer. Scavengers are another member of a complete system, eating what other animals leave uneaten, or are not simple enough for decomposers or plants to use. Decomposers break down what remains into minerals or other substances useful to plants. They utilize what other organisms leave uneaten, the waste of animals and dead organic matter, thereby contributing to the soil. These types of life forms interact with each other and with the environment in the maintenance of the habitat.

If it were to be expressed metaphorically, it might go something like this. Primary producers are the "daily bread" of primary consumers, who are the "meat" of secondary consumers. Scavengers eat the "crumbs" and "scraps" left over, and decomposers clean up the "mess." A neat, self-reliant banquet.

2.2 Enormous Cells

A cell is one of the fundamental structural units of which all living things are composed. Likewise, the habitat is one of the fundamental structural units of the Earth's life zone, the biosphere. Microscopic units that maintain a cell's life functions are comparable to the interrelationships of life and the environment within the habitat. For this reason, habitats can be understood as a likeness or analog of the cells of what we normally call life. Could these analogues of enormous cells constitute one of the outer layers of an enormous organism? This analogy may seem somewhat tenuous to some, but the remainder of this book confirms that this analogy is real.

Biologists studying the cell have observed a number of parts called organelles, performing the different life-sustaining activities of the unit. In fact, organelles are now being understood as bacteria and algae that evolved to be incorporated into the cell. They are populations composed of vestiges of organisms that interact within the boundaries of the cell membrane. From single-celled organisms to complex organisms, plants and animals are communities of smaller organisms (*as stated in the Endosymbiotic Theory*).

Few have seen that the interrelationships between organelles in the cell reflect the interrelationships of plants, animals, and the environment. Perhaps the reason is because most scientists are specialists in a given field, or because none wants to venture a theory based on an analogy, possibly losing the respect of colleagues. Still, one of the first and continuing methods of distinguishing new life forms is through analogy. Let's examine that analogy.

The most primary organelle of the cell is the mitochondria. Liberating energy for the cell, it is active in the transport of water and ions, and forms oxidized phosphates (*ATP*). Similarly, soil and mineralized waters (oceans, lakes, etc.) transport water, ions, and phosphates to plants, which liberate energy for all other members of the habitat. The mitochondria regulates the electrochemical gradient of the cell, especially with regard to positively charged protons. The soil is similar by interacting more with positive ions (i.e., the Earth's surface is negatively charged). Furthermore, the mitochondria has cytochrome, which is composed of iron-sulfur proteins. Soils, in fact nearly all rocks on the Earth's surface, are abundant in iron, while sulfur is a major ingredient of the healthier organic soils. In analogy, it might be said that the mitochondria are the soil of the cell, or inversely, *soils and mineralized waters are the mitochondria of the habitat.*

Another constituent of the cell is called the endoplasmic reticulum. Absorbing ions and phosphates, it produces protein and carbohydrates. Its other functions include bringing energy to all parts of the cell and eliminating waste. Plants, likewise, absorb mineral ions for growth, utilizing them in their structures as protein and carbohydrates. These plants feed some animals who in turn are food for other animals. Plants, either directly or indirectly, feed or transfer energy to all members of the habitat. Waste products of the decomposer are so mineralized that they are of benefit only to plants. Therefore, plants also remove the final waste product of the habitat. In conclusion, *primary producers (or plants) are the endoplasmic reticulum of the habitat.*

Another member of the cell in contact with the point of protein synthesis on the primary producer (endoplasmic reticulum) is the golgi apparatus. It produces and concentrates more complex compounds by utilizing these proteins and carbohydrates from the primary producer (endoplasmic reticulum). Similarly, primary consumers of the habitat, plant-eaters, feed on primary producers, utilizing their proteins and carbohydrates for growth and metabolism. Therefore, concentration of these proteins and carbohydrates takes place in the cellular makeup and energy (metabolism) of the plant-eater (primary consumer). *Mutually complimentary in function are primary consumers or plant-eaters, and the golgi apparatus of the cell.*

Produced from the proteins and carbohydrates of the golgi apparatus (primary consumer) are lysosomes. These mobile members of the cell consume invaders, such as microorganisms (disease bacteria, etc.) and viruses. Scavenging particles which have become unusable to the cell, in order to maintain energy needs and repair cell structure, is another function. All secondary consumers have one thing in common: they always prey upon weak and sick animals. Predators, such as the lion, help to eliminate disease, and by keeping herds of animals moving, they prevent overgrazing, and death in winter, thereby regulating energy transfer within the habitat. They maintain their own energy while mainly eliminating diseased or weak animals who are easier to catch. Just as lysosomes are produced by the golgi apparatus, secondary consumers are made up of the substance of their food (primary consumers). Scavengers consume unused material in the habitat, and upon passing it through their systems, make it usable to decomposers. As this material is broken down further by these decomposers, minerals are released into the soil for use by plants, and the transfer of material for maintaining the habitat (Earth cell) takes place. *The functions of the secondary consumers (predators), scavengers, and decomposers reflect those of the different types of lysosomes.*

Superior to all other cell components is the nucleus. It is indispensable for regulating metabolism, growth and reproduction. Within its "memory" (chromosomes) is the information necessary for replacing cell components and for creating other cells. This is also the exalted position we humans occupy within the habitat. Unlike other creatures, humans possess the capacity and potential to fully understand and interact with all other life. *Humans, as the nucleus can and should regulate the metabolism, growth, and reproduction of all habitats* (though we have not).[2]

In the following chapters is a look at the microcosm (the molecular and quantum world), in Volume Two, the macrocosm (the planets, Sun and Universe as a whole), and the prehistory and history of *In Defense of Nature—The History Nobody Told You About* will demonstrate this relationship of life and humans with the Earth. Also, shown is that

---

[2]* Humans are also the collective mind of the Earth, as will be discussed in the **Conclusions** in **Volume Two.**

altruistically working with the biosphere is not only our most fulfilling endeavor, but creates very different physical and social worlds. Our most enlightened perspective involves nurturing *all* life, and in so doing we will bring about the perfection of both the microcosm and macrocosm, and their linkages.

As demonstrated in quantum mechanics, we are not just observers, but participators in the physical world. Perceiving the interrelationships within the habitat and nurturing them will also lead to a peaceful and tranquil Earth, and bring about a world the likes of which we have *never* experienced, and in which some of today's scientific "truths" appear to be nothing more than illusion. All of this will become clear with the observations and realizations presented here, in this book.

CHAPTER 3

Living Waters
(Are Sea Water and the Hydrosphere Analogous to Blood and a Circulatory System?)

All life needs water. Not even a single cell can function without it. Molecules necessary for biochemical reactions are suspended in it, or it is an integral part of these reactions. No life on the Earth's surface (and probably anywhere else) can exist without water.

We are not sure how the oceans came into being. One theory states that the oceans formed out of the Earth when it was first fashioned into a planet. However, no evidence exists prior to the first fossils, the first evidence of life. It has been suspected that the oceans once covered the entire Earth: no continents, no shallow water. If so, our current theories on the origin of life must be revised. Its average depth is 3,200 meters (approximately 2 miles), its volume is 1.2 billion cubic-kilometers (about 300 million cubic miles) and it weighs 1.3 quintillion tons (metric). Still, that tremendously vast enigma has *never* been unsuitable for life!

3.1 Ocean Water

Blood and ocean water have much in common. Both are liquids with dissolved minerals (*electrolyte solutions*), which make them excellent for electrical activity and chemical reaction. They share the same major positively charged particles (*cations*), which are sodium, potassium, calcium and magnesium. Dissolved gases in both are oxygen and carbon dioxide. Blood plasma—blood minus the red and white cells—has a specific gravity of 1.026 compared to sea water's 1.03, minus the ocean life. This shows they both have a similar capacity for holding solids suspended in solution. An animal's body and the biosphere's largest single constituent is water. Water absorbs excess heat without much temperature change, and therefore, helps to maintain an ideal temperature (*high specific heat*). Average pH, which designates acidity, is not all that different, either. The average pH of blood is 7.4 (7.35-7.45), and that of sea water is 7.6 (7.0-8.2).[411] Both are ideal for the maintenance of life, and anyone who doubts this need only observe any one-celled creature in the sea whose blood supply is literally the ocean.

3.2 Circulatory Systems

Creatures of the higher orders typically have circulatory systems. The water of life must be pumped to cells for renewal and to supply them with nutrients for energy. Because the Earth has cell-like structures, it could be suspected that something like a circulatory system would also exist, if indeed it is a living entity. The hydrologic, or water, cycle and the hydrosphere, or water-zone of the Earth, conform to this analogy.

When looking at the transport of water over the globe, it is easy to see grandiosity, but less obvious is its minute intricacy. If one begins in the largest component of the hydrosphere, the ocean, some of this intricacy becomes apparent. Bubbles are created mostly by wind action against the water's surface, causing breaking waves, and thereby, trapping gases from the atmosphere. Also, plants and animals in the ocean release gases, and rain aerates the ocean as it breaks through the water's sur-face-tension.

Being lighter than water, gas rises to the surface in the form of a bub-ble. It has been observed that the bursting of a single bubble sends three droplets into the air. Then a wind takes hold of them or they evaporate into the atmosphere, causing them to rise where they combine with atmospheric elements to ultimately form clouds. It may sound trivial, but millions of bubbles are involved in making the whitecap of a *single* breaking wave, and three times as many droplets are sent skyward. If you have ever stood at a shore, you have felt the mist of this continuing process. This is no mundane realization when the vastness of the oceans covering 71% of the Earth's surface (361,700,000 sq. km.; 139,616,200 sq. mi.) is considered; nor is this limited to the oceans.

Also, the gaseous envelope of the Earth, its atmosphere, plays an extremely important role in the transport of water. Infrared radiation emitted by the Sun is shielded partly by the ozone layer of the atmos-phere with its ability to absorb certain wavelengths of heat. Similar con-ditions exist for carbon dioxide and water vapor, which absorb different wavelengths of the infrared spectrum. Each of these gases also warms the Earth by preventing infrared radiation, emitted by the Earth's sur-face, from escaping into space. Providing these gases remain at natural levels, they prevent the Earth from becoming overheated or too cold to sustain life. The Greenhouse Effect is what science has labeled this effect, because the same situation exists in a greenhouse, whose win-dows can be compared with these gaseous layers. The balancing of con-tained heat sets an ideal situation for life and the evaporation of water.

A myriad of mechanisms exist in Nature for turning water into water vapor. For example, as water seeps down to certain depths in the soil or porous rock, it becomes heated due to the increased temperature and pressure, thereby rising up again through evaporation. This process is known as percolation. Its results ranges anywhere from the morning dew and fogs to an invisible vapor at other times.

Even celestial bodies get into the act by pulling on water through gravitational attraction. This brings about a greater separation of water particles, making it easier for heat and gases to combine with it, and draws up more vapor from the soil, combining with the percolation process. The greatest effects are exhibited by the Moon, which brings about more plentiful dew and high tides when closest to the Earth.

Comparatively simple is the contraction of muscles pumping life-sustaining fluids to all parts of the body. This process is, however, analogous to the operation of the more complex hydrosphere. Collectively, the hydrosphere consists of the various parts of the atmosphere and Earth that involve the transport of water.

A heart and the continual "pumping" up and down of water are very much alike. The right ventricle of the heart pumps blood through the pulmonary artery to the lungs to obtain oxygen, and then enters the left ventricle through the pulmonary vein. Similarly, ocean water is "pumped" up by evaporation, bursting bubbles and other mechanisms, rising through the atmosphere mixing with oxygen and other gases, and then, through condensation, transforms into clouds. From the left ventricle, blood passes through the aorta to arteries, which eventually bring blood to all parts of the body. Clouds rain, bringing water to all parts of the globe. After the body has received blood, it is sent on its way back through veins, and finally, enters the inferior and superior vena cava, which empties into the right ventricle, and the series of events starts all over. After the Earth's surface receives water, it is sent on its way back through waterways, and finally, enters the major rivers, which empty into the ocean, and the series of events begins again. The grand and complicated hydrosphere can be considered analogous to a giant circulatory system.

To illustrate this more clearly, let's review the two, event for event. The ocean (the right ventricle) pumps water (blood) by way of evaporation, bursting bubbles and other means (the pulmonary artery) through the atmosphere (the lungs), and into clouds (the left ventricle) by way of condensation. Precipitation (blood), emptying everywhere (into the aorta and its branching arteries), then collects into waterways (veins) that flow into the major rivers (the inferior and superior vena cava), and ends up back in the ocean (the right ventricle). Complimentary functions exist in both systems.

Visually, as well as functionally, the tributaries (brooks, streams, etc.) of major rivers appear almost identical to capillaries connecting the arterial and venous systems. Anyone who has ever been in a cavern or cave has a good understanding of how a waterway can resemble a giant vein. Caverns and some caves are ancient underground waterways, which through Earth movements, have been lifted above the flow of groundwater. Evidence exists that many waterways should be below ground, but because of a restless Earth are not, and would then bear likeness to veins even more so than they do presently.

CHAPTER 4

Internal Balance
(Homeostatic Characteristics of the Biosphere)

Homeostasis is a word used primarily to describe equilibrium within an organism. It constitutes the regulation and adjustment of vital functions that maintain a steady state of the blood, tissues and body functions. The objective of this regulation is to sustain internal constancy. Through higher levels of control, a dynamic equilibrium is achieved independent of the fluctuating environment. This control takes the form of reactions to compensate for imbalance, and mechanisms that maintain checks and balances. With the exception that this type of process exists in some man-made mechanical regulatory devices (which are made by life), whose operation is rigid and not so self-sustained as biological types (which are more flexible and adaptable), it is found nowhere else, but in living things.

4.1 Ocean Salinity

The Earth, however, displays this capability, for one, with the salinity of the oceans. Salt enters the oceans at the rate of 540 megatons each year, yet the average salinity is 3.4%. And it has been monitored long enough to make it obvious that it is fairly constant. This figure is only of the annual runoff of mineralized water from the continents, and does not include what salt is produced by seafloor spreading. Both calculated together would make the oceans no more than 60 million years old, if no means for removing it existed. However, geology with radiometric dating says the oceans are at least 3.5 billion years old (the oldest fossils). This inconsistency indicates that there is a means for establishing the equilibrium, a homeostatic mechanism.

Oceanographers have long known that a mechanism for removing salt from the oceans as fast as it is added is needed to explain present salinity. If chemical equilibrium was considered the figures would be something like 63% water, 35% salt, and 1.7% sodium nitrate, but instead, it is 96%, 3.4% and a trace, respectively. Inorganic, nonliving mechanisms that have been proposed fail to establish the answer, and it remains one of the major unsolved mysteries of chemical oceanography.[258]

To top it off, fossil evidence indicates it has been fairly constant since life began. Also, geologic evidence shows that the salt content of the sea has not changed very much since the birth of the oceans and the origin of life. The internal and external environment of a living cell must never exceed 6% salinity for even a few seconds or it will literally fall to pieces. The oceans have never been inhospitably salty—Why? The answer can only be that some homeostatic mechanism exists for maintaining ocean salinity.

4.2 <u>Temperature</u>

The Earth's (mean) surface temperature has also been very cordial to life throughout its history. Surface temperature has never varied more than a few degrees, and it has never been too hot or too cold for life to survive. According to a typical life history of a star, the energy output of the Sun would have been 30% less at its earliest stages. This means that the Earth's climate would have been in a frozen state, but fossil evidence and life's persistence show this never happened. In contrast, if the Earth

were a solid inanimate object, its surface temperature would have fluctuated with the Sun.

In a certain sense, the Earth as a whole sweats. A major portion of the
Earth's surface is covered with water. In the equatorial regions the
oceans evaporate due to more solar irradiation. The vapor produced
does not remain in the tropics, but drifts away from the equator into
cooler regions. There it condenses into precipitation, allowing the
energy spent in evaporation to be released when the vapor condenses.
As a result, the colder regions of the Earth are warmed and the warmer
regions are cooled.

If water did not have such immense heat of vaporization and the
ability to absorb great amounts of energy, while undergoing only mild
increases in temperature, the Earth's climate would be very harsh. The
tropics would be a desert wasteland, the mid-latitudes would be frozen
tundra, and the poles would be devastatingly frigid. Nighttime would
be freezing, while daytime would be scorching.

In addition, the Sun's thermal radiation bombarding the Earth's surface amounts to six thousand, million, million, million (sextillion) calo-
ries annually, but are readmitted into the atmosphere. The outcome is a
constant average temperature of 8° Celsius (46.4° F) on the Earth's surface. Again, the Earth exhibits an organism-like homeostatic characteristic: a constant and fairly uniform "body" temperature.

4.3 Atmospheric Composition

The strange composition of our atmosphere is also optimum for life.
Had it been established by chemical equilibrium alone its composition
would be close to 99% carbon dioxide and 1% argon, but instead it is
0.3% carbon dioxide, 78% nitrogen, 21% oxygen, and 1% argon. For
each molecule of carbon dioxide removed in buried plant residue, one
molecule of oxygen is left behind. Were it not for this process, oxygen
would be steadily withdrawn from the atmosphere through reactions
with decaying matter. If the present level of oxygen were to increase by

just 4% so that it was 25% of atmospheric composition, lightning or an explosion would cause a raging fire, a conflagration, that would literally destroy all vegetation.

Without methane produced by (*anaerobic*) microorganisms, oxygen would increase to that concentration. A counterbalance to methane is nitrous oxide, which is (*catalytically*) destructive to ozone (3 oxygen molecules combined; 0^3). And the ozone layer protects the Earth from receiving too much ultraviolet radiation. How is this balance maintained? The only possible way is that all life, existing as a single entity, balances these gases by sensing too much or too little ultraviolet is getting through the ozone layer.[258]

According to our present understanding of the Earth, helium should exist in quantities that exceed its present level by at least one thousand times. That is, the present levels of helium would require less than one thousandth of that of the Earth's supposed age. A scientist examining the sources of helium and its output says that it would have taken a "catastrophic event" to bring about the present level.[31] Of course, this is not the only interpretation, but again it is evidence of another homeostatic mechanism on Earth. In addition, more recently it was discovered that some areas in the Pacific Ocean show excess quantities of helium, indicating another primary input of helium.[100] This only makes the situation even more difficult to explain without considering the existence of a homeostatic mechanism of a living Earth.

4.4 Vital Organs

Lakes, too, are natural control systems. They are maintained at a fairly constant level regardless of the flow of the river emptying into them. The currents in a lake are more varied in direction and less swift than those in the river. Larger particles suspended by the high velocity of the river lose momentum and settle at the lake's bottom. When the river empties into a lake, its surface area is increased. As a result, it is affected more by sweeping winds and exposed to more heat (*infrared radiation*), increasing evaporation. Through varied currents and evaporation, the lake increases minerals which are more easily utilized by aquatic plants and are more chemically active.

Lakes also regulate nitrogen more efficiently, because denitrifying bacteria dwell in mud, absent of oxygen (*anoxic mud*), at their bottoms. These bacteria use the oxygen bound in animal waste and decaying matter (in this mud) for maintaining their own existence. By doing this, they release new compounds, which are easily taken up by roots and utilized by plants. Nitrogen gas is released into the atmosphere, which is dissolved by rain, bringing an important nutrient to plants outside the lake.

Lakes also undergo what has been called "succession." This term describes the process where by a lake eventually becomes filled in by sediments. Then brushy growth, and finally, tall trees and large grazing animals follow. In the process, lakes bring about the renewal and maintenance of habitats (Earth-cells). All in all, life is enhanced by the effects of lakes.

In many ways this reflects the function of one of the vital homeostatic organs of an animal. Kidneys regulate fluid volume, and adjust water quality and quantity to chemical content. More useful and chemically active minerals needed for bodily functions are increased. Harmful nitrogen compounds are eliminated from the system. Kidneys regulate the volume, acidity and composition of body fluids by filtering the blood and excreting waste products. Those biological functions that bring about the renewal and maintenance of the cells of a given living organism are partly regulated by the kidneys. Lakes and kidneys clearly operate toward the same objective, which is purifying a life-giving substance (again, a parallel between the hydrologic cycle and a circulatory system exists).

There are also the element cycles. Each of these cycles hoard nutrients in one form or another. Carbon, for example, is stored in limestone after becoming the calcium carbonate of calcareous shells of shellfish or reefs. It is also reserved in coal deposits, buried plant residue, soil, the hydrosphere and the atmosphere before being put to use in the makeup of plant and animal life. The carbon cycle is one of the more active and well-known cycles, along with nitrogen, phosphorous and sulfur. These cycles often exist by and always for life.

Throughout human history, these cycles have often become imbalanced for a period of time. The imbalances were always caused by humanity, either directly or indirectly. For example, phosphorous (from soaps) was dumped into waterways, which caused an overabundance of plant growth. Eventually the waterways were cleansed as the plants died and the toxic amounts were stored once again. Another example involved the overabundance of metals released by mining, which were absorbed by organisms who, on death, removed the toxic amounts. In all cases, the element was stored once more in the environment, as originally, not causing any threat to life.

Again, these element cycles reflect another vital homeostatic organ of an animal: the liver. The liver stores the nutrients: vitamins A, D, E and F, and other fatty acids. It is essential for the detoxification of toxic substances, which it eliminates through the intestines or in fatty tissue. The liver is indispensable for metabolism of all body functions, just like the elements are essential to all life on Earth.

Many other examples of homeostasis exist. However, many are too intricate and time consuming to be dealt with here. Others will become apparent as this book unfolds. One thing must be said here, though. There must be a set of interlocking mechanisms on Earth that regulate this homeostasis as occurs in other organisms. Life is an integrated whole that maintains and supports *all* life on this living Earth.

CHAPTER 5

A Huge Nervous System
(Telluric Currents, Geoelectricity, Geomagnetobiology and Superconductivity—The Unity of Electrostatic Time-Varying Characteristics of the Biosphere and the Earth's Electrical Environment)

All living things have a nervous system. Various complexities in nervous systems exist, ranging from the simplest forms in single-celled plants to highly complex ones in mammals, such as humans, dolphins, and whales. The web of life on Earth both creates and is affected by electrical currents on the Earth's surface and in the atmosphere. Termed telluric currents, geoelectricity and terrestrial electricity, many things contribute to this overall network. For the sake of simplifying the terminology these will be referred to as earth-currents in the following discussion.

5.1 <u>Generation</u> <u>and</u> <u>Flow</u> <u>of</u> <u>Electricity</u>

The soil with its electrically conducting gases, metals, semiconducting mineral crystals, water-soaked organic matter, and electrolytes offers a good medium for maintaining and producing electrical currents.[99,302,303,422,423] Water transports and renews these components and is an excellent electrical conductor itself. Pores, spaces, and other voids in the main surface rocks of the Earth (*sedimentary, and fractured crystalline and metamorphic rocks*) contain relatively large amounts of water, making them moderately good conductors of electricity. Radioactive decay of elements and radioactive gases in the air and soil produce charged particles (ions).[175,199, 222,394] Moving water and breaking water-films, such as waterfalls, rain and breaking waves, also produce charges.[52,61,62,139,142,144,191,198,226,256,272,306] Studies conducted on sea water disclose the fact that the entire ocean has a high-frequency conductivity.[61,136,139,144] These charges are emitted into the ground and water, or within the first few meters (several feet) of the atmosphere. Furthermore, the earth-currents are greatly increased by electrically grounded tall objects, such as vegetation and animals.[175,199,394] Thereby, the charged particles (ions) are draw to the ground (*planetary boundary layer*) and high concentrations (*10 ion pairs cm^{-3} sec^{-1}*) increase the intensity of earth-currents (*electrode effect*).[199] As a result, static electricity flows parallel with the ground (*orthogonal quasi-static electric field*).[394] These and other influences create a pathway for generating electricity, which travels to and along surface layers of the Earth. All things considered, the Earth's surface is a good conductor, and as a result, charges are observed to be distributed worldwide in only a short time.[175,199,205,394]

Interrelationships between life and the environment are essential for maintaining chemical activity, and producing electrical currents. Plants, with their chemically active roots and root-hairs, interact with the soil, aiding in the production of electric currents. These currents eventually reach water-saturated rock surfaces, establishing a basis for migrating electricity. Roots, particularly those of trees (especially arid and desert species), can also create gaps in hardened soil layers and underlying rock. The gaps would then serve as reservoirs for water and organic matter, establishing a pathway for the movement and production of electrical currents. Furthermore, the respiration of both plants and animals releases electrically active molecules (ions) into the environment.[364]

In contrast, the human-made environment often does not produce electrically active ions due to pollution or artificial currents, especially in urban-industrial areas.[140,175,199,394] Moreover, the soil is largely affected by what is called the state of hydration, where water increases conductivity and some human-made substances decrease conductivity (*e.g., clathrate hydrates*).[239] This will be discussed in Tome Two.

Animals and tall vegetation develop free charges from the environment and the atmosphere, and these charges travel to the ground. We humans have experienced something similar to this when static electricity builds up and a spark is discharged by touching a metal object, such as a doorknob. Other animals and plants, however, have no insulating shoes, and the charge is continually grounded. Animals which are able to move from one habitat to another, as in migration, exhibit one of the more active transports of electricity from one Earth cell (habitat) to another. All things being interrelated, the web of life works its way from near the tops of the highest mountains to the depths of the oceans, taking with it electrical, as well as, magnetic, energy.

Due to the fact that the importance of life in producing the earth-currents has nearly been overlooked, it has led scientists to make statements like these: "The simple laws of electromagnetic induction do not fully explain the cause of geoelectric and geomagnetic activity."[332] "Either the current circuit is in the horizontal plane or the currents are not the result of the induced [electromagnetic field]."[416] Those simple laws do not include life's influence (*a bioelectrostatic-generator, and biological superconductivity, conductivity and semiconductivity*), which generates a current horizontally across the Earth's surface.

Undoubtedly, an abundance of life is extremely important and elementary for the production of strong electrical currents. The organized network of all nerve cells in any organism branches out from a central system, and research in terrestrial electricity formulates a worldwide system of electrical currents or eddies (see Figure 1.1). These currents are geometrically arranged, centering along the 30° to 40° latitudes and both poles (four eddies around each pole), focusing on a number of phenomena previously thought to be unrelated.

Observations of these electric currents have revealed some very interesting results. Plant life needs light to function, and its fundamental importance in generating electricity is noted by currents of greater strength on the daytime side of the Earth. Contributing to this daytime effect is the fact that there is more daytime animal life, producing a greater supply of electricity for the system than does nocturnal life. Also important in this regard is the photoelectric effect, which is basically (electromagnetic) radiation, mostly from the Sun, striking rock and releasing subatomic particles, thereby producing an electric current. This same process is at work in the solar energy device known as the photovoltaic cell. These currents are fixed in relation to the Sun, shift with season, vary with sunspot number, and display a daily lunar variation.[111]

Frequently the geomagnetic field is altered by changes in these currents, and these currents fluctuate inversely with changes in the magnetic field, demonstrating a definite mutual relationship.[53,141,157-160,203,327,422,423] Though we often claim that the Earth's magnetic field effects the earth-currents, which in turn effect life, it is also the reverse.[97,123,126,127,310] All observations show the mutually reinforcing nature of life, the electric flow and the geomagnetic field (GMF); this will be demonstrated further as this book unfolds.

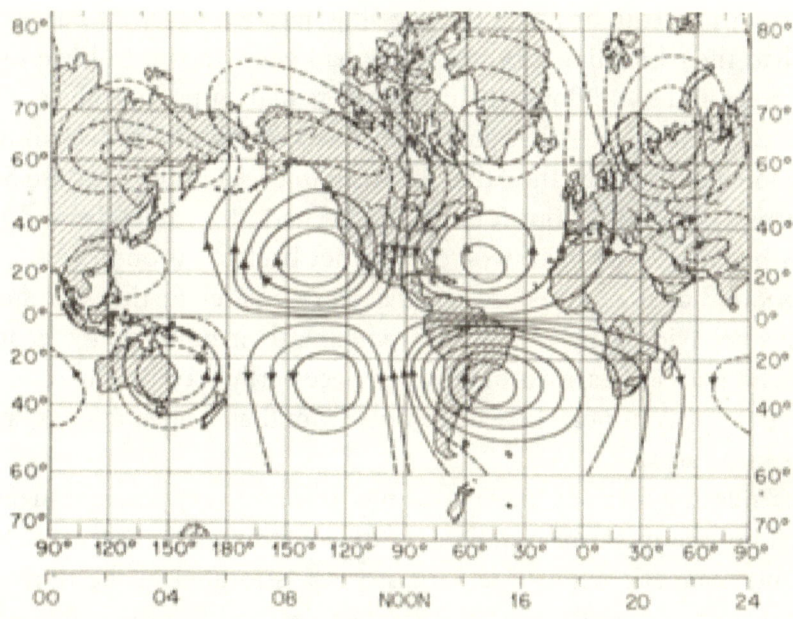

Figure 1.1. Telluric Currents. Earth-currents (*telluric currents*) are shown here in a pattern derived from observatories' data. The basic discernment is that the currents spiral into the Field areas. Summarizing completely accurate maps of these currents' patterns is a somewhat unresolved problem.[245] One of the basic reasons for this problem is that few stations exist for the observation of these currents, and solar linkages, and particularly, human activities constantly offset the currents' patterns. All of the Fields are clearly represented with the exception of the Mediterranean and South African Fields, which are only hinted at, because there were no observatories in Africa.[by permission of the *Journal of the Washington Academy of Sciences*; from reference 245].

5.2 Electrical Sensitivity of Life

One of several mysteries connected with one of the areas these electric currents flow into, and one that has intrigued ancient as well as modern observers, concerns the spawning of eels. Aristotle (384-322

BC) was the first to bring up the puzzling question of where European Eels had their breeding grounds. They had been known to leave their waterways and swim down the major rivers that empty into the ocean. It was 2,500 years later that they were observed to unite and swim in a great shoal clear across the Atlantic Ocean to breed in the Sargasso Sea, a journey that took about four months. There the adults die and the newly born eels start their two-year journey back to Europe.[279,375]

No one has ever proposed a satisfactory answer to this seemingly strange behavior. Why would eels, reducing their numbers by feeding gulls and packs of carnivorous fish, unnecessarily traverse the Atlantic to breed? The American Eels also empty from their waterways to spawn in the same area.

Considering some of the facts brings at least a very possible suggestion, if not the actual answer. Eels are known to be very sensitive to electric and magnetic fields. The Sargasso Sea is located in the midst of possibly one of the strongest electrical eddies, and is also an area with an unusual magnetic pattern or anomaly (east of the southeastern United States).

Eels are the only fish that have hemoglobin, the iron-containing pigment of red blood cells, and not even the carcasses of the adults are ever seen again.[279,375] In the sediment of the depths of the Sargasso Sea there are sulfate-reducing bacteria that take putrefying matter, and precipitate iron and other heavy, electrically conducting metals that do not dissolve in water.[43,63,183] These bacteria also orient themselves to the Earth's magnetic field, even at death. Could these eels unknowingly be making their sojourn across the Atlantic so that these bacteria can precipitate the iron in their hemoglobin and make the sediment a better electrical conductor?

When the geographic distribution of eels is examined (see Figure 1.2), they display definite relationships to the areas that electrical currents flow into. All of these areas contain sulfate-reducing bacteria, as well. This might be just coincidental to the environmental parameters of eels. However, the European Eel with its transatlantic voyage tells us that it is a very good probability that eels contribute to the electrical characteristics of the system.

Figure 1.2. Geographic Distribution of Eels. Here we see the areas in which eels are known to exist (biogeography of eels). For the location of the Fields in respect to this distribution, see the other figures. Recently, young urban professionals, Yuppies, have acquired a preference for eel skin wallets, which due the eel skin's electromagnetic properties have erased the magnetic strips on credit cards.

The Submarine Alvin, built to withstand great pressure (recently used to observe the sunken Titanic), was sent into the great depths in order to study erosion processes at the bottom of the Florida Escarpment, and found that this precipitation of iron occurs under extreme conditions. This area consists of a range of underwater cliffs in the Gulf of Mexico, which reach such great heights that even the Grand Canyon seems dwarfed. At such great depths there is no sunshine, and the pressure is extremely intense. A graduate at Scripps Institute of Oceanography in La Jolla, California comments on a discovery there: "Rather unexpectedly we came across communities of abundant organisms."[40] The variety of creatures included bacterial mats, crabs, bivalves, fish, limpets, and various other organisms. These same types of communities are also known to exist in other inhospitable ocean areas, such as the Eastern Pacific Rise where the Earth's interior spews forth, creating very hot temperatures.

When drilling the Florida Platform, hydrogen sulfide was found, and the Escarpment sediments yielded large amounts of an iron sulfide (*pyrite*). In fact, its abundance far excesses what would occur if the ocean were the only source of the sulfur and iron. Furthermore, core samples of the hydrogen sulfide sediments disclosed that water in its pores contained ammonia (NH_3), but was low in sulfate, suggesting that organic mater was decomposed by sulfate-reducing bacteria.[40,209] Metal sulfides do not dissolve in water.[289] This is the observation of the formation of an electrically conducting sediment, and the organic matter includes eels. In addition, insoluble metal sulfides maintain a balance of low sulfur and more iron in the oceans so that ocean water becomes more electrically conducting, too.[112,238,342] These sulfur bacteria were one of the first organisms on Earth, and iron sulfides (*pyrite or pyrrhotite*) can be found in nearly all kinds of rock on Earth.[124,209,238,342]

More recently, exploration of the ocean floor and deep within the Earth's surface has revealed more of the unforeseen. Bacteria were located 500 meters (1,640 feet) below the ocean floor itself. It seems that the bacteria feed on organic matter in the sediments. Moreover, drilling at a location west of the Mid-Atlantic Ridge (*Leg 82*), where there is no sunlight and extreme pressure, uncovered microbes deep beneath the ocean floor where they live in and on volcanic glass, and may even "eat" the glass. Likewise, a borehole in Virginia, about three kilometers deep, revealed that bacteria were everywhere. Moreover, the deepest drill hole on Earth, the 12-kilometer (7.5-mile) deep hole in the Kola Peninsula (Russian Arctic Circle), also uncovered bacteria. The KTB hole (to be discussed in a later section) penetrated to a depth of 7.5 kilometers (4.7 miles) and found fluids at nearly all depths, which might be utilized by bacteria. Another bore hole drilled to the

6.8-kilometer (4.2-mile) depth in the Siljan Ring, Sweden yielded micrometer-sized grains of magnetite of bacterial origin.[111] These discoveries make the biosphere, along with its electrical characteristics, much larger than originally thought.

5.3 A Global System

All the various nerves throughout an organism channel into main nerves that bring impulses to a central system. Life on Earth displays a similar state with fishes that have what has been termed electric organs. At least seventeen species of fish have been recognized as belonging to this type. One of them, the Electric Eel (not a true eel, but related to carp), emanates a shock powerful enough to stun animals as large as a horse or a human. These fish species are geographically closely associated with at least four of the electrical currents previously discussed and illustrated. For example, the Northern Star Gazer's (*astroscopus y-graeum*) main range is within the Sargasso Sea, and its entire range barely reaches outside. Six species in South America are within the proximity of the electrical currents which center off the east coast of Brazil. Europe and Iceland have one species by the mouth of the Mediterranean Sea, and another that stretches from there to the Sargasso Sea (much like the European Eel's migration route). Africa has six species linked to both the current centering east of South Africa and at the mouth of the Mediterranean. Each of these types of fish is extremely sensitive to weak electric currents (*0.01 microvolts/sq. cm.*). All species continuously emitting electricity have been shown in studies to be sensitive to changes in the conductance of water.[219,220,252-254,376]

When life is destroyed on land (the Earth-cell, the habitat destroyed) topsoil is lost and more sediment flows into waterways and the oceans. This enhanced erosion not only changes electrical production on land, but also changes the conductance of the waterways and oceans (*due to the presence of more electrolytes*). Could these fish, sensing the change, somehow send impulses to a central system? Our current knowledge on the subject cannot answer this question, but it seems worth contemplating, because electric fish respond to very weak magnetic and electric fields.[219,220,252-254,376]

These fish are not entirely unique in this respect, because all life on the planet is sensitive in this way, though not quite as much. Superconductivity, a term used to describe the movement of an electrical current with little or no resistance, is the only known physical mechanism which can explain this sensitivity (it has also been termed the Josephson Junction Effect).[217] This super-sensitivity to electric currents (biological superconductivity) occurs in organic material, and living things.[106,107,168,181,182,255,266] For example, current voltage characteristics measured in dry and water-swollen wheat seeds reveal a (negative) resistance that indicates the existence of this superconductivity.[126,127,229] Even animals in greater numbers would increase the superconductivity of soils by releasing waste (defecating and urinating) due to the fact that bile acids and salts in that waste are superconducting, and enhance chemical reaction rates, which produce currents.[60,181,182,255] In addition, acids and salts, including decaying organic matter in the soil, will also further the corrosion of metal, which has been shown to produce a magnetic field.[60,62] The Earth's surface would then receive the magnetic charge with its accompanying electric activity. In fact, corrosion of buried metal is significantly enhanced by earth-currents, and typically, power failures and railway corrosion often take place within a month of the equinox (vernal equinox in the Northern Hemisphere).[239] Almost a perfect likeness of biological membranes, many of today's superconductors are being made of layers of biological molecules consisting of a metal and a semiconductor separated by organic molecules.[152,255,266,329] Furthermore, they mimic the process behind nerve conduction.[106] The unique aspect of the biological superconductors is that they are functional under physiological temperatures, while manufactured types require very low temperatures.

Superconductivity is also evident in plant and animal life's orientation to the geomagnetic field (GMF). As previously stated, the electric currents and the magnetic field fluctuate in unison so well that they may be considered a single unit. In addition, life and the environment fluctuate with it as well. Animals under natural conditions select compass directions along these electric currents during migration.[47,77,282] Plant roots' and tubers' growth also orient along compass direction.[47,126,127] Even flower symmetry, leaf rotation, and germination are oriented.[126,127,145]

Birds migrate along GMF lines, obtaining direction and position as if they possessed a map of the magnetic field.[3,59,240,275,277,401,414] Responding to changes in the field's direction, they are thrown off course during magnetic storms when the field changes direction.[275,277,399,400] During an eclipse, birds are known to become totally disoriented, because the field strength becomes imperceptible at that time. Within their physiology are magnetic minerals, such as magnetite, which aid in this ability to orient fight in relation to the GMF.[313,400]

Other animals are also known to have tiny magnets used for orientation, including fish[77,185,245,253-255,281,282], honeybees[171], shellfish (chitons, mollusks, etc.)[259], arthropods (insects, etc.)[171], dolphins[424], many species of mud bacteria[65], Yellowfin Tuna[400], moths[48] and others.[284] Human central nervous system, physiology, biorhythms, sleep, disease states, and even extrasensory perception and DNA synthesis (*fibroblasts*) are affected.[125,126,250] The soil's temperature, mobile forms of iron, dynamics of soil bacteria and water, and chemical activity are also affected.[126,127]

The entire biosphere, from molecules both biological and inorganic to entire organisms including humans, fish, insects, plants and other life-forms, is oriented to the Earth's magnetic field. This is no accidental arrangement. Organisms have their entire physiology oriented to their central nervous system. It seems the Earth is no exception. This is especially apparent in the fact that all living things, including humans, react three to four days *prior* to changes in the magnetic field.[96,127,310] It is also apparent in the fact that the Earth's magnetic field became strong and consolidated at the same time that a life system became well established (Cambrian). An actual physical connection exists between the magnetic field, these electric currents and life on the Earth (*GMF-telluric-superconducting-electrostatic interrelationships*). More will be discussed on these and related topics in Chapter 15 and Volume Two.

CHAPTER 6

Journey to the Center of the Earth (Seismological and Experimental Observations of the Earth's Interior)

Understanding the internal features of the Earth is an exceedingly difficult task, because humans have barely scratched the surface of the first layer. If the Earth were to be reckoned as an egg, we have not even reached beyond the shell. Like fleas on the back of an elephant, our perception of what we live on is extremely limited. Being unable to probe to any appreciable depth, we must turn to other sources for our knowledge of the thing beneath our feet.

6.1 Earthquake Waves

The study of earthquake waves, seismology, is about the most important contributor to comprehending the terrestrial orb's interior. Two types of earthquake waves stimulate our perception of the divisions within. These are "P" waves, or primary, push-pull waves, and "S" waves, or secondary, shake waves. They are designated this way to describe their time of origin and action. "P" waves are first in occurrence (primary), moving the materials they affect back and forth (push-pull). "S" waves are second in occurrence (secondary), moving the materials they affect from side to side (shake). As these waves pass through the various divisions of the Earth their speed changes, thus distinguishing these divisions one from another.[69,73]

According to the study of earthquake waves, the Earth can be depicted as an object with three major divisions: crust, mantle and core. The major divisions are understood to extend to certain depths, and to possess certain characteristics that distinguish their density. Shared by humans, other living things, and the oceans is the crust, which has been observed to be a solid descending to a depth of about 32 kilometers (approximately 20 miles). The mantle, next in descending order, extends to a depth of 2,900 kilometers (about 1,800 miles) and possesses the characteristics of a solid. Actually it is claimed to be molten, which is to say it is more or less liquid, but being contained by pressure it behaves like a solid.[69,73]

A property of "S" waves is that they can only travel through solids, and the final division, the core, does not transmit "S" waves. For this reason, the core offers a special problem to the seismologist and geophysicist. Most geophysicists presently insist that the core is liquid with a solid inner core, while others have proposed that it might very well be composed of undifferentiated solar matter.[234] The concept of a core consisting of undifferentiated solar matter is closer to the truth, despite the fact that most seismologists and geophysicists are rather skeptical about the idea, and have argued against it. The only seemingly credible argument against a core of solar matter (rich in hydrogen) is from the standpoint of chemical thermodynamics, which states that an important intermediate stage in the cooling of the Earth during its cosmological birth caused silicon (Si) and hydrogen (H) to combine into silicate rocks (SiH_4).[132] However, this is not a valid analysis, because they have not pondered the possibility of a renewable source of hydrogen—solar matter (i.e., solar wind plasma)—continually being involved in the chemical reactions.

6.2 <u>Discontinuities</u>

In order to understand this a bit more in depth, it is necessary to look at the Earth's interior. The interior is not quite as simple as the three concentric layers discussed previously. Earthquake waves have indicated further (sub)divisions marked by areas the seismologist labels discontinuities. A discontinuity is an area where the chemical composition of main divisions, or layers within a division, changes composition (and bear the name of the scientist who convincingly presented evidence of its existence).

For example, the crust can be divided into two subdivisions, the upper and lower crust. Between these two is what has been called the Conrad Discontinuity, where the upper crust is made of one material (*granite*) and the lower crust is believed to be made of another (*basalt or gabbro*). The Conrad Discontinuity, however, is a combination of these chemical compositions (*granite and basalt*), and thereby, shares the properties of both. A discontinuity is referred to as such because chemical composition, density, and the speed of earthquake waves do *not* continue to be the same; that is, they are discontinuous.

However, some of this reasoning has been brought into question. The Conrad Discontinuity is generally between 7.5 and 8.6 kilometers (4.7 and 5.3 miles). According to present theory, one would expect to find a significant change in rock type when drilling through such a strong discontinuity. The granitic rocks of the continents should suddenly change to basalt. Yet, Soviet drilling below the Kola Peninsula found no such switch over. In fact, they had drilled to 12 kilometers (7.5 miles), more than three kilometers below the depth of where the discontinuity should have been. This brings into question what the Conrad Discontinuity, and in fact, what any discontinuity represents, and therefore, whether our models of the Earth's interior are realistic.

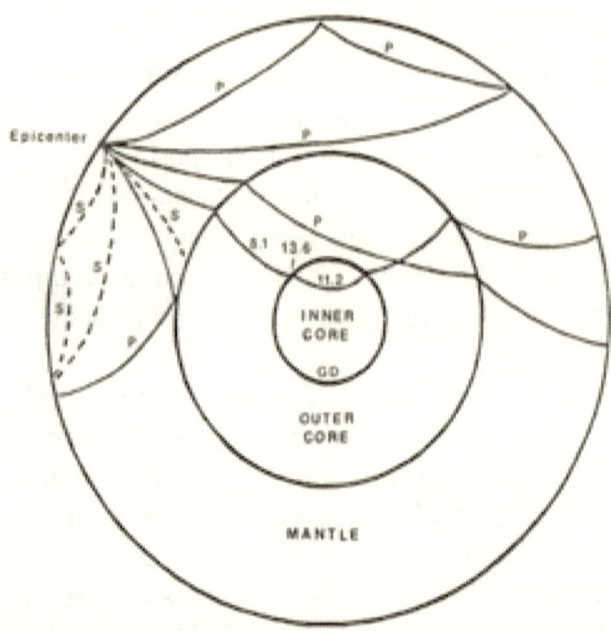

Figure 1.3 Main Divisions of the Earth's Interior. The interior of the Earth has been contemplated on the basis of the behavior of earthquake waves shown here by typical 'S' waves and 'P' waves (marked by S and P, respectively). Note how the 'S' waves do not travel into the core. Figures marked, on the 'P' wave that passes through the inner core, are the different speeds that the wave travels in kilometers per second. Note the higher speed the wave takes as it passes through the Gutenburg discontinuity (marked here by GD). .

Notwithstanding, a true understanding of the Earth comes from the facts derived from these finer divisions. These subdivisions (discontinuities) are important in comprehending the interior, the magnetic field, and the process by which the hydrogen-rich core is maintained (see Figure 1.3). Confirming the existence of a different core than hypothesized are recent observations that reveal earthquake waves travel faster north-south than east-west. Such observations disclose the presence of a magnetic field along the outer core that is as strong as a star's. Other geophysical phenomena confirm the existence of such a core (this will be discussed)

CHAPTER 7

Spiraling Forces
(Is the Dynamo Theory a Tenable Solution to Geomagnetism?)

Arriving at an acceptable understanding of how the Earth achieves its magnetic field has stirred the human imagination nearly to a state of frenzy. Usually new theories embrace theories in adjacent disciplinary fields, and speculation about the magnetic field is no exception.

7.1 <u>Dynamo</u> <u>Theory</u>

Current rationale begins by embracing the theory that the core is liquid iron-nickel. It then hypothesizes a number of possible ways in which the fluid core is set in motion so that the magnetic energy can be drawn from the kinetic energy of the fluid motions. In addition, in this supposed state, the core is a good conductor of electricity, and coupled with these motions could generate a self-sustaining field.[90,208,304,404] If you don't understand, don't agonize over it, because the geophysicists really don't either. Thus goes the dynamo theory, presently the most popular philosophy on the origin of the geomagnetic field.

7.2 Dynamo Theory's Failure

Dynamo theory has many drawbacks that make one doubtful of its soundness. It has *not* been proven possible to demonstrate dynamo action in the laboratory using a sphere similar to the Earth's proposed core. In contrast to a sphere, engineering dynamos have a coil that has two directions of rotation (two fields) that are not equal (two unequal spirals), and bear the shape or symmetry of the face of a clock. This arrangement causes a current to flow in such a way that it generates a field (electric and magnetic) that reinforces the first field. A spherical shape, such as the core, does not possess these characteristics, because it is not in the shape of a disc or face of a clock, and it is unknown where the initial field reinforced by the dynamo came from, and whether there are two directions of motion.

The question then arises: Can disc-like motion occur in a sphere, creating a magnetic field, or is it also necessary to have a disc shape in order to produce it? (*Does asymmetry of motion allow for dynamo action, or is it necessary for there to be an asymmetry of structure also?*). The problem is so complicated that it will probably never be solved, even with the aid of the most advanced computers. Geophysicists are still at work trying to make the dynamo theory a tenable solution to the Earth's magnetism, and in fact, basically insist that it is the only answer.

Even some observations contrast those conditions that should exist if the dynamo theory were fact. During eclipses the geomagnetic field becomes sufficiently weakened to cause the disorientation of migrating birds. The gravitational forces operating at this time are not capable of causing the core's supposed dynamo action to slow to this extent. Furthermore, studies using a pendulum disclose that gravity changes during eclipses, revealing the existence of an unknown field (this will be discussed).

Other observations also contradict the dynamo scenario. The inner core seems to be structurally unsuitable for dynamo action as well, as it is either a prolate sphere (its axis is about 100 kilometers or 62 miles longer pole to pole), or it has hexagonal symmetry.[111,129] Moreover, the geomagnetic field is tilted about 11° from the rotational axis, which is, again, unsuitable for dynamo action. Drilling into the crust of the Earth reveals temperatures rise too quickly, and by extrapolation, the temperatures at only 100 kilometers (62 miles) would be too hot (*beyond the Curie point of known materials*) for magnetism to exist. As will be discussed in Volume Two, variations in the geomagnetic field (polar reversals, wandering and jerks) occur too rapidly as well, and show correlations with solar activity that cannot be accounted for by the solid, permanently magnetized materials proposed for the core. Also unresolved are the source of the initial magnetic field needed to start the self-excited dynamo, and the energy source that keeps it going.[111] Moreover, as will be discussed in Volume Two, the tilt of others planets' magnetic fields (notably Uranus and Neptune) and the existence of magnetic fields where they were thought not to exist (Mercury) seriously challenges the dynamo theory.

Increases in solar activity (i.e., solar flares, prominences, etc.) cause changes in the length of day, and make the magnetic field shift direction and intensity (*i.e., geomagnetic storms*). The solar wind does *not* have the physical properties to cause this if the dynamo of an iron-nickel core were producing the magnetic field. Moreover, earth-currents (*i.e., telluric currents*) are understood to have an influence upon the braking or acceleration of the Earth's rotation. In addition, electric currents and areas with unusual magnetic patterns, known as geomagnetic anomalies, on the Earth's surface are too persistent in strength and geographic location for this hypothesis (*spherical harmonics vary too smoothly*).[239]

Incidents in which the magnetic field flips north for south, or south for north (magnetic reversals) are unexplainable with the current theories and mechanisms of the dynamo. Comets approaching near the Earth could cause reversals with a dynamo mechanism. However, reversals are not as frequent, observations have been made of comets passing without a reversal, and there are times of proposed asteroid or cometary impacts without a reversal.

Methane has always been known to be produced by bacteria and was always thought to be biological in origin. Now it is known to be coming from a primary source: the core. This is not possible with an iron-nickel core, but is possible with a core of hydrogen fusion. Furthermore, there is an excess of helium-3 with regard to the ratios of helium-3 to helium-4 in rocks, which can also be accounted for by hydrogen fusion.

Recent evidence shows that the Earth's interior is not as previously envisioned. The core-mantle boundary (*is not like the thermal boundary layer theorized*), and the inner and outer core are not as was once theorized (unexpected anomalies exist). Furthermore, earthquake waves were observed to bend like that caused by a lens, which was also confirmed by stresses and gravity measurements.[113,407,408,417] A fact such as this indicates a magnetic field boundary (a discontinuity) that helps to create the Earth's magnetic field.

Experimental evidence is beginning to indicate that an element lighter than iron exists in the inner core.[210,408] One of the suspected elements is hydrogen (*other possibilities are S, O, or Si*). Such light elements make the dynamo theory less possible, because they are not good conductors or generators of electric or magnetic energy.

An expert on magnetic fields indicates one of the dynamo theory's failures: "The dynamo mechanism does not provide for the outright creation of a magnetic field, but only for the reproduction, that is, the amplification, of existing fields. **Where then did the fields the dynamos amplify come from?**"[300]; p.54 (bold added)

Meanwhile, geophysicists have recently made experimental observations of iron at the super pressure of the core.[413] They say that the Earth's core is thousands of degrees hotter than the Sun's surface! Therefore, the forces that drive the continents into "drift" (plate movement), and give rise to volcanoes and earthquakes have their origins in the Earth's core. Previously, it was thought to be the mantle's heat that produced plate motion. In addition, recent space probes reaching the outer planets have shown that magnetic fields exist on planets thought to be too cold, and made of material too light to produce a magnetic field, indicating the need for new theories on planetary geology. At best, the dynamo theory is an educated guess, though an incomplete one.

CHAPTER 8

Lights in the Sky
(Recent Observations of Solar, Magnetospheric and Ionospheric Plasmas—Implications for a New Model of the Earth)

It is winter in the Northern Hemisphere. The circumpolar region of this part of the world has been plunged into six months of nearly perpetual darkness. The scene is set for one of the great mysteries of modern science. Suddenly a stream of light cuts the cover of darkness in the neighborhood of the North Pole. Awe envelops the population. Nature has sent her most spectacular show straight to the hearts of her audience. It's the Aurora Borealis, more commonly known as the Northern Lights. Such beauty could even make those six months of darkness a cherished thing.

This scene also occurs during the six months of light but goes unobserved by the naked eye. The South Pole has a similar display called the Aurora Australis or Southern Lights. Both bring evidence of what the Earth's interior is really all about.

8.1 <u>Auroras</u>

The real unraveling of what produces the magnetic field lies in our observations of the auroras. Auroral activity is *always* associated with the Earth's magnetism. This association is due to the electrons present in the aurora (solar plasma), which in association with the Earth's electric currents, are accelerated into the Earth's interior generating a magnetic field. Such a relationship existed during the formation of the Earth from the spiraling planetary nebula that produced a dipole magnetic field, as well as mid-latitude fields. This is the result of the Coriolis and centrifugal forces in the plasma, and the pre-planet (protoplanetary) nebula's interaction with the Sun's Interplanetary Magnetic Field (i.e., due to the mechanical forces generated, as described in the Einstein Equivalence Principle). As the nebula condensed, it lead to the alignment of minerals due to their magnetic properties (this will be discussed more in Chapter 10 and "Churning Up Creation" in the **Conclusions** in Volume Two).

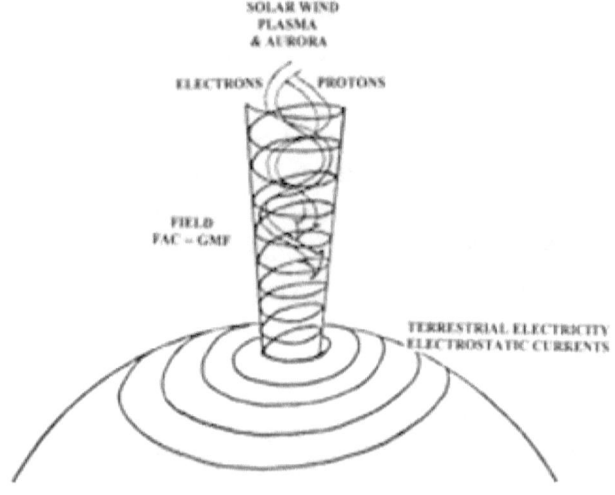

Figure 1.4. Diagram of the Field at the Pole. This crude representation of the Field at the pole shows the earth-currents (*biogeoelectrostatic*

currents) that help to reinforce the Earth's magnetic field. The Field then directs the electrons and protons of the solar wind plasma by accelerating them toward the Earth in oppositely directed spirals (i.e., a helix). [FAC is Field-Aligned Currents]

Electrons rotating at an angle and with speed are equivalent to a current. Such a current will produce a magnetic field at right angles to its plane of rotation. Curtain-shaped auroras exhibit the best example of how this is accomplished. They are caused primarily by an intense sheet beam of energetic electrons that spiral down into the polar regions. This spiraling vortex (actually a helix of electrons and protons), shaped like a super-elongated funnel, produces the magnetic field as would the wire wrapped around an electric magnet or the antenna in a radio; only here there is no wire. Consequently, there would be no need for an iron-nickel core, complex fluid motions, or for that matter, a dynamo theory to explain the geomagnetic field (see Figure 1.4).

Ah! Delicate green lines with a crimson red-purple lower border embellishing a black velvety background. That may sound like a painting in a gallery, but actually it's a red aurora, type B. And no painting could compare with its beauty.

When aurora particles make contact with the atmosphere, reactions occur which emit various colors, leaving a telltale sign of their composition to the trained observer. With instruments known as spectrometers and photometers, science is able to understand what the variety of particles are in auroral displays. On close examination it becomes obvious that one of the main constituents causing lighted displays is energetic hydrogen protons (nuclei). Any nuclear physicist will tell us that those particles could be used as plasma in a fusion reactor. Of course, its source is a colossal reactor, the Sun.

One of the properties of plasma is that it can be repelled by a magnetic field. In fact, using magnetism to confine hydrogen plasma (i.e., magnetic confinement) represents an ideal solution to nuclear fusion.[92,173,221,357] Certainly the neutral sea of outer space would offer no repulsion to the plasma, a highly ionized hydrogen gas, yet it sends its remarkable art-form earthward (see Plate 1.1).

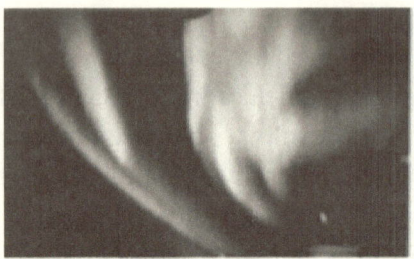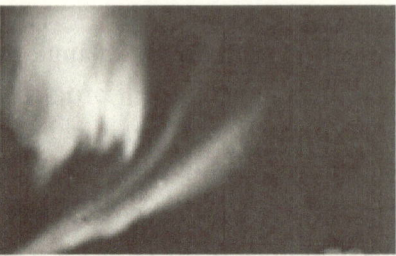

Plate 1.1. Simultaneous Aurora Over Each Pole. Here we see both the Aurora Australis and Aurora Borealis. These photos were taken simultaneously by planes flying along the same magnetic field lines in both the Northern and Southern Hemispheres. Note how they basically display a mirror image of each other. [From Eather, R.H. Majestic Lights: The Aurora in Science, History and the Arts. Copyright 1980 by the American Geophysical Union].

8.2 A New Earth Model

When one begins to interpret the information about the manifold phenomena affecting the Earth, it becomes evident that the most widely accepted theories about the interior are, at the least, inadequate. The terrestrial orb's interior is ideally suited for thermonuclear fusion. There maybe some, or even most, geophysicists who will resist this idea outright, but rejecting this understanding now must surely be labeled "hasty closure," for there is much supportive evidence, and science must remain an open-minded discipline. The major portion of this book lends supportive observations that make it so obvious we will wonder why it wasn't seen before.

Underlying it all is that there is a long overlooked reality about gravity's role in the formation of the Earth, particularly the core. Downward from the Earth's surface, primary differentiation with depth is accompanied by the progressive decline to zero of the force of gravity at the center of the Earth. Normal gravitational force, downwardly directed, is replaced by a reversed upwardly directed force at depths greater than 2,700 kilometers (1,678 miles), which would pull matter away from the core. This is especially true because the core is now known to be much hotter than previously thought.

This is why research involving deep mines and boreholes has shown different values for gravity with depth. In fact, a person deep within a mine weights less than that same person on a mountain top. Moreover, those scientists studying the phenomena have called for a new model of the Earth, but long-held theories are not easily dismissed. As a result, the phenomena has been virtually ignored.

Meanwhile, there is more to the story. During the protoplanetary stage there is more to reckon with than the force of gravity. According to the Einstein Equivalence Principle there are also the mechanical forces, which are put on an equal level with the gravitational force. The Earth, whether in the protoplanetary stage or at present, has a velocity gradient from zero at the core to progressively higher velocities as the surface is reached (particularly along the equatorial bulge).

Likewise, the centrifugal and Coriolis forces are also an aspect of the Earth's structural arrangement, and its earliest differentiation. There is a velocity gradient from near zero along the Earth's axis to progressively higher velocities as the equatorial bulge is reached. The equatorial bulge has always been perpendicular to the geomagnetic dipole, even during polar wander and reversal, and cannot be considered to be a hold over from an earlier, faster rotation. Furthermore, the Coriolis force is at zero in regions along the 30° to 40° latitudes, making the mechanical forces more active around these latitudes.

These mechanical forces are rarely, if ever, contemplated when theories of the Earth's formation and dynamics are envisioned. Yet, there is another factor to ponder. As is typical of solar system formation, the

Sun formed into a star before any of the planets began to form. Furthermore, the Sun was and is sending waves of solar plasma along its Interplanetary Magnetic Field (IMF), which would have hydrogen-plasma (*magnetohydrodynamic*) effects beginning with the protoplanetary nebula. This interrelationship can be seen even today in solar-terrestrial physics, and in the various solar cycles reflected in geological and meteorological cycles, including the auroras, weather, earthquakes, volcanics, the Chandler (polar) wander, and polar jerks, wander and reversals, among other things (these will be discussed in Volume Two).

When all of these factors are considered, they force a more elaborate Earth model than can be deduced from gravity, geochemistry and chemical thermodynamics alone. For example, the centrifugal force would have aided in the formation of the Earth's dipole field. That is, according the tenets of ferromagnetism, molecular arrangement evolves symmetrically under influence of an applied field—the centrifugal force and the IMF in combination. After all, the dynamo theory is seriously flawed if we merely consider the fact that dynamos only reinforce existing fields, and do not provide for the outright generation of a magnetic field.

This whole scenario leads to an Earth model that is quite different and more dynamic than the present model. The combination of forces pulled matter away from the core, and as a result, created a "chamber," for want of a better word, with very dense metals, such as iridium, producing a magnetic confinement chamber at the Gutenberg discontinuity (GD), at a depth of about 2,900 kilometers (1,802 miles). This depth is sufficient for the gravitational force to be primarily away from the core. Because the solar wind plasma and pulsating IMF are time-varying phenomena, the fields that developed became time-varying fields, ideal for accelerating hydrogen plasma into and away from the core. According to the Special Theory of Relativity, such fields should exist.

Another indication of a structure at the core-mantle boundary involves the extensive analysis of earthquake data. Both seismic tomography and the analysis of whole-Earth vibrations have revealed an unknown structure at the boundary. Both of these techniques are independent studies, yet they reveal the same understanding.[111]

Furthermore, a core of hydrogen fusion would make sense of the aurora (*hydrogen plasma*) within the Earth's magnetic field lines, and the spiraling of electrons and protons into the polar regions. Polar-glow type auroras, with their separated electron- and proton-excited streams, are high in hydrogen plasma, an ideal fusion material. And the electron-excited portion is ideal for helping to maintain a magnetic field (and particle accelerator). The spiraling electrons would offer a perfect pathway for the protons to enter the Earth's core where pressure and magnetic confinement are quite sufficient to cause thermonuclear fusion. In fact, some present-day fusion reactors are in its likeness, though we have remained unaware of the fact.

One of the arguments opposing a core of hydrogen is that pressure is not sufficient to condense the core to the point where it would behave like a liquid, as earthquake waves indicate. This reasoning does not take into account the possibility of hydrogen, in the form of plasma, being contained by a magnetic field, which would cause the observed effects.

The Gutenberg discontinuity (GD) displays the characteristics of a magnetic confinement chamber. It is unlike other discontinuities which cause seismic waves to speed up slightly on their sojourn from one subdivision to another. Typically, earthquake waves never move faster than they do in the next inner division, but this discontinuity causes a drastic speed-up. Actually, the acceleration experienced makes it the most unique portion of the globe's interior (13.6 km/sec., or 8.45 mi./sec.). Even in the supposedly more dense and solid inner core the same wave is much slower (11.2-11.7 km./sec., or 7.0-7.25 mi./sec.).

According to the GD's position, it should exhibit the characteristics of the lower mantle and the outer core. That is, the GD should have a

"P" wave velocity of somewhere between 8.2 kilometers or 5.1 miles per second (lower mantle), and 8.1 kilometers or 5.0 miles per second (outer core). Considering the velocity of primary waves of the order of 13.6 kilometers or 8.45 miles per second, the GD should be very dense material.[69,73] It could easily be the only portion composed of an iron-nickel compound or possibly even more dense, such as iridium or gold. However, it is doubtless that the GD possesses the qualities for high electrical conductivity.

Furthermore, the electrons that produce the aurora eventually make contact with the GD and generate an intense magnetic field around the plasma. Coupled with pressure, a factor aiding fusion, which at this depth is 21.1 million tons per square centimeter (136.1 million tons/sq. in.), would make the hydrogen-rich plasma core yield characteristics that would cause the observed earthquake-wave behavior. This interpretation is especially evident when we consider the recent observations that earthquake waves travel faster north-south than east-west, which reveals a magnetic field the strength of a star slowing the east-west waves (see Chapter 29 for discussion).

What is actually at the core is hydrogen plasma, confined by a magnetic field in a constant state of thermonuclear fusion. A sun-like structure exists in the core. The iron-nickel core theory has no plausible explanation for the observed primary wave behavior in the Gutenberg discontinuity (GD), nor the different north-south/east-west earthquake wave velocities.

This model may be hard to visualize, but it goes something like this. First, the Earth's surface (including life) generates an electric current which spirals around specific areas. Two of the areas are the North and South Poles, where this current reinforces a funnel-shaped pathway of electric and magnetic forces (*a time-varying, biogeoelectrostatic particle accelerator that reinforces an electromagnetic field*). Then a highly ionized hydrogen gas from the Sun (solar wind plasma) enters the polar regions as high-energy electrons and high-energy protons. The electrons reinforce the initial field, causing the protons to be accelerated (i.e, a time-varying accelerator) to speeds approaching that of the speed of light (*relativistic velocities*). Furthermore, that same portion, reinforcing the initial field (electrons), quickly reaches the Gutenberg discontinuity, reinforcing an intense magnetic field that confines the protons so that they constantly collide. When protons collide they let off tremendous amounts of energy by undergoing thermonuclear fusion. The end result is a sun-like structure in the Earth's center.

This model has been somewhat revealed in drilling programs. One such program is called the Kontinentales Tiefbohrprogramm der Bundesrepublik der Deutschland, in German. A research drill penetrated to a depth of 7.5 kilometers (4.7 miles), which is the second deepest man-made hole, and is often referred to as the KTB hole. One of the surprises was that temperatures rose much faster than predicted by present models. Another unexpected find was that there were crevices and pores, filled with fluids, at almost all depths, though theory indicates that there should be no such thing, as the pressures are supposed to be very intense. Likewise, when a drill hole in the Kola Peninsula reached 4,500 meters (14,800 feet) a sudden decrease in density occurred.[111] Again, this evidence reveals that gravity changes with depth.

Temperature changes also encourage the idea of an energy source deep in the Earth. When drilling in the Kola Peninsula reached the 10-kilometer (6.2-mile) depth the temperature was expected to be 100° C (212° F), but was instead 180° C (356° F). At the bottom of a 3.5-kilometer (2.2-mile) hole in Oberpfaiz Forest, German, the temperature was 118° C (244° F), not the expected 80° C (176° F). In this hole there was also circulating brine rich in helium-3.[111]

Another observation in support of this new Earth model is the anomalous abundance of helium-3. Helium-3 has been observed outgassing from the Earth's crust, especially from the mid-ocean ridges, but also from the continental crust. The imbalance with respect to helium-4 means that new helium-3 is being generated by nuclear reactions within the Earth.[111]

8.3 Aurora Dynamics

Further observations of the aurora during the Dynamics Explorer missions yield further evidence of what the Earth's interior is all about. An unknown plasma wind has been observed coming *up* from the pole when previously it was only thought to flow down. Called the "polar wind," scientists do not know which is more important to auroral displays, this polar wind or the solar wind.[4,174,318,328,335,344,345,350] The ions that were discovered to flow upward are hydrogen and helium, just as experienced if one were in the vicinity of the Sun. In fact, helium ions were found to be several times more abundant than previously thought, making the two major components helium and hydrogen. Meanwhile, oxygen ions were also discovered, and a part of the radiation belts (*plasmasphere*) has more oxygen ions than other areas originally thought to have more (*ionosphere*), even greater amounts than the solar wind. Nitrogen ions, totally unexpected, were observed high above the Earth (*at 3 Earth radii*), restricted to regions directly over the pole.[174,344,345] Again, this is something that has been observed with the Sun and the occasional occurrence of the combining of two helium ions. Helium and oxygen at high levels and nitrogen at any level are not common to our present understanding of the Earth, but would have been predicted with the new Earth model presented here. The temperature and speed of the ions indicate that "additional acceleration or heating would be required."[344; p.401]. Furthermore, these ions are correlated with auroras (*substorms*), GMF changes, and the solar cycle.[344,345]

Another previously unknown aurora (*theta arc configuration*) frequently forms in the center of the oval where auroras commonly occur (*aurora oval*), although auroras are normally and theoretically supposed to be formed at the edge (see Plate 1.2). It can develop on the edge of the oval, but it is rapidly centered over the pole.[143] Auroras are also intimately related to changes in the direction and intensity of the Earth's magnetic field (*geomagnetic storms*), indicating that the Earth also triggers aurora. The reason for this is unknown, because scientists think only in terms of a solid Earth with an iron-nickel core (dynamo theory) interacting with the solar wind.

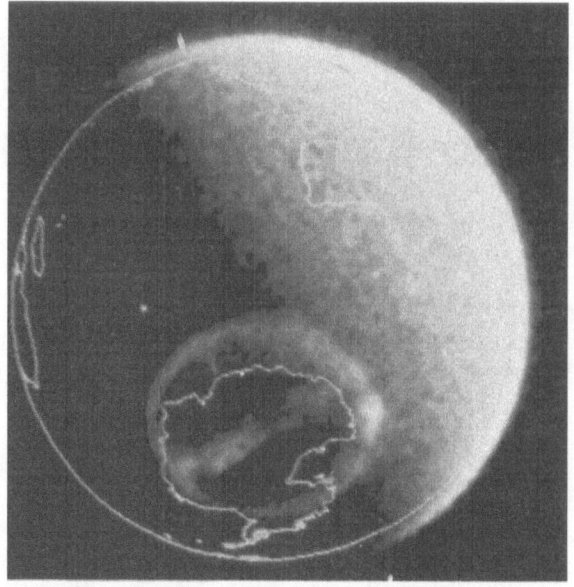

Plate 1.2. Theta Arc Aurora. This photo shows a Theta-Arc Aurora. Note how the aurora forms a circle with an auroral line dissecting the circle. Theta arcs form at the center of the auroral oval, a place that has no magnetic lines of force. This shows that there is more to auroral formation than just solar plasma interacting with the geomagnetic field's lines of force. [NASA/JPL, Dynamics Explorer image].

In terms of presently accepted theories of the geomagnetic field, not one of these observations is explainable, but with the understanding presented here they are completely predictable. All of these data and more quite clearly demonstrate that a sun-like structure dominates the

Earth's interior. Also, fields accelerate these ions both up through the polar region into space and down into the Earth's center. This new model of the Earth will be referred to as the Field-dynamical Earth Model, abbreviated FEM, which also brings to mind the feminine Earth Mother that has permeated various cultural beliefs throughout history.

8.4 Field Observations

Space probes, particularly the Dynamics Explorer, have examined the polar region, and found various fields and currents there. A pair of currents were observed to flow along the direction of the Earth's magnetic field (*field-aligned currents; FAC*). These two parallel electric fields accelerate solar wind particles toward the Earth, producing the aurora.[78,358] When auroras occur, a strong electric field was detected at the edge of the auroral arc. A parallel electric field is necessary in order to explain the existence of this first field, and therefore, two fields exist.[78,164,265,412] "It is evident that the upward current above the arc must be fed by horizontal currents."[78; p.81]. While scientists look above the Earth in space for the answer to the origin of the horizontal currents, the answer lies below on the Earth's surface.

The electrical properties of the Earth's surface and the life on it are responsible for the horizontal currents, but scientists have not been trained to look there. In fact, this is why as the aurora occurs it disturbs the magnetic field (*substorms*), with the event taking place both on the ground and in space simultaneously.[204,206] The absence of a delay from space to the ground does not fit current scientific thought, and in addition, indicates the presence of a field current that connects the two regions.

Other observations at the polar region bring forth more information about the character of the force operating there. A "twist" in the magnetic field was observed (*flux transfer events*), but again the source of this field is not known.[330,334] Later observations confirmed that this twist does exist and was also noted to conform to the electric currents flowing in the direction of the geomagnetic field (*FAC*).[96,261,332,412] It was also noted that the aurora formed a spiraling "funnel" (*plasma vortex motion*).[96,334,412] This spiraling vortex moves in a clockwise direction in the Northern Hemisphere and counterclockwise in the Southern Hemisphere.[80] In the midst of all this, an electrical change (electrical potential drop) was observed that rotates with the Earth (corotational) and aids in producing the aurora (i.e., its acceleration processes).[78,80,349]

Some of the magnetic lines extend indefinitely into interplanetary space.[190,261,293,318,353] Electrons flow into the polar region (*electron precipitation*) and are called "polar showers."[206,416] Furthermore, the magnetic field, in combination with electric fields, is connected to outer layers in the environment around the Earth (*plasmasphere, ionosphere and magnetosphere are coupled*) by some "unknown" mechanism that is electromagnetic.[149,353] Reports claim an accelerator is needed to produce the distinct variety of features of the aurora (*electron distribution: spectral peaks and field aligned pitch-angle distribution, and the upward flow of terrestrial ions*).[345,412] Finally, after all is considered, the mechanism is said to be "unknown" and follows magnetic field lines, but in order to produce all the observed effects it requires a field like that which is produced on the Earth's surface (*a field-aligned electrostatic accelerator*).[412]

A combination of the aurora particles and the spiraling magnetic field produces something similar to a device used by plasma physicists in the laboratory to accelerate particles. Called a time-varying field accelerator, it uses surges of electricity to push plasma to higher speeds, and must be done in particle groups or "bundles." Bundles of solar plasma that enter the pole (*auroral substorms*) are observed to occur every few hours and peak in ten minutes, producing electrical surges just as could be predicted if the Earth's magnetic field were like this type of accelerator.[363]

Observations at the South Pole are intriguing for the same reasons and a few others. The former Soviet Union's Molodezhnaya station, using laser radar, detected an "unknown" energy source 95 kilometers (60 miles) above Antarctica.[39,146,341,423] This source causes a number of phenomena in the atmosphere, such as rapid temperature changes, density and pressure changes, and other physical manifestations that could be expected of controlled particle flow.

In 1973 and thereafter, Nimbus 5 of NASA used microwaves to take pictures of the Earth's surface.[162,235,425,426] During winter in the Southern Hemisphere, when the Weddell Sea in Antarctica is frozen solid, there lay an unfrozen 482,000 square-kilometer (186,000 sq. mi.) lake. That's an area greater than the state of California, unfrozen in the middle of thousands of kilometers of ice during six months of darkness with the lowest temperatures and strongest winds on Earth! Many more "mysteries" at the South Pole will now begin to make sense with this new model of the Earth.

8.5 Field-dynamical Earth Model (FEM) and Life

If the Polar Fields are indeed reinforced by the electrostatic properties of life's generation of static electricity, then we should observe changes as we destroy life in favor of the lifeless components of our modern world. This is exactly what we find. First of all, the geomagnetic field has been becoming continually weaker as time progresses. In fact, from the first observations (1830s by Gauss) to the present we find that this is the most pronounced feature in geological observation.[53,118,119] This interpretation is supported by the facts that the Earth's magnetic field first reached near its usual strength when life began to blossom (Cambrian), and its strength has fluctuated with life crises throughout the geologic and archeological records (this relationship will be confirmed).

Furthermore, it is only in the older literature that we find auroras with a very different character. Only in the 19th and early 20th Centuries do we find observations of "pillar," "fan-shaped," and very low altitude auroras. Pillar and fan-shaped auroras, observed by earlier scientists, were known to occur for a few minutes to as much as half an hour, often with waves of light that rushed up and down.[16,27,68,70,105,128,175,178,216,243,324,331,387,419] Low-level auroras occurred below mountain top levels, and some even at ground level.[21,25,57,58,75,88,89,97,108,109,116,133,150,151,212,347,348] However, today several long photographic and optic series have yielded observations that have led to theories that state auroras occur around 80 kilometers (50 miles) above the ground. The most interesting correlation in regard to this is that pillar auroras were last seen in 1937 and low-level auroras in 1941. This basically coincides with increased levels of natural resource exploiting industry around the time of World War II (1939-1943) when much more wilderness was beginning to be destroyed the world over.

In addition, swishing and crackling noises that have accompanied auroras are also contrary to present theories. At 80 kilometers (50 miles) there is a near-vacuum in the atmosphere, and therefore, audible sounds should *not* be transmitted to observers on the ground, but yet they have been repeatedly.[55-58,83,88,89,116,133, 148,150,151,224,275,315,316,346,359,362] However, reports of this sort seem to occur only up to 1975. Meanwhile, since that time the number of auroras have increased significantly. Sulfur smells have also occurred during auroras, which can be caused by a discharge of earth-currents (*terrestrial electricity*) into the atmosphere.[58] All of this shows that an electrostatic field, that is maintained by life's generation of static electricity, has become progressively weaker, and that it is an integral part of the aurora's formation and dynamics. Additional support for this interpretation will be presented throughout this book.

CHAPTER 9

Time, Space, Matter and Energy (The Geographic Distribution of Geophysical Phenomena and Anomalies—Insights for a Unified Theory)

In order to distinguish the fields that make up the Field-dynamical Earth Model (FEM) from other types of fields, the fields of FEM and other Field-dynamical Models of other objects, will be referred to as Fields. Capitalizing will differentiate them from other fields in physics.

9.1 Oceanic Features

Many other phenomena are associated with the Field areas, in addition to the terrestrial electricity that spirals into them. Some of the deepest parts of the oceans are within them. For example, there is the Puerto Rican Trench off San Juan in the southern tip of the North Atlantic Field, which is the deepest part of the Atlantic Ocean (about 9,100 meters or 30,200 feet below sea level). Almost on the opposite side of the Earth in the southern corner of the Japanese Field is one of the world's greatest ocean depths (approximately 10,900 meters or 36,200 feet). Even something as landlocked as the Gulf of Oman, at the mouth of the Persian Gulf, shows a deep depression beneath the watery abyss. Not only are the deepest parts of the ocean located in these regions, but ocean currents also swirl around these Field areas (see Figure 1.5).

The High Energy Benthic Boundary Layer Experiment, with sensors approximately 4.8 kilometers (3 miles) underwater, has disclosed that vast storms rage undersea, clouding the water with mud. These bottom currents, totally unexpected, travel at approximately 1.6 kilometers per hour (1 mi./hr.), moving along with the surface currents. This is a mystery, because it requires a very, very strong force to move water at that depth due to the intense pressure. Furthermore, the presence of huge ring currents (2.5-3.5 km. or more than 2 miles in diameter) was a great rarity, so much so, that none were found only a decade ago. However, today they are quite abundant. The source of this powerful energy is unknown to the scientific world, but it can move water from the Mediterranean to the eastern Atlantic (from the Mediterranean Field to the North Atlantic Field). That is a distance of more than 6,000 kilometers (3,720 miles) and the same migration route taken by the European Eels.[196,227,269]

The capability of the Fields to displace water is, in fact, what is being overlooked, and a map showing the areas in which they develop again discloses the areas of the Fields (see Figure 1.5). Their recent development is the result of the destruction of life, thereby weakening the electric currents, so that the release of fusion by-products in the Field area "churns" (highly ionizes) the water when previously it had not done so. Again, this interpretation will be supported further in this work.

Figure 1.5. Ocean Currents and Surface Salinity. Major, spiraling ocean currents (black arrows), and the highest values in salinity (grey areas) are centered in the Field areas. The deep abyssal currents observed by the High-Energy Benthos Boundary Layer Experiment kicked around coarse sand on the ocean bottom, shown here by the locations of beds of coarse sand (black areas). In addition, all the Field areas, with the exception of the Field in the North Pacific, west of the United States, are areas of high kinetic energy.[201]

9.2 Luminous Phenomena

In addition to these observations, these areas are also noted for producing strange, lighted displays and milky seas. One type of lighted-display involves long, parallel bars or ripples of greenish-white light that move swiftly across the sea. They were observed and reported for areas that are near the North Atlantic Field (Gulf of Mexico)[280], the Persian Gulf Field[24,85,200], the Japanese Field (South China Sea)[8,9], and near the North Pole (Greenland).[288] The majority of sightings occur around the vernal equinox (20-21 March), disclosing the solar linkage, and often precede and follow the wheel-type.[111]

The wheel-type of lighted displays involve an arrangement of greenish-white lights that appear like the moving spokes of a wheel. Again the majority of sightings occur near the Persian Gulf and Japanese Fields.[15,18,22,33,297,308] Unlike the parallel bar type, which scientists claim is produced by waves of "glowing" phosphorescent bacteria (*bioluminescence*), this wheel-type has a geometry that cannot be explained by luminous bacteria in waves. Some sort of applied field is essential. Again the majority of these sightings occur around the equinoxes, especially the vernal equinox.[111]

Other luminous phenomena involve milky seas and luminous fogs, some of which occur in the depths of the sea, while others appear above the sea surface. Once again the majority of sightings have occurred near the Persian Gulf Field (Arabian Sea and Indian Ocean) and also near the Brazilian Field that is east of Brazil in the south Atlantic.[11,12,19,23,54,79,131,314] Here the peak occurs around the equinox (September), and less so, following the solstice (21 December).[111] As has often occurred in the North Atlantic Field, there are incidents when luminous fogs appear; it is as though the sea and sky join together and all sense of distance is lost.[17]

The effects of luminous organisms (*bioluminescence*) have been used as the standard explanations for these lighted-displays. However, this explanation falls short of describing the geometry of some types, their restricted geographic locations, and their timing in relation to the equinox and solstice (i.e., a solar-terrestrial linkage is apparent). These luminous phenomena result from ionizing radiation producing thermoluminescence, and the Fields' electromagnetic properties create geometric patterns.

9.3 Audible Phenomena

Explosive sounds along shorelines that are distant and muffled, often picked up by sensitive barometers, have been called "water-guns." Occasionally triplet booms occur, more of these sounds take place on warm calm days, and on some occasions they are associated with flashes of light.[111] Water-guns have been heard near the North Pole Field by people along the coast of the North Sea, and near the Mediterranean Field, at the mouth of the Mediterranean Sea, by those in Europe.[185,340,367,377,389] In Japan, especially near the Japanese Field, they are fairly common and are called "uminari."[1,14,370] Booming sounds are the most frequent near the North Atlantic Field, and predictably are reported by people along the East Coast of North America.[20, 37,38,104,120,134,167,228,257,336,369] One of the more pronounced cases of this phenomenon, reported from this area, involves the so-called "mystery booms" that continued from late 1977 to 1978, and were occasionally accompanied by flashes of light.[37,38,167,228,343] Some have also occurred in an area between the Japanese and Persian Gulf Fields with reports coming from the Ganges Delta, the Bay of Bengal and the East Indies.[28,115,242,290,361] Reports also come from areas around the Fields off the east and west coasts of Australia, from residents in Victoria and New South Wales (East Australian Field) and western Australia (West Australian Field).[101,103,291,367,396]

The majority of these sounds occur around both of the solstices (21 June or 21 December), particularly the winter solstice (21 December; most reports were from the Northern Hemisphere).[110] They are produced by the creation of a vacuum due the release of ionizing radiation through the Fields, and when the Field deactivates (i.e., it is time-varying) the air masses smash together, creating the "boom" sound. These phenomena result from a process that it is very much like lightning, and the way in which it can produce thunder.

Another peculiarity that occurs in these areas are sounds that resemble a harp, a bell, or an organ. Again, reports of these phenomena come from the Field areas, and mostly occur during the night. They have frequently been heard in the region around the North Atlantic Field, such as the area along the Gulf of Mexico, Florida and Central America.[7,13,93,120,230,271] They have also been heard near the Persian Gulf[10,291], near the Japanese Field (Cambodia and Malay)[13,28,29,] between both of these Fields (Tenasserim)[134], and near the West Australian Field (Java and Sumatra).[28,29]

One scientist claims that these sounds are the result of whales singing or conversing.[401] Whales are sensitive to magnetic fields and use magnetic maps of the Earth to guide themselves while migrating. The Field areas do have magnetic anomalies, but the description of these sounds do not seem to fit that of whales, especially because they are so audible inland. Instead, these sounds may be the result of changes in earth-currents, characteristics of the Fields, or changes in conductivity or resistance in the area. This seems to be suggested by the fact that most occur near Full Moon, which is a time when electric currents are more active, lunar gravitational forces are strongest, and there is a minimum in auroral activity.[111,189,396]

9.4 <u>Weather</u> <u>Centers</u>

The unique characteristics of these areas do not end here, either. Centers where weather phenomena develop characterize them also. In each there are high and low pressure systems constantly forming. A worldwide map of storm tracks discloses the fact that most storms originate in or near the Fields (see Figure 1.6). Tornadoes, hurricanes, and typhoons are propagated and display the shape of the forces present; that is, they are spiral-, funnel-, or tube-shaped patterns.

Figure 1.6. Weather Centers and Warm-Water Ocean Areas. Areas of high and low pressure are designated here by dashed lines. Warm water ocean areas are represented in grey. The arrows indicate the paths and relative frequency of typhoons and hurricanes.

For example, there is the Bermuda High in the North Atlantic Field, an area which is particularly noted for spawning hurricanes. Of the entire United States, it is the southeastern section that experiences the greatest percentage of both tornadoes and hurricanes, as well as the most lightning. The Japanese Field area produces more typhoons than any other area on Earth, and the Aleutian Low is a weather formation in the region. Of course, warm surface ocean water is important to the formation of both typhoons and hurricanes, so some begin away from the actual Field.[114,190,195,233,283,323] Weather, though, is most definitely an associated phenomenon and fluctuates along with changes in solar activity, as could be predicted from an understanding of FEM and it's solar linkage.

A combination of various eyewitness accounts reveals the forces involved:

> "There was hard wind. I was outside. Suddenly there was no wind. My eardrums felt like they would burst, very intense. Then a great wall of white came. There was hard wind and all

white. I could not see through the white. Balls of orange and lightning came from the cone point. Lights darting around in the clouds were sort of luminous and appeared to be more round in shape than anything else. Also they were quite large. [There was] a surface glow, some three or four feet deep, a general brightness, static on [the] radio, [and] we saw one black funnel with a slight glow coming from the sides. The sky was really black. All at once a big hole opened up in the sky with a mass [that had] a yellow tinge in the center and the edges darker cherry red with black spots in the edges."[393]

These are the observations of residents in Toledo, Ohio as it was struck by a tornado on 11 April 1965.[393]

Theories on the formation of tornadoes suggest the effects of a particle accelerator and ionized particles are involved. Electromagnetic acceleration of ions and heat released from lightning strokes are theorized to be the reasons tornadoes form.[283,392,393] When one visualizes the spiraling electric and magnetic force of the particle accelerator, shaped like a funnel, it seems as though the mechanism has indeed shown itself indirectly. What happens is that the Field accelerates particles into the upper atmosphere and magnetosphere, but being time-varying acceleration it ends, and the a vortex of particles re-enters the lower atmosphere, producing tornadoes (also hurricanes, typhoons, etc.). Some weather phenomena display both the time-varying nature and vortex formation, such as the previously puzzling vortex streets (see Plate 1.3).

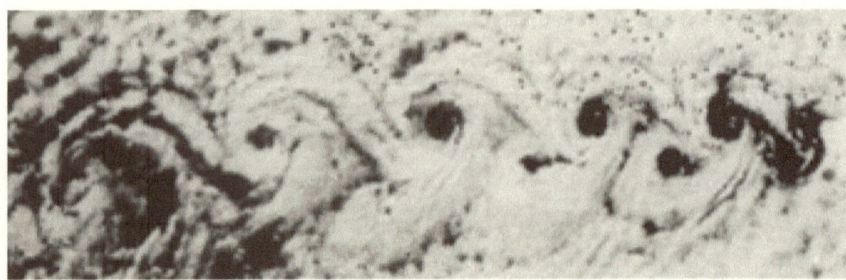

Plate 1.3. Vortex Streets. Vortex streets are another unexplained weather phenomenon. Shown here are the earliest observed over the North Pole. A time-varying Field, in this case the North Polar Field, accelerating particles in bundles can explain vortex streets.[image courtesy of the NERC Satellite Receiving Station, University of Dundee]

All lightning occurs as the result of imbalances in pockets of positively and negatively charged particles (ions). Some who have survived a passing tornado and looked up into its center have noted that the walls were constantly illuminated by lightning that "zig-zagged" from side to side. The white glow, very low pressure ("ears feeling as though they would burst"), lights, dark masses, static on radios, and various colors suggest that ions and neutrons play a part. The southeastern United States, near the North Atlantic Field, and west of that area (due to the Earth's rotation), experiences the greatest number of tornadoes.

It is obvious that similar factors play a prominent role in the development of hurricanes (and typhoons). Existing theories on hurricanes present what seem to be some important contributors that are unique from what generates tornadoes. The effect of the Earth's rotation on wind currents, known as the Coriolis force, give added energy to these cyclones. But, then again the effect of the planetary nebula's rotation also produced the Coriolis force and was instrumental in the formation of the Fields as the Earth formed. Also, a warm-water surface area sufficient to supply large amounts of vapor plays a fundamental role in generating such storms. However, hurricanes have, in a number of cases, been accompanied by one or more tornadoes. Unlike tornadoes, though, they have a wide radius and a center, the eye, characterized by light breezes or a calm. In fact, the extremely low pressure of the eye has caused cans, bottles, and even air-tight crystals on watches to burst. Heat is often a by-product of reactions, so frequently the eye is 8° to 10° Celsius (46° to 50° F) warmer than the storm's main body, because more reactions take place in the center.

Hurricanes are not fully understood by scientists who do not consider the Fields and ionizing radiation.[273,283] The Fields, which are particle accelerators, are not yet known to exist, and therefore, are not considered. However, the wind speeds of hurricanes must be uniform from top to bottom, or wind shear would destroy its structure. A field that controls particle flow seems to be required in order to prevent wind shear. All the dissimilarities between tornadoes and hurricanes designate a wider diameter to the accelerator in hurricanes, and the reason for this lies in FEM's interrelationships with the Sun (a solar-terrestrial or Interplanetary Magnetic Field (IMF)-FEM linkage).

9.5 FEM's Solar Linkage

One of the most revealing observations of FEM's interrelationship with the Sun involves those periods when peaks in the numbers of tornadoes, hurricanes, and other phenomena occur. The North Atlantic Field offers some of the best statistical records, and the results are particularly enlightening. Tornadoes and their water-born relatives, waterspouts, are most frequent from April to June, while hurricanes have their peak in the period which extends from July to October. These observations lead to establishing an understanding that a gradual increase in the release of the by-products of hydrogen fusion occur from tornadoes to hurricanes to a winter peak in the coupling of atmospheric layers and upward planetary waves. As will be shown, a winter peak also occurs in other geophysical phenomena controlled by the Fields (as discussed in Volume Two). This alone suggests the relationship is one of the Earth's position in orbit in relation to the Sun (see Figure 1.7).

The magnetic field of the Sun, which is called the Interplanetary Magnetic Field (IMF), and Geomagnetic Field (GMF) interact to produce these effects. The IMF's arms extend beyond the planets.[44-46,81,202,248,299,378] It is shaped like an Archimedes spiral, which is to say it appears something like the water coming off a spinning-armed lawn sprinkler (see Figure 1.7). In the oval where auroras occur, this magnetic field always cuts along the line between the Earth and Sun. When there is a particularly strong merging of the IMF with the GMF, the Earth's field shifts and the solar wind plasma blows across (*auroral substorms*). Observations of the aurora particles show the electrons moving in one direction and the protons in the other. When these magnetic field lines converge at the poles, they produce a spiral of electrons which looks enough like a tornado's funnel cloud to have been labeled an "electron funnel."[72,204,245,247,296,318,378]

The curve or arc of the IMF is greater the further the Earth is from the Sun. In its interaction, the accelerator at the Pole (North Polar Field) is opened less (smaller diameter). When the Earth is nearest the Sun the reverse is true; that is, the accelerator is opened wider (greater diameter). Herein lies the reason for the observations of these peak periods.

This relationship can be noted in other observations. First of all, the Northern Hemisphere experiences more tornadoes, hurricanes, and lightning than does the Southern Hemisphere. Also there are more cyclones and thunderstorm activity in the north. The reason for this is that when the Earth is nearest the Sun (*perihelion*; early January), the Northern Hemisphere (North Polar Field) is tilted away from the Sun in winter. Furthermore, being nearest the Sun there is less curvature to the IMF at this time, which allows the IMF to appreciably affect the North Polar Field. In contrast, the Southern Hemisphere is tilted away from the Sun in winter when the Earth is farthest (*aphelion*; early July) from the Sun (more curvature to the IMF), so the South Polar Field is less affected. Due to the orbit and tilt of the Earth, the relationship between the GMF and the IMF cause the Northern Hemisphere's Fields to be activated more, and thereby, results in more drastic weather and other phenomena in that hemisphere (see Volume Two for other discussions).

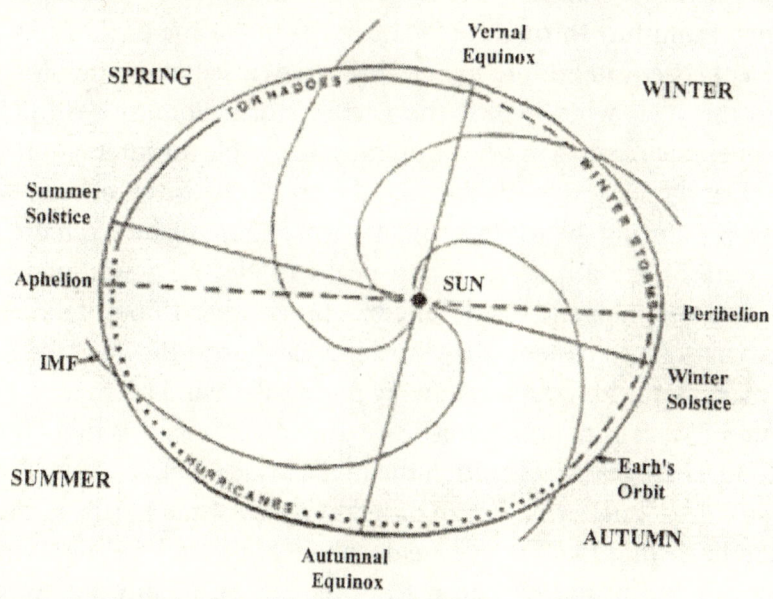

Figure 1.7. **Peak Periods.** Peak periods of tornadoes, hurricanes and winter's enhanced weather disturbances are displayed here for the North Atlantic Field. The peak periods are shown with respect to the Earth's interrelationship with the Sun's Interplanetary Magnetic Field (IMF) as takes place in the Earth's orbit (i.e., a solar-terrestrial linkage). All Fields display similar periods.

The mid-latitude Fields are activated similarly along with (3-4 days after) those at the poles, and again using the North Atlantic Field an example is illustrated. During the tornado peak period of April to the end of June, the Earth is moving away from the Sun and the Northern Hemisphere only begins to tilt toward the Sun. At this time, solar wind and the IMF are moving in the same direction as the Earth, and thereby, the North Pole Field is affected less. In July the Earth is its farthest from the Sun, with the Northern Hemisphere tilting toward the Sun during summer. From July through to October, the hurricane peak, the North Pole Field is activated more by the IMF (less curvature of the IMF), as well as the solar wind, which the Earth's North Pole moves into. By November, the Earth is in a position most favorable for interaction with the IMF, and is closest to the Sun and the greater densities of solar wind plasma. It is during the winter that an intense coupling between the various layers of the atmosphere and peaks in electric fields, planetary waves and geomagnetic disturbances, among other things, take place. To put it in one statement, the closer the Earth is to the Sun, the more their magnetic fields couple, allowing more solar wind plasma to interact with FEM, causing more drastic weather phenomena as more fusion by-products are released through the mid-latitude Fields.

Even weather patterns shift to different zones of the Earth, and these shifts indicate the effects of the Fields.[163,184,190,197,311, 354,380,386] When a hemisphere begins to point away from the Sun, as in autumn, weather patterns begin to shift toward the equator. In winter, weather patterns move to even lower latitudes. In contrast, when the same hemisphere is pointing toward the Sun, as in spring and summer, weather patterns shift toward the pole.[163,190,311] One scientist noting the temperature disturbances in the Tropical Zone witnesses the fact that it takes an "immensely powerful force" to alter temperature so much (*several standard deviations*).[311] Observations from the Dynamics Explorer have revealed planetary waves move upward from the Earth, and that there are much stronger disturbances in the upper layers of the atmosphere in winter.[354] An important link of air exchange between upper layers of the atmosphere (*tropo- and stratospheres*) is in the 30° to 54° latitudes, corresponding not only to the jet stream axis, but also the Fields' latitudes (the Fields spread with altitude due to their funnel shape).[386]

As a result of the Fields and their interaction with the IMF, various cycles have been shown to be correlated. Studies have long shown a strong correlation between cycles in solar activity and weather.[162, 190,270,269] But now there is the unexpected: a lunar influence on weather.[86] Earth-currents are also known to be correlated with solar activity and lunar phase.[239] In combination, these observations indicate the effects of this new model of the Earth, the Field-dynamical Earth Model (FEM), while the lunar influence involves particle flow along Field lines (*i.e., electrostatic repulsion, bow shock, plasma torus, etc.*).

The unusual and other natural phenomena discussed previously show relationships to orbital position, too. Around the vernal equinox (21 March) there is a sudden and abrupt change from winter to summer conditions. These changes include: the reversal of upper atmospheric winds, an end to upward moving planetary waves, lowered absorption in one region, a pressure maximum in the middle atmosphere, and earth-currents (*GMF Sq-currents*) are half as strong.[363,401] It is during this same time that these currents (*Sq-currents*) and high-altitude spirals in the atmosphere (*ionospheric vortices*) clearly show the positions of the Fields at the 30° to 40° latitudes (see Figure 1.8).[335]

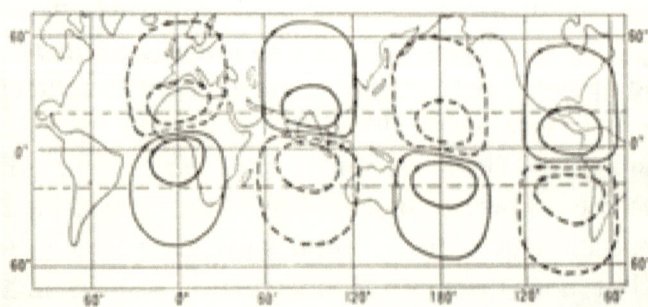

Figure 1.8. Currents in the Upper Atmosphere. Upper atmospheric currents (*ionospheric Sq-currents*) are, in some cases, slightly shifted west from a given Field's center (due to the Earth's rotation and magnetohydrodynamic flexibility), but disclose the relative position of the Fields.[from reference 391; Fig.49, p.140]

One study indicates that the majority of whistlers, which are lightning-induced high-frequency waves, also occur more often at that time.[379] Both the "pillar-shaped" (vernal) and low-level (autumnal) auroras have been observed much more often around the equinoxes. The majority of the wheel-type (vernal and autumnal), milky (autumnal) and parallel bar (vernal) luminous seas occur around the equinoxes. Around the time of the solstice, we see a maximum in upward moving planetary waves, low-level auroras take place slightly more often (equinox more so), and the booming "water-guns" occur more often (especially in winter).[110,111] Again, all of this is the result of the interaction between the GMF and IMF at times when the GMF is at right angles (equinoxes: vernal: 21 March; autumnal: 21 September) or the polar Fields are pointed towards, and mid-latitude Fields away from, the IMF (solstices: 21 June and 21 December).

Observations of electric currents yield similar results. During the summer, these currents are at their strongest with respect to season, which is the time hurricanes are at their peak. In January, these electrical eddies are suddenly intensified (for example, as the Tucson observatory reveals), upward moving (*propagating*) planetary waves develop a maximum, and the Earth is closest to the Sun.[354,391] These currents are always fixed with respect to the Sun and shift with the seasons. They also vary with solar activity (sunspots and solar flares) and lunar position, as do the magnetic field and auroras.

The Moon affects these currents, causing daily variations and greater effects during certain phases of the Moon (New, Full and closest approach to the Earth—*perigee*—especially, and less so during quarter phases). During the Full Moon there is also a minimum in aurora activity, and the sounds emitted near the 30° latitudes are at a maximum.[110,189] The effect of the Moon, as will be shown in Volume Two, involves the triggering of particle flow along Field lines (*electrostatic repulsion, plasma torus, bow shock, etc.*).

Furthermore, earth-currents are known to produce power surges that can destroy transformers or produce power failures.[155,239] Some of the worst power failures occur near the equinox as, for example, on 24 March 1940 when New England, New York, Pennsylvania, Minnesota, Quebec and Ontario all had blackouts. **All evidence shows that changes in electric currents, atmospheric electricity, the geomagnetic field, auroras, weather phenomena, orbital position, lunar phase, the biosphere and solar activity occur in *unison*!**

The interaction between the IMF and the Earth demonstrates the operation of the Earth as described here. The ways in which the Fields at the poles operate (*dynamics of the polar cusp*) are related to its interaction with the IMF.[71,72] Solar wind plasma is injected into the region where the IMF and GMF lines merge.[71,189, 296,378] An electric field is needed to produce this, but its actual character is "unknown" (it is the electrostatic field produced by life and the currents under the ice cap in conjunction with the Polar Field). This field helps to produce the polar-cap arcs or the more continuous auroras (*i.e., theta-arc configuration, etc. along the y-component of the IMF or northward IMF*).[143,187]

A particularly strong disturbance of the Earth's magnetic field, increased electric potential in the polar region (i.e., the polar cap), and an enhancement of auroral currents occur whenever the IMF is strong and pointed southward. Furthermore, all of these observed effects end if the IMF's polarity becomes northward.[170,389] The fields and currents in the polar region are known to be sensitive to the IMF and are understood to possess the characteristics of the life-system on Earth as occurs with FEM (*electrostatic and time-varying*).[190]

Whenever the IMF points northward, it not only ends the aurora, but the electric field around the equator (*equatorial electric field or electrojet*) becomes enhanced and the high-latitude electric fields decrease.[170,225] In fact, a reversal of the electric field at the equator occurs whenever the IMF points northward from a previous steady south.[225] All observations show that the IMF, and the electric and magnetic fields at the equator are strongly interrelated.[170,225,391] Moreover, the depth of the radiation belts at the equator becomes reduced (vernal) or enhanced (autumnal) when the Earth's axis is at right angles to its plane of orbit around the Sun (ecliptic plane), during the equinox and when the IMF is directed towards the Sun (negative IMF).[333]

The reason for these observations is unknown, because this model of the Earth is not yet known. However, it demonstrates that the Fields along the 30° latitudes are activated when the IMF becomes northward. Also, it shows that these Fields repel away from each other along the equator inward toward the Earth as they spread out with altitude (i.e., due to the funnel shape). That is, the solar wind plasma enters the Earth (polar Fields are activated) when the IMF is southward, and then the waste products are released (30° to 40° latitude Fields are activated) when it becomes northward; this will become clear as we proceed.

9.6 The Radiation Belts

Surrounding the Earth are belts of radiation with inner and outer regions shaped something like donuts. Since their discovery in 1958, space probes have gathered a wealth of information about the Van Allen radiation belts. Particles in these belts are what would be predicted from hydrogen fusion. To some, though, this seems to be nothing unusual, because particles such as these are continually sent off into space by the Sun. As with a number of things associated with the Earth it is indeed fortunate it did form and renew. For without these belts, cosmic radiation (x-rays, gamma-rays, etc.) would have impinged upon the atmosphere and biosphere, and most, if not all, life-forms on Earth would have been impossible.

The formation of these belts is the result of the release of the by-products of hydrogen fusion from the center of the Earth, as well as solar wind plasma. The inner region of the belts is composed of protons and some electrons, which scientists indicate are produced by the decay of neutrons. After only 11.7 minutes (*the neutron's half-life*), half of all neutrons present decay into protons and electrons, forming this inner region. In the outer region, helium nuclei (*alpha particles*) are found, and direct measurement shows the amount of trapped particles increase at the time of auroras. Ideally located with respect to the accelerators' release of waste products, the belts are not at the poles where it would neutralize the solar wind plasma producing the aurora, but instead allow solar plasma to enter the Earth's interior. The outer region of the Van Allen radiation belts farthest from the Sun (the tail of the magnetosphere) undergoes a considerable change when affected by solar winds.[338] This alteration of the belts' outer region is so effective that the particles return to a plasma state, rejoining the solar wind (see Figure 1.9).[92,130,137,215,249,250,286,287,306,338,339,380]

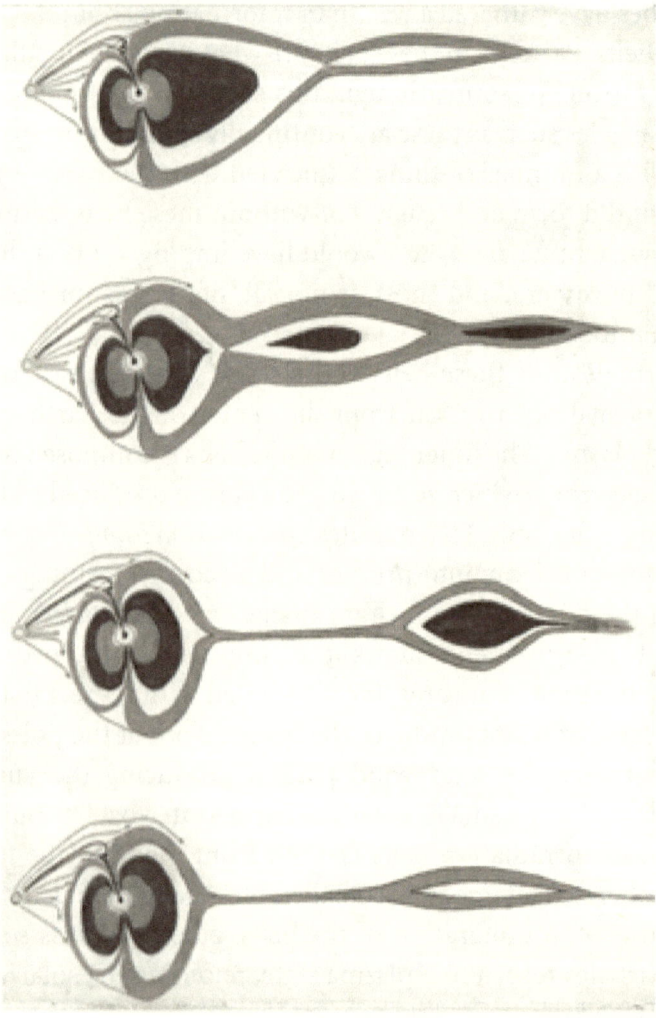

Figure 1.9. Phases of the Magnetosphere. The Earth's magnetosphere undergoes these alterations after receiving solar wind plasma. The lines and arrows show the flow of solar plasma around the magnetosphere, and the flow of plasma into the Earth's interior. Grey indicates the inner, and black the outer, radiation belts, and a lighter gray is the outer layer (mantle) of the magnetosphere. From top to bottom the outer

radiation belt grows, and particles (by-products of hydrogen fusion) are eventually released to rejoin the solar wind. This illustration of the phases of the magnetosphere bears a likeness to the release of waste by an animal. [after reference 344]

Electrons of the inner region of the belts are recycled with the help of x-rays. Energetic electrons during an aurora generate strong x-rays when they collide with upper atmospheric particles. Particle accelerators, which are similar to the Fields, also routinely produce x-rays. X-rays are known to interact strongly with very low frequency radio waves, called whistlers, which are generated by lightning.[368] Some of the electrons are scattered out of the radiation belts by these waves and are precipitated into the upper atmosphere. This mechanism is so efficient that it limits the maximum number of electrons trapped in the inner region of the belts. These recycled electrons, allied with those in the aurora, produce a greater flow of electrons to be utilized by the Fields.

During intense auroras (*magnetospheric storms or substorms*), very frequently a large number of protons are accumulated and the inner region grows abnormally intense (*labeled the equatorial-ring current*). At the same time the GMF is disturbed near the 30° latitudes where fusion by-products are accelerated out through the mid-latitude Fields. This is a result of their magnetic properties, because they are very similar to and interconnected with those at the poles. It is believed that these protons leak out of the polar regions where they too are recycled.

The following section brings forth evidence, obtained from satellite observations, which shows the effects of the 30° to 40° latitude Fields. Observations indicate that the Fields control the conditions of the upper layers of the atmosphere and the magnetosphere. Scientists have noted the existence of the Fields, but do not know where they originate. The data, however, shows that they do exist and are releasing the by-products of hydrogen fusion. These by-products include those ions that were also observed at the poles, such as hydrogen, helium, oxygen and nitrogen. Dynamics of the Fields reveal synchronous alterations in the earth-currents, the magnetic field, auroras, solar activity and the position of the Earth in its orbit.

The Fields are activated by the Earth's relationship with the IMF, especially during the seasonal change that marks the times of the equinoxes and solstices. This is particularly evident during the equinox which shows abrupt, sudden changes in spring and autumn. Earth-currents double, ion counts increase, and the upper layers of the atmosphere become disturbed as a given hemisphere points away from the Sun (autumn and winter). Shortly after the March or vernal equinox, earth-currents lose half their strength, ion counts decrease, and upper layers of the atmosphere are no longer deeply affected (spring and summer). Furthermore, changes in the earth-currents and the magnetic field occur along the 30º to 40º latitudes and the equator following auroras, which demonstrates a coupling between these Fields and those in the polar regions. The dynamics of FEM disclose that there is an activation of the polar and 30º to 40º latitude Fields at the same time.

Gravity anomalies, airglow and radio waves are noted within the Field regions. Recyclable ions flow back into the solar wind due to the dynamics of FEM. All of the observations prove the existence of FEM. The observations are presented in the following discussion, but it may be somewhat complex for the average reader, though it is written with them in mind. Therefore, some of you may find it easier to pass by the section marked by the asterisks (* * *) and should consider this paragraph a review.

*   *   *   *   *

9.7 Observations of Mid-Latitude Fields

A number of observations provide evidence for the existence of the mid-latitude Fields situated along the 30º to 40º latitudes. During conditions of enhanced activity in the Earth's magnetic field, the electric field at the equator (*equatorial zonal electric field*) is transformed by "direct penetration fields."[170,225,260,334] Furthermore, a coupling exists between the electric fields along the equatorial and auroral regions during these same conditions.[334]

In winter, the middle atmosphere shows the effects of an upward movement (propagation) of planetary waves that reach a maximum in January. Also, in winter there is an upward extension of some layers of the atmosphere (*stratosphere and mesosphere*), unusual patterns in some regions (*D and E*), and a doubling of ion counts (*in the ionosphere*). These data reveal the time-varying effects of the mid-latitude Fields of FEM as they become activated.

Shortly after the equinox, in early April, a time when the IMF and GMF interact the most, there is a sudden, abrupt transition from winter to summer conditions (at all levels between 30 and 150 km., or 19 and 93 mi.). This sudden change includes winds that reverse from west to east in the upper atmosphere (*stratosphere*), the disappearance of the upward moving planetary waves, a pronounced minimum in one region's (*D-region*) absorption, a pressure maximum in the middle layer of the atmosphere (*mesosphere*), reversal of the meteor wind, and changes in geomagnetic field currents (*GMF Sq-currents are half as strong as in winter*).[353] Because of the limitations of scientific theory and instrumentation, statements like these arise: "The evidence shown for an intense coupling between the basic states in the strato-, meso- and thermosphere during late winters is physically not yet explainable."[353]; p.302

Another scientist examining the electric field in the atmosphere notes a strong field in a layer of the atmosphere (*lower mesosphere*) that is "most startling." Commenting further he states that there is "no satisfactory explanation of their existence [and that it] must involve [a] generation mechanism."[320]; p.192

Because of the sudden abrupt spring equinoctial changeover in the upper atmosphere (*stratosphere*) and the decoupling of the upper middle atmosphere from below, a question arises: "Why are the internal atmospheric processes so well reflected in the upper region during winter?"[353]; p.302. The answer to this question is that the mid-latitude Fields are more active in winter, because that hemisphere is pointed away from the Sun, which allows for interaction with the IMF (*solar wind/magnetospheric coupling is noted in convection processes, especially in the magnetotail*).

During some events (*excitation of kinetic Alfven waves*) there is a strong density of warm plasma near the equator and a large-scale magnetic field variation that coincides with the onset of strong irregular pulsations on the ground.[230] Evidence indicates that the electric field at mid-latitudes is influenced by currents in the polar region (*FAC*). The electric fields are sometimes reversed at mid-latitudes, indicating a definite mutual, time-varying relationship.[260]

Energetic hydrogen, helium, oxygen and nitrogen ions (*and neutrals*) were observed to flow along a zone centered along the magnetic equator.[71,98,334,394] This zone has a very limited spread and a maximum of ionization, when theory suggests a much broader spread, and no maximum, but a reduction of ions (*due to backscattering*).[394] These particles rapidly spiral around field lines, "bouncing" between mirror points on the opposite sides of the magnetic equator, and have a relatively slow drift around the Earth (electrons eastward and protons westward).[161]

During the main phase of a magnetic storm the area extends from 15° to 45° latitudes.[98,394] Here we see a spread of 15° on either side of the Fields at 30° latitudes, which reveals their cone or funnel shape. It is important to keep this spread in mind, because it is observed in various phenomena on Earth and other celestial objects. The observations noted along the magnetic equator are due to the Fields turning inward towards the Earth along the equator. All of these data are evidence of the existence of the Fields that release fusion by-products. In fact, it is understood that some "unknown" acceleration mechanism in the region is needed to explain the evidence.[391]

Observations indicate an electrical coupling between the magnetosphere and an outer layer of the atmosphere along the equator (*equatorial ionosphere*), and also that this coupling is strongly influenced by the IMF.[137,147,192,225,287,360,380] During magnetic storms, electrons are injected into the radiation belt (*slot region*). This results in the development of a low-latitude radiation belt, producing a zone of ions along the equator (*equatorial zone ionization*).[395] Activity in the geomagnetic field is the feature most correlated with these changes, along with ions being "injected into outer regions" (*ring current and equatorial injection boundary*).[2,71,344, 394] As a scientist states: "The 'sudden' appearance of heated ions in this region has been correlated with the expansion phase of magnetic substorms. The actual electric fields and wave-ion interactions experienced by the ions along their paths are not known."[98; p.415 & 418]. The "unknowns" are merely the result of not knowing the model of the Earth presented here, but the observations confirm its existence.

The global character of changes in the equatorial, low and middle latitudes show the effects of the Fields, too. Magnetic disturbances (*small-scale transverse*) occur more often in the hemisphere experiencing winter, and mainly in an east-west direction, but this does not occur around the pole.[43] Observations disclose the fact that there are global zones of energetic particles precipitating towards Earth in the middle, low and equatorial latitudes.[394] The mid-latitude zone is the northern end of the Fields in the Northern Hemisphere, and the southern end of those in the Southern Hemisphere. The low-latitude zone is the southern end of the Fields in the Northern Hemisphere, and the northern end of those in the Southern Hemisphere. The equatorial latitude zone is the result of the Fields of both hemispheres deflecting inward toward Earth along the equator (see Figure 1.10).

Electron concentrations are observed to either increase or develop contours in relation to the Fields.[232] During geomagnetic activity a worldwide electric current around the equator, known as the equatorial electrojet, changes intensity (*variability of the zonal component*).[170] These effects, as well as the unexplained maximum of ionization at the equator[394], are the result of these Fields spreading with altitude then descending inward along the equator. An occasional reversal of ion flow along the equator (*counter-electrojet*) occurs as the result of solar and lunar influences.[391] In the low-latitude zone, energetic electrons were observed to precipitate towards the Earth above the 20° latitudes, and this coincides with a low-energy radiation belt. In the mid-latitude zone, protons and other heavy ions were observed to precipitate earthward.[394] At times of very disturbed magnetic activity there is an increase of particle precipitation, enhanced electron temperature, reduced electron counts in certain regions (*i.e., density of the E and F regions*), intensified airglow, and a greater flow of electrons, protons and heavier ions.[170,394]

Their relationship with the aurora is predictable with FEM. When the aurora takes place electric fields and currents decrease, and there is a triggering of changes (*in amplitude*) along the equator and at the 30° latitudes. Furthermore, this change is greater than predicted from present models of the Earth (*from a constant E-field*), indicating that there is a need for the type of field that is found in the new model (*i.e., time-varying electrostatic auroral conductivity*).[170]

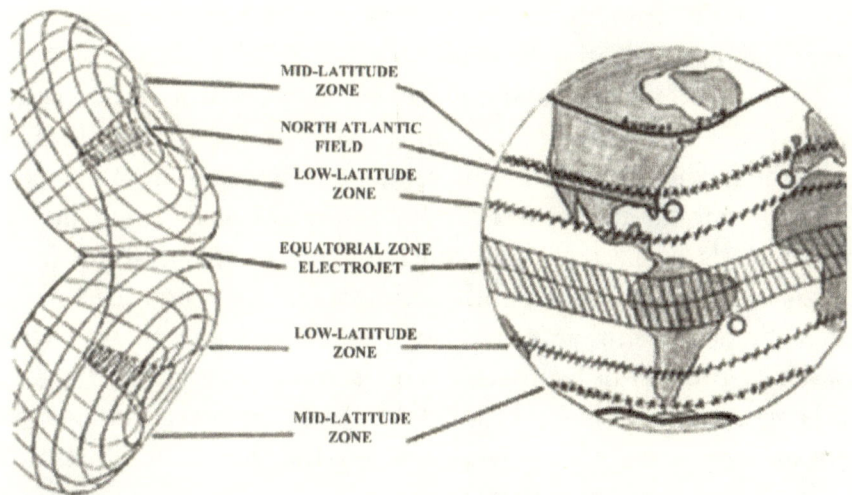

Figure 1.10. Zones of Observed Particle Flow. The illustration on the left shows what the Fields would look like if we could see them. However, they are not quite so perfectly shaped as shown here, but take on contours according to FEM's interaction with the IMF, the intensity of solar activity, and wildlife's presence or absence. The illustration on the right is the zones of observed particle movements as illustrated by scientists.[394] The North Atlantic Field is labeled, and the Mediterranean Field and the Brazilian Field are also shown as circles. Take note of the sine curve arrangement (basically an "S" on its side) of the Fields and the zones, as this will be evident in other celestial objects.

There have even been direct observations of these Fields aside from what has been mentioned previously. An "excessive electron density" occurs at night, and higher levels of oxygen ions are noted in the North Atlantic Field (i.e., the east coast of North America) than anywhere in the Northern Hemisphere.[149,161] As could be predicted, there are also occasional rapid changes in the earth-currents in the Field area (Florida Straits) that are tied to effects observed in the GMF.[240,410] The Field just east of South Africa shows particle loss in a "slot region" between the inner and outer radiation belts.[120]

The Field off the east coast of Brazil has been particularly observable, and is referred to as the South Atlantic Geomagnetic Anomaly (SAGA). Here the magnetic field has low values, and particles are observed to "dip low" into the atmosphere (i.e., into the Field's contours). There are higher levels of electrons and ions, and higher temperatures in the Brazilian Field (*SAGA*) than other comparable areas, especially during magnetic field disturbances (*L-shell height is at a minimum*). Helium ions, a by-product of hydrogen fusion, show a pronounced "bulge" in this area. At times this Field (*SAGA*) shows an extremely complex pattern on some maps, while other maps show no irregularity whatsoever, indicating time-varying characteristics.[161] A strange airglow is also noted in the area, similar to the aurora and the lights noted in the other Field areas.[102,165,172,264] In addition, an enhancement of particle precipitation is noted in the area.[156,165, 379,394] Recent observations show what has been called a "hole" in the area, and a scientist comments that "observations prove for sure it exists."[165]; p.214.

As if all of this evidence is not enough to prove the existence of these Fields there is more. Anomalies of magnetism and gravity exist in each of the Field areas, with typically low values for each. For example, (*free-air*) gravity values in the North Atlantic Field area (Caribbean Region) are the most extreme in the world, with the lowest in the Puerto Rican Trough (*-355 mgal*), and high values in the Greater Antilles (*+200 mgal*). In fact, the Puerto Rican Trench has a gravity anomaly so strong that the ocean above it is actually depressed. There is a very complex zone and a prominent gravity low at the mouth of the Mediterranean, in the Straits of Gibraltar. The Persian Gulf shows a strong negative value just outside its mouth, and in the Gulf of Oman is a strong positive value. In fact, there is a belt of negative values that stretch across parts of Europe and Asia for which the Persian Gulf and the Gulf of Oman mark the southern boundary. Low values occur within the

Japanese Field, South African Field, East and West Australian Fields, and within the Fields in the Pacific, as well.[74,111,135,161,301]

Other observations also lend credence to the acceleration processes of these Fields. Photomultipliers in the Fly's Eye telescope (100 kilometers southwest of Salt Lake City, Utah) recorded the most energetic "cosmic-ray" particle in the upper atmosphere ever detected. When the particle struck atmospheric molecules it caused the atmosphere to fluoresce. This indicates that there is a particle accelerator (the Fields) orders of magnitude more powerful than our largest earthbound atom smashers. Furthermore, observations of the upper atmosphere (especially over thunderstorms) have revealed mysterious light flashes, radio emissions, and even gamma-rays, all of which are indicators of acceleration processes. The gamma-rays, detected by the Compton Gamma Ray Observatory, came in brief intense flashes, and must occur above 30 kilometers in order to escape absorption by the atmosphere and reach the detectors. Even though the gamma-rays are associated with thunderstorms, lightning is not energetic enough to produce gamma-rays, and lightning does not occur above 30 kilometers anyway, indicating high-altitude electrical discharges. Mysterious double radio bursts have been recorded more than 100 times (since 5 November 1993) by the radio receiver on the satellite Alexis, designed to spot nuclear blasts. These radio emissions are not typical of those generated by lightning and are not accompanied by flashes of light. The bursts are separated by forty microseconds, indicating the time-varying acceleration of the Fields, and the frequency dispersion indicates that they have traveled through the ionosphere. All of these observations are considered mysteries, but clearly display what would be expected of the effects of the Fields.

During the equinox, spiraling disturbances in the upper atmosphere and currents on the Earth (*ionospheric current vortices and Sq fields*) disclose the position of the Fields quite well (as shown in Figure 1.8). When considering electromagnetic Fields on the Earth, scientists note that there are at least two oppositely directed spirals (*vortices*), one in the South Hemisphere and the other in the Northern Hemisphere, with foci at the 30° latitudes.[325] More and more information is becoming available that demonstrates the Field areas have many unique characteristics.

9.8 The Global Field System

Other observations indicate that the mid-latitude Fields are very similar to those in the polar regions. A delay was noted during events when eastward moving electrons entered the Brazilian Field (*SAGA*), indicating that an electric field exists there. Increases in electron counts occur at the southerly end that decrease during geomagnetic disturbances as the Field activates. In fact, during geomagnetic storms, fluxes of energetic electrons move into the outer regions of the radiation belt. After this enhancement, they defuse to a lower level (*L-shell*) over a few days. Persistent ionization (*in the D-region*) and radio waves are absorbed over several days after these storms occur. In addition, greater quantities of electrons are precipitated in the mid-latitude region during these storms.[161]

An electrical coupling between an area of ionization at the equator (*equatorial ionosphere*) and the magnetosphere is noted.[161] This coupling involves a penetration mechanism that is "unknown," but it is observed during increased magnetic activity; a time when the Fields activate. As could be predicted by the Field's effects, electromagnetic coupling between the high and low latitudes took place nearly simultaneously (*in the plasmasphere*).[170] The physical characteristics of the Field (*fluctuating electric fields and double layers: an electrostatic/time-varying accelerator*) are needed to explain the electrons' parallel and antiparallel movement in relation to geomagnetic field lines (*counterstreaming*).[200]

Furthermore, irregularities in the stream of electrons along the equator (*equatorial electrojet*) are the result of electrostatic waves from the Fields.[312] Hydrogen, helium, oxygen and nitrogen ions (*terrestrial ions*) are noted to appear along the equator worldwide, but the means for their acceleration is said to be "unknown."[161,344] Speculation has led to the belief that these ions would occur only in the lower energy range of the magnetosphere, but instead it is throughout. In order to explain this density a "second population" of ions, other than the polar regions (solar and polar winds), is needed.[344] The interconnected character of the Fields is also evident in the observation that electric field disturbances occur worldwide during periods of intense auroral zone, magnetic field and electric field activity.[170,274,363]

Other observations show interconnected relationships between the Fields at the poles and those at mid-latitudes. Auroras have effects on the atmosphere's electric field (*even the fair weather field*) that cannot be explained by simply considering solar plasma entering the polar region.[180] Vertical fields and gradients (*electron density*) are coupled to and fluctuate along with the horizontal electric fields and gradients (*electron density*).[312] With an understanding of those Fields at the 30° to 40° latitudes, these observations and the basic problems underlying the maintenance of the atmospheric electric field are solved.[125]

The precipitation of energy input is affected by parallel electric and geomagnetic field lines. Also, electrostatic fields become enhanced along the equator (*in the resonance region of the magnetosphere*). Furthermore, the effects of strong localized electrostatic fields directed upward from high-latitude regions were observed ($1R_E$). The most significant wave-particle interactions occur in the equatorial region of the magnetosphere, and the precipitation of energy input changes along with parallel electric field shifts.[349]

Within the auroral ovals of both the winter (Northern) and summer (Southern) hemispheres, sunward winds occur. Likewise, sunward winds occur in the equatorward regions with a westerly direction before midday (*suppression of the equatorial electrojet*) and an easterly direction after midday.[161,319] Furthermore, there are strong changes (*spatial gradients*) in the earth-currents (*telluric*) that are related to the auroral and equatorial electric currents (*electrojet*).[239] The mutual character and interconnected relationships of the polar and 30° to 40° latitude Fields are quite evident, but virtually ignored.

*   *   *   *   *

9.9 FEM's Dynamics

All these observations demonstrate the dynamic equilibrium of matter and energy flow (more evidence of homeostasis). The evidence shows limits are set to the number of particles in the radiation belts, just as there are limits to the amount of digested food in an animal's intestines. In addition, the way in which the excess particles are released is very similar to an organism's release of waste (as noted in Figure 1.9).

When the interaction of the GMF and IMF is considered, it becomes apparent that these regions are both intensified and altered by the solar wind so that the radiation belts do not grow abnormally large nor too thin, thereby counteracting its protective capacity. It might even be likened to an outer layer of "skin" that is somewhat transparent. Considering the orbital position of the Earth in relation to the flow of solar wind plasma, and the effects on the radiation belts, a dynamic state of equilibrium maintains a consistent amount of matter undergoing fusion. As more solar wind reaches the Earth, more fusion by-products are released into the radiation belts and more of the radiation belts are altered by the solar wind.

Space exploration has yielded an interesting observation in relation to the release of fusion by-products. Spiraling gaseous nebulae were observed in photos taken by the cosmonauts on one of their earliest space flights.[218] This, and other observations, brought about a grid system which reveals each of the Fields (see Figure 1.11).[64,169] The release of fusion by-products into the radiation belts had been observed in these photos.

Figure 1.11. Grid System of the Earth. This grid system was produced by professionals from the now defunct Soviet Union who were experts in the areas of geochemistry, ornithology and meteorology. They noted that certain geologic features, bird migration, and weather phenomena occurred along these grid lines. Unknown to them was the fact that at the center of each pentagon is the location of a Field of the Field-dynamical Earth Model (FEM); the descending limbs are along the edge of the pentagon (basically). This was only reported in scientific literature in Russia and later in magazines in the U.S., but was presented here to show that others have seen the manifestations of the Field-dynamical Earth Model, even though they had not realized it fully. [from references 64 and 169]

Other terrestrial and biological phenomena demonstrate the unique features of the Field regions. Seventy percent of all life on Earth is situated between the 40° latitudes. The importance of life in producing electrical currents has already been discussed, and this ensures that relationship.

Ocean surface salinity is also at some of its highest values in these areas. The electrolytes causing this salinity are similar to those that produce the electrical charge in a battery. Drops that emerge at the collapse of the jet that forms when a bubble breaks at the ocean's surface also carry a charge.[66,67] Measurements made at sea near Barbados, in the West Indies (the North Atlantic Field area), for example, show that whitecaps, where many bubbles are bursting, have a charge at the surface that is greater than in other areas. Whitecaps are directly related to wind speed (*approximately proportional to the square of the wind speed*), and therefore, oceanic charge should be greatest where winds are greatest. As already discussed, weather centers in each hemisphere occur in these same areas, stronger winds are observed, and as a result, oceanic charge is greater.

Moving water also creates electrical currents, and ocean currents spiral in the Field areas.[52,196,225,306] Ocean waves produce a magnetic field, and ocean contact on the coast has been observed to generate currents on land.[139,144,191] Shoreline also helps generate bursting bubbles, and all but two Fields (in the Pacific Ocean) are in the ocean near some land mass. The largest conductivity values (*electrostatic and VHF*) occur at the bottom of ice sheets, especially at the bottom of a grounded ice cap, and both poles are ice covered.[51,211,262,298,321]

There are even conditions that counterbalance the effects of winter hibernation and dormancy in some organisms, which generate electrical current. The point charge of leafless trees is greater than the same trees with leaves.[87] During winter, deciduous trees, which shed their leaves seasonally, release more point charge into the atmosphere. Deciduous trees are more abundant above 30° latitudes, and their origin might be considered as an adaptation to the maintenance of electrical environment of FEM. In Earth history when a full-year growing season existed (*Cretaceous climate was uniform*), these trees did not exist, but when winter came (*Cenozoic*) these trees (*dicotyledons*) are first noted in the fossil record.

Later discussions will confirm the existence of these Fields by noting their effects on biological and physical phenomena. These effects will be shown in the impact of weather on mental and physical health (biometeorology; Chapter 14), extinction and speciation (biogeography and paleobiogeography; Chapter 22, 23, 27 and 28), continental drift (plate tectonics; Chapter 29, and Chapter 16 of Volume Two), weather (solar-terrestrial linkages and geographic restrictions; Chapter 15 of Volume Two) and more. Also, the importance of life in maintaining the stability of FEM's dynamics is supported throughout.

9.10 FEM: A System of Unified Fields

On 18 April 1955, one of the greatest scientists of the 20th Century left this world and behind him an array of theories for future generations of scientists. Many mathematical models and experiments each day undergo tests which examine an elite logic. One of Einstein's theories yet to be proven is that of the Unified Field.

The basis of this theory is that our limited perceptions of time, space, matter and energy are not separate and distinct in existence, but are changeable under the influence of electromagnetic forces. In practice, it concerns electric and magnetic fields as follows: An electric field generated in a coil produces a magnetic field at right angles to the first field. Sound familiar? Each field represents a plane of space, but since there are three planes of space, there must be a third field other than these electric and magnetic fields.

Just as theorized, time, space, matter and energy are transmutable by the Fields of FEM. The hydrogen-rich solar wind produces aurora displays and is transformed into helium nuclei, electrons, neutrons, and even occasional nitrogen and oxygen ions. Energy is changed from high-velocity plasma to fusion energy, and high-speed electrons become a magnetic field. As discussed, many aspects of the physical world, from terrestrial electricity to weather centers, are affected by the Unified Fields of FEM, which are the particle accelerators along the 30° to 40° latitudes and the poles. More will be revealed throughout this book that confirms the existence of Unified Fields, not just with FEM, but other objects throughout the Universe, as theorized by Einstein.

9.11 <u>Organism</u> <u>FEM</u>

In review of this section, again observations of organism-like qualities characterize the Earth. Life has a hierarchy of components and processes. When we consider the interrelationships of life on Earth with the internal structure, it must be concluded that there are centers in vital processes.

The Earth takes in a food-like substance (solar wind plasma) and utilizes the energy for work or potential work. One could even consider that the energy liberated in earthquakes, geothermal energy, volcanics, the hydrologic cycle, plate tectonics, and other geophysical and biosphere activities as the metabolism of a giant organism. The release of this energy, especially through time, totals such a vast amount that it cannot be generated by today's image of an Earth with an iron-nickel core.

As occurs with all living things, all conditions on Earth operate in response to the necessities and in support of the preservation of the whole. A living Earth becomes inescapable in light of a single understanding: **"With very little qualification, one could say that life is transformed sunlight."**[337] (bold added).

CHAPTER 10

Growth and Reproduction
(The 'Expanding' Earth and
Observable Reproduction
as Criteria for Life)

All living things grow and reproduce. This, of course, presupposes that both these processes can be observed within a period that is somehow relative to a human lifetime. With the Earth, this obviously becomes a difficult task, because its life span is far greater.

10.1 Growth of the Earth

However, evidence of the fit of the continents against one another in the distant past shows that the Earth may have once been much smaller. Far back in the history of the Earth, today's continents were grouped together in one super continent (*Pangea in the early Jurassic and prior*). From that point, the continents we know of today separated, and through time, took their present positions. If the continents were taken and reconstructed into that original super continent to fit as well as possible on a globe with the dimensions of the modern Earth, gaps (*gores*) would appear between different parts of the Earth's crust.

These gaps can be eliminated by shrinking the dimensions of the globe to an Earth with a mean diameter that is 80% of the present-day value. Then the continents would have a direct fit and correspond with the available geological evidence of the connection. At these dimensions, the Earth was only 51% of its present volume, and the surface area was 64% of that of the present day.[294,295] The implications for FEM can be observed in this statement: "Perhaps we are completely wrong and the inner core is in some state nobody has yet imagined, a state that is undergoing a transition from a high-density state to a lower density state, and pushing out the crust, the *skin* of the Earth, as it expands."[295]; p.29 (bold and italics added). If it had been something else in the natural world other than the Earth, the word "expands" would have probably been replaced with "grows," but the insight is there because the crust is referred to as the "skin," and grow and expand are synonyms in any thesaurus.

10.2 Formation of the Oceans

Another indication of growth is afforded by the oceans. No one really knows how the oceans formed, but it is believed that they derived from the interior some time after the Earth formed as a planet. The large quantities of water in the oceans is itself an enigma, because water molecules separate (*dissociate*) in the upper atmosphere with hydrogen being lost to space. Beyond this we have also observed hydrogen (terrestrial ions) being accelerated away from the Earth in the polar wind and throughout the magnetosphere (even during quiet times).[249,344] Therefore, the loss of hydrogen is more than was previously considered, making this a mystery that is even harder to explain.

More water must be produced by hydrogen release (*outgasing*) from the Earth's interior. In fact, this may have been witnessed in the North Atlantic Field area during the Barbados Oceanographic and Meteorological Experiment. During the first few weeks of May low salinity water was observed in the area.[35] Heavy spring rains or water from South America's Amazon River were the possibilities suggested as an explanation. However, these suggestions do not adequately explain the evidence.

One of the more probable explanations is that hydrogen and oxygen ions, released through the Field, combine to form water. The combination of oxygen and hydrogen under the influence of electrical energy to produce water is an established fact. Moreover, rocket and microwave detection of water in the upper atmosphere have been noted. As we will see in Volume Two, the planets and the Moon also reveal a Field-dynamical Model and some show evidence of episodes when water swept there surfaces.

Another possible explanation is that sulfate bacteria produce iron sulfide at a depth that is deficient in oxygen, which could yield water and free sulfur.[112,342] Both of these possibilities require the presence of life in association with geologic change throughout Earth history, either chemically or electromagnetically. This is strong evidence of a changing Earth, if not direct evidence of growth and regeneration.

10.3 Reproduction?

Because all living organisms known to humans reproduce, should reproduction be a criterion by which we can determine whether the Earth is alive? This is not a simple question to answer. It could easily be concluded that there are other Earth-like planets throughout the cosmos, therefore, the Earth has reproduced. However, there is the question of whether quantity means something has reproduced.

Possibly, planets are reproduced in a form of asexual reproduction. In asexual reproduction new individuals are produced by budding from a parent system. There is no requirement for one individual to fuse with another.

Stars are produced in the spiral arms of galaxies and planetary systems seem to be the inevitable outcome of star formation. Protoplanetary nebulas will be discussed in Volume Two. Recent instrumentation in space research has also revealed, through spectral analysis, that water is on at least one nearby planet, suggesting other Earth-like planets exist.

CHAPTER 11

Adaptation and Responsiveness *(Animal Response to, and Phenomena Associated with, Earthquakes and Volcanic Eruptions, Including the Coevolution of the Biosphere with Geophysical Phenomena)*

11.1 Responsiveness

Responsiveness is readily observable between life on the Earth's surface and the Earth as a planet. *All* biological organisms, from bacteria to whales, react in response to changes in the Earth's magnetic field. Life, every single organism, shows what has been called a "preperception" three to four days *prior* to changes in the GMF.[127] Could it be that life's interrelatedness somehow helps to cause these changes? After all, how does something as simple as bacteria or a single-celled plant know about changes that are yet to happen? Humans, who have the most independence from the environment, not only show this response, but also have everything from biochemistry and biorhythms to accidents and disease states correlated with these fluctuations.[163] As will be shown, this responsiveness is apparent in many ways and throughout time (e.g., see Chapters 14, and 15, and Volume Two). Responsiveness is always known to exist with a living organism and is one of the criteria for establishing whether a thing is alive or not.

Communication between living things on Earth is becoming increasingly apparent. For example, trees can communicate with one another in order to warn of possible predators. When under attack by insects or animals they release a chemical into the air. This distress signal is received by nearby trees, which then step up their production of tannin, a mild poison in their leaves that gives insects and some animals indigestion. Its production is determined according to the duration and intensity of the attack. Some tree species in late summer, with fall nearby, take no steps to defend themselves with tannin.[49] They will soon lose their leaves anyway, but how do they know when and when not to defend themselves at just the right time?

Many of us are familiar with experiments conducted with brine shrimp and plants. Brine shrimp were dumped into boiling water, and a plant nearby, with electrodes attached, nearly always responded. Other experiments, some criticized for not being scientifically performed, have suggested that other types of communication may exist, especially between human and other life. Diverse types of responsiveness exist, including whole regions of organisms responding to geophysical phenomena; such as the following.

At first the day seemed like a fairly ordinary one, when unusual things began to happen. Normally in the mountains, flocks of deer came down and crowded together near the village, and seemed uninterested in grazing; a sight never seen before. Cats left their homes and not one could be found in the village for the two days that followed. Kittens were taken outdoors by their mothers and bedded in vegetation. Mice and rats left their hiding places, and in some locations they could be seen aimlessly scampering around. Fowl refused to roost, but scattered about noisily. Cattle and other livestock panicked in the barns and about fifteen minutes before the event showed clear signs of fear. Dogs seemed to bark without reason. Birds became restless and emitted calls at times they were normally more likely to be inactive. It was on that day, 6 May 1976, that the village of Friuli, in northeastern Italy, was struck by an earthquake.

Unusual animal behavior preceding earthquakes is so consistent that it has been used to predict them. In 1975, Haicheng in China was hit by a quake that was successfully predicted partially as a result of this knowledge.[94] Even an illustrated booklet, *Earthquakes*, which has been compiled by the Seismological Office of Tientsin, in China, says both historical and recent surveys prove animals react before the event. Additional evidence from the Chinese indicates that 58 species are aware of approaching earthquakes, and there undoubtedly are more.[154]

For example, a Japanese scientist noted that quakes in the Idu peninsula and the number of fish caught near the end of Sagami Bay were correlated. In the spring of 1930, swarms of quakes hit Ito on the east coast of the peninsula, and it was around that time that abundant catches of horse mackerel and other fish occurred at the Sigedera fishing grounds. On the other end of the biological spectrum, falls of camellia flowers were also correlated with quakes, by this same scientist.[26,374]

Even we humans are affected with disorientation, giddiness, nausea, uneasiness and feelings of impending calamity prior to and during a quake. Scientists suggest that this is the result of human sensitivity to ground waves[188], and to electrostatic effects (*including the Serotonin Irritation Syndrome or Serotonin Hyperproduction Syndrome*) and electromagnetic forces.[381] In other words, observations have shown that we humans are sensitive to the Earth's nervous system impulses, too.

Knowledge of this sort extends back at least 2,000 years to the time of the naturalist and writer, Pliny the Elder, who designated animal response as one of four signs of a threatening earthquake.[309] Thirty-three independent reports from various parts of the world have been compiled by the United States Department of the Interior[388], one author has collected 78 reports from folk tradition and mythology[383-385], and various other reports also exist.[5,176,250,365,370-374,409]

Probably one of the most interesting observations dates from 373 BC in a region of ancient Greece. Helice, in Achaia, bordering the Gulf of Corinth, was hit by such a violent earthquake that it sank beneath the sea where it remains to this very day. However, five days prior, swarms of animals, including rats, weasels, snakes, worms, centipedes and beetles, migrated across a connecting road toward the city of Koria. They had successfully predicted the earthquake, saving their lives (the components of the Earth-cell). The reason for this type of behavior has most scientists baffled.

Yet, one researcher states what could be expected of FEM: "The ground gives off static electricity before an earthquake."[381] In addition, increases in the intensity of the earth-currents (*telluric*) are considered one of the warning signs of an impending quake.[198] This coincides well with the understanding that earth-currents are maintained by static electricity and the superconductivity of living things. The physiological effects on animals may also be due to air ions offsetting biochemistry (*Serotonin Hyperproduction Syndrome*).[381] The outcome is that life, the components of the Earth-cell, is preserved. Like an organism with a damaged cell, an electrical message (nerve impulse) is sent, the cell is saved, and that which is damaging the cell (i.e., the city or disease) is destroyed.

Aurora-like glows often accompany earthquakes, and this plus the animal response has led to a theory. It is said that the effect may be due to quartz microcrystals in rock under high pressure. However, there must be a fairly high proportion of crystals, but this is not sufficient in itself. They must not be randomly arranged, but, rather, in the same direction, so that the electricity produced by one is not canceled out by another. Then and only then, with sufficient pressure, will an electrical discharge be produced. This theory is known as the piezoelectric effect.

A number of problems with this theory are known. For one, it is inapplicable in at least some situations. These lights (aurora-like glows) have also been observed over the sea. The sea floor is not solid, and currents constantly rearrange the crystals. Also, sea water is high in conductivity, which would neutralize (*buffer*) the forces. Earthquake lights are most frequent when the Moon has passed its closest approach, and thereby, occur during a decrease in the lunar tide.[397] If the piezoelectric effect were producing these lights, then the opposite would be true (i.e., quakes would occur with an increase in lunar tide). One report correlates luminous seas, discussed previously, and earthquakes.[30] This observation cannot be explained by either the bioluminescence theory nor the piezoelectric effect, but can be readily explained by electrostatic forces produced by particle flow along the Field lines of FEM (discussed further in Chapter 28).

There are also many lighted-displays that quartz could never produce, such as the following. It was a clear, cloudless, morning sky when a strange rainbow appeared, attracting attention. Being out of season and unlike anything ever seen before, that unique occurrence wedged its way into the memories of many. As the next morning progressed, with it came the rumbling and quaking of the ground. In the eyes of awestruck faces one could see reflections of flashes of light, bluish flames, and aurora-like afterglows that painted the heavens. Where the ground shook most, there came bewildering bright beams, fireballs, funnel-shaped lights, and moving luminous columns. Toward Manpukuzi Temple, a straight row of radiant round masses revolving with considerable splendor could be seen. On that day, 30 November 1930, the residents of Tango, Japan had much more to remember than the previous day's unusual rainbow.

Spectacular ostentations and a variety of wonders are commonly a part of the earthquake scene. Rain often occurs before, during, or after the shock, attended by thunder, lightning and wind. Globes of fire, illuminations, extraordinary lights and ball lightning, often claimed to be meteors, are seen. Dark fogs, red and blue suns, and gray and red lurid skies, to name just a few phases of the colored atmosphere, are other associations.[110,361] Magnetic fluctuations at or near the quake area always occur. In fact, changes in magnetic field characteristics during and after quakes can be local or even global.[32,34,36,110,117,193,194,213,214,223,278,351] The atmosphere also manifests aurora-like incandescence, fire, smoke, electrical activity, cold air, tempestuous winds and/or total calms. Added to the list is an array of indescribable sounds or total silence.

Anomalies in the earth-currents near an earthquake's epicenter demonstrate that FEM's electrostatic forces are involved in the event.[372,381-385] Sparking, electric shocks, and the mutual attraction and/or repulsion of objects also show these electrostatic effects.[381-385] Radio emissions can be caused by electric currents along field lines, and have also been observed. For example, radio emissions occurred during the Chilean quake of May 1960, and were picked up by cosmic radio noise monitors across the United States.[166,263,402] No earthquake exhibits all of these somewhat ambiguously described displays, but each strange occurrence adds another detail to the potpourri of facts that indicate electric currents and magnetic fields (the Earth's nervous system), and the by-products of hydrogen fusion play their part. Illustrating the effects of this new understanding of the Earth we find correlations between earthquake activity and the polar or Chandler Wobble, the length of day, the Moon's position and solar activity, as well as a global system of quakes (these will be addressed in Chapter 29 and Volume Two).

Luminous and electrical phenomena are also part of the volcanic scenario. Some of the most extraordinarily marvelous lightning wonders are those that scintillate amidst a column of smoke and vapor rising from an active volcano. The bizarre show, for example, that attended the eruption of Pelee (8 May 1902) brought flashes of lightning that ended in star-like outbursts. Three days after an eruption of

Vesuvius (4 April 1906) there were sharp, sudden and powerful explosions occurring from one every three or four seconds to at least three per second. At the instant of each, a thin, luminous flashing arc moved outward and upward from the crater. This display was repeated hundreds of times, each arc disappearing into space.[110]

An undersea volcanic eruption off the southern coast of Iceland (14 November 1963) produced clouds that rose vertically at twelve meters per second (40 ft./sec.) to an altitude of nine kilometers (5.5 miles). Within ten days an island was formed from the debris, and for an additional seven days, clouds were still rising, this time with an intense, almost continuous light. During a period with no lightning, an aircraft flew through the volcanic cloud where huge electric fields were measured, occasionally exceeding 11,000 volts per meter. Undoubtedly the charge was stronger when lightning was present. Again, electricity and ions were observed.[110]

One prominent neurologist and biometeorologist, studying the biology of the nervous system and the influences of weather on humans, has made an interesting observation. When earthquake activity, no matter how slight, increases in a region, reports of UFOs increase as well. Increases in UFO reports between the years 1820 and 1971, and the number of earthquakes during that same time, when laid out on a map of the United States, coincide quite well. Seismic activity often produces magnetic fields, which could be responsible for central nervous system (neurological) effects. Hallucinations or amnesia, he says, could account for some of the odd encounters reported.[307]

Earth-movements are not exactly what we have been led to believe. Sure, the majority of earthquakes and volcanic eruptions are the Earth bringing equilibrium to its shape and surface. These occur along plate boundaries in the process known as plate tectonics. Science does not know exactly how the Earth gets the energy to do this, or for that matter, why. Today the continents are basically situated along the edges of the Earth's rounded-pyramid shape, yet these events continue. Now it can be said that growth (expansion) is part of the reason, because this seems to have been occurring through time. The remainder of earthquakes, which mostly take place on the continents, are in response to an imbalance in a life-system (a disturbed or destroyed habitat). This is why they center in or near urban areas, agricultural fields, reservoirs, dams, and other artifacts of civilization that remove wilderness. Other disasters, such as floods, hurricanes, and the like always destroy these same things. Most "natural" disasters are actually the equivalent of a damaged or dead cell being restored, or an attempt at restoration. It may seem to offend common sense to consider the idea that major upheavals of the Earth's crust are being manipulated in the interest of the biosphere. However, as we examine history in Volume Two, this is most definitely a sound and substantiated conclusion.

11.2 Adaptation

New perspectives on primeval conditions on Earth are also available. Nitrogen in its 78% abundance in the atmosphere, in contrast to a hypothetical chemical equilibrium, which yields almost no nitrogen, no longer needs an elaborate hypothesis. With hydrogen fusion there is the occasional combination of helium atoms to produce nitrogen, and this has already been unexpectedly observed over the poles. Likewise, the atmosphere and oceans with ammonia as a source of nitrogen are not a necessary scenario for the production of protein and the formation of life. We no longer need to search for a renewable source of hydrogen to keep the amount of water constant on Earth, because there is a primary source within the Earth's interior. In fact, this becomes the most likely candidate for the long sought origin of the oceans.

One of the first things to develop in any living organism is a nervous system. In the geologic record, one of the first conditions we observe is vast quantities of surface crustal rock containing ferrous or a more reduced form of iron that was oxidized during the early stages of life. The oldest fossil rocks are those that show the presence of bacteria that produce iron-sulfides (*sulfur bacteria and anaerobic organisms*). Later, plants, through photosynthesis, brought about large deposits of iron oxides, which were so vast that they make up today's iron-ore deposits of the world. In fact, so much did both of these conditions occur that nearly every type of rock on Earth contains iron (*pyrite or pyrrhotite*).[112,124,210,238,289,342] Like other life forms, life on Earth, as a whole, established an electrically conducting surface—a nervous system—as one of its most primary objectives.

Adaptation of the Earth as an organism is the primary force behind evolution. There is what is known as resistance adaptations, which are changes that permit survival at extremes of some physical condition in the environment. This is evident in life's changing to adapt to changes in solar activity, which also brought about fluctuations in weather, resulting in ice caps and deserts, among other new environments, with new life-forms.

Capacity adaptations are changes in cellular functions to maintain normal activity over a range of internal states. This is observable in life's adaptations to a growing Earth and continental displacement. Life's adaptation is viewed as a response to physical laws and the physical world, leading to a new scientific discipline called biophysics.[384] The following chapters And Volume Two will clarify that this relationship is a two-way street, where the laws of physics and chemistry conform to life, as well as life conforming to physics and chemistry.

In the later stages of the Earth's existence (*Carboniferous to the present*), the situation could be viewed as a state of disease and its adapting to maintain health (*disease homeostasis*); episodes or cycles of extinction and speciation. A biological system is an open system; a steady state in dynamic equilibrium. An open system has two principal characteristics: (1) it is inseparable in form, and constantly interacts with the external environment; and (2) it does not maintain unique stationary levels. Both the Earth and life are inseparable in form, both interact with changes in the environment (i.e., solar activity, lunar cycles, cosmic cycles, etc.), and stationary levels evolve into other stationary levels (new life-systems/land-forms take over for extinct life-systems/land-forms).

Though there are an infinite variety of normal homeostatic open systems, all have one common feature, which is to keep balance along with constant change. Evolution is the Earth's constant change, and it even occurs at regular intervals (cycles or episodes). There are cycles of biological and geological changes that occur in unison. The Earth and the life upon it evolve together at the same time, and each new life system becomes more stable (i.e., more diversity through time). Bodily substance has broken down and been built up through reactions—irritability has led to adaptation and reaction—leading to extinction and changing land-forms, followed by new species and land-forms. Our observations of life progressing from simple to more complex forms can be described as the life history (*ontogeny*) of a developing organism. Evolution is literally the checks and balances, the hierarchy of controls, and compensatory reactions of a living entity, FEM, bringing into existence and maintaining its cellular functions.

CHAPTER 12

Electricity and Magnetism
(A New Model of the Earth and
Human Influences on Geophysics)

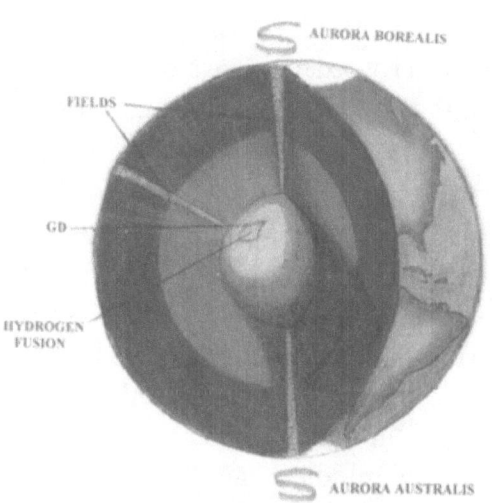

Figure 1.12. The Field-dynamical Earth Model (FEM). After considering various observations this illustration shows the new model of the Earth (interior only). The Fields are somewhat exaggerated for purposes of visualization.

In review of this new model of the Earth it is plainly observable that there are many interrelated natural phenomena. Here we find the relationship of electricity to magnetism, and the beginning of new perspectives on the Universe and our place in it. Like a pane of glass out in space, the Unified Fields have remained transparent, "solid" yet allowing light to pass through them. Our apparatus for measuring the physical world cannot detect something that we do not know exists, cannot be directly experienced by the senses, and that transcends those physical laws we do know something about. However, by witnessing other natural phenomena they have become visible, as would the pane of glass if an observer were situated so that a beam of light reflected its position in space. This is just an introduction, let us not forget, and in the pages which follow is further confirmation of the existence and influences of FEM, and the electromagnetic forces throughout the cosmos. Let us not confuse the point. Those subjects that follow are quite different, but the physical world has been and continues to be trained or shaped by these Fields and life. In other words, we have begun to touch upon a grand unifying theory of the physical basis of the cosmos.

12.1 A Unified Whole

The Fields (time-varying particle accelerators), which are situated just above the bulge along the equator at about the 30° to 40° latitudes in both hemispheres and also the poles, are associated with other phenomena. Each forms a sinusoidal or "S"-shaped relationship to another Field. Each is on the exact opposite end of the Earth from another, so that if a rod were put in one and through the center of the Earth, it would come out in the middle of another Field. Terrestrial electricity spirals, strange sounds, and magnetic and gravity anomalies occur in each. The deepest parts of the ocean floor, ocean current patterns, strong storms deep on the ocean floor, some of the highest values of ocean surface salinity, and the most plentiful locations for bursting bubbles, producing stronger atmospheric charge are other associations.

These same areas are high and low pressure systems in weather and storm centers. Spiraling gaseous nebulae have been observed from space centering in each Field. The Earth's surface layer (*lithosphere*) and the physical aspects of plate tectonics (*subduction zones, plate boundaries, rises, rifts, stress patterns, earthquake-wave anomalies, etc.*) are related to the Fields' locations (further discussion is in Chapter 29). Lighted displays (ionized particles) and dark masses (neutrons) have been observed repeatedly. Patterns and changes in the upper atmosphere, radiation belts, and the magnetosphere show relationships to the Fields. Eels, iron-precipitating bacteria, and electric fishes are geographically located in these areas. More than 70% of all life on Earth is situated in relation to the Fields (between the 40° latitudes), which is ideal for the generation of electrical energy by life. A partial model of this Earth is depicted in Figure 1.12.

Some things that seemed unexplainable now find answers, such as the Chandler Wobble. As the Earth spins on its axis it wobbles, like a top losing its momentum. A seventy-foot shift occurs every fourteen months and was once believed to be due to tidal effects. However, with our current theories about the Earth, with its elastic interior, this should disappear gradually. As scientists say, some "unknown" force seems to reinstate it periodically.[236] That force is the Fields at the poles which, in a weakened state (decreased by more than half)[50, 118,119,285,355,356], caused by the destruction of life, does not keep the axis upright as it interacts with the Interplanetary Magnetic Field (related discussions can be found in Volume Two).

All extremes in the shift of the GMF (*declination, inclination and total intensity*) occur over a wide area together. Also, the magnetic field has weakened since 1850, the beginning of the industrial revolution. Furthermore, the pole can shift position (wander) by 80 kilometers (about 50 miles) in a single day as a result of solar wind particles.[118,119,405] Again, characteristics such as these cannot be explained by a magnetic field produced by an iron-nickel core (*dynamo*) and the effects of gravity.

The grid system, discussed previously and illustrated in Figure 1.11, leads to some very curious observations. Near the edges of each grid—the descending limbs of each Field—are a variety of physical phenomena on the Earth. There we find the edges of the tectonic plates, which are the basis of plate tectonics. Volcanic and earthquake activity are at their peak in these areas, and electrical conductivity, and magnetic and gravity anomalies are observable, as well.[74,83,179, 244,352,418,420] The ribs of the grid system fall along the mid-ocean ridges, core faults, zones of active rising, magnetic anomalies (*along the vertices of the polygon*), depressions on the ocean floor, and mark the paths of hurricanes, winds and water currents.[64,169] In addition, ore and hydrocarbon deposits can be found along these lines.[237]

Another observable relationship involves those areas where one plate rides under another in continental drift (*subduction zones*), showing different earthquake wave speeds (*seismic velocity anomalies*). Regions where these waves travel the fastest are related to the Fields. One area stretches from the North Polar Field to the North Atlantic Field (Aleutian Islands to the Caribbean), and another from near the Persian Gulf Field to the West Australian Field (Red Sea to New Zealand). Other Fields show either fast or slow speeds that are more geographically isolated to the Field area (*findings are the result of a tomographic map of the mantle to the 1,200-kilometer or 745-mile depth*).[129,155,403,407] As observed in photos of the Earth taken from space, structures below the surface, the upper rock portion of the Earth's crust (*lithosphere*), are said to "shine through" accumulated sediments (sand, soil, clay, etc.); however, this observation may be due to particle flow (i.e., electrostatic forces). At other times a barely noticeable network of nebulous streaks are seen, as noted previously.[218] Heat-flow contours and the thickness of the upper portion (*lithosphere*) also show a relationship to the Fields, and heat flow increases upward movements (*warping*) that help to generate electric currents.[239] Also, heat flow (see Figure 1.13) and the thickness of the Earth's crust (*lithosphere*) conform to these Field areas.

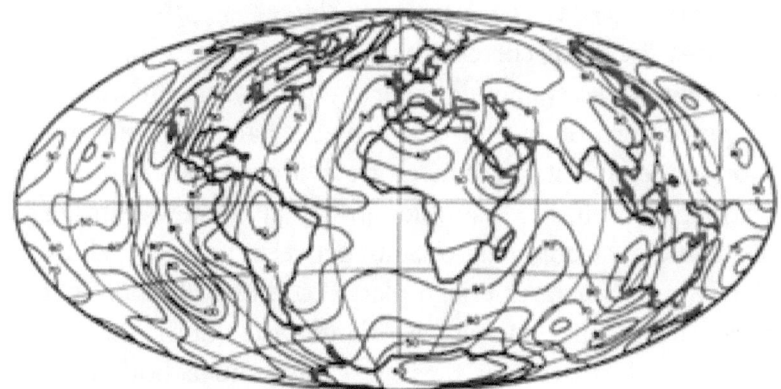

Figure 1.13. Heat Flow of the Lithosphere. On close examination this figure reveals that contours of heat flow from the Earth's rock surface (lithosphere) conform to the Field locations. The thickness of the rock surface (lithosphere) displays similar patterns. [from Champman, D.S. Pollack, H.N. (1977) Regional Isotherms and Lithospheric Thickness. *Geology* 5:265-268]

12.2 FEM's Electromagnetic Forces and Life

Human-made structures, which destroy life, a number of scientists have noted, disturb the natural earth-current environment by creating different, imbalanced or severely weakened charges. For example, observations taken near Stuttgart, Germany indicate this influence on the vertical electric field. West of the city, a disturbed field was noted near an industrial area. East of the city, near a rural area, a quiet, "fair weather" field was observed. The reason for the disturbed field of the industrial area is that the atmosphere has built up a charge that cannot be grounded, because asphalt, concrete and buildings, insulators, cover the ground. In contrast, the rural area has much more open ground and vegetation so that the charge is grounded. Urban areas, like industrial areas, also produce the same effects, and in addition, prevent the production of charges when pollution is present in the air and/or through waste disposal (plastics, solvents, etc.).[140,175,199,239,391] The result of all of this is that the earth-currents become weakened so much that they take on a new character, literally, altering the physical environment.

The importance of life for producing a balanced electromagnetic field worldwide, and its influences are evident in a number of other observations, as well. For example, gravity anomalies[***] in the North Atlantic show a strong relationship between positive values and a boundary of the North American plate (i.e., the Mid-Atlantic Ridge in plate tectonics). Unlike the North Atlantic, the South Atlantic has no strong relationship between positive values and a boundary of the South American plate.[74] This boundary (ridge) runs along an edge of a grid, and at the center of each grid is one of the Fields.

The reason for this difference between the North and South Atlantic can be seen in the human exploitation of wilderness. North America has had much more wilderness destroyed, especially along the coast near the Field. Therefore, strong, positive relationships exist with the ridge due to a weakened Field bowing down on the plate's edges (this will become clear in later chapters; particularly Chapter 29). In South America we have a relatively strong wilderness dominated by the Amazon's rain forests and also ocean waters that are less polluted, therefore no strong positive relationships exist with the boundary (Mid-Atlantic ridge).

In fact, this is also why, as discussed previously, the Brazilian Field (*SAGA*) is more observable from space. It is unique from other Fields by

---

[***] Later discussionss will show that there is an electromagnetic coupling with gravity that is the result of forces at work in the planetary nebula that eventually formed the Earth. Evidence indicates interaction between gravitational and electromagnetic fields in accord with general relativity.

observations of lower values in the magnetic field, and more particles flow earthward. There is also a bulge of helium ions, more airglow, a "hole" in the area, higher temperatures, and higher levels of electrons and ions are noted. However, all of these characteristics are due to change because 20 hectares (50 acres) per minute of the Amazon jungle are being destroyed *day and night*! That is, more life leads to less drastic anomalies, because the Field becomes strong and balanced with life's presence.

The South Pacific offers another example. Of all places on the face of the Earth, human habitation is at a minimum there, and so, relatively little exploitation of the wilderness occurs. Life's electrostatic properties in the region are the least disturbed, and therefore, we should see a quiet, smooth area if, in fact, the life-sustained Fields control physical aspects of the Earth. A scientist's comment on this area does show this relationship, "the impression gained from contouring the existing data is that this part of the world probably has the smoothest gravity field of any part of the Earth."[74]; p.5

Another human effect upon the environment due to life's destruction involves changes in weather. For example, thunderstorms are produced most frequently in the area around the Mediterranean, in both Europe and Africa.[207,322,392] Presently, this area is very heavily populated and has been the region of many of the oldest civilizations with very little untouched wilderness, thereby producing weak Fields, and as a result, ions are generated in the air, leading to thunderstorms (i.e., particle acceleration is inefficient). America is next, then southern Asia, then southern Africa, followed by Australia and South America in order of both thunderstorm frequency and the extent of destroyed wilderness. Those regions that are particularly hit by thunderstorms are areas near the Fields: southeastern North America (North Atlantic Field), southeast Asia (Japanese Field), southern Africa (South African Field), New Zealand (East Australian Field), Australia (East and West Australian Fields), western Europe (Mediterranean Field), and eastern

South America (Brazilian Field). Another confirmation of this effect is reflected in the fact that the Northern Hemisphere has more thunderstorms, as well as more human exploitation of the wilderness, than the Southern Hemisphere.

Even during fair weather, but especially during thunderstorms, currents flow from the atmosphere to the Earth, neutralizing the atmospheric charge. Meanwhile, the continuing electric field suggests a generating mechanism to oppose this "leakage" into the global electric field. This fact is again confirmed by the observation that winter shows stronger electric field measurements than those produced by the peak of thunderstorm activity in the summer.[322] Precipitation, winds (*convection currents*), blowing snow or dust, and thunderstorms are now considered to be a global generator of the electric field by a number of scientists.[175,392] As stated previously, these Fields are storm centers, and thereby, at a very basic level influence this electric field.

Presently, there are more storms (heavy precipitation; floods, snowfalls, etc.), more strong positively charged lightning (10-100 times more), and a greater number of record-breaking storms. Considering the Fields and the growing human exploitation of wilderness brings about an understanding of these recent observations. This is confirmed further by an examination of history, and the geography and timing of distinct weather phenomena (as discussed in Volume Two and *In Defense of Nature—The History Nobody Told You About*).

The recently observed hole in the ozone layer above the polar regions demonstrates another effect. It is shortly after the autumnal equinox, as the Northern Hemisphere moves into its fall and winter seasons, that the hole develops over the Arctic.[406] Likewise, it is shortly after the vernal equinox, as the Southern Hemisphere moves into its fall and winter seasons, that the hole develops above the Antarctic.[84,138,406] As a hemisphere points away from the Sun (IMF-GMF relationships change) the Field at that pole becomes one in which fusion by-products are released rather than taking in solar plasma. The Polar Fields take on characteristics like those of the mid-latitude Fields at these times. Observations disclose that the maximum extent of the hole is correlated with temperature drops.[122,406] This is due to the charged particles (ions) that are released into the atmosphere creating a vacuum into which the much colder upper atmosphere flows.

This is also the reason that the holes form at times of solar maximums.[84] In fact, nitrogen compounds may be responsible, but not necessarily as the result of the Sun, as suggested. Rather, contributing to this effect is the sun-like structure in the Earth's interior, and the release of nitrogen up from the pole, as observed by the Dynamics Explorer missions. This is why a NASA scientist at Goddard says: "The ozone is being blown from one place to another by a disturbance whose source we're not very sure of."[406] The effects of particle flow by the Field are what is being overlooked.

Furthermore, the Southern Hemisphere is less disturbed by humans, and as a result, the hole in the ozone layer is immediately above the pole. However, the Northern Hemisphere has its hole in the ozone layer shifted 1,125 kilometers (700 miles) towards northern Europe's more devastated wilderness, rather than above the pole. In addition, it is during the south's fall-winter that the Northern Hemisphere's Field interacts with the Sun's Field (GMF-IMF interactions), during times of greater solar activity.[84,122,267,268] As a result, more release of fusion by-products are accelerated through the South Pole's Field, yielding the larger hole over the Antarctic.[122,406] Further confirmation of this relationship is indicated in the fact that the holes in the ozone have only recently been noted as the human exploitation of Nature reaches its maximum (further discussion is in Volume Two). This is only a few of the examples of this relationship of life to electromagnetic and electrostatic forces on Earth, which also include the weakened magnetic field of the Earth and other observations (to be discussed).

12.3 Paradoxes Lead to New Paradigms

The Earth as a planet has brought before the scientific community an array of paradoxes (contradictions between theory and observation). Not only is it impossible to demonstrate and thoroughly explain how the magnetic field works with an iron-nickel core (dynamo theory), but also why it has changed position. The poles have suddenly changed north for south (polar reversals) after long periods of relatively unchanged (*static*) pole positions. There are also periods of rapid movement of the pole from its position (polar wandering) and sudden changes in these paths of wandering. These shifts in the poles take place far too rapidly form the proposed dynamo (see Volume Two, Chapters 18 and 19 for discussion).

Likewise, there are long periods of continental stability followed by the rapid breakup and "drift" of the continents. All with respect to the location of the poles, there has been shifting of faults (rifting), the melting of continent or ocean floor that has been pushed under another continent (subduction and melting), the uniform shape (symmetry) of elongated elevations on ocean bottoms (ridges), and zones where the Earth's crust has fractured. Finally, volcanic activity, cratering, extinctions, and all these other events are correlated in time and in cycles or episodes.[6,41,317] In fact, recent observations of plate tectonics show spurts of movement that are accompanied by increased volcanic and mountain-building activity.[390] These topics will be addressed in Tome Three and Volume Two.

The American Geophysical Union, where this information was presented, reached the conclusion that the Earth's inner workings are not capable of producing these phenomena, and therefore, new theories are required to explain the Earth's interior. In fact, recent evidence involving earthquake waves indicates that waves travel faster east to west than north to south. This demonstrates that the core is not in the proposed state, while such observations are completely resolved by a strong star-like magnetic field at the core. None of this (and more) can be explained by present theories of the Earth, but is plainly and completely explainable with the model presented here.

Various scientists do realize that today's model of the Earth does not explain numerous observations or that contradictory theories exist for many observed facts. One scientist examining the contradictory characteristics or paradoxes expresses the need for a new model of the Earth in an article titled "The Earth as a Planet: Paradigms and Paradoxes": "A paradigm is the set of rules and assumptions that guide workers in a given field. As facts accumulate, paradoxes multiply to the point where they topple the paradigm. The standard way to eliminate a paradox, aside from ignoring it, is to change rules and assumptions, thereby incorporating the paradox into a new paradigm. Thus, today's paradoxes are tomorrow's paradigms."[6] So far the choice has been to ignore the paradoxes, but "tomorrow" is *here,* in the knowledge presented in this book.

CHAPTER 13

Cybernetics
(Human Participation in Physical
Laws—Our New Found Responsibility)

Unfortunately most scientists, regardless of the talent present, go through an educational experience that limits their perception of the facts. It is all too easy to take that fatal step of selecting only that data that fits a preconceived assumption. And soon science has built an image, not of the real world, but of an illusion that has been maintained through shear repetition and belief. As a result, there is the perception that this planet is an "earth" of lifeless stone instead of the *organism* Earth; FEM, the Field-dynamical Earth Model.

13.1 Future Directions

Cybernetics is a Greek word that means "the art of steersmanship." This word is used to describe self-regulating systems of communication and control in living organisms, and those machines made by living organisms (humans). The primary function of cybernetic systems is to "steer" the best course through changing conditions towards a desired goal. Such a system functioning on a global scale, and having as its goal the establishment and maintenance of optimum life, and internal structure is the best evidence that the Earth is an organism. This will be illustrated throughout this book.

A general approach used by cyberneticians for recognizing control systems is called the black box method. In electrical engineering, it is used to teach students who must describe the function of the black box by testing a few wires that protrude from the box without opening it. A student is allowed to utilize power supplies and instruments to test the wires, and from the observations obtained she or he must deduce what the box contains.

The black box, or its equivalent, in cybernetics is assumed to be functioning normally. The box is then tested by changing some property of the environment that is controlled by the system. From this change would come a compensating action, demonstrating the capacity for the control system to balance itself. By observing the changes under its control, it would lead to the understanding of the control system.

For humanity, scientific exploration, prehistory and history (geologic and human) is the black box. Quite unconsciously, humanity has been interacting with the control system of life on Earth. So unconscious is this, in fact, that we have not included our interaction with the natural environment in our look at prehistory and history, until recently and then only by a few authors. Likewise, in science we have only recently considered our participation in the physical world we are examining. Still we have unwittingly tested, so to speak, the "wires" protruding from the black box, but have not deduced the characteristics of the system.

If a control system exists, there should be certain aspects that are observable. The objective behind understanding this control system would be to set a goal for obtaining specific conditions, and find the easiest route to that goal, utilizing all practical means and methods. This is known as purposeful behavior. To have knowledge of these results and using these results selectively is humanity's duty, not only to all life, but for all of humanity as well.

If right in action, we must know we are right. If wrong, we must know we are wrong. Knowledge, then, is a very prime commodity. If there is no choice, then there can be no control. And let us not fool ourselves, we do have control.

There is only one greater truth about history than that expressed by Santayana: "Those who do not learn from history are condemned to repeat it." And that greater truth is the lesson not yet learned. history and science, then, are humanity's struggle to enact, with free will, the correct choice, and its greatest truth is that life nurtured perfects the physical world.

The following is that story (see also Volume Two and *In Defense of Nature—The History Nobody Told You About.*

TOME TWO

How Small is Small If One Principle Connects it All?
The Microcosm and Reality

"We have to cross out the old word 'observer' and replace it by the new word 'participator.' In some strange sense, the quantum principle tells us that we are dealing with a participatory universe." John A. Wheeler (Universe as Home for Man. *American Scientist* 62:683-691, 1974)

Thou canst stir a flower, without the troubling of a star.
Francis Thompson

Our most basic link with the outside world is visual. At the same time, our eyes have the greatest range of capabilities of all living species. At a glance we can gauge size, record movement, make sudden shifts in focus, determine depth, and in a moment, minute detail can be distinguished. Of course, we cannot see an object at a great distance with the sharpness of the eagle, nor the tiny world which microorganisms experience. Regardless of our eyes remarkable capabilities and versatility there are infinite sights to which we are totally blind. For us, the myriad of minute details that make up the microcosm often go completely unnoticed.

Whose eyes or even imagination could have pictured the tiniest of worlds? Before the invention of the microscope it was very near impossible to visualize a world of plankton in a single drop of water. At this very moment there are as many creatures on your skin as there are people on Earth; yet, who would have guessed that? The imagination did picture matter divided into molecules and atoms. However, it took experiments and the electron-microscope to reveal the fact that atoms were made up of smaller particles: neutrons, electrons and protons. Now with our complex physical theories and apparatus we have uncovered still smaller "quantum" particles, labeled with colorful names like gluons, quarks, baryons, leptons, and so forth. These particles may be composed of even smaller ones and we still do not know where, or if ever, the world of the small will stop. There may well be a universe within the atom itself, it is merely our limited perception that prevents us from seeing deeply enough. We only have our imagination and inventions for exploring this hidden realm. We grope in the dark, using only educated guesses and contrived apparatus to "see" the unseeable.

Understanding such a tiny world may appear of no importance to daily life. However, we do need to comprehend that world in order to know the most fundamental physical laws and be certain our actions are making the best of life. What does our limited understanding do to that minute world and daily existence? Are sickness, disease, mental ailments, aging, birth defects and other similar phenomena an indication of an unmerciful world with which we are at odds against? Is entropy (decay, disorder, etc.) really an immutable physical law, or are there certain situations when it is so because of our participation in the physical world? The answer to resolving some of the world's greatest problems lies in our individual actions, and it involves our influence on the microcosm to create a different and better world.

Experiments have revealed how interconnected the quantum world is. Creating pairs of particles, with each flying off in opposite directions, has shown that each "knows" instantaneously what is happening to the other. The information about all of the Universe is encapsulated in every part of it. Everything from gluons to galaxies, including you and I, illustrates all the details of the grander picture. This is not merely theory, but has been proven in several experiments that have been performed within the last two decades.

In some strange sense, all objects constitute an indivisible whole of which we are a part. Our actions influence all those objects, so how can we say, with any certainty, how it actually works if we do not take into account what we do to it? And, therefore we must ask: How small is small if one principle connects it all? It is something along this line: you cannot disturb the smallest flower without it unsettling the farthest star. Let's take a look at that world, and examine our awesome power to shape what goes on.

CHAPTER 14

Its All Up In The Air: Weather Or Not It Affects You
(The Weather Matrix, Biologically Active Ions, Biometerology and Psychobiology)

Weather phenomena can have a profound impact on physiological and psychological balance. Temperature extremes, lightning generated extremely low frequency waves, and air ions all interact with living things. Plants, animals and human beings all respond in some way to the influences, as uncovered by a new discipline called biometeorology.

For humans the effects are represented in a wide range of statistics and confirmed in numerous studies. Extreme conditions are known to trigger violence and aggression, and increase disease rates. On the less extreme end of the effects, there are increases in a variety of phenomena such as asthma attacks, colds, mental confusion, increased quarreling, and a myriad of other mental and physical outcomes.

The Field-dynamical Earth Model and the role of life are represented as powerful influences. FEM's fields release energetic particles that ionize the atmosphere producing what have been referred to around the world as "evil winds." Evil winds are known to increase violent crimes, civil unrest, war, disease rates, and suicides, among other things. Moreover, wilderness balances these influences in a beneficial way, while urbanization and industrialization imbalance them.

14.1 Physiological Effects

Human sensitivity to weather, known as biometeorology, is a fairly new science that is only in its infancy. One effect is that sudden cool and very humid air accelerates heat loss from the skin and lungs causing quick drops in (*peripheral*) body temperature. Any extreme deviation from the normal 37° Celsius (98.6° Fahrenheit) body temperature can cause very different mental responses, from poor thinking to self-destructive thoughts.

Prolonged extremes of hot and cold (hot or cold waves) cause the greatest difficulties because of the compensating effects of that part of the brain which controls body temperature, the hypothalamus. The hypothalamus also regulates hormones, hunger, moods, and numerous other psychobiological functions. Cold fronts are known to effect the part of the nervous system that controls involuntary (*sympathetic*) functions. Warm fronts affect another nervous system that controls other involuntary (*parasympathetic*) functions. Changes in barometric pressure, which typically accompany fronts, influence both physiology and behavior, but the basis of the effect is not well understood. However, pressure fluctuations are correlated with various human problems from heart attacks to arthritic complaints.[460,937,939,1192]

Lightning or thunderstorms, which generate extremely low-frequency (ELF) waves, are able to penetrate buildings with their effects. The influences can change respiration rate, blood clotting, and internal biological (*circadian*) rhythms, which at times are referred to as biorhythms. Typically, the effects lag behind the weather by one to three days. Some of the observed changes are a decrease in water retention, along with sodium decreases and potassium increases (the opposite occurs with shielding). Those parts of the central nervous system that control conscious thought (i.e., the lower, front part of the brain; *temporal lobes*), parts of the limbic system (*hippocampal-amygdaloid complex*), and the higher brain (*occipital cortex*) are affected. Also, a part of the central nervous system that regulates internal balance (*thalamus*) is functionally altered.[460,937,939,1192]

An estimated three out of ten people are very weather sensitive, while all of us are sensitive to extremes.[460,937,939,1192] Women tend to be more easily affected, because women's involuntary nervous system is synchronized with the hormonal periods of the menstrual cycle. Overall a snowballing effect occurs as the weather sensitive affect those who are not weather sensitive possibly leading to social, economic, mass biological and/or mass psychological phenomena.

14.2 Weather and Disease

As could be predicted from the above discussion, weather has an influence on the occurrence and severity of disease. Pressure changes are correlated with several diseases, ranging from heart attacks to arthritic pains.[937-939] For example, decreased pressure several days after a cold front is associated with decreased fibrin formation, leading to slower blood clotting. Changes in pressure, as a result, are correlated with ruptured blood-vessels (*embolism*).[526]

The hypothalamus controls body temperature, heart rate and blood pressure, while it is acted upon by temperature changes in the environment. Heart disease and temperature shifts were correlated with the highest death rates in New York (1959-1962) and London (1960-1964) due to cold temperatures. Tokyo's population disclosed that most perished as the result of hot temperatures.[1065] Stroke also displays a similar relationship to temperature.[937] Other studies provide evidence for a relationship with heart attacks (*myocardial infractions*), brain blood-vessel (*cerebrovascular*) accidents, and diseases involving blood-vessels and the respiratory system.[198] Correlations exist between the lightning generated ELF waves (before minimums and after maximums) and heart attacks[181] and arthritis.[393] Weather phenomena are also important factors in cancer causation[1241], ulcers[527], and heart[203], infectious[1242] and rheumatic diseases.[725] The influence of the weather is noted in various other physical disorders and diseases as well.[1238,1243,1246] Seasonal changes affect asthma, bronchitis, dental cavities, and numerous other physiologically related symptoms.[1238,1243,1246] Weather itself can be considered more than just a "dis-ease" when extreme.

Mental balance and brain chemistry are swayed by weather. Deviations from normal body temperature cause complex behavior patterns where thinking, concentration and overall mental clarity are disrupted or suppressed. Weather extremes that fluctuate from normal body temperature have been known to create panic, intense anxiety, a generally intense negative tone, and heightened feelings of suicide or self-destructiveness.[937] The hypothalamus controls body-temperature, and is defective in those who attempt suicide or are self-destructive. Suicides display a seasonal maximum in autumn and early winter, as well as at times of pressure changes when temperature shifts often occur.[1248]

Even though temperatures were controlled in housing, increased restlessness in schizophrenics, especially females, took place at times of decreased atmospheric pressure.[1238] An analysis of 50,000 rapes in the United States (14 locations) disclosed a maximum in summer. Assaults also peaked in summer, and robberies in November and December.[842] Psychiatric facilities show the highest degree of unrest occurs in November, December and January, and some display a second peak in July and August—typically, the coldest and hottest months.[1248] Schizophrenia, serious neurotic disorders and personality all demonstrate seasonal relationships.[667]

Various aspects of psychology and sociology, from child restlessness to road and industrial accidents, are influenced by weather fluctuations.[1245] In fact, ELF waves (*type-I*), generated by lightning, are strikingly similar to alpha brain-wave patterns.[688,689] The altered mental states roused by weather undoubtedly involves the adjustment of internal balance to extremes (homeostasis), modified biochemistry, and electromagnetically induced nerve firing.

14.3 Civilization's Impact

Urbanization and industrialization, aside from altering the Earth's electrical environment, have more local effects on weather, called microclimates. Mean temperatures are elevated in both summer and winter. An overall decrease in humidity, a 100% increase in fog (especially in winter) and greater precipitation are noted. Dust is increased by more than ten-fold, and along with greater cloudiness, decrease the total radiation that reaches the horizontal surface. Tall structures particularly contribute to this microclimate creating lower wind speeds and a greater frequency of calms (5% to 20%).[187] Air electricity is also rapidly modified particularly by pollutant levels in industrial centers.[1192] To sum up the effects, it could be said that the transformation created by urban-industrial centers attracts weather that is destructive to those centers (i.e., floods, acid precipitation, ion imbalances, etc.), and contributes to various human problems (e.g., disease, mental illness, and crime).

Aside from this, agriculture has even more profound effects on climate. The most widespread modification of the climate has been created by the conversion of wilderness into arable land and pastures.[96,808] Over the past eight millennia approximately 11% of the world's land area has been converted into agricultural land, and 31% of the forests are no longer in a natural state.[425] In central Europe about 60% of the land has been converted from forest to farmland during the last 1,000 years alone.[322] The total figure of primary or tropical rainforests cleared or degraded every year is an astounding 200,000 square kilometers (77,000 square miles).[78,876] If this continues there will be *no* tropical forests in 50 years.[78,493,876,1134] The climate changes that will result include massive flooding, droughts, degradation and desertification.[78,876, 1107] According to the experts, the droughts created in the belt extending from Sahel to Ethiopia, along with the devastating famine of the late 1980s, originated by cutting down western Africa's forests.

Without the forests, the tropics would be hotter and drier, and the temperate areas would be colder. Forests are a major sink for carbon dioxide, and without extensive forest, carbon dioxide would accumulate increasing the so-called "Greenhouse Effect." The burning of fuels continually makes, and has made, significant contributions to this problem. The increase in temperatures would melt ice caps and raise water levels worldwide, threatening many cities and agricultural areas; the regions most responsible for the problem.[78]

Forests are also major producers of ozone. Without them the ozone holes would become significantly worse. Most importantly ozone protects all life on Earth from the damaging effects of cosmic radiation.

Involved in the water (hydrologic) cycle, forests are effective traps for solar radiation, and provide the greatest surface area for the interception and re-evaporation of water.[933] Merely altering ground cover affects surface runoff, and the amount of light reflected (*albedo*), changing heat transport, and greatly modifying surface winds. In turn, these variations would bring about alterations in soil temperature, moisture and erosion rates.[1062] Forests, in contrast, stabilize the atmosphere and weather, more so than can be stated in a few paragraphs.

14.4 Biological Effects of Air Ions

Air ions are essential to the health of all living things. Like most biologically active substances, air is composed of molecules that are either positively or negatively charged. Animals, plants and microorganisms are all affected to the extent of bringing about balance or imbalance within individual organisms or entire communities of living things. When balanced in proper levels air ions promote health, disease resistance, mental alertness, enhance growth, and prolong life. As with all things in the physical world, air ions offer an understanding of the grander picture, and our place in it and how we influence it.

Our first knowledge of ions' biological significance was obtained from experiments with plants. As far back as 1748 there was the Frenchman, L'Abbe Nollet, and in 1775, Father Gian Battista Baccaria at the University of Turin, Italy, who used electricity to stimulate plant growth. Mostly within the last thirty years, more specific and controlled experiments were performed.

In the 1960s the United States Department of Agriculture grew cucumber seedlings which yielded cucumbers that were 46 centimeters (18 inches) longer than normal. The University of California using high densities of positive and negative ions produced barley, oats, lettuce and peas that were significantly bigger and healthier plants.[1163] Tomatoes, cucumbers and other crops can have their yields doubled with significant densities of ions.[700] Even, fungus in a state of inactivity (*stasis*) can be rejuvenated with negative ions.[1117]

Negative ions increase a plant's uptake of soil nutrients, and aid in the production of iron enzymes (*i.e., cytochrome, etc.*) important to the formation of chlorophyll, and hence, photosynthesis.[702,1132] Plants, unlike higher organisms, demonstrate no particular biological preference for the charge of the ions, only that they should be present in sufficient quantities.[660,1323] An exception to this is the positively charged carbondioxide essential to photosynthesis, which plants obviously have an increased affinity for. Ions increase growth by 10% to 30% and dry weight by 30% to 60%, while also stimulating seed germination. This overall enhancement is due to a slight increase in chlorophyll, a considerable increase of iron containing proteins important in cell function (*cytochrome*), stimulation of a compound that releases energy for cell growth and reproduction (*adenosine triphosphate*), and increasing the uptake of iron (*chlorosis*) from iron-depleted soils.[962,1323] The enhancement of growth may also be due to elevated levels of a growth-stimulating hormone (*auxin*).[1098] A change in electrical activity (*potential gradient*) between the upper plant, the root and the soil takes place. The uptake of nutrients is a purely electrical phenomena and involves the flow of electrical currents.[151,414,698,843,962,1323]

The Earth's surface is negatively charged attracting positive ions to the ground, thereby enhancing the root environment (*positive potential*).[462,463] Numerous studies have shown that the electrostatic fields produced by ions, particularly fluctuating ones (time-varying), enhance plant growth. Similar alterations of charge distribution (*potential gradients*), as occur in the atmosphere, or the Earth's electric environment in total (*telluric currents, geoelectricity, biogeoelectrostatic currents, etc.*), can affect the germination of seeds, and the respiration and growth of plants, including trees.[871,873,874,1132,1240,1334]

Animals, including insects, are also affected by air ions. Similar to the effects of the Earth's electrical environment on plants, electrostatic fields produced by ions affect insects (honey bees, drosophila, aedus, aphids, diptera, etc.), including their pupation, molting, oviposition and growth.[247,379,385,386,517,742,814,1091-1093,1263,1317] Silkworms displayed greater growth, earlier onsets for the hatching of eggs and the spinning of silk, a marked increase in larval growth, and a significant increase in cocoon weight.[660]

One of the first experiments using animals exposed mice, rats, guinea pigs and rabbits to totally deionized air, and within two weeks most had perished. Diagnosis revealed various problems were responsible for each death, such as, fatty liver, kidney failure, heart degeneration, anemia and other physical disorders.[1163] The same experimenter demonstrated that negative ions increased sexual activity, and caused an increase in weight and overall health.

In another experiment, guinea pigs with tuberculosis (TB) were divided into groups, and one group was given negative ions. After eight weeks the control group, which was positive-ion exposed, perished. Meanwhile, every one from the negative-ion group lived, was free of TB, and was even healthier than before contracting the disease.[260]

A number of other effects have been noted. The Turkey's fertility rate can be increased by 10% with an environment supplemented with ions.[1192] Others have shown that negative ions destroy disease bacteria, and aid in resistance to diseases, such as influenza.[699] Lowered bacterial counts or resistance to infection occur with negative ions, as well.[702,1192]

Continuous exposure to ionized air increased maternal attention, which elevated the survival rate of newborn rats. High ion counts increased the weight of newborns, and more growth took place after weaning. When high levels of ions with only one charge were present detrimental effects occurred, such as the regression (*involution*) of the thymus gland, which maintains the immune system (*T-cells*), and the male reproductive organs (*testis*). The most profound effects took place prior to weaning, and may be due to the inhibition of the immune system.[133] For animals, ions, particularly negative ions, are essential to the maintenance of disease resistance, sexuality, fertility, and general mental and physical health.

14.5 Air-Ion Effects on Humans

Human physiology and well-being are regulated by the effects of air ions, as well. Air ions, regardless of charge, are an essential nutrient in maintaining human health. Depletion to 2,000 ions per cubic centimeter (cm^3) reduces reaction time.[247,937] Higher concentrations ($10^4/cm^3$ or greater) increase levels of alertness, the ability to think, rapid decision making, and subtle positive-mood changes, among other effects.[247]

Completely ion-depleted air has resulted in the death of many laboratory animals. "Female" (*estradiol, estriol and estrone*), "male" (*testosterone and adrenosterone*), and thyroid (i.e., *parathyroid and thyrocalcitonin*) hormones are stimulated by air electricity. At times, extreme changes in brain-chemicals (*catecholamines*) are observed. One hypothesis states that negative ions stimulate, and positive ions block, the action of a biochemical that regulates brain-chemical production (*monoaminoxidase*).[702] Imbalances in air ionization are known to trigger mental and/or physical disorders and diseases, as well as criminal behavior and civil unrest.

A close interrelationship exists between cell growth, cell metabolism, and the state of the central nervous system, and air ions. This influence extends to the enzyme mechanisms' actions, tissue function, antibody formation and other physiological functions. Without ions any of these physiological activities may malfunction, thereby triggering or accelerating any disease, including cancer.[1241] An estimate suggests that 25% of the population are significantly affected, and 50% are considerably affected, while 25% are not affected by air ions.[1191] Numerous experiments have demonstrated the role of negative air ions in helping to cure and prevent disease.

Both positive and negative ions retard the growth of bacteria, but negative ions are more effective.[702,1192] Other beneficial effects of negative ions include increases in body capacity and normalizing vitamin metabolism.[690] With few, if any, exceptions, negative ions are important in combating most physical diseases and disorders, and promoting general health.

Negative ions in high concentrations, or concentrations of mixed ions with negative predominating, reduced the death rate from infectious diseases.[1242] Fewer perish from respiratory infections with a substantial amount of negative ions present, while positive ions caused many to perish.[701,1192] Aside from destroying disease bacteria, negative ions also increase the effectiveness of antibodies.[158]

For example, cystic fibrosis is noted for having abnormally high electrical levels in air passages, causing mucous to clog the passages by inhibiting the cleansing hair-like structures, known as cilia.[682] Negative ions increase and positive ions decrease the effectiveness of the cilia.[247] Likewise, in multiple sclerosis certain aspects of the immune system are sensitive to negative ions.[470]

Negative ions ameliorate various diseases and disorders, such as, sickle cell anemia, heart disease (*e.g., coronary insufficiency, myocardial infarction, etc.*), asthma, pneumonia, influenza, bronchitis, tuberculosis, other respiratory diseases (*e.g., chronic rhinitis, sinusitis, coccidioidomycosis, etc.*), emphysema, migraines, insomnia, hay fever, eye diseases (*e.g., conjunctivitis, etc.*) and more. In addition, negative ions can prevent the blocking of blood vessels (*thromboembolism*), stimulate wound and burn healing, and reduce pain.[247,660,1192]

Negative ions are attributed with boosting vitality and resistance to infection, and enhancing the metabolism of carbohydrates.[702,1192] Stimulation of electron carriers that aid in oxidation reactions, the oxygen consumption of the liver, and the carbon dioxide combining power of blood-plasma are other biological effects.[1192] The resultant increased efficiency decreases respiration, blood pressure, skin temperature, and metabolic and blood sedimentation rates, while increasing immunity. The frequency of mitosis in cell division is increased, and so wound healing and growth are improved. Agility, endurance, and reaction time are enhanced as the result of less time required to electrically excite muscles. Cilia in the windpipes are stimulated, increasing the effectiveness of respiratory defenses.[1192]

Currently the most lethal disease in the United States, arteriosclerosis, has also been linked to the effects of air ions. The (*endothelial*) cells of arteries are strongly negatively charged, because of certain biochemicals (*sulfate groups of mucopolysaccharides*). As a result, arteries normally repulse negatively charged blood-cells and proteins. A negative charge is maintained by the presence of these biochemicals free on the arterial wall, the acidity of the blood, and the electrical charge of those things that are consumed or absorbed into the blood, such as air ions. Positively charged foods, such as red meat, can lead to the loss of negative charges on the artery walls, and eating habits that include red meat consumption have been linked to the disease. High blood-pressure (hypertension) is the most important variable for predicting the development of arteriosclerosis. Experimentation has shown that negative ions decrease, while positive ions increase, blood pressure.[1039] Negative air ions are essential to the prevention of arteriosclerosis, as well as proper diet.

The chemistry of the brain is also greatly affected by air ions, as illustrated in the case of one very important brain-chemical, known as a neurotransmitter, called serotonin. A predominance of positive ions or a depletion of negative ions triggers what is called the Serotonin Irritation Syndrome, or more accurately, the Serotonin Hyperproduction Syndrome (SHS). Symptoms include migraines, nausea, vomiting, impaired vision (*i.e., scotoma, amblyopia, vasomotor rhinitis and conjunctivitis*), electrified hair, irritability, tension, laryngitis, vertigo, sleeplessness, irregular heart-beats, and tremor. Other symptoms include reduced function of the digestive tract, congestion of the respiratory tract, and flushes with sweats and chills.[247,702,1192] Other signs of SHS include dryness, burning and itching nose, nasal obstruction, difficulty in swallowing, itching eyes, dry scratchy throat, and breathing is reduced from 35 liters per minute to 25 liters per minute (almost the same in quarts), which can trigger asthma.[1192]

Exposure to positive ions increases serotonin and its major metabolite (*5-HIAA*) in blood-plasma and urine, as levels become lowered in the brain.[247,702,937,1192] Negative ions also produce a similar effect, but this is due to increased production and metabolism, while positive ions increase serotonin's breakdown.[247] Serotonin constricts blood vessels in the brain, influences hormone levels (*melatonin, norepinephrine, steroids, etc.*), and regulates metabolism, including food intake. Low levels of serotonin result in suppressed mating, obesity, depression, impaired learning, suicidal urges, aggression, anxiety, fear, and more. Furthermore, there are influences on protein metabolism and the regulation of cellular processes due to effects on the higher brain (*cerebral cortex*), and the liver (*hepatic protein synthesis via tryptophan*).[344] Other neurotransmitters and adrenal hormones (*catecholamines*) are also affected, resulting in influences on learning, emotional stability, agility, stamina, and other psychobiological factors.[702,937,1192] Both behavior and physiology are significantly altered by these influences.

Observations of the electrical activity and growth of the brain, as well as changes in behavior, serve to confirm these effects. Negative ions were noted to increase weight in those areas of the higher brain that control the interpretation of sensory input (*somatosensory and occipital lobe areas of the cortex*).[660] They are also known to advance the alpha rhythm from the hindbrain to the forebrain, which is indicative of increased conscious thought. Observations of brain-waves (*EEG*) show that alpha waves are reduced in amplitude and frequency, as occurs when a person is relaxed. Negative ions also stabilize a (*10 Hertz*) frequency that designates relaxation, increase amplitudes that denote improved work capacity, and stimulate the synchronization of the left and right brains.[88,247,1192]

Negative ions produce good sensations, euphoria and feelings of well-being.[330,411,549,937, 1163,1192] They enhance performance and learning, while decreasing conditioned emotional responses, such as fear.[247,660] Heightened physical and mental capabilities arise from the effects of negative ions.

One study demonstrates how negative ions improve performance. Increases in physical endurance were observed in weight-lifters, and women athletes. Those who were given negative ions displayed a 240% increase in performance, while those who were not supplemented with ions only showed a 53% to 132% improvement. A mock "Ion Olympics" showed that the negative ion team improved by 370% to 393%. The utilization of vitamin C was increased and may account for the some of the observed effects.[852] Vitamin C is highly essential in both physical and mental biochemistry.

Unlike negative ions, which create mostly positive effects, positive ions lessen physical and mental capacity. High doses ($10^3/cm^3$ to $10^6/cm^3$) are associated with complaints of itching nose and/or eyes, dryness of the mouth, difficulty in breathing, and inflammation of the region in back of the nose, mouth and larynx.[937] Positive ions are also known to produce headache, sore throats, fatigue, nausea, nasal obstruction, and dizziness. Normal infants experience rapid breathing, and spasms of the windpipes or bronchial tree. In adults, breathing capacity was suppressed by 30%, activity of the cilia was reduced, and the respiratory rate quickened. Furthermore, metabolism, blood pressure and skin temperature rose. The fatigue experienced is probably due to an extended time-period required to electrically excite muscles, and the reduced metabolic activity of the adrenal glands. Positive ions were noted to inhibit the growth of tissue cell cultures, which suggests an inhibition of growth and wound healing in general.[1192]

Decreased learning and performance, slowed reaction time, and less spontaneous activity are other effects. The reduction of serotonin by 20% to 60% due to positive ions, may suppress mating and learning, or bring about depression, obesity, aggression, anxiety, fear or compulsive behavior.[247,660] In spite of all these impairing effects, positive ions are necessary at natural levels.

In contrast to negative ions, positive ions reduce the efficiency of the (*autonomic*) nervous system controlling involuntary functions. Such an effect has been known to be associated with disorganized responses to verbally threatening communication, increased defensiveness, emotional disturbances associated with threats, and a lessened ability to adapt.[248] Young children's IQ's are affected by disturbed protein metabolism important to brain chemistry (*disturbed oxidation of phenylalanine and tyrosine due to imbalances in MAO*).[1245]

The world at large demonstrates the impact of what might otherwise be considered trivial: "Epidemiological data indicate that increased small positive air ionization due to changing weather conditions is associated with increases in industrial and automobile accidents, suicide, and crime, as well as depression, irritability, and interference with central nervous system function."[248]

Positive ions are known to create or trigger other phenomena. They are associated with negative mood swings, slowed reaction time, increased fatigue, and decreased task performance, attentiveness and sociability.[248,931,1192] Observations in hospitals revealed that depressives, epileptics, hysterics, psychopaths, neurotics, schizophrenics, alcoholics, and drug addicts react to positive-ion winds.[411,1192,1248] Intelligence and restlessness of children is also affected under these conditions.[1248] Other influences are associated with weather and those seasons when positive ions are present in greater quantities.

14.6 "Evil Winds," and Human Aggression and Disease

Winds that are composed of a predominance of positive ions, often referred to as "evil winds," are recognized throughout the world. There is the *Chinook* of the North American plains; *Autan* and *Vent Du Midi* of France; *Tramontana* in Northern Italy; *Norther* of Texas, Southeast Australia and Portugal; the *Norte* of Northern Mexico, Chile and Spain; the *Koebang* and *Gending* in Java; and *Bohorok* in northern Sumatra. All of these evil winds are merely examples of *numerous* others that stretch across the globe.

The biological impact of evil winds can be understood from what transpires when the *Sharav* blows in Israel. Thirty percent of the population is affected by this "evil wind," causing an observable increase in murders, suicides, attempted suicides, asthma attacks, depression, unbearable tensions, and aching joints.[1192] In fact, its influence is so well known that judges are more lenient in their sentencing if a crime occurred when the wind is blowing. Positive ions increase by ten-fold when the *Sharav* blows.[1193] Worldwide, similar winds are noted to increase crime, murder, suicide, tension, irritability, sleeplessness, migraines with vomiting and nausea, and automobile and industrial accidents.[248]

Another well-studied evil wind is the *Foehn*. Common in central Europe it is a very warm, dry and electrified wind that often blows in the mountain valleys of the Alps in Switzerland and Tyrol (Tirol). In Switzerland it is associated with distress, depression, migraines, irritability, dizziness, decreased self-control, and slow reaction times, leading to increases in accidents, crime and suicide. Significant increases are also noted for perforations of ulcers, suddenly ruptured or blocked blood-vessels (*embolism and thrombosis*), hemorrhages, bleeding in hemophiliacs, deterioration of treated psychoses, and even, birth frequency.[1282]

The *Foehn* is characterized by everything that could be expected of the Field-dynamical Earth Model (FEM) ionizing the atmosphere. There are changes in atmospheric pressure typified by strong gusty winds, a rise in temperature, a fall in humidity, and atmospheric electrostatic fields are intensified, becoming unstable to strongly fluctuating (time-varying). Increases in electromagnetic long waves, the electrical conductivity of the upper atmospheric layers, sudden strong atmospheric radioactivity, ionization, and colloid and dust content are observed.[1282] As discussed in Tome One, these observations are what occur when ionized by-products of hydrogen fusion (ionizing radiation) are released along the Field lines of FEM, and here they are shown to be responsible for producing the evil winds (also the latitude locations of these winds indicate this).

The impact of evil winds on murders can be seen in the effects of the Santa Ana Winds, or Santanas as they are called locally, in California. Within 53 days there were 58 homicides in Los Angeles, compared to the 50.8 average for such a time period. In 1965, during the week of 20-26 October, the longest period was noted for the wind within the two years of the study. The author of the study comments: "The total of 10 reported homicides was 47 percent above the 6.8 for a normal week."[848] These winds are also notorious for their effects on brush fires in the Hollywood Hills, and often precede or follow earthquakes.

The ionized by-products of hydrogen fusion not only create evil winds, but change weather in general, and the interrelationships have been observed. Positive ions precede weather fronts by one to two days, because electricity moves faster than air, such as observed for the *Sharav* and *Foehn*. In the United States, storm centers that move down from the northwest or up the Mississippi and Ohio River Valleys leave behind increased frequencies of respiratory attacks and appendicitis. Marked changes in the mental state of patients occur with low-pressure centers approaching. Most patients experience feelings of futility, an inability to accomplish difficult tasks, or poor mental efficiency. Children display restlessness, irritability and petulance. Adults are more quarrelsome, fault finding, have a tendency for pessimism, and as a consequence, more marital disharmony.[849,1248] Each of these weather phenomena and the associated behavioral problems are the result of air ionization.

More than 5,000 publications link suicides with weather and weather fronts. Peaks in suicides occur with low pressure and rising temperatures. Even those who migrated from the South to a more stormy section of the North had a marked increase of suicide.[1248]

For the new model of the Earth, FEM, presented in Tome One, it was shown that due to solar linkages seasonal relationships exist in the release of ionizing radiation (this will be illustrated further in Tome Five of Volume Two, and *In Defense of Nature—The History Nobody Told You About*). In Tome One it was shown that the position of the Earth in its orbit brings about various interactions between the Sun's Interplanetary Magnetic Field (IMF) and the Fields at the Poles. This interaction then activates the mid-latitude Fields, which release fusion by-products, creating changes in weather. These Fields are activated more when the Earth is farthest from the Sun, and when the Earth is tilting on its axis away from the Sun. For either hemisphere it is the fall, and particularly the winter, that more of the ionized by-products of hydrogen fusion are released. The vernal equinox (20-21 March) is a time when both poles are at right angles to the ecliptic plane along which is the IMF, and this interaction activates both poles. The Earth is closest to the Sun at that time and traveling its fastest into the solar wind. The end result is greater storms in many or all of the Field areas, during winter, fall, and the months of March and April.

Ions and weather go hand in hand. Therefore, weather changes should produce ion effects on human behavior and disease. Summer should also display a similar effect, because hot winds produce positive ions, and there is typically greater human activity and pollution, which depletes negative and produces positive ions. As discussed, more deep-seated weather changes occur in winter and around the vernal equinox (also see sections 5.3 and 5.5 for further discussions). Therefore, effects on human behavior and disease should occur mostly in the winter, somewhat in the fall (near the solstice), in March and April (around the vernal equinox), and during summer's heat.

Numerous studies do show these relationships. Unrest in schizophrenic patients occurred most often during November, December and January with a secondary peak in July and August. More schizophrenics are born in the Northern Hemisphere in January through April and more so if the preceding summer was a hot one.[667,1065] According to nine independent researchers studying a total of about 30,000 cases of schizophrenia, the highest birth rate is during January and February.[1243] Another study of 15,000 cases disclosed that it was the months of February and March. Meanwhile, Australia, in the Southern Hemisphere, displays the predicted six-month shift by exhibiting a peak that is also during the winter.[1248] Another way of looking at these facts is that schizophrenics were either born or conceived during the winter, around the March equinox, or in summer. In addition, many schizophrenics developed critical brain structures (fourth to seventh month of gestation) during the summer's heat.

Other studies disclose a similar relationship. Britain has more serious neurotic disorders in those people who were born during the summer. In North America more of the emotionally disturbed are born in the winter.[667] The number of abnormalities in pregnancy are highest in January, February and March. Suicides are at a maximum during March and May in the Netherlands. For Southern Germany, suicides peak in women during May and October, and in men during October (just following the September equinox in fall). In the Hague, while most births occur in May, those born in February and March (winter and the vernal equinox) are the most prone to suicide.[1248] The death rate for all causes in the United States from 1950 to 1965 is highest from November through March.[937] The United Kingdom, France, Denmark, Germany and Japan display a winter peak for all causes.[1065] Cancer is highest in those born from December through March, with Australia again displaying the six-month shift.[1243] Excess mortality occurs during heat waves and cold waves in the United States.[1251] Deaths from heart attacks occur more frequently when temperatures are very low and during times of snowfall or hot, humid weather.[102] Even stripe rust epidemics on winter wheat are correlated with low temperatures in December and January, and high or low temperatures from April through June.[273]

Low temperatures show worldwide trends that produce changes in liver function and inorganic chemicals (*colloids; i.e., Piccardi Tests*) that may explain long-term alterations in mental function and behavior, as well as physiology.[1239,1247] The months of extreme *Sharav*, the evil wind in Israel, are March through May and August through September, when acts of aggression are two times higher.[715] Statistical representations bear out the influence of positive ions on humans in relation to the solar-FEM linkage that result in times of peak ionization.

Correlations with weather in general also disclose the impact of positive ions. Weather effects (*i.e., the Biometeorological Index*) are correlated with total admissions to hospitals and significant prognosis.[1020-1025] Atmospheric electricity slows reaction times, and shifts in atmospheric electricity are correlated with traffic and industrial accidents as well.

14.7 <u>Geographic</u> <u>Distribution</u>

Another way of confirming this relationship is to consider the geographic location of the Fields, and note the differences in behavior and disease near those regions. The areas near the Fields should display higher rates of disease, especially during the winter, and more violence or civil unrest. This relationship is unmistakably clear in the data.

The Field near North America is to the southeast, east of Florida. Total mortality is highest in the southeastern United States, particularly the more southerly locations. The lowest mortality occurs in those regions furthest from the Southeast.

Other areas of high mortality take place where storm tracks occur or the descending limb of the Field is located. These regions are around the Great Lakes and the West Coast.[1065] The North Atlantic Field produces low pressure when the atmosphere is ionized. As a result, a wind corridor is created that flows through the Great Lakes. Hence, for example, Chicago is called the "Windy City." Furthermore, the Great Lakes are affected by the dynamics between the North Atlantic Field and the North Polar Field, because it is a region where both descending limbs interact. Other winds flow along the West Coast (i.e., the area along the descending limb of the Field), producing positive ion winds, such as the Santa Ana Winds. The high mortality localities are also urbanized and industrialized, which have a predominance of positive ions and biologically inactive ions (*Langevin ions*).

This relationship holds true for other regions with these same factors governing the situation. Another Field, referred to as the South African Field, is located east of South Africa and south of Madagascar. Studies conducted in South Africa examining the incidence of asthma indicate that a greater frequency of asthma occurs along the east coast of South Africa. Furthermore, a group of patients' symptoms improved or disappeared when they moved inland, away from the South African Field. Other patients experienced asthma for the first time when they moved to or while vacationing along the East Coast.[1244] Positive ions are known to trigger or aggravate asthma.

Another Field is located east of southern Japan and has been referred to as the Japanese Field. Here we find a winter peak for most diseases and infant mortality, particularly blood-vessel diseases of the brain (*cerebrovascular*) in adults. Furthermore, disease rates are more pronounced in southern Japan.[667]

A comparison of three cities, Tokyo, London and New York, also discloses this relationship. Considering the relative position of each city to a nearby Field the three cities can be categorized as such: Tokyo is closest, London is median and New York is farthest. A comparison between the death rate and changes in temperature reveals the predictable. A biometeorologist summarizes: "The correlation between mortality and temperature among the three cities is the largest in Tokyo and the smallest in New York."[667]; p.126. The ionization that occurs in the Field areas creates a partial vacuum, which the much colder upper atmosphere flows into, resulting in temperature drops and positive air ions, producing these results.

The relationship between the Fields and violence or civil unrest is also quite discernible. More evil winds are found throughout the Near East and Middle Europe than anywhere on Earth. This region is between the two most weakened Fields; the Persian Gulf Field and the Mediterranean Field. This area has been civilized for the longest time (ancient Egypt, Mesopotamia, Indus, ancient Greece, etc.), life systems have been destroyed and their electric contributions (bioelectrical, superconducting, etc.) are mostly lost. With Field-strength weakened (reduced particle acceleration) the atmosphere undergoes more ionization, and hence, more positive-ion winds result in increased violence, civil unrest and war. This region has had more wars than any other area on the face of the Earth. Nowhere else has there been as many countries that have instigated international conflict nor been used so often as a battleground.

Today, all the hot spots of civil unrest take place in the vicinity of the Fields. For example, there is the Persian Gulf Field around which we find the Iran-Iraq War, the Iraqi invasion of Kuwait, terrorism in Beirut and Jerusalem, Palestinian-Israeli conflicts, Arab-Israeli wars, Libyan aggressions, the Soviets in Afghanistan, the Persian Gulf War, Serb-Croatian conflicts, Kosovo and the U.S.-Afghanistan war. Near the South African Field there are the South African race riots and civil war. Due west of the North Atlantic Field there are the repeated riots in Miami and the wars in Central America. Along the descending limb there were the Watts and Los Angeles riots. In the vicinity of the Mediterranean Field, at the sea's mouth, there are the wars of northwest African countries, civil war in the British Isles (IRA), and riots in Spain, England and Ireland.

The effect of the Fields in producing air ionization can be seen on maps of thunderstorm activity which occurs mostly between the 40° latitudes. The impact of human activities is manifest in the greater frequency of thunderstorms in the Northern Hemisphere where human population is much greater. All the Field areas display similar effects on aggression and civil unrest, while Volume Two, Tome Five will disclose a correlation with solar activity and lunar cycles, which enhances the effects of air ionization.

A study of suicides reveals an association with the latitudes of the Fields. The latitudes just north of the Fields (45° to 55° North Latitude) have two to eight times more suicides than other latitudes.[371] The Fields are situated on the equatorial bulge in such a way that they point away from the bulge.

Solar activity and sharp fluctuations in the Earth's magnetic field (geomagnetic storms) have a direct relationship with the activation of the Fields and increased ionization, and are also correlated with disease and human behavior. Solar flares are unquestionably followed by increased ionization, and are associated with peaks in pathological conditions and suicides.[1192] They are also correlated with increased epidemics, plagues, pestilence, heart attacks, hypertension, brain diseases, mental disorders, crime, accidents, and disease in general.[474] Solar activity and magnetic storms are also correlated with war and civil unrest in general.[929] Together, these findings confirm the existence of the new model of the Earth with its Fields, and their role in air ionization with its biometeorological effects.

14.8 Urbanization and Industrialization

The impact of human participation in the physical world can be observed in air ion imbalances and their effects. Many human-created conditions and human actions produce positive ions. The friction of shoes on a carpet can dissipate into the air, if not grounded by touching a door knob or other metal object. Clothes, especially synthetic fabrics, rubbing against themselves can produce similar results. Only natural plant fibers, such as cotton, do not do this. Heavy, positive ions accumulate in closed spaces, such as offices and stores, especially during business hours when carbon dioxide increases.[937] Buildings with centralized heating and air conditioning remove negative ions and add positive ions. Even more positive ions are created as the result of friction when the air flows through vents. In addition, air brushing past a metal surface or through a pipe also creates positive ions. Fire or heat from anything, such as a match, a candle, a stove or central heating are rich sources of positive ions.[1192] Cars and other vehicles, with their metal surfaces and heat rob the air inside of negative ions, while adding positive ions. The exhaust robs the air of negative ions and increases positive ions, as well. Part of the health risk of smoking is that smoke depletes negative ions, and thereby, reduces natural defenses. These are merely examples of everyday activities most of us engage in, but we do not realize that they increase positive ions and decrease negative ions affecting our mood and health.

Meanwhile, urbanization and industrialization have far more of an impact. Air pollution from industry leads to small, air-ion depletion by attaching to molecules, thereby producing the heavier, biologically inactive (*Langevin*) ions. Though both charges of ions are transformed into these biologically inactive ions, more negative ions are involved. The reason for this is that negative ions zig-zag through the air at great speeds and are more chemically active. Sources of pollution that neutralize negative ions range from particulates (dust), high voltage lines and carbon dioxide to coal and oil combustion.[660,1192] In rural areas particulates may reach 6,000 per milliliter, and are predominately pollen and some dust. However, in industrialized cities in North America and Europe particulates reach several million per milliliter, and are predominately dust and soot.

Main intersections in cities average 50 to 200 ions per cubic centimeter (cm³) at mid-day, when there should be about 10,000 ions per cm³. Scientists in Zurich and Munich took an ion count at noon on a sunny day and found only 200 ions per cm³. Worse still, during the San Francisco rush-hour the smaller biologically active ions were non-existent.[1163]

Reaching far beyond the immediate environment, large ions result in a reduced conductivity of the atmosphere. In contrast, the Earth's electrical environment is maintained by a high potential gradient (large ions move more slowly).[1192] In the overall picture, this ion imbalance produces all the biological, mental and social ills discussed previously, as well as, weakens the producing effects that literally reach beyond the Earth (this will be discussed).

Weather is profoundly influenced by humans, which in turn has its effects on ion balances. Friction between air masses, between the air and the ground, between particles of ice and/or snow, or hot, dry wind blowing over land, particularly dry desert sands, produce positive ions.[937] Deserts are, and have been, increasing due to the human misuse of land, and this compounds the effect. Covering the land with asphalt and concrete, or maintaining conventional agriculture (especially with chemical use) prevent or impede the production and/or grounding of ions. Add to this the impact of industry, pollution, the components of indoor environments, population growth, desertification and so forth, and the effect is *profound*.

In the better environments there is typically a preponderance of positive ions, usually six positive to five negative. Truly natural settings have a preponderance of negative ions, especially forests with running water. Meanwhile, the bigger the city, the more industrialized and polluted, the greater is the imbalance with either or both greater quantities of positive ions and the large biologically inactive (*Langevin*) ions. The resulting environment would inevitably bring about more physical, mental, social and economic disorder.

Urban-industrial conditions and the ion-weather matrix display a relationship that may explain some unanswered questions. As discussed in Tome One it was noted that urban and industrial areas have an accumulation of ions in the atmosphere (*i.e., a disturbed weather field*). This was due to insulating materials preventing the grounding of ions, which are predominately positively charged in urban and industrial areas. This is an additional factor in the higher rates of disease, crime, accidents, civil unrest and so forth, in urban-industrial areas.

Under natural conditions, the Earth's surface is negatively charged, and with life positive charges are grounded producing positive earth currents. In contrast, urban and industrial areas do not allow the grounding of positive ions, which are also in greater quantities than in a natural setting. As a result, the negative surface retains its charge continually attracting positive ions, which typically accompany storm fronts. Consequently, these areas attract storms, and as will be discussed in *In Defense of Nature—The History Nobody Told You About*, more and more people are affected by disasters, such as extreme weather, floods and landslides. This brings an overlooked aspect of the destruction of civilizations throughout history to light. It is following the eradication of wildlife that wars, disease, violent weather, and other disasters increase, striking the urban centers.

Positive ions are also important in triggering plagues. One researcher comments on this: "It is these that cause animals to be restive and insects to erupt suddenly with an explosion of energy and become a plague instead of just a nuisance."[1163]; p.22. History shows that plagues erupt after widespread urbanization, the destruction of wildlife, climate shifts and increased solar activity (see discussion in *In Defense of Nature—The History Nobody Told You About*).

Urbanization and industrialization, by preventing the grounding of ions, weaken the earth-currents (*telluric currents and the Earth's electrical environment*). In turn, this weakens the Fields, increasing air ionization, and a correlation with increased disease is conspicuous in the data. In Japan, from the latter part of the 19th Century up to the 1930s, diseases took their greatest toll in summer. Following World War II and the intensification of industrialization the summer peak was replaced by a winter peak. A biometeorologist takes note of this transition: "Mortality became concentrated in winter, in striking contrast to its previous tendency to be centered in summer."[1065]; p.15. Originally, it was summer's hot air that had produced the positive ions, but with increased industrialization the Japanese Field weakened, and being more active in winter, ionized the air more at that time, thereby shifting the season for the highest mortality.

Due to the increase of positive ions, and the presence of insulating materials preventing their grounding, disease also shows a relationship with these areas. Advanced countries, such as the United Kingdom, France, Denmark and Germany, display a winter peak, unlike the "underdeveloped" countries which still show peaks in the hot season. The United Kingdom was industrialized before other countries and displays a winter curve in the 1840's, while the summer peak also disappeared in the 19th Century. The United States and Japan, only show a slightly higher peak in the winter with little seasonal variation in recent decades. This has been ascribed to a controlled environment indoors with heating, cooling and food refrigeration along with the institution of medical services, social security, welfare and advanced medical practices.[1065] Meanwhile, and in spite of all these measures to prevent disease: "It should also be noted that the adjusted rate stands higher in industrialized countries like Japan (6.0), the United States (5.9) and England (5.7)."[1065]; p.33 (figures are the adjusted death rate per 1,000 of the population).

One reason behind such correlations is that indoor environments offset ion-balances towards a predominance of positive ions. Americans, like other industrial countries, spend 80% to 90% of their time indoors, with the higher figure in the more urbanized areas.[247] For example, air-conditioned offices typically have three positive ions for each negative ion, while in Nature the balance is equal for the most part with a slight predominance of negative ions. Meanwhile, negative ions clean the air of dust, smoke, particulates, bacteria and pathogens. Everything in the artificial environment produces positive ions, with the exception of showers.[247,660,1192] Negative-ion generators designed to correct the imbalance produce ozone, which can be destructive to biological tissues.[660,1192] Requiring electricity, they produce imbalances in other places such as in the vicinity of high voltage power lines, mining and industrialization to produce natural resources used in energy production and in manufacturing the ionizer. Meanwhile, plants produce negative ions, but can be damaged or killed by high levels of ions produced by the generators.[474,871,872,1323] House plants are also capable of removing pollutants and particulates, as well as utilize positive ions and produce negative ions. The more artificial the environment, the greater are the quantities of positive ions, and biologically inactive ions, with all their ill effects.

14.9 Wilderness

Wilderness is in clear contrast to the urban/industrial picture. Plants, trees, active soils and radioactive substances (*nuclides*) produce ions. As usual, forests are particularly active in this sense, and are known to generate negative ions. Furthermore, the Earth's surface is negatively charged, attracting positive ions and repelling negative ions. This brings about a one- to two-meter thick atmospheric layer with a preponderance of negative ions near ground level.[1192] As can be easily noted, the vast majority of the human population is within this two-meter (6.5-foot) height. Furthermore, we are the only creature on Earth that has its nostrils pointing downward in a position perfectly adapted to breathing in the upward-flowing negative ions. Negative ions zig-zag pattern conveys its charge to bacteria, dust, smoke particles, water droplets and other particulates cleaning the air. Finally, mental and physical functioning are thereby enhanced, including disease prevention, increased intelligence, greater stamina, better moods and the various other benefits of negative ions. Obviously, it is in caring for the natural world that we care for ourselves.

Another facet of this involves environmental stability and enhancement. More life forms produce electrical activity and ground positive charges, resulting in stronger earth-currents (*telluric currents and geoelectricity*), which are positively charged. The Earth's surface is negatively charged, and therefore attracts positive ions, which in naturally balanced quantities increase microorganisms in the soil. Consequentially, there is greater degradation of organic matter, or in other words, enhanced natural waste-disposal occurs, and nutrients are released more quickly and made more available to plants. Furthermore, increased chemical reaction rates occur in the soil due to the heat released by the decomposing matter, and the greater chemical combinations as a by-product of decomposition. More free negative ions are produced by the chemical reactions and the enhanced photosynthesis. The negative ions are then repelled by the negative surface of the Earth cleansing the air above the soil of bacteria and pathogens, and enhancing the physiological and mental functions of animals and humans. A greater abundance of healthier animals is stimulated who ground more positive ions strengthening the earth-currents (*telluric currents and geoelectricity*) even more.

Finally the whole scene brings about a high state of balanced functioning that produces an abundance of healthier plant and animal life, the likes of which we have ***never experienced***; yet, this is only the beginning of the story.

One of the major atmospheric changes that results from the destruction of wildlife, typically with the establishment of civilization, is the increase in the ratio of carbon dioxide to oxygen, and in the ratio of positive ions to negative ions. The biological effects ascribed to positive ions appear to be the result of positively charged carbon dioxide, and negative ions are due to negatively charged oxygen.[347] This immediately implicates the importance of vegetation in balancing air ions, since plants utilize carbon-dioxide in photosynthesis and release oxygen.

A closer look suggests what are very probable realities, but have often been thought of as dreams, fantasies or even religious dogma. Aggression is controlled, for the most part, by the limbic system, and less so by the hypothalamus.[346,1085] Oxygen and negative ions increase, while carbon dioxide and positive ions decrease, the major brain-chemicals that are found in these brain structures (*serotonin and catecholamines, and indirectly, via vitamin C, acetylcholine*). Low levels of these brain chemicals are known to trigger violence and aggression. In contrast, negative ions enhance brain functioning, promote good mood and stimulate relaxation.

The question is: Could balanced air ionization, with a preponderance of negative ions, defuse violence and aggression? Violence, which is often considered part of the natural world, might be eliminated in both animals and humans. The results of studies suggest that this is very possible, particularly with animals; humans are a little more complicated with factors grounded in psychology and diet, too.

Another consideration involving aggression and violence in animals is the health of animals of prey. Negative ions promote disease resistance and physical health, and therefore, preyed upon animals would be healthier and stronger. It is a thoroughly confirmed observation that most predators prey upon weak and sick animals, because it requires much less energy to catch less healthy animals. Typically, strong, healthy animals cannot be caught by predators. Therefore, the predator would be encouraged, if not find it necessary to seek easier prey, such as high-protein vegetable foods or carrion.

With a life-replenished Earth then, we might find that predator and prey peacefully co-exist. Until we do replenish the Earth with life we cannot conclude with any scientific validity that it cannot happen. It could very well mean the end of violence in the natural world!

14.10 <u>Our</u> <u>Future</u> <u>World</u>

When we consider the full extent of the possible impact of balanced air ionization it is quite significant. There exists the potential to at least reduce, if not end, mental and physical diseases and disorders. Other factors show promise of being eliminated on a life-abundant Earth (i.e., more nutritious foods, less genetic defects, enhanced environmental quality, etc.; also, see following sections). Likewise, there is every indication that mental and physical functioning will be more efficient and healthful. Stability in the Earth's electrical environment and general environmental enhancement are definite.

Feasiblely, an end or reduction of violence in animals and humans, including civil unrest and war, could be accomplished. As we see violence and terrorism increasing, we need to consider that this influence is part of the reason behind the escalation of aggression. This is partly why we are seeing more violence as more of Nature is destroyed, and this holds true of our past history, as well (as discussed in *In Defense of Nature—The History Nobody Told You About*).

Plants and animals would grow in greater abundance and health. Even if any organism were injured, healing and resistance to infection would be more easily achieved. Ions can aid in the production of electric and magnetic fields, which, as will be discussed in the following section, promote better cell function, growth, reproduction, healing, regeneration of bone and tissue, and physical and mental functioning, including a larger brain size. Here is one way in which we are reminded of the power of our participation in the physical world.

CHAPTER 15

Intimacy of a Different Sort:
The Earth and the Life Upon it
(Geomagnetobiology and
Biogeomagnetology)

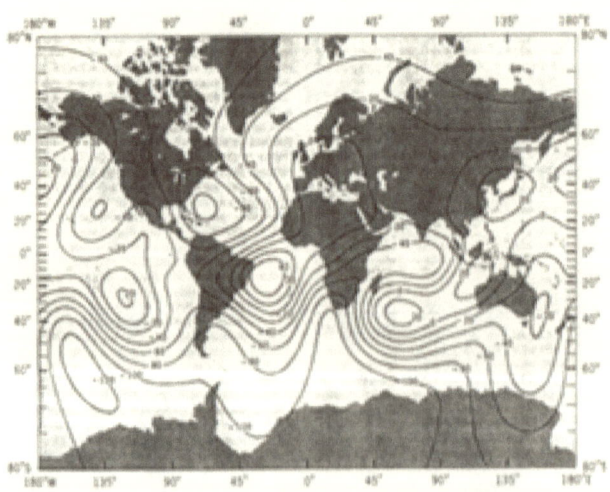

Figure 2.1. Geomagnetic Contours. Again, the Field locations are evident in this map of the Earth's magnetic contours. Animals and plants are very sensitive to these contours and align themselves to these contours during

growth, migrations and other behavior. Life also helps to reinforce the magnetic (and electric) forces. Weather systems, along with their accompanying air ions, flow along these contours, as well. [source: US Naval Oceanographic Office and British Geological Survey].

The Earth's magnetic field and life exhibit such unity that a fairly new science has surfaced called geomagnetobiology.[365] In fact, so interrelated are they that it could be just as effectively be termed biogeomagnetology. After all, the fossil record reveals a common time of origin for life and the full strength of the Earth's magnetic field.[349,1257] This correlation is the result of a mutually reinforcing relationship where the magnetic field helps regulate life processes and life contributes electromagnetic energy to geomagnetic field intensity. Figure 2.1 shows the geomagnetic contours, and their relationship to the Fields of FEM.

15.1 Life's Sensitivity

Any living thing can distinguish the intensity of the geomagnetic field, and sense the direction of those magnetic lines of force which pass through its body. Many methods have been employed in order to comprehend this sensitivity. Biochemical ions moving across cell membranes, bioelectricity, crystals in biological materials (*pyroelectricity and piezoelectricity*), and superconductivity, all of which are interrelated, have been deemed responsible for the sensitivity.

Observations of the biological activity in and around cells discloses that most of what takes place is electrical. Natural alternating current electric fields exist in and around the cells of every living thing, from primitive bacteria to mammals.[962] Moreover, cell function is regulated by weak electromagnetic fields that modulate electrical and biological windows.[334,952]

A charge difference (i.e., potential gradient) exists across cell membranes, especially when metals (*Ag, Cu, Hg, etc.*) are present.[513] Ions that diffuse across cell membranes in response to temperature or concentration differences (gradients) produce currents. As a result of these electromagnetic phenomena, electric or magnetic signals from the environment can trigger or modify cellular activity.[1358] The sensitivity of living things to the geomagnetic field is partly due to enhanced activity across cell membranes (*i.e., their permeability*).[365]

Bioelectric materials, especially crystal types (*piezoelectricity and pyroelectricity*), may also contribute to this sensitivity. Proteins (*biopolymers*) can create pathways for electrons, ions and particles that are attracted to a trapped particle (*not polypeptide chains, but tunneling between impurity centers due to polarons*).[944] This type of (*piezoelectric*) material is found in very common biological substances, such as cellulose, the main constituent of starch (*amylose*), proteins, nucleic acids (DNA and RNA), and a biochemical (*hyluronic acid*) found in connective tissues and body fluids, bone and tendons, known as collagen.[448] Other materials that exhibit the same properties include shells of shellfish (*chitin*), silk, teeth (*dentin*), ivory, aorta, trachea, intestine, elastin and wood, among others.[448,1359] These materials may also be responsible for an organism's detection of electrical and magnetic stimuli.

Living things and organic materials also have a physical property known as superconductivity. A complete absence of electrical resistance or any inhibition on the flow of electrical current exists in superconductivity (*also referred to as the Josephson Junction Effect*). Biological superconductivity takes place at physiological temperatures, unlike the electronic superconductors, which require extremely low temperatures.[288,293,813] It probably plays a role in nerve conduction, and could easily explain the sensitivity of organisms to electric and magnetic fields in the environment.[288-293,365,634]

Other probable influences involve the magnetic properties of minerals in living organisms. Minerals within living tissues can have magnetic properties (*biomineralization and ferromagnetism*), such as the mineral magnetite. Some animals are known to have magnetite or some other mineral in their central nervous system encased in a bundle of nerves.

These animals use the magnetite as though it were a compass, and thereby, find direction while migrating.[675,676]

Every living thing responds to changes in the Earth's magnetic field and electrical environment, regardless of whether it is daily, weekly, monthly, seasonal, annual or any other fluctuations (i.e., solar and lunar cycles, etc.). Interaction at this level is a well established and a generally recognized fact.[128,129,363,365] A geomagnetobiologist expounds: "Confirmation of the non-accidental nature of these phenomena is provided by the scale of their occurrence (from global, where they are reported at geographic points separated from one another by thousands of kilometers, to distances of a few centimeters on the same plant). All these phenomena are affected by a sole factor and there is a common reason for their periodicity, rhythmicity and cyclicity."[365]; p.33

For example, one experiment measured the air, the Earth and two trees about 25 kilometers (40 miles) apart for electrical variations. All were noted to vary at approximately the same time.[209-211] Connections between what is normally called the "non-living" world and life are clearly unmistakable.

The relationship is not as we typically perceive it, as only the geomagnetic field's influence on life. It is a mutually influential interrelationship, and life often creates the changes. A scientist declares: "Organisms themselves contribute to their environment's electric and magnetic concomitants of their own physiological processes."[194]; p.238. In fact, living things might be considered a sheet of magnets (*dipoles*) bathed in controlled external energy fields.[9,11,89,90,502,1358] Together they channel electric and magnetic energy at the atomic, subatomic and quantum levels.[9,11,502] The microscopic electric fields and potentials of living things (*heterogeneous media*) can be taken as the macroscopic electric field and potential of the Earth.[777] While this is known by the scientific community, it is virtually ignored both in science and by society in general.

Interconnections extend far beyond the Earth, including as far as the Sun. The Sun's Interplanetary Magnetic Field (IMF) extends from the Sun to beyond the planets and couples with Earth's magnetic field. A 26.875-day cycle exists in the variation of the IMF polarity that shifts from away from the Sun (positive) to toward the Sun (negative). The period is reflected in variations in the intensity of the Earth's magnetic field (*vertical component*). Biological processes also display this same period. Solar cycle variations, such as the 11-year solar cycle and solar flares, cause strong storms in the Earth's magnetic field and are manifest in cycles in the biological world, as well.[365] Demonstrating the real impact of life is the fact that the earthly cycles peak *before* the solar cycles (this will be discussed further in Volume Two, Tome Five).[342]

15.2 Plants/Geomagnetic Field (GMF) Unity

Plant growth and respiration are intimately related to the Earth's magnetic field. Plant respiration, growth, biochemical reactions, algae reducing nitrate, metabolic variations and other physiological reactions are synchronized with cycles in the geomagnetic field (GMF).[365] Decreased GMF intensity (*shielded GMF*) caused pine (*conifer*) seedlings to undergo a prolonged dormant period, as well as reduced seed germination, oxygen uptake and dry matter content by 30%.[377,977] Under similar conditions mustard plant growth was retarded. Likewise, barley, vetch, pea and millet seedlings' embryonic roots were shorter, while the plants also underwent various changes in growth.[268,709] When exposed to a magnetic field stronger than that of the Earth's, dandelions' flowers delayed their opening and closing, and when a more lengthy exposure was enacted they wilted and died.[895] As we will see, these responses work in favor of plant survival and evolution under the conditions present during major evolutionary transitions.

Electric fields similar to the Earth's electrical environment have an impact too. Enhanced growth and increased germination is one effect electric fields have on plants.[961,1323] Lettuce seedlings displayed a 35%

increase in certain root structures (*internodes*), a 425% increase in potassium, and a 50% decrease in sodium. Most of these effects are due to increased levels of a growth hormone (*auxin*).[698,1097] Others have shown that electric currents at physiological strength modify plant growth and ion transport.[843,1119] Ion uptake is purely an electrical phenomenon, and therefore, electrical currents in the plant or ground affect nutrient uptake. This electrical effect leads to increased nutrient uptake of nitrate, phosphate, calcium and potassium, yielding more vigorous growth.[151] Electric field effects involve the interrelationships between the Earth's electrical environment, nutrient availability, the bioelectricity of plants and air ions, as an integrated whole.

Bioelectricity in plants is a well-known phenomena. Charge differences (*potential gradients*) within the plant and root environment, wound recovery period, and the amount of light (*photoperiodicity*), each create electrical changes.[462] Plant stems have extremely low-frequency (ELF) electrical properties, as well.[536-538] A variety of plants are reported to generate radio frequencies and microwaves in their leaves and branches.[189] Electric fields around individual cells, plant organs and tissues are also known.[708,1097] Ripening fruits, such as avocados, under go electrical changes as they ripen.[117] Organic substances that are mobile within cells and tissues also carry charges (*e.g., organic acids, amino acids, adenosine phosphate, etc.*).

Light enhances ion transport, producing electrical phenomena as well. Hydrogen-ion concentrations in the soil and air affect the electrical (potential) properties of plants internally.[558] The amount of solar radiation and the leaf surface temperature also produce electrical changes.[730,731] These electrical variations could be due to responses to light, evaporation, and/or heat (*transpirational effect via photocontrol of stomata*[731], *effect on evapo-transpirational pyroelectric effect*[89,718], *and/or the photoelectric effect*[1209]). Other electrical effects are due to particle or liquid mobility (*electrokinetic transport mechanism*)[414], or particle movement across a membrane (*an electro-osmotic mechanism*).[1165]

Spores on plants are ejected by like charges of static electricity on leaves and spores (i.e., the electrostatic liberation of spores). The triggering and size of these electrostatic charges is closely related to (*red-infrared*) light exposure and atmospheric humidity.[500,599,729,731] In general, various plants display electrical (*biopotential and electrostatic*) properties.[320,414,500,599,616,708,729,731,733,950,951,1097,1099,1253,1312]   Meanwhile, few comment on the fact that bioelectricity in plants plays a role in contributing to the Earth's electrical environment.

When a plant is injured, such as when an insect or animal consumes part of it, the electrical charge increases. Significant fluctuations in electrical potential take place when a leaf is severed or injured.[708,730,1253,1268] As a scientist reveals this: "Severance caused a rapid increase in negative potential. It was concluded that severance of leaves from plants profoundly affects their electrical behavior, and that this behavior is dependent on plant-soil-water relations."[730] Therefore, when a wilderness area is complete with insects and animals that eat plants, there is greater electrical activity in the environment due to these effects on plants alone.

Plants may also communicate with each other when they are injured, much like nerve impulses between cells. One doctor of physics spent more than two decades uncovering an understanding of why sap flows up a tree. Attaching the electrodes of an instrument sensitive to electric charges, known as a strip-chart recorder, to a tree, steel needles were used to repeatedly stab the specimen. Observations disclosed that the voltage varied up and down the sample, suggesting a possible explanation for the upward flow of sap. Later he attached electrodes to two trees separated by 30.5 meters (100 feet). Striking one tree with an ax, a response was noted on the strip-chart recorder. Surprisingly, seconds later the strip-chart recorder on the other tree also showed a response. The doctor concluded that when one tree is under attack it sends out a "W-wave" as a warning to surrounding trees.[1287,1289] Similar electrical warnings occur between an organism's cells that are under attack.

Interestingly, Native Americans appear to have known intuitively that this communication between trees existed. They used sticks to beat those trees surrounding the one which they would fell for a totem pole. The belief was that the cut-down tree would suffer less by going into shock first.

What might be more surprising is that the electrical charges produced by plants contribute to the Earth's electrical environment. The daily or diurnal changes in a plant's electrical charges are like those observed elsewhere in the environment. "The diurnal cycling of fair leaf potentials is similar to the daily changes of atmospheric potential gradients of the Earth's surface."[730] In fact, the variation of the atmospheric electricity resembles the daily variation of atmospheric water vapor content, and indicates an integrated whole.[605] Likewise, the so called "Sunrise Effect" of atmospheric electricity may relate to the significant increase in leaf potentials that occur at sunrise.[605,730] Furthermore, seasonal and daily changes in long electromagnetic waves reflect plant growth periods. The daily number of impulses is relatively small during winter, but increases suddenly with spring's plant growth (May and June), reaching a maximum in late summer (August).[1246]

In fact, here we see the reason why the atmosphere is disturbed more in fall, and especially, in winter. Less plant growth during these seasons produce less electricity, weakening the Fields of FEM. When the Fields are weak they do not accelerate particles efficiently, and thereby, disturb the atmosphere. This is not the full effect, which also involves FEM-solar linkages. Daily cycling of leaf potentials is a common characteristic of all plants, and demonstrates that the Earth and the life on it are intimately connected.

This intimate connection can be seen in the orientation of a plant and its roots, which align with the Earth's magnetic field. The side roots of beets, for example, were noted to grow along a uniform compass direction.[1087] When a plant embryo is oriented toward the North or South Pole different growth responses occur. Depending on the plant, growth was greater in one direction or the other. Red silk-cotton flowers displayed different shapes (*dissymmetry*) according to the geographic location and its relationship to the angle (*dip*) of the GMF. A similar phenomena was observed with sugar beets, corn, wheat and radish root-creases, and overall plant response and separation into new functional types. Orienting the seed-embryo's root toward magnetic north produces females and toward south produces males in plants, such as hemp and cucumber.[365] Space, germination, growth and root development of any plant is oriented with respect to the GMF. Growth is greatest when seeds were oriented toward the South Pole when planting in the Northern Hemisphere.[365,1212]

Growth response according to the GMF is referred to as geomagnetotropism and can be found in all plants in one form or another.[365] For example, pine (*conifer*) seeds with their embryo roots (*radicles*) oriented toward the south, germinated four to five days sooner than those oriented towards the north.[960] After examining the root systems of 175 wild oat plants, sampled from four geographic regions of central North America, it was discovered that the roots lay predominately in a north-south direction. Meanwhile, rye that was sown in a field had roots that lay in various different directions, which demonstrates agriculture's abnormal influences.[365] Again, FEM displays organism-like characteristics with life's orientation in relation to the main dipole field, just as an organism has its cells oriented in relation to its central nervous system.

Lunar phase is known to affect the Earth's magnetic field and electrical environment, and likewise, plants respond to lunar phase. The levels and rhythm of root excretions by plants depend on the time of soaking in relation to lunar phase.[362] Growth was at its best during Full and Quarter Moon phases.[362,365] Undoubtedly a strong stable geomagnetic field is the best environment for the healthy growth of plants, and plant abundance strengthens the GMF by producing magnetic, electric, electrostatic and electromagnetic fields.

Soil activity and chemical uptake by plants are also involved in this strengthening and stability. Plants produce electric fields as the result of nutrient utilization (*ionic charge carriers*), and their coupling with soil and air ions.[538] Changes in the GMF are correlated with the rhythm of root excretions and root uptake of nutrients.[362,730] Soils with organic matter produce weak acids (*i.e., humic substances*), and this effects the internal electricity of the plant, as well as the electrical activity of the soil. Electrical potentials exist between root excretions (*exudates*) and the soil, which includes its water holding capacity.[558] Organic acids, either from organic matter (*humic and fulvic acids*) or animal excretions (*bile acids*), contribute to the electric fields in the plant, soil and air. Fluctuations in the Earth's electrical environment is correlated with the distribution of a plant hormone (*auxin*) responsible for stimulating growth and increasing electrical activity.[558,1098] Electromagnetic fields produce changes in the amount of dry matter and ash assimilated by plants, which increases electrical activity and trace mineral uptake, which is also electrical.[670,730,771,956,957]

The seasonal changes in various types of soil, regardless of water availability and temperature, are very similar when considering chemical activity and mobile forms of iron.[107] For example, soil bacteria, fungi, algae and chemical activity undergo significant changes in winter in various geographic regions (again, a winter effect is noted).[984] Soil bacteria play a major role in making trace metal ions available, particularly iron, increasing electric and magnetic activity.[304] Organic salts, in general, are superconductors which expel an external magnetic field (*Meissner Effect*) and increase electrical activity.[813] The interrelationships between life, the soil and the GMF are apparent in these observations, and are reflected in the similar cycles of the chemical composition of underground (*striatal*) waters.[730]

15.3 Bacteria/GMF Unity

Sulfur (*sulfate-reducing and sulfur-oxidizing*) and other types of bacteria are both affected by, and contribute to, the Earth's electrical environment and magnetic field. In an environment that has a magnetic field with a strength less than that of the Earth's, bacteria's biological activity was altered. Cultivation of some bacteria (*staphylococci*) were reduced by fifteen fold in their number and size of colonies.[123] Another experiment produced considerable changes in metabolism and physiology so much that it could be considered an evolutionary change (*phenotypic and genotypic*).[364,926] Numerous experiments yielded reduced growth, mutant strains and alterations in metabolism, antibiotic resistance, and morphological and cultural characteristics.[365] Stronger than normal magnetic fields also inhibited growth.[334] Both electric and magnetic fields influence bacterial growth in a number of ways.[28,334]

Bacteria that can be found in soils, sediments and water make otherwise inactive compounds active.[304] This is especially true of iron compounds, the most abundant metal on the Earth's surface and essential to photosynthesis and all life (cytochrome).[304,608-610,683,844,845,1214,1215] These bacteria are especially good at corroding metals, including nickel, nickel-based alloys, galvanized and stainless steel, and cast iron and carbon steel. [683,844,1214,1215] Corrosion of metals produces constant electrical activity and magnetic fields.[134,304,734,1184] This electric and magnetic activity contributes to the Earth's electrical environment and magnetic field.

Bacteria in water of various types orient and swim in the direction of the GMF.[154,635] Furthermore, they swim north and downward in the Northern Hemisphere, south and downward in the Southern Hemisphere, and in both north and south directions along the equator. Directed toward the mud-water interface they contain enough of an iron mineral (*single-domain magnetite*) to passively align with the geomagnetic field, even at death.[675] Rates of reproduction change along with the interaction of the Interplanetary Magnetic Field (IMF) with the GMF (*passage of the IMF solar sector boundary*).[7,8] Even something as small as bacteria display the interconnectedness of life with FEM in an organism-like organization.

15.4 Animals/GMF Unity

All animals are sensitive to and have electric and magnetic fields. Physical (*pyroelectric and piezoelectric*) properties in all animals can be considered permanent electromagnets (*dipole electric moments*). These electromagnets consist of rod-like or prism-shaped molecules, and liquid crystals that possess an electromagnet in a lengthwise direction.[21,90,891] These molecules and crystals can align in a parallel direction spontaneously without the application of any external field or force.[90] A direct interaction with other electromagnets (*component dipoles*) exists and can account for some of the animals' sensitivity to their environment.[14]

Insects respond in various ways to both electric and magnetic fields. They orient flight direction according to the strength and direction of the GMF.[658] For example, flies (*i.e., diptera*) orient their flight at 20° angles on either side of the north-south and east-west axes of the magnetic field (*i.e., the horizontal component*).[121,122] Bees can more accurately communicate the location of pollen bearing flowers using the GMF as a reference frame.[761] A sharp increase in the weight of beetle larvae takes place during unusually strong disturbances in the geomagnetic field.[1147] Even attraction to a lamp or light, depends on the state of the GMF. During magnetic storms the daily (*diurnal*) rhythm is greatly altered, when great numbers of insects take to flight.

Planarian worms directed north by light, display a (*synodic*) monthly variation in their turning left. They show a maximum in turning during New Moon and minimum during Full Moon. When directed south there was right turning, and a reversed field caused the phase to be shifted by exactly half a cycle. This experiment disclosed that the movement of planarian worms is related to geographic location and lunar phase.[194] All insects are influenced by and contribute to the GMF and electrical environment, including the effects of lunar phase.

All insects react, though somewhat differently, to the GMF and magnetic disturbances.[256,257] For example, termites orient their nests in an east-west direction.[121,122] Rhythms in the termite's feeding behavior correlate to magnetic and solar activity. This is also true of the feeding behavior of some beetle species (*e.g., Hylotrupes Bejulius L.*).[129] The Earth's electric environment influences the activity of flies, and insects' growth, development from larvae, shedding and egg-laying (*i.e., pupation, molting and oviposition*).[379,384,386,814,1240]

Fish and other sea creatures are some of the most sensitive. The (*electroreceptive*) sensitivity of fish responds up to the strength of earth currents (*telluric currents, etc.*) and then falls off rapidly. In addition, they generate and respond to extremely low- and very low-frequency electric fields as part of normal social communication.[935]

Eels are particularly sensitive to the GMF.[365] The European Eel, as discussed in Chapter 5, employs this capability to traverse the entire Atlantic Ocean. They then spawn in the Sargasso Sea, which is located in the area of the North Atlantic Field.[669]

Disturbances in the Earth's magnetic field are associated with changes in earth currents (*telluric currents and geoelectricity*) in the oceans and seas, as well as the distribution of fish.[976] Fish orient themselves in relation to the GMF, particularly its horizontal and northern (*X*) components.[1229] GMF orientation is involved in fish migration, and as a result, data show a correlation between magnetic activity and the number of fish caught.[643,853] The greatest catches occur in autumn and winter when earth currents double in strength (see Tome One).[976] In fact, all marine animals with backbones (vertebrates) use extremely low-frequency electric fields for orientation when navigating and attacking prey.[199]

The response of a marine shellfish (*mollusk*) to magnetism is modulated by a lunar (*or circa-lunar*) rhythm. During Full Moon the animals turned left, and likewise, during New Moon in a reversed field. One nerve displays enhanced electrical activity synchronized with Earth-strength magnetic fields. Mud snails and flatworms also vary subtly with lunar phase.[770]

Magnetic field pulses are known to correlate with nerve firing in the crayfish.[334] Similarly, snails display orientation responses that are integrated with the Earth's magnetic field and electrical environment (electrostatic fields).[1306] In some instances this sensitivity may be due to the presence of magnetic (*ferrimagnetic*) material, such as magnetite, somewhere in the animals physiology, usually wrapped by nerves.[100,101,676]

The orientation of some organisms may even involve relativity. Electromagnetic orientation could entail interaction with the Earth's magnetic field, and tidal and wind-driven currents. This, in turn, informs sharks, skates and stingrays about their drift in relation to the water. Swimming movements induce electric fields that may provide the physical basis for magnetic compass sense. Ocean currents, by their interaction with the Earth's magnetic field (*vertical component*), induce an electromotive force both in the water and in the fish.[644] The main magnetic field of the Earth is generated from electric currents that flow in concentric circles from east to west around the north-south axis.[1175] A fish that is swimming or drifting in waters where these current loops exist would find that the currents are negatively charged on the left and positively charged on the right in the Northern Hemisphere, and vice versa in the Southern Hemisphere. The production of global electric fields directed horizontally at the fish results, and the fish utilize them for orientation.[644] The author of this study concludes: "By sensing the horizontal as well as the vertical component of the Earth's magnetic field, the fish could determine the magnetic inclination, and thereby, the local magnetic latitude."[644; p.37]

15.5 Bird and Mammal Migration

The beginning of studying how animals orient themselves in migration started with birds. The Earth's magnetic field is utilized so well by birds that they can be blindfolded and still reach their destination (see Tome One, and Chapter 15). In contrast, during eclipses and other times of disturbance in the Earth's magnetic field, such as geomagnetic storms, birds lose some or all of their homing capabilities.[935] Bird navigation, as well as mammals' biorhythms, utilize the extremely low-frequency electric fields of the Earth's magnetic field and electrical environment.[192,935,1321]

Birds contribute to the Earth's electrical environment and magnetic field as well. A flying bird becomes charged with static electricity (i.e., electrostatically charged), and its flapping wings become sources of alternating electric fields. Furthermore, the "V" formation taken by flocks of birds increases the electrical build-up and transfer. Finally the birds land, grounding the charge on the Earth's surface, a tree or other object.[1299] The atmosphere has less positive ions, because the birds ground them. The final result is a strengthening of earth currents, which stabilizes the Earth's magnetic field, and produces a potential gradient between the Earth's surface and the upper layer (*electrosphere*) of the Earth's electrical environment.

Mammals are also sensitive to the Earth's magnetic field and electrical environment. The central nervous system of mammals is very sensitive to environmental electric and electromagnetic fields, and even the (*intrinsic*) fields in brain tissue.[9] The spinal cord is a dipole magnet with the negative end near the head.[90] This dipole magnet could be the means by which an animal detects the environment's fields.[14,1359] The effects on physiology can alter growth, activity, behavior, nerve function and the immune system.[334,365,1125] Hormones, blood composition, the heart, circulatory system, and brain waves of the higher brain are influenced.[365] Even capillary action and pupil contractions can be involved.[1064] Daily and annual biorhythms (*circadian and circannual variations*) in growth and cold resistance in rats correlate with GMF activity.[556] Rodent and mole activity is also correlated, as well as that of the sable.[927,1204] Effects on mammals appears to be a general phenomenon.[100,101,364]

15.6 Human Sensitivity to the GMF

A widespread misconception is that humans do not have a magnetic sense or are capable of overcoming any influence.[100,101] Yet, this is assuredly not so. The brain and spinal cord are among the most potent generators of extremely low-frequency electromagnetic fields of higher life forms.[275,928,935] Changes in this field have been observed for different parts of the brain and during various states of consciousness, learning and emotion.[275,928,935] Furthermore, the human spinal cord is a permanent electromagnet (*dipole*) in the direction of its (*longitudinal*) axis with the negative pole near the head (*cranially*) and positive pole near the posterior (*caudally*).[90] Humans, like all mammals, are particularly sensitive to environmental electromagnetic and electric fields, and built-in (*intrinsic*) fields in brain tissue itself, but often repress and deny that sensitivity, either intentionally or as the result of social influences.

These GMF influences are so obvious that the device used to record the electrical activity of the brain, the electroencephalograph, which is abbreviated EEG, cannot be considered as merely the record of the brain's electrical function alone. With regard to this understanding a scientist emphasizes: "I certainly do not believe that we can dismiss the EEG as the noise of the brain's motor."[10]; p.75. Furthermore, outer (*peripheral*) nerve bundles in the body are attracted to magnetic fields, which was unexpected.[939] The higher brain structures display very strong magnetic fields for biological systems.[225] Therefore, like any two magnets, such as the central nervous system and the Earth's magnetic field, there are influences on each other.

Numerous experiments disclose that magnetic fields do have a wide range of influence on the human central nervous system. Evidence demonstrates significant effects on nerves, biochemistry and behavior.[112,750] There is a rise in a brain-chemical, known as a neurotransmitter, and hormone (*norepinephrine*), which potentiates emotional reactions and exhilarates one physically and mentally.[1018] This increase may lead to lower levels subsequently. Electromagnetic fields alter the functioning of calcium-dependent systems and calcium-ion binding in the higher brain (*cerebrum*).[12,13,115,286,334] This may indicate a reduction in certain biochemicals utilized in nerve transmission, known as neurotransmitters (*catecholamines*). An indication that this may be true is afforded by magnetic field activation of the central nervous system, which causes the subject to be more relaxed, be less reactive to external stimuli, and experience increased sleep.[334] Elevated activity of the B-vitamin, choline, in the brain stem takes place, which may effect memory and reproduction (*i.e., the effects of acetylcholine*).[896]

A barrier in the brain (*blood-brain barrier*) that prevents some substances in the blood from reaching the brain is affected by electromagnetic fields, which reduce the effectiveness of this essential barrier (*increased permeability*).[112,439,440,915] Changes in the synthesis of hormones and neurotransmitters, and altered brain waves (*EEG*) also take place.[12,13,112,458,1281] These effects may be due to the pineal gland, which is at times referred to as the master gland, because it regulates most physiological functions. It is very sensitive to magnetic fields, is capable of altering the central nervous system's function, and controls numerous vital organs.[128] As a result, studies disclose that electromagnetic fields regulate the central nervous system, cell physiology and nerve regeneration.[12,14,180,199,313,461,523,758,815,899]

Experiments with humans demonstrate that the ability to detect the geomagnetic field (GMF) also exists in humans. Magnets worn on the head for only ten to fifteen minutes can effect one's ability to orient to the normal state of the GMF. This effect can last for 1.5 to 15 hours after the magnet is removed. Magnets worn on the back of the head have little, if any, influence, while those worn on the forehead or between the ear and the eye had a significant influence on orientation and navigation. When worn behind the ear they have an intermediate effect.[100] This suggests that the influences are mostly related to the pineal gland and the higher brain structures that involve conscious thought (*temporal lobes and limbic system*). However, magnetic material has only been found in human sinuses (*in remanence studies*).[100] These facts suggest that human conscious-processes, involving the higher brain, possibly coupled with magnetic material, are responsible for human sensitivity to the GMF.

Anatomical, physiological and behavioral data indicate that the effects trigger influences throughout the central nervous system. The internal (*intrinsic*) communication system in higher brain (*cerebral*) tissue permits nerves to "whisper together." This is unlike the traditional concept of an all-or-none, spike discharge from single cells. Nerves affect adjoining nerves. As a result, imposed electromagnetic fields that mimic the frequencies of brain waves (*EEG rhythms*) can modify behavior, even conditioned responses. These effects operate at the cellular or nerve levels, affecting the higher brain, hormones (including the hypothalamus and pineal gland), and biochemistry.[12,14,152,365,439,523,750] Finally, classical psychophysics and the detection of extremely low-frequency magnetic fields are systematically related to the intensity of the Earth's magnetic field and electrical environment, including weather.[938]

Observations disclose the effects of this direct relationship between the activity of the Earth's magnetic field and the central nervous system. Variations in the IMF are correlated with fluctuations in the GMF near the Earth's surface (*particularly the vertical component at high latitudes*). At such times geomagnetic activity is enhanced and the hypothalamus is disturbed.[365,1279] A dysfunctioning hypothalamus is typical of depression, eating disorders, suicides and self-mutilation. The hypothalamus regulates hormones, homeostasis, and the hunger and thirst drives, revealing an overall influence on physiology.

An amino acid important to brain and body chemistry is tryptophan (*i.e., involved in the production of serotonin, melatonin, endogenous opiates, hepatic protein synthesis, and indirectly catecholamines via norepinephrine*). In solution, tryptophan shows changes along with disturbances in the GMF, known as geomagnetic storms, and lunar phases and solar activity.[365] As a result, sleep, mood, sexual activity, aggression, energy levels, and hunger and thirst drives, among others, can be readily affected.

Neurotransmitters and hormones (*norepinephrine and epinephrine*) decrease with an increase in the intensity of the GMF. Stress hormones (*corticosteroids*), for which there is a great density of sites in the higher brain (*limbic system*), are observed to peak about 32 hours after a disturbance in the GMF.[340] An important characteristic of the operation of sensory interpretation (*critical flicker frequency*), revealing the functioning of the higher brain, changes rapidly along with the geomagnetic field.[130,212] These effects lead to distinct changes in the central nervous system (CNS).[130,131]

Extremely low-frequency electromagnetic waves effect certain neurotransmitters (*dopamine and endogenous opiates*) that are typically imbalanced in mental disorders and diseases. Even short-period variations of the Earth's magnetic field can directly affect the CNS.[124] During geomagnetic storms two-thirds of the healthy people observed showed a moderate increase in sympathetic nervous system operation (*tonus*). Meanwhile, the other one-third displayed an increase in parasympathetic nervous system operation (*tonus*).[1050] The parasympathetic and sympathetic nervous systems regulate all involuntary bodily functions.

Something as simple as orienting the body at a 45° angle in relation to the main compass points can reduce the amount of deep sleep (*13.3% reduction of theta and 6.8% increase in delta amplitudes*).[365] An improved sleep and psychobiological well-being are achieved when the body is oriented in a north-south direction.[365,1192] Demonstrating this overall influence is the fact that psychiatric hospital admissions are correlated with the state of the GMF.[444]

The electrical and magnetic or electromagnetic fields around humans have been used to diagnose mental disorders. In one experiment, this so-called "electro-dynamic" field was observed and its intensity was postulated to reflect the state of a person's mental health. Independently, a group was studied by different psychiatrists who used an "electro-metric exam" and uncovered individual diagnoses that were correlated. As a result, it was surmised that emotional state is capable of altering the electric and magnetic energy emanating from a human being. The physiological basis of the alterations involve CNS controls, particularly the higher brain influencing hormone levels and biochemistry. Extremely low-frequency magnetic fields, such as the Earth's, were noted to produce fluctuations in tissue weights and function, water consumption, oxygen consumption and positively charged ions (*cations*), altering psychobiological potentials. Changes in direct current potentials influenced learning, memory and consciousness.[211] Perception was also noted to influence the electromagnetic emanations from the brain.[654]

Due to the fact that magnetic fields control each other, perception and the mind can be influenced by changes in the GMF. This control is evident in the fossil record, which shows increased brain size (*cranial capacity*) along with the increased strength of the Earth's magnetic field.[607] When minds are connected they can influence our environment by the collective state of mind, or in the words of the old adage, it becomes a case of "mind over matter," as well as matter over mind.

Other studies confirm the impact magnetic fields have on brain chemistry. Various behavioral changes, associated with an under functioning thyroid (*hypothyroidism*), are correlated with magnetic field influence.[939] During increased geomagnetic field activity, the number of epileptic psychoses and their equivalents (schizophrenia, paranoia, etc.) increase.[646] Maximums and minimums of GMF activity lead to a greater number of neurotic reactions, complaints and difficulties.[505] A greater number of schizophrenics are born at times when there are more magnetic storms (*indicated by solar indexes*), during the first trimester.[604] Though the area of study is controversial, a correlation was observed between the monthly GMF disturbances and monthly data on extrasensory perception, including telepathy, clairvoyance, and psychokinesis (the ability to levitate objects with the mind).[83,936] The human mind and the Earth's magnetic field are, in a sense, a single unit. Our nurturing of life strengthens and stabilizes the GMF, which in turn, enhances the functioning of the human brain and mind.

Lunar phase influences the Earth's magnetic field and electrical environment, which in turn controls some aspects of behavior. Changes in direct current potentials between the chest and head varied along with lunar phase.[1006,1007] A five-year study totaling 11,613 cases of aggravated assault disclosed that peaks occurred during Full Moon and First Quarter with a secondary peak during Last Quarter.[756,757] Two thousand murders were analyzed for Dade County, Florida, between 1956 and 1970, disclosing peaks during the Full Moon and a few days following New Moon. Likewise, Ohio County Florida murders, from 1958 to 1970, revealed peaks a few days after New Moon and around Last Quarter.[758] Self-inflicted poisonings through drug-overdose in Winnipeg and Manitoba, Canada during February to August 1971, indicated significant increases around the First Quarter.[916] Furthermore, suicide attempts in Melbourne, Australia during 1970 to 1971, disclosed the same phenomena.[1217] Between 1971 and 1972, emergency room admittance in a mental health center increased by 30% during Full Moon.[153] Admissions and patient contacts increased during Full Moon at a psychiatric hospital in Toronto and Ontario, Canada, as well.[459] An exacerbation of symptoms in mental patients were associated with New and Full Moons in other studies.[1006,1007] The hormone changes that accompany menstruation can alter mood and emotion, such as during PMS. A study of 810 high school girls, who had 7,500 menstruations, revealed that most had occurred during the First and Last Quarters, with the least occurring during New Moon.[792] Numerous theories attempt to explain the influence of the Moon and some even deny that any influence exists.[219,937] A biometeorologist comments on this conflict of opinion: "The existence of a small but persistent relationship between lunar phase and aggressive behavior, birth incidence and psychiatric admission rates is not disputed by knowledgeable scientists."[937]; p.15

Brain waves (*EEG*) may be partly the result of the effects of low-frequency electromagnetic radiation. Many biological cycles or biorhythms are the result of light's effect on the pineal gland, which regulates hormones. However, removal of the pineal gland does not put an end to the biorhythms (*circadian rhythms*). Light is known to penetrate the skull and consequentially these rhythms are controlled by some unidentified light-receptors in the brain. There is also a physical effect due to cycles (*Schumann Resonances*) of standing waves in the Earth's electrical environment, or at least a correlation exists. Scientists concede that brain waves and the physical world cannot be considered two totally separate phenomena.[349] This alone indicates that collective mind has an impact on the physical world, and that our participation in the physical world has an effect on the mind (including perception).[617]

15.7 Effects of Electric and Magnetic Fields on Human Physiology

As could be suspected when something affects the higher brain, the entire body is functionally changed, but there are also direct effects on the body. Cell growth, wound healing, hormone balance, muscle tone, and immunity are regulated by bioelectricity, which is influenced by electric and magnetic fields.[286,496,777] Electric fields modulate cell physiology and the healing of bone and cartilage.[98,111,182,286,443,723,744,802,1035,1304] In fact, bones' vital state is related to its electrical conduction.[1035]

Magnetic fields can also change the rate at which ions flow across a plasma membrane.[750] Very basic biochemicals essential to cell metabolism (*creatine phosphate, ADP, ACTH, etc.*) are regulated by magnetic fields.[112] Altered hormone production (*corticoids, ACTH, ACTH-RF, catecholamines, insulin, etc.*), oxygen metabolism and biochemistry also take place.[112,364,692,935,1321,1322] Within the cell there are modifications in enzymatic and metabolic sequences, such as changes in receptor mechanisms for antibodies, hormones and nerve transmission.[12,13] White blood cells and other immune factors (*neutrophil, lymphocytes, leukocytes, etc.*) are also affected.[112,286,365,496] Electric fields are known to regulate bone, nerve and limb regeneration.[313] Blood cell composition, clotting time and suspended biochemicals (*sedimentation rate*) vary along with the GMF.[365,1240]

Vital organs, and particularly the central nervous system, produce their own magnetic fields. One of the strongest magnetic fields in higher life forms, other than the brain, is in the heart.[935] Biomagnetic fields have also been observed for the eye[650], liver[410], muscle[276], torso, limbs, scalp and breast.[277] As a result, magnetic fields can influence these organs, such as changing the heart rate, and the functioning of the retina.[586] It is easy to imagine that external magnetic and electric fields, such as those of the Earth or human-made objects, would influence physiological and psychological functions.

As a result, studies correlating the fluctuations in the Earth's magnetic field with disease rates are plentiful. The recurrence and spread of epidemics have been correlated with disturbances in the GMF (also indicated by solar activity).[261,360,365,474,1129] Global patterns involving the dynamics of the onset of some diseases are apparent, such as childhood measles.[865]

General spontaneous variations of the GMF, particularly intense, prolonged geomagnetic storms, have proven to be a leading factor in cardiovascular disease.[30,475,722,795,897,937] In fact, irregular heart beat is due to a loss of electrical conductance.[409,413] Moreover, extremely low-frequency electromagnetic fields cause irregular heart rates.[586]

Correlations exist between the course of several diseases' death rates and the degree of GMF disturbance.[365,368,369,474,863] Different levels of magnetic activity are reflected in different death rates for blood vessel diseases in the brain, as well.[16] Other diseases that display a similar correlation include eclampsia, pulmonary hemorrhages, tuberculosis, eye diseases, cancer, infectious diseases and various others.[364,365,474] This is understandable, because electromagnetic fields affect the growth of bacteria, metabolic activities, genetic expression, plaque formation, immune factors, and tumor incidence, among other disease-related factors.[126,286,334,399,427,762,1152] Reflecting this understand is the fact that magnetic activity, including the influence of 27- to 28-day IMF polarity shifts, is correlated to the rhythms of daily human death rates.

A neurosurgeon noticed that severe bleeding occurred in surgical patients during Full Moon and for two days thereafter. After studying data from blood banks all over the country it became apparent that the demand for blood was highest at that time.[1163] The Earth's magnetic field and electrical environment, which are affected by lunar cycles, have influences on disease and health in general.[474,863]

Reproduction is also regulated by these influences. The Earth's magnetic field and electrical environment take part in regulating the development of the fetus.[127] They also play a part in genetic expression[1261,364], sex differentiation[202], DNA synthesis[751,752], and cell function and growth[1035], among other psychobiological functions. The daily rhythm at the start and end of birth depends on the daily rhythm of the magnetic field and times of geomagnetic storms, which can also intensify labor. This is why premature births occur more frequently during geomagnetic storms, and the number of births increase near the end of such disturbances.[508] A study over a ten-year period, involving 510,000 births, in all municipal and ten private hospitals disclosed a predominance of births during the Full Moon, when the Moon has the most influence on the Earth's electrical environment.[837]

There are also effects on the menstrual cycle. During reduced magnetic activity the onset frequency of menstruation was higher, and during increased magnetic activity onsets were less frequent. The length of the menstrual cycle was also longer with more intense GMF activity.[149,792]

Even the orientation of the head of the embryo in relation to the GMF can determine the sex of the offspring. A geomagnetobiologist comments on this surprising discovery: "The statistical significance of the obtained results confirms that if women in the first two months of pregnancy sleep with their head towards the north there is a predominance of girls among the children born, and if they sleep with their head toward the south, boys predominate."[6]; p.8

Results of studying 824 births disclosed that 81.5% were male if the mother's head was oriented toward the south, and 86.3% were female if the head was oriented toward the north (in the Northern Hemisphere). This effect was also noted in fruit flies (*drosophilia*), hens, cows and horses.[365] Here is a clear example of the fact that genetic information is not the only basis of selectivity, and that fluctuations and reversals in the geomagnetic field influence evolution.

Growth is also controlled by magnetic fields. Normally occurring electromagnetic forces regulate growth. They regulate internal electrical forces that control the regeneration of tissues, fetal development, and hormones that govern overall metabolism (*ACTH, etc.*).[112,127,753,782,1210] Cell physiological properties, including bone growth and regeneration are also modulated.[111,442,724,744,920,1035,1153,1164,1295] Development of the central nervous system and brain size are enhanced as well.[180,607] If the Earth's electrical environment were at its optimum strength and stability it might be possible to regenerate lost limbs, bone, tissue and nerve.[126] A scientist studying the influence of the GMF on growth concludes: "The

Earth's normal magnetic field exerts an overall regulatory effect on many basic life processes including growth in general."[127]; p.112.

A number of direct observations have illustrated this phenomenon quite well. The average growth of all age groups from newborns to adolescents is increased, girls mature earlier and sexual development is speeded-up. At times this accelerated growth is known to occur around the entire globe.[365,535,647] The most plausible explanation for this phenomenon is an increase in the mean level of electromagnetic fields throughout the Earth's biosphere.[365,972,973]

Variations in the local strength of the Earth's magnetic field also have a similar effect. The lowest value of the magnetic field is in Chile, South America, and a reduction in body length of males was noted for the period between 1920 and 1960. Furthermore, North American children living in Rio de Janeiro differed from their American-grown counterparts by having a smaller body length.[365,1273] The westward drift of the GMF is also reflected in changes in body size and the migration of the human population.[1272] The weight of breast-feeding infants and the height of males in Europe were exceeded by those in Africa where GMF intensity is greater.[1273]

Brain size (*cranial capacity*) for all mammals, including humans, displays a parallel course along with the intensity of the GMF. The stronger the field the larger the brain size, which may reflect the effects of the growth hormone (*STH*).[607] Our growth and intelligence are intimately linked to our living Earth. Overall, what we do to the Earth affects our health, development and intelligence; we are intimately connected to the Earth.

15.8 Effects of Human-made Devices

Our participation at this level is quite substantial. Some human activities create extremely low- and very low-frequency (ELF and VLF) electromagnetic waves. The sources of ELF and VLF include radar, remote sensing in satellites, CRT screens in computers, microwave ovens, radio and television signals, waterbed heaters, electric blankets, power lines, appliances, cellular phones, and numerous other devices. Much of the research has focused on electric fields, magnetic fields and microwaves, and is recent. Strengths at or above natural levels have been shown to offset the functioning of the body and the central nervous system.

In the very low-frequency or VLF range we find microwaves. Radar is one such source, and has been reported to increase cancer. For example, Wichita, Kansas has radar stations and more cancers occurred on terrain crests than in valleys, where microwaves cannot reach.[739] Nationally it was reported that counties with Air Force bases, which use radar, have greater cancer rates than other counties.[740,741] The results of this study has been denied by the Air Force and radar manufacturers, evidently to protect their liability.

Some aspects of behavior are altered and related to the higher brain (*dopamine and endogenous opiate related; i.e., the limbic system*).[440-442,834,1103] The first phase of exposure is active inhibition, followed by the second phase of protective inhibition. In this second phase there is a weakening of central nervous system function, a decrease in biochemical activity, low blood-pressure, and the suppression of development in general.[586,1111]

Microwaves can cause damage to the nervous system, developmental abnormalities, and alteration and damage to the eye.[439,529] A filter that prevents certain substances from reaching the brain, called the blood-brain barrier, becomes less effective due to increased permeability. Those parts of the brain that are most affected are in the higher brain, including the limbic system (*midbrain, diencephalon, thalamus and cerebellum*).[435,529,900,1103,1104] This reduced effectiveness is probably due to effects on the pineal gland, pituitary, hypothalamus and enzymes.[1103,1104]

Sperm abnormalities, irregular heart rate, changes in immunity, and genetic mutation that was often lethal, were other outcomes of microwave exposure.[584,586] Exposure is also capable of increasing an intestinal yeast (*candida albicans*) that is typically overproduced in those people with diseases.[323] Microwaves produce an audible sound (*4.8 KHz +0.8 KHz*), which has been known to cause hearing loss.[438] The egg production of hens was greater (13.7%) with microwave exposure, but they suffered twice as many deaths. They also exhibited two major groups of disease and "profound deterioration in health."[1211] Microwaves are also produced by microwave ovens, remote sensing in satellites, telecommunication systems, television transmission by satellite (dishes), cellular phones and medical imaging, among other devices.

The most damaging effects of microwaves are on the central nervous system. There are significant changes in the structure, function and numbers of nerve cells, at all ages. Permanent damage seems inevitable when there is fetal or newborn exposure.[529,900] Studies also show that "even humans occupationally exposed to microwaves of moderate to high intensity could be at risk of brain damage."[529]; p.131.

Microwaves have also been used for weed and pest control. They kill plant roots, fungi and soil organisms (*nematodes*). Germination is inhibited or prevented, and seeds and plant tissues are destroyed. All forms of insects, both pest and beneficial, are killed. The soil along with everything in it is rapidly dried. A scientist examining this possible application of microwaves states: "Microwave treatment would be very expensive and pose health hazards."[348]

Electric and magnetic fields in the extremely low-frequency range effect mental and physical health. One reason for this is that cells operate by means of an electric flow across cell membranes (*potential gradient*). Negatively charged surface receptors on the cell membrane, and electrically controlled "little fluid gutters" between cells, are sensitive to minute changes in both electric and magnetic signals.[383] Blood is transformed, with red blood cells increasing their volume, and cleansing substances and immune factors change levels.[243,287] As could be suspected, there are changes in heart rate and hormones, as well as decreased growth.[202,250,251,287, 540] Fetal development and sexual differentiation are altered, including diminished sexual response.[202] Chromosome abnormalities (*breaks, aneuploidy, polyploidy, deletion and fragmentation*) that are produced could lead to birth defects.[316,334] The pineal gland is sensitive to electromagnetic fields, which may result in mental or physical disorders.[128]

The communication system in brain tissue permits nerves to "whisper together," and imposed electromagnetic fields that mimic brain waves (*EEG rhythms*) modify behavior.[14] Behavioral changes have been noted that exceed the effects of microwaves.[14,152,334,834] A scientist comments: "Since the central nervous system is intimately involved in the detection and resultant functional expression of these field parameters, abnormal field exposures may result in the production of disturbances ranging on the psychiatric."[128] Electric and magnetic fields with these frequencies are common place in almost any artificial environment.

Electromagnetic fields are known to produce physical diseases. Power lines have been linked to an increased risk of cancer, including childhood leukemia and brain cancer.[80,381,383,450,657,947] Electrically heated beds, especially electric blankets and heated waterbeds, have been linked to miscarriages, birth defects, and an increased length of pregnancy.[381,383] Electrical occupations are known to have increased health risks, and in some cases, wives of workers have had abnormal pregnancies, and children with birth defects.[657] Reaction time and daily biorhythms (*circadian activity*) also can be thrown off.[152] Some studies even indicate the possibility of increases in the occurrence of suicides.[934]

Irregular heart beats (*arrhythmia*) are caused by conduction disturbances, and heart failure is electromagnetic.[409,410,413] Electric fields change heart rate, and blood flow, and magnetic fields change blood volume.[250,251,287,540,243] Due to these effects, there is an increased risk of heart disease. Moreover, arteriosclerosis involves bioelectrical imbalance, so there is also the probability of increased risk of arteriosclerosis.[762]

Appliances with motors, particularly large motors, produce strong magnetic fields. This is especially true of portable devices, such as saws, drills, vacuum cleaners, and so forth.[457] Homes near power lines are in double jeopardy, particularly with appliances and many windows.[221] Electric and magnetic fields localize in houses and under streets, especially with nearby wiring particularly from power stations, welding equipment, subways and movie projectors.

ELF is the abbreviation used to represent extremely low-frequency electric and magnetic fields. A scientist discusses the possible effects of ELF:

> "The introduction of novel frequencies by man's activities, particularly in the ELF spectral range may be expected to be productive of general, subtle alterations in the biological cycle patterns of the entire population exposed. Such alterations, if long continued, may be further expected to be productive of a wide variety of functional psychological disturbances and actual pathological conditions in the human population."[127]; p.115

Human activities also alter the Earth's electrical environment and magnetic field. The main magnetic field of the Earth, according to geophysical theory, is the result of electric currents that flow in concentric circles from east to west around the north-south axis.[1175] This is especially true of the new model of the Earth presented in Tome One. Buildings, concrete and asphalt are typically insulators, and prevent the grounding of electricity.[710] A biometeorologist comments: "The complexity of the fields are often enhanced by the idiosyncratic contributions of local man-made architecture [which] can either enhance local geomagnetic field levels or reduce them to near-null levels."[937] The mental and physical effects of these human-created imbalances were made evident in the previous sections and contribute to *all* human problems, but few of us realize this effect, and even fewer are concerned.

15.9 The Natural Environment and Its Potentials

The electrical contributions of plants depend on root excretions (*exudates*), the type of soil, which ideally should be organic (*humic substances, soil organisms, etc.*), and whether animals are present (*bile acids and urea*).[151,538,558] Meanwhile, humans typically destroy plants, use chemical methods when plants are grown, and prevent animals from being free to roam. Herbicides change a soil's conductivity, usually lowering it, as well.[22] Light enhances electrical activity (*ion transport*) in plants, while pollution and dust inhibit it.[558] High-voltage electric fields, such as power lines, damage and kill plants.[348]

Life contributes to the intensity and stability of the Earth's electrical environment and magnetic field. As a scientist states: "Organisms themselves contribute to their environment's electric and magnetic concomitants of their own physiological processes."[194]; p.238. Living things also produce electricity, semiconductors, conductors and superconductors, and change optically inactive material to an optically active medium (i.e., a chirowaveguide).

In plants, both respiration and photosynthesis obtain their primary energy source by mobile electrons. These electrons flow in a stream through conductors, such as proteins, and can be measured as a current. The rate of transfer depends mostly on its coupling with the environment and occurs at temperatures too low for ordinary chemical reactions.

In animals, cell components (*mitochondria*) and other bioelectric reactions produce electron transfer.[341] Cell function is controlled by electromagnetic fields that regulate electrical windows.[952] Biological superconductivity takes place at physiological temperatures, unlike the electronic superconductors, which require extremely low temperatures.[288-293,365,634,813]

Electrical and magnetic activity involving living things is widespread and profound. Flying birds pick up charges from the atmosphere, their

'V' formation transfers electrical charges, and when they land they ground the electricity.[1299] Plants, particularly tall and deep-rooted trees, and migrating animals also ground electricity. Organic carbon, present in all living things, and organic soils are electron donors for microorganisms that reduce sulfate, and corrode iron and other metals.[304] The corrosion of metal produces a magnetic field.[608-611,1291] According to quantum mechanical theory, microscopic electric and magnetic fields contribute to the larger macroscopic fields, such as those of the Earth.[9,341,502,688,689,777,1175] Living things can be considered part of the physical properties of the Earth, including the electrical environment and geomagnetic field.

With a stronger electrical environment and magnetic field of the Earth, there would be an increase in the near 100% quantum efficiency of photosynthetic electron transport.[341] Life contributes electromagnetic (*bioelectric and quantum electrodynamic*) forces to the Earth's electric environment and magnetic field, which are strengthened and stabilized, and in turn, enhance life's abundance and health. An abundance of life contributes more to the Earth, and so on, until a dynamic balance is achieved, producing a world we have *never* experienced that is superior to anything heretofore.

We humans will benefit in a number of ways, one of which is an increase in intelligence and enhanced central nervous system functioning. Nerve regeneration, repair and reestablished nerve connections is one possibility from an enhanced and balanced electrical and magnetic environment.[875,899] The stimulation of "internal pain-killers" (*endorphins*) under these conditions would not only increase feelings of well-being, but enhance the functioning of higher brain-structures, and relieve stress and anxiety.[112,924] A hormone and neurotransmitter (*norepinephrine*) that exhilarates one's mood and physiology is also increased.[607]

Fossil evidence discloses that human brain size increased along with an increase in the strength of the Earth's magnetic field. This increased strength occurred during the ice ages when grounded ice caps increased the Earth's electrical currents, enhancing the magnetic field, and in turn, increased human intelligence. There are even interactions between the Earth's electrical environment and brain-waves (*EEG*), nerve conduction and cellular activity. Furthermore, all of these benefits are also available for all mammals and other animals.[607] Enhanced intelligence and brain function also means the capabilities of mind over matter are more readily available, and have been known to increase with strong, undisturbed geomagnetic field conditions.[83,365,936,940]

Aside from the increased mental capacity to alter physiology there are also direct effects. Electric and magnetic forces play a role in the formation of tissues and organs (*morphogenesis*) by communicating with cells and altering their genetic expression. This can allow for regenerative growth, such as growing back damaged organs, nerves, bone and muscle, and even lost limbs.[25,125,127,129,398,780,1142,1153] There is every indication of increased immunity, and combating diseases that already exist, such as cancer.[112,127,285,365,399,427,497,863,892,946] Growth and a number of physiological processes are enhanced by the Earth's magnetic field and electrical environment.[112,127] This enhancement of growth, physiological functioning and disease resistance is not limited to humans.

These mental and physical effects have been confirmed. Mentally related problems such as pain, nerve blockage, partial deafness, mental depression, memory relapse and schizophrenia have improved with the application of one understanding. Likewise, physical disorders, such as arthritis, osteoarthritis, chronic bronchitis, earache, infections and others also improve with the same knowledge. All of this enhanced rejuvenation was accomplished simply by orienting patients during sleep to the north-south component of the Earth's magnetic field.[365] How much more would mental and physical functioning be enhanced with a complete and plentiful life system on Earth enhancing the magnetic field?

Again, throughout this section we have seen the unity of life and the physical world. Here we see the mechanism behind the events that occur in cycles of human history and evolution (these will be discussed). Simultaneous increases in the numbers of animals or insects, which yielded the recurrence and spread of epidemics and plagues, occur throughout history. Mass migrations, swarms of insects out of season, climate changes, war and civil unrest, "natural" disasters (volcanic eruptions, earthquakes, etc.), and solar and geomagnetic activity are correlated in cycles. All of these phenomena are known to fluctuate together.[365,474,929] Furthermore, each of these events can be triggered by the Earth's magnetic field and electrical environment fluctuating (this will be discussed in the following sections, and Tome Five of Volume Two, and *In Defense of Nature—The History Nobody Told You About*). All in all, we are at the controls in such a way that we could make our existence more peaceful, prosperous and plentiful simply by nurturing life, and not destroying it.

CHAPTER 16

You Reap What You Sow (Plant-Soil Interrelationships and Human Participation)

Plants are the most primary food for all of the higher organisms on Earth. Aside from needing water, plants are intimately dependent upon the elemental components of the soil. When humans are reduced to a very simple basis, the importance of the soil's constituents in plant life, the primary food source, begin to be appreciated.

The human body is about 5% ash or about 5% soil (e.g., 7.5 lbs. of 150 lb., or 3.5 kg. of 70 kg. person). About 1.6% of normal body weight is calcium, 0.9% is phosphorous, 0.3% is potassium, 0.3% is sodium, 0.2% is sulfur, 0.05% is magnesium, 0.004% is iron, and the list goes on. Certainly if the soil does not contain these elements, or in their proper proportions, then imbalances in the plant, and thereby, the animal that consumes it, are inevitable.

A distinct contrast between the trace element requirements of plants and those animals that use them for food is obvious. Aside from the major nutrients, nitrogen, phosphorous, potassium and calcium, there are also minor elements. These minor elements are iron, manganese, zinc, sulfur, copper and boron. The total number of elements required in plant health is far less than those required in human nutrition.

The number of essential elements for humans is at least twenty-seven.[840] More recently the total has reached 30 with every potential for this to increase. Requirements for the elements range from 50 micrograms to 18 milligrams per day for humans. Other elements may be necessary in very small quantities, but our limited knowledge has yet to reveal them. When considering the contrast between plant requirements and those of humans, the possibility of human needs being unmet becomes immediately apparent if the soil is deficient.

Throughout our experiences with plant cultivation we have discovered that it is even difficult to know when plants are experiencing a deficiency. Latent deficiencies, those that do not cause obvious deficiency symptoms, are often ignored and can cause a great deal of damage before they bring about visible symptoms. There may be problems with water utilization (*transpiration rate*), and decreases in reproductive capacity, yields, disease resistance, pest resistance, and overall quality.[1095] Just how much damage these deficiencies cause for humans is not easily determined, especially when considering human needs far outweigh those of plants.

16.1 Importance of Organic Soil Constituents

Soil, and the factors that govern its quality, affects nearly all living things. Natural organic components of the soil increase soil fertility and health more than any other type of fertilization, such as chemical applications. Manures, and the decomposing plant and animal debris called compost, are slow release fertilizers.[93] Heavy manure applications increase the most important plant element, nitrogen, in both the soil and subsoil.[188] Organic matter increases yields beyond what chemical fertilizers can, because chemical fertilizers reach a certain saturation point and then decline.[93] In fact, "it helps to maintain the maximal yield even when conditions are at either side of the optimal range."[93];

p.8. The use of organic matter enhances the quality of the soil, and crop health and yields are directly related to the soil.

Soil scientists make these comments about soil organic matter: "It encourages granulation, increases water storage, nutrient supply and soil organism activity, and improves soil fertility and productivity."[1016] "The added organic matter induced changes in the physical properties of soils. A marked improvement in the structure of the soils is often detected. The biological activity induced by the organic matter has the potential to antagonize root disease and supply different growth factors."[93]; p.8. "Soil organic matter has been recognized for centuries as the key to soil fertility and productivity. Organic matter plays a major role in the chemical, microbiological and physical aspects of soil fertility."[910]; p.29

Studies have shown the numerous benefits of organic matter. Water holding capacity is increased, especially in sandy soils.[93] Higher crop yields are realized, particularly with manure (dung and leguminous plants, such as clover, soybeans, etc.).[93,910] Leaf mold and decomposing plant matter, called humus, is a soil conditioner and stores available plant nutrients.[554] Nutrients are released gradually and this makes the soil less sensitive to nutrient loss (*leaching, volatilization or fixation*). Improved soil structure, and increased availability of minerals and trace metals occurs (*through chelation, organic and inorganic ligands, ion exchange, absorption, precipitation and dissolution of solids, and acid-base equilibria*).[93,254]

Organic matter plays a key role in enhancing nutrient availability, but also reduces the toxic effects of some substances (*free cations*). Trace metals (*Fe, Mn, Cu and Zn*) remain active, allowing increased utilization (chelated forms are more readily used by plants). Iron, utilized in chlorophyll, plays an important role in the form of soluble organic complexes. Diffusion of metal complexes (*cations of Fe, Mn, Zn, Cu, etc.*) are facilitated by the presence of other ions and organic matter enhances this relationship. Not only does it aid in the transport of nutrients to roots, but creates an activity gradient of ions.[254] Generating electric activity, this gradient contributes stability to the Earth's electrical environment and magnetic field (*bioelectricity contributes to telluric currents, geoelectricity and the geomagnetic field via electromagnetic induction*).

Weak acids, called humic substances (*humic and fulvic acids*), have beneficial effects on plant growth. Humic substances interact with metal ions, oxides and hydroxides, and more complex soil minerals (*they form polynuclear complexes*).[254,1012] The accumulation of humic substances is higher under undisturbed (*anaerobic*) soil conditions than tilled soil. Higher levels of humic substances increase the solubility of the most essential nutrients, carbon and nitrogen, as well as, other nutrients.[1012] A team of scientists expound: "Organic matter plays a key role in the behavior of micronutrients in soil. Substantial evidence has accumulated to indicate that complexing agents (e.g., defined biochemical compounds and humic substances) play a prominent role in the dissolution of micronutrients and their transport to plant roots."[254; p.108]

Organic matter's role in soil fertility is particularly illustrated by a few facts. The correlation between organic matter, soil microorganisms and soil fertility is made clear by the decline in soil fertility following the removal of the surface, humus-rich soil layers.[554] In fact, the temporary increase in soil fertility which attends chemical and physical treatments of the soil is partly due to the enhanced degradation of the life that was destroyed by its application. Not utilizing chemical treatments produces longer-lived greater fertility, as well as, the direct elimination of plant diseases.[554]

Organic matter is the key to all aspects of soil fertility including chemical, biological and physical properties. It is the main recipient of plant nutrients particularly carbon, nitrogen, phosphorus, sulfur, and exchangeable calcium and magnesium.[23,167, 255,769,902,1016,1156] An organically-farmed soil had higher levels of soil enzymes (*urease, phosphatase and dehydrogenase*) and microorganisms (*microbial biomass*).[167,1016] Polysaccharides, which serve as active binding agents in soil aggregate formation, bring greater stability in organically-farmed soils.[1016] The surface most fertile layer of the organically-farmed soil is thicker (by 3 cm. or 1.17 in.), and yet, seedlings could emerge more easily.[1016] The electrical and chemical qualities of the soil (*cation exchange capacity*) and physical properties were also enhanced.[216,224,1200]

Microorganisms are also more abundant in organically farmed soils and are essential to plant health. A soil scientist comments: "Soils differ from a heap of inert rock particles in many ways, but one of the more important is that they have a population of microorganisms living in them which derives its energy by oxidizing organic residues left behind by plants growing on the soil, or by animals feeding on the plants."[1055; p.150]

Most microorganisms cannot survive without organic matter, while plants cannot utilize organic matter nor effectively utilize nutrients without microorganisms. The rate at which microorganisms degrade organic matter depends on soil moisture, acidity (pH), temperature, available plant nutrients, microflora, soil texture and structure, oxidant supply and structure, and those factors that inhibit or enhance microbial activity.[1012] All of these conditions are at optimal ranges in organic soils.

Microorganisms play a major role in humus formation and decomposition. A slow rate of decomposition brings a steady supply of nitrogen, phosphorus, potassium, carbon and micro-elements to plants. Soil microorganisms include nitrogen-fixating bacteria, fungi, protozoa, actinomycetes, yeasts, algae, worms and beneficial insects, which thrive under the optimum moisture and temperature. Living animals excrete waste that is incorporated into the soil by microorganisms forming more readily utilized substances (*chelates*), and organic acids and inorganic acids, such as nitric and sulfuric acids. They are also known to be essential for the elimination of plant diseases (*pathogens*).[554] Organic complexing agents in the soil are produced by microorganisms, which produce various substances (*humic acid, fulvic acid, polynuclear complexes, polyphenols, and aliphatic, amino and sugar acids*).[254] All of these factors increase the availability, solubility and active state of most elements.[254,554,1012]

Most significant are microorganisms' ability to balance soil conditions under extremes. Referred to as "homeostasis," as occurs with any organism, checks and balances maintain equilibrium. "The microbial population has the ability to maintain community stability and integrity in a variable environment."[514] Once again we witness an example of life bringing stability in contrast to the decay and disorder expected (i.e., as indicated by the Second Law of Thermodynamics).

In fact, the difference of a manured plot can mean fifteen times more energy than an unmanured plot. At least twice the number of bacteria and protozoa are present. Fungi of the beneficial sort also increase. Organisms live much more actively than in other plots, spend less time in the resting stage, and live right up to the income of matter and nutrient supply.

A general rule in natural conditions is when a greater number of organisms exist there is also a greater number of species. Any increase of organic matter at once increases the number of microorganisms.[1055] Microorganisms increase activity in the region close to the plant's roots (*rhizosphere effect*), and accelerate rock weathering bringing about nutrient release.[254,279,554,775,1055] Bacterial fertilizers alone can increase yields by 10% to 25%, or more.[694] Microorganisms are the single most important factor (other than water), and they require organic matter in order to make nutrients available to plants.[279]

Fungi are some of the most important soil microorganisms. Fungi (*i.e., mycorrhizal and ectomycorrhizal*) increase growth and yield through improved phosphorous uptake, resistance to drought and salinity, and increased resistance to disease (*tolerance to pathogens*) and root infections. They also activate root excretions (*exudates*) that stimulate other growth factors. By creating mats (*hyphae*) across root hairs, root-absorption surface area is greatly increased aiding in the roots capacity to absorb charged particles or ions.[93,514] For example, plant growth on phosphorus-poor soils is greatly improved by the application of fungi.[554] Their mutually beneficial relationship (*symbiosis*) with plants trades plant nutrients for carbohydrates. One of the most important nutrients released by fungi is a hormone (*auxin*) that stimulates plant growth. Inoculation with fungi, as well as bacteria, have been shown to produce greater, healthier yields.

Adding cultures of fungi and bacteria has proven to enhance plant growth in ways that no chemical application can. When fungi were added to soil in England, potato yields increased by 20%. After inoculation, corn, barley and wheat were grown on an otherwise infertile field in Pakistan.[1052] Parasitic invasion is also prevented, and when the fungi are disturbed, such as in tilling, bacterial or fungal parasites begin to destroy the plant. For example, root-rot fungi do not attack healthy plants nor root cells, but form a mutually beneficial (*symbiotic*) relationship with the plant.[1055] Bacterial cultures (*phosphobacterin*) added to the soil have been able to increase crop yields in some instances, such as in eastern Europe.[188] Fungi and bacteria tend to congregate around particles of decaying plant and animal debris, called humus and soil organic matter.

The earthworm can bring ten or more tons of castings to the surface per acre per year. These castings are composed of a neutral colloidal humus, which yields nutrients that are immediately available to plants. Earthworms stimulate nitrogen release and aerate the soil, making it permeable to rain, and they mix organic matter throughout the soil, both of which lessen the tendency for soil erosion by wind and rain. Burrowing animals, including earthworms, transport nutrients by mixing and turning over the soil (*i.e., bioturbation*).[93]

Animals living in, on or above the soil contribute to nutrient distribution, too. Manure aids in the uptake of plant minerals, such as iron, copper and zinc, among others. Herding animals and birds, especially those that go on extensive migrations, distribute minor and trace elements. Without migrating animals mineral distributions would tend to be localized and imbalanced. Animals are involved in all major nutrient cycles, such as nitrogen, phosphorus and carbon, which are major plant nutrients.

In the past, animal manure was highly valued as a source of plant nutrients. Chemical fertilizers were considered adequate substitutes and have been used extensively since 1950 (post-World War II). Meanwhile, the use of manures increases crop yields, particularly with root crops, corn and grasslands. The problems encountered with chemical fertilizers are typical of conventional agriculture: leaching and surface runoff of nitrogen, phosphorus, potassium and organic matter into waters, and the introduction of heavy metals.[1285] In contrast, the addition of organic matter produces bacteria that transform and detoxify heavy metals, and prevent leaching and runoff.[1133]

Manure is a much better source of phosphates and potassium than chemical fertilizers. Rapid improvement of plant nutrition is unmistakable.[1148] Manure is an important source of nitrogen, phosphorus, potassium, calcium, magnesium, zinc and other elements (Na, B, Cu, Co, Mn and Mo).[1144] Moreover, these nutrients are more chemically active.

Without organic constituents any imbalances transfer to the plant consumer. For example, manure on a pasture increased the potassium and lowered the sodium in cows' blood (*serum*).[704] Due to chemical fertilizers, which are cheap, easy to use and pushed by salesman, the present practices of manure handling and application often lead to environmental problems instead.[1266]

Another virtue of combining soil organic matter and microorganisms is the formation of carbohydrates. Soil carbohydrates are essential to plant growth, while they also stimulate soil aggregation. Both major and minor nutrients are more available when soil carbohydrates are available (*contributes to macronutrient cation exchange capacity, and micronutrient interaction with anions*). Microorganisms release simple carbohydrates with binding (*chelating*) properties that accelerate mineral weathering, consequentially making even more nutrients available. Other natural constituents, as yet unidentified, are enhanced by decomposing herbage and tannic acid.[775] The region around the roots (*rhizosphere and rhizoplane*), where the roots release substances (*exudates*), is 90% carbohydrate.[1012] The final result leads to improved plant growth (*including nitrogen fixation*), uptake of micro-elements (*chelates*), and the elimination of plant disease organisms (*pathogens*).[554] In contrast to chemical applications, a scientist asserts: "In conclusion, there is little doubt that carbohydrates in soil have an influence on physical properties, and are likely to be involved in a wide variety of biochemical processes which influence the soil as a medium for plant growth."[554]; p.88

Organic matter and microorganisms balance the acidity and enzymes of soil, too. Plants grow best in a pH that ranges from four to eight, or the neutral to mildly acidic range. A pH of three or below is acid, which injures plant roots and releases toxic elements that plants then absorb. Such is the effect of acid rain, which has a profound effect on human disease rates. A pH of nine or above is basic, which prevents the utilization of essential nutrients, such as phosphates and trace minerals.[1055] Organic matter and microorganisms maintain an ideal pH.

The uptake of trace minerals depends on the soil's pH, some minerals increase and others decrease in relation to pH level. For example, red clover absorbs half as much cobalt, nickel and manganese, and six times more molybdenum when the pH shifts from 5.4 to 6.4 (from mildly acidic to almost neutral).[855] The decomposition of humus creates weak acids and animal waste produces uric acid (lowering pH). Balanced acidity is accomplished by microorganisms, which increase carbon-dioxide concentrations, thereby raising the pH.

The best soil condition has fine particles (i.e, clay), and abundant organic matter and microorganisms, yielding a pH between 6.0 and 7.5 (neutral is 7.0 and acidic is below 7.0). Plant materials in the soil increase soil enzyme activity, which prevents the formation of inorganic sulfur. Plants find only organic sulfur useful and inorganic forms are usually acid.[1]

16.2 <u>Soil:</u> <u>The</u> <u>Basics</u> <u>of</u> <u>Health</u>

Soil and the factors that affect nutrient availability are essential to the health of all life on Earth. At the most fundamental level, the vitamin and mineral content of foods depends on the humus content of the soil. For example, organic fertilizer enhances vitamin content, while compost causes a striking increase.[20,685,694,1094] Composted crops contained higher levels of vitamin C, potassium, phosphorus, iron and usable protein; a 50% increase in some cases.

Sodium and nitrates were lower, and fiber is higher in organic foods, indicating disease preventive characteristics for humans and animals. High sodium is a factor in the development of hypertension and heart disease. Organically-grown foods are low in sodium, disclosing one of the possible benefits. Also higher fiber and vitamin C, and lower nitrates are disease-preventive, particularly with regard to cancer.

Pesticides, not used in organic farming, can cause both mental and physical disorders and diseases. Problems created by pesticides can range anywhere from depression and schizophrenia to cancer, heart failure, birth defects, and even death.[519] Numerous other factors affect the quality of a plant's nutritional status and safety, and those animals, including humans, who consume it.

The practices of conventional agriculture have a profound effect, for one, by lowering trace minerals. Minor elements are often neglected when using chemical fertilizers. Deficiencies are often not immediately apparent in the crop itself, and because the main objective is profit, deficiencies are often overlooked.

Chemical fertilizers can also interfere with the absorption of other nutrients. For example, when ammonia or ammonium is added to the soil it displaces other positive ions, typically metals, by leaching them into waterways. Potassium in excess is absorbed at the expense of magnesium, calcium and iron, which are lost.[374,902] An excess of either nitrogen or potassium lowers the uptake of calcium, phosphate and iron.[902] For example, nitrogen fertilizers have brought about copper deficiencies in cereals, and phosphate has caused zinc deficiencies.[855]

Chemical fertilizers also decrease soil microorganisms, because most require organic matter and these chemicals are too simple to break down for energy requirements. Part of the increase in fertilization following chemical applications is the destruction of microorganisms, which then decompose releasing nitrogen. In addition, humus nutrients become unavailable, causing the loss of metal and non-metal ions.[24,26,641,1052]

Compounding the whole problem, each time a crop is harvested those trace elements that were available are shipped to market never to return to the soil. The entire situation makes subsequent harvests more and more deficient in trace elements. The impact is especially pronounced for those soils that have been farmed for one or two centuries.

Chemical fertilizers have profound effects on the plant and soil. Mycorrhizal fungi, which form a mutually beneficial relationship by aiding plant roots in nutrient uptake, are either substantially reduced or absent. An infertile field in Pakistan, which grew corn, wheat and barley when the fungi were added, had all the beneficial effects nullified by the addition of a chemical phosphorous fertilizer.[1052] Humus is rarely added and often washed away by rain or irrigation, thereby the consequent loss of nutrients and their availability ensues.

Most microorganisms require organic matter to provide energy, and thereby, soil life plunges to a bare minimum. Bacterial fertilizers, in contrast, can increase yields just as effectively or greater than the typical 68 kilograms (150 lbs.) per acre of commercial nitrogen that is applied. And, there is better nutrient content, storage life and flavor in produce with the use of organic fertilizers.[351] When heavily fertilized with chemicals, corn has lower quality, including less usable (*not chelated*) and more difficult to digest protein.[854,1076] Due to the effects of chemical fertilizers, the average protein content of grain has consistently declined in the United States, and undoubtedly elsewhere.

The impact of chemical agriculture does not end with food either. Nitrogen and phosphate run-off causes algae to multiply in waterways. Being short-lived, the algae quickly perish and decay, thereby starving the water of oxygen, and the aquatic life that depends on it perishes. Normally dormant, pathogenic organisms become active and outbreaks of previously rare diseases can occur. Chemical nitrate fertilizers are so soluble that they seep into groundwater, and high amounts have resulted in hazardous effects, including infant deaths.[70,108,283,423,1269,1324]

Some of the nitrogen fertilizer converts to nitrous oxide, which is destructive to the ozone layer and is, at least, partly responsible for the ozone holes.[1114] In addition, some of the nitrous oxide transforms into nitric acid and contributes to the acid-rain problem. Compounding the problem are multi-stomached cattle and sheep who reduce nitrate in feed grains to toxic nitrite, contributing to the detrimental effects of animal protein in cancers.

Baby foods show excessive levels of nitrite when improperly stored and children under four months are susceptible to its toxic effects. The part of the plant which did not convert to protein, due to chemical fertilizers, is the source of the nitrites.[1344] The full scope of the problems created can only be guessed at, because the first year of life is the greatest determinate of mental and physical health throughout life (other than the gestation period).

16.3 Nature's Way of Telling Us Something is Wrong

Plant disease and infestation are Nature's way of telling us something is wrong with the health of the soil. Pesticides do not only kill the so-called "pests," but many other living creatures, particularly those that consume the pests. In fact, they destroy many of the soil components that bring a fertile, healthy soil. Soil fumigants or other pesticides reduce mycorrhizal fungi. Also affected are earthworms, actinomycetes, protozoa, yeasts, insects, and the other soil organisms and animals that feed on them.[24,26,498,1356] Herbicides reduce mineralization and nitrification of the soil.[218] Pesticides and herbicides predictably alter the ecological balance of the soil population by reducing some, and stimulating other organisms, such as parasites (*saprophytes*).[498,805,1356]

The accumulation of (residual) chemicals increases the solubility or concentration of manganese, copper, zinc, phosphate, chlorine, boron, sulfate, nitrate and others. This shift can result in the formation of toxic substances or the loss of mineral uptake.[27,805] Enzymes, nitrogen (*N-fixation*), and fungi, algae and other organisms (*microflora*) are altered in ways that do not promote soil health.[498,1356] Bees, essential for pollination, are killed by the sprays, costing millions annually in federal compensation.[882]

The entire effect is partially summed up in the fact that, in the United States from 1950 to 1970, there was a 53% increase of yield per acre that required a 700% increase in chemical use to accomplish. The increase of herbicides is equal to or ahead of insecticides as major environmental contaminants. Furthermore, there is the future probability of sub-lethal amounts leading to new lethal substances.[105,868] Making matters worse, a scientist comments: "Consequently, the synthesis, within the last 50 years, of a wide range of organic compounds that have no counterpart in Nature, poses an enormous problem for the decomposers, which are often unable to breakdown these chemicals."[560]

Often pesticides backfire and do not accomplish the proposed task. Many examples are known of "pests" becoming immune to the pesticides designed to destroy them.[238,258,677,816,1274] Such was the case when "super rats" plundered newly sown wheat and barley on farms in southern England. Making matters worse, once an insect becomes resistant many other pesticides are also ineffective.[400] Meanwhile, the pest's predators have become severely reduced or destroyed due to the increased levels of the pesticides created by consuming the pests.

An example involving malaria demonstrates how unsound pesticides can be for humans. By 1968, 38 species of malaria carrying mosquitoes in India were resistant to pesticides. In 1956 it was only five resistant species. A major portion of the 30 to 50 million cases of malaria in India alone in 1977 were obviously the result of the loss of mosquito predators destroyed by the pesticides. Likewise, Central America was sprayed in the 1960s, and by 1975 there were three times more cases of malaria than a decade earlier.

When stressed by air pollution, chemicals or a deficient soil, plants release a toxic substance. This has lead to increased attacks on plants by insects who use the plant's own chemicals for detoxifying both natural and synthetic insecticides. For example, soybeans that were stressed were attacked by Mexican Bean Beetles who normally shun the plant.[988]

Furthermore, food and chemical demands increase every day. In three decades there was a ten-fold increase in insecticide use, yet crop losses nearly doubled.[953] The increase in food production by the year 2000 will be accompanied by at least an 180% increase of chemical fertilizer use than was used a decade ago. Yearly, 80 million metric tons were used between 1973 and 1975. Meanwhile, in 2000 it will be 225 million metric tons.[300] A five-fold increase in pesticides will probably be "required."[34] Adding to the overall effects, these chemicals will necessitate extensive mining and industrialization, devastating wildlife, the environment, and undoubtedly, humans. Simply by buying chemically grown foods, we are literally removing mountains with the life on them, and jeopardizing our health, well-being, and the Earth.

When considering the fact that 60% (by weight) of all herbicides, 90% (by volume) of all fungicides, and 30% (by volume) of all insecticides used in the United States can cause tumors in animals, the threat to human health is evident. Such chemicals become concentrated in processed foods, making them particularly hazardous.[77,297,318,401,408,601,807] Pesticides and other chemicals are not only present in an area due to direct or aerial spraying on food crops, but are also in running waters, rain, fog, adjacent lands, accidental spillage and industrial waste.[76,161,228,473,804,820,868,954,1031] Pesticides have been found in humans, water, soil, animals, crops, air and sediments.[226] Other studies show that very hazardous pesticides (*DDT, DDE, carbamate, organochlorine, organophosphorus, chlorophenoxy, etc.*) are stored in our bodies (*in lipid tissue*).[60,142,195,301,501,642,645,870,1143] "The data also suggests that most people are not occupationally exposed, they come in contact with these substances through other sources."[870; p.120.]

Meanwhile, the dangers of pesticides and other chemicals is unmistakable in the statistics. Of all cancer as much as 90% is triggered by some agent in the environment. The cancers we see today are from agents in the environment 20 years ago. Latency periods of various cancers range from 15 to 40 years, so recent developments are not yet observable. Those developments include tens of thousands of new chemicals.

From 1950 to 1969 a geographic analysis of the United States disclosed that cancer mortality was higher among those males, in 139 counties, that were in the regions where the chemical industry was most concentrated. Nationally and internationally migrating people acquire the predominant cancer of the area they move to, and also lose the cancer of the area they came from.[419] Much of this cancer-susceptibility is undoubtedly due to chemical imbalances in those areas.

Many diseases and imbalances are linked to pesticides, such as depression, schizophrenia (Korsakoff's Syndrome), Parkinson's disease, infertility, cancer, vital organ damage or failure, and just about anything.[27,42,58,62,95,214,233,244,249,366,387,477,511,541,553,687,746,877,1108,1208,1324, 1349,1354] Recently, leukemia, particularly during childhood, and especially during pregnancy or nursing, has been tied to pesticide use in the home.[71,424,453,632,764,776,931,992] The increases in cancer and other diseases, including AIDS and mental illness, could stem from the widespread presence of chemical hazards in the environment.

Two-thirds of those humans conceived never make it to the end of the first year of life. Most of this group perished in the early stages of pregnancy with a stillborn rate at nine per thousand. In over 60% of the aborted embryos a chromosome abnormality is found, and more than half of the spontaneously aborted embryos and stillborn fetuses have gross developmental abnormalities.[464] Pesticide-induced mutation is clearly evident in a number of studies.[178,498,1269,1341,1356] An Environmental Protection Agency (EPA) study of vital organs of 100 fetuses disclosed that all tissue samples contained "significant amounts" of pesticides. In the United States, as well as on the other side of the Earth, in Moldavia, pesticides were discovered in stillbirths, placenta and fetuses.[60,142,301,507,642,645,1143] This is only a sample of the hazards that stretch far beyond threatened human welfare!

> "Inert ingredients, which are not usually treated in discussions of pesticide health effects, frequently comprise a large percentage of a commercial pesticide product, and their adverse effects may exceed those of the active ingredients. Due to the widespread use of pesticides, low-level exposure to numerous pesticide ingredients are common. Pesticides are present in most commercially grown food, are used extensively for residential and commercial building pest control, and may be found in most water supplies."[519]; pp.1 & 2

Predictably, changes in food quality also occur. Pesticides affect the flavor and nutritional quality of foods. This is especially true for root crops, such as potatoes and carrots where the pesticide is absorbed into the plant.[754] Petroleum-based sprays on fruit trees retard fruit development and even photosynthesis. Sprays such as these are known to lower the vitamin, amino acid and mineral levels of fruits.[902] Any animal ingesting a food with pesticides, fungicides, herbicides or other chemicals have their nutritional requirements for detoxifying vitamins increased (B-complex, especially B-1, and vitamins C, A and E).[215] The destruction of life, we will see in numerous examples, leads to the weakening or decay (entropy) of those things responsible: people via commercially grown food crops.

A statement discloses the underlying basis of the problem: "The policy of protecting crops from pests by means of sprays, powders, and so forth is unscientific and unsound for even when successful, such a procedure merely preserves the unfit and obscures the real problem—how to grow healthy crops."[581]

16.4 Disturbing Consequences

Disturbing the existing wilderness life-system of an area is considered an essential part of conventional farming. Yet, one of the most basic plant nutrients, nitrogen, is lost immediately after an ecosystem is disturbed.[1280] Under average farming conditions 25% of the nitrogen is lost in the first 20 years, and losses continue for another 40 years.[1089] Cleared areas no longer have an input of organic matter, which leads to changes in morphology, particle size, clay mineralogy and chemistry. Upper soil layers are either destroyed (*O horizon*) or very modified (*A and B horizons*). The usual granular structures, so prevalent and well-developed in the best of soils, such as in forests, are destroyed, humus is depleted, and clay in plowed horizons is greater. Soil studies show that the physical and electrical properties (*e.g., cation exchange capacity*), organic matter, organic carbon, and nitrogen are significantly lower in plowed fields. Less retention and more evaporation of soil moisture takes place as well.[719]

Under optimal moisture and temperature, soil life thrives, and "the soil may be regarded as a living system composed of many individual creatures."[554]; p.159. Tilling and harvesting reduce (*ectomycorrhizal*) fungi, which grow on the substances released by roots (*exudates*).[554] Agricultural soil, even left unsown for one to two growing seasons, shows a sparse population of fungi near organic matter, and a small number of isolated bacteria (*commonly short rods, and long rods are scarce*).[1050] Decomposition of organic matter is also disrupted.[1160] The accumulation of humic substances, so vital to plant health, is higher under untilled (*anaerobic*) soil conditions than tilled (*aerobic*). As a result, the untilled soil produces greater levels of soluble organic carbon and nitrogen, the two most essential plant nutrients.[1012]

Continuous cultivation compacts soil (raises the bulk density), thereby reducing infiltration, aeration and root growth, raising energy needs and causing the loss of nutrients. Manured soils show less compaction (lower bulk density), and a better subsoil structure.[93,1016] One soil scientist comments: "Modern agriculture is almost equivalent to the use of heavy machinery, leading to the compaction of the soil."[93; p.9.] One can imagine what a combination of the two create. Neither necessity nor wisdom, but very short-term economic interests, are at the bottom of utilizing chemical agriculture.[65,1220,1232,1255] In fact, today grocery stores that offer both varieties refer to chemically grown crops as commercial produce.

Meanwhile, the worst of conventional agriculture is soil erosion. And, "soil erosion may well be the world's most serious environmental problem."[1232] Lost topsoil reduces organic matter, fine clays, water-holding capacity, plant rooting depth, productivity, and crop yields.[93,418,818,1016,1258] In the United States alone 6.4 billion tons of topsoil is eroded *each year*.[138] At least a half metric ton of soil is lost each year for every man, woman and child![376] Meanwhile, all countries are contributing to the erosion of the soil, and many more so than the U.S.[84,193,259]

In 50 years all of the topsoil will be lost, exposing the denser, less fertile subsoil (*argillic horizons*).[1016] The entire problem can be summed-up by recognizing that it takes 500 to 1,000 years to develop a mere 2.5 centimeters (one inch) of topsoil, empires have collapsed throughout history partly due to erosion, and recently, since 1914, more topsoil has been lost than in all of previous history![234]

"Grave though the loss of topsoil may be, it is a quiet crisis, one not widely perceived. Unlike earthquakes, volcanic eruptions or other natural disasters, this human made disaster is unfolding gradually. It is not widely recognized because of the intensification of cropping patterns and the plowing of marginal land that leads to excessive erosion over the long run can lead to production gains in the short run, creating the illusion of progress and a false sense of food security."[193]

It is quite ridiculous for us to continue with such a devastating practice as world hunger and populations grow.

Erosion is not limited to agriculture either. Grazing, timber harvesting, mining, construction and even recreation contribute to the problem.[376,1232] Leaching, wind-blown erosion and desertification are also taking a toll.[376] For example, the Sahara Desert has expanded by about a kilometer (0.6 miles) and the Turkana Desert by more than four kilometers (2.5 miles) each year. In Roman times the Congo Forest nearly reached Khartoum, but today it is more than 2,400 kilometers (1,500 miles) away, separated by long stretches of desert and semi-desert. The entire Mediterranean region has been undergoing a progressive desiccation, the worst of which has taken place during the past 300 years. In Asia, what was once the hunting grounds of Genghis Khan, a wooded grassland, is now the Gobi Desert.[234,376] A long history of humans despoiling lush productive natural areas, causing their civilizations to collapse, should have awakened us, but we too have not heeded the lessons (see *In Defense of Nature—The History Nobody Told You About*).

Wilderness and soil organic-matter reduce erosion. Throughout the literature, there are examples of mulch, humus, plant residues and plant cover fighting erosion.[84,332] No-tillage land-use fights erosion, especially in fine-particle soils (*i.e., loess and loamy areas*) and uplands.[259,339] Organic farms usually employ green manure, which involves the planting of crops that have pea-pods (*legumes*). These plants have a mutually beneficial relationship with bacteria that take nitrogen from the air and bring it into the soil (*i.e., symbiosis with nitrogen-fixing bacteria*). A significant reduction in soil erosion occurs when crops of this type are periodically planted (*legume-based crop rotations*).[620,846,1016,1050,1102]

The best use of computers ever conceived of is to aid in the translocation and reestablishment of life-systems (ecosystems) for food production and other uses.[625] There is a dire need for improved environmental management and the quality of life, which can be met by forest and other ecosystems, because erosion is threatening future productivity and human welfare.[259,587,706,1150,1176] Yet, erosion is only one aspect of the entire picture.

Contemplating the knowledge obtained about these hazards is no trivial matter. Food production must increase *25 million tons each year* just to keep up with population growth! Furthermore, new strains of crops usually require additional fertilizers and pesticides, along with a greater degree of water control. More industrialization is then "required," which requires mining and releases waste products that affect the soil, plants, animals and humans. Climates are changing around the world with desertification, droughts, floods and other disasters exacting their toll on food production.[376] In contrast, ecosystems overcome all of these problems, and do not require the endless maintenance and energy input of conventional agriculture.

Pesticides, chemical fertilizers, farm machinery, factories, transportation, mining, and other environmentally threatening components are required in conventional agriculture. Cleared areas show reduced organic matter input, and changes in morphology, particle size, clay mineralogy and soil chemistry.[791] A study of urban Scotland disclosed that parks and gardens contained five to ten times more lead, copper, zinc and boron than rural areas.[978] This study was conducted in the 1960s indicating that concentrations are greater now due to population growth. Furthermore, rural areas have greater concentrations than wilderness so that the real contrast between industrialization and wilderness is far greater.

Mining and emissions from transportation and industry increase element input into the environment. Mining production alone usually exceeds the rates at which many metals (*Cd, Cr, Cu, Hg, Ni, Pb and Zn*) are normally transported to the oceans by streams.[455] The same can be said of metal input from industrial activities and the combustion of fuels. Many of these metals are toxic to plants, animals and humans.[87,679,1226] Fish, particularly shellfish, contain high levels of metal as a result of this pollution.[141,868] Heavy metals reduce microorganisms and animals due to their toxic effects, and also increase pest infestations of plants.[847] Agricultural chemicals, mining and industrial wastes are very significant sources of heavy metal input into the environment.[679] Health effects on humans can be mentally and/or physically debilitating, though there are critical gaps in our knowledge of the full extent of the health problems created.[215,898]

Air pollution is one major source for producing damaging effects on plants, animals and humans. It is known to cause injury and retarded growth in plants.[15,548,735,932,1268,1310] Soil interactions also occur, and plants stressed by air pollution are more easily attacked by pests, who even use the plant's own defense against it.[847,988] Chemical and microbial activity in the soil becomes altered when exposed.[673,735,1290,1310] Heavy metals from motor vehicle emissions accumulate in soil, water, vegetation and crops, affecting humans through air, water and foods. Most accumulate near highways, but heavy metals (*Cd, Cr, Cu, Ni, Pb and Zn*) are carried by weather systems to remote areas.[844] It is clear that considerable contamination of plants and the soil has taken, and continues to take, place in areas near smelters.[431,768] Dust, produced by cleared land, mining, vehicles, and so forth, decreases the rate of photosynthesis in plants, obviously retarding growth, development and health.[1221] The Congressional Office of Technological Assessment reported that at least 50,000 premature deaths each year are attributable to air pollution alone, and the facts considered did not include food toxicity and food crop health.

Major elements are offset by human activities as well. About 30% of nitrogen-fixation is produced by humans, when it is normally accomplished by bacteria, often in a mutually beneficial relationship with plants.[1102,1138] Naturally, elements occur from high to lower levels as follows: calcium, magnesium, sodium, carbon, nitrogen, phosphorous and potassium. Conventional agriculture, however, has upset that balance by putting nitrogen, phosphorous and potassium, in that order, as the most available elements.

One example occurred when the United States dumped at least 436,000 metric tons of phosphorous into the environment. This excess of phosphorous increased plant growth (primary productivity) by 50 million metric tons. The problem is that this runaway growth occurs with mostly single-celled aquatic plants, which are short lived and then decompose, starving waterways of oxygen, killing other aquatic organisms.

The carbon dioxide content of the atmosphere has been steadily increasing by two-tenths of a percent each year since 1958 (*an increase of 0.67 x 10^{16} grams/year*). A 13% rise in carbon dioxide has taken place since the industrial revolution (about the late 19th Century) to 1977, or about one century, with 5% of that 13% in only the last 15 years of that period (1962 to 1977). From 1958 to 1968 the annual increase was 0.7 ppm (parts per million), while more recently it has risen to more than 4.0 ppm. Destruction of forests, such as the Amazon and other wilderness areas, as well as increases in human activities are sure to make this a major problem. The results of blocking heat (infrared) could lead to an ice age, or global warming, which could raise sea levels, in what is known as the Greenhouse Effect.[522,658] Of course, these are only a few examples.

Industrialized and urbanized societies are introducing heavy metals to the environment at rates that far exceed natural (sedimentary) cycles. The combustion of fuels and production of cement alone release metals into the environment at greater levels than do rivers to the ocean. For example, more than a decade ago, 24 tons of lead had been deposited *every day* in Los Angeles alone.[1345]

The amounts of heavy metals that have been mined and ultimately wasted in the biosphere up to 1979 is astounding. In millions of tons they are: 0.5 cadmium; 20 nickel; 240 lead; 250 zinc; and 310 copper. And this discharge increases in tons each year by at least: 7,300 cadmium; 400,000 lead; 56,000 copper; 214,000 zinc; and so forth.[898] Industrial pollution dumped into waterways also includes lead, cadmium, zinc, nickel, copper, mercury and other potentially hazardous metals (*Cr, As, Co, etc.*).[429]

Baltic Sea sediments illustrate how coal fly ash alone has increased heavy metals. Cadmium is seven, lead is four and copper is two times greater than a century ago. Comparable levels are discernible on the United States coastal plain.[1345]

Mining releases many times the stream load or atmospheric rain-out of lead, copper, nickel, chromium, cadmium, zinc, silver and antimony, all of which are toxic in low concentrations.[455,885] Various historical records exist that show metal pollution from mining has been toxic to plants, animals and humans.[1226] Mining destroys wildlife and results in major instabilities that threaten those who perpetrated the problem, we humans.

Sewage of industrial cities often pose a disposal problem because of toxic levels of heavy metals: zinc, copper, nickel, cadmium and lead.[87,254] The most important sources are air pollution, sewage, agricultural chemicals and industrial waste, which end up contaminating plants and soils, thereby entering our foods and water.[679] Making things worse, hazardous waste has been used as fertilizer.[114,1019] In addition, sewage and other pollutants enter the oceans and shellfish accumulate heavy metals. For example, cadmium is 30,000 and lead 9,000 times greater in shellfish than in the surrounding waters.[868] The resultant changes in natural biological cycles and mobilization of toxic elements by advanced industrialization are increasingly detrimental to *all* life on Earth.[174,215]

Mercury, a heavy metal, serves as an example of this toxicity. This element is so toxic that a spill from a thermometer, those tiny drops, can seriously contaminate an entire room. Mining and mercury utilization have increased the mercury content of rivers by about four-fold or more. The "Interference Index"—an indication of how much humans have disturbed an element's natural levels—for mercury is the highest of those metals examined, being placed at 1,000% or ten times its natural level.[455] The Greenland ice sheet displays a three-fold increase of mercury in the period between 1952 and 1965 when compared to 1951 and prior; it is undoubtedly worse now.[1315] High concentrations have been detected in shellfish particularly, and other fish (with tuna and swordfish as the main species involved).[455] Soil concentrations have also increased making its presence in all foods. This does not include pesticides, paint, some batteries, fungicides, fabric softeners, air conditioning filters, furniture polish, barometers, antiseptics, thermometers, floor wax and anti-mildew agents, all of which contain mercury as an active ingredient.

Metals pose one of the greatest threats to human welfare in very subtle ways. Toxic or even sub-toxic levels could be responsible for numerous mental or physical illnesses for which we are unaware of heavy metals as causative agents. Some are known to cause illnesses, such as lead causing learning disabilities or mental insufficiency (*neuropsychological deficit*), even at low doses.[215,885] Hair analysis of black female hypertensive patients revealed an excess of cadmium, lead and zinc.[827] One researcher comments on the history of known problems: "However, because of man's activities, local concentrations of some of these elements have been multiplied many times, particularly in water. Illness and even death have resulted."[455]; p.112. We need to consider the future possibilities we are creating because, "The fingerprints of man's technology are readily seen in the enhanced metal contents of his air and water."[1345]; p.149. Our soils and food, undoubtedly, intimately interact with both the air and water.

It is obvious that these metals in the air, water and food are being absorbed by we humans. This is evident in municipal sewage, which has been considered for use as an organic fertilizer, but its use poses a hazard and so it has not been utilized. The reason is that there are heavy metal accumulations in those soils treated with sewage. High levels of cadmium, chromium, copper, nickel, lead and zinc have been detected.[242,395] This is obvious testimony that heavy metals, with attendant mental and physical effects, are entering our bodies.

16.5 Little Things Mean A Lot

In contrast to the heavy metal problem, trace elements important to our health are rapidly disappearing from our foods. Trace element deficiencies were first reported in the late 1800s. Today we are still learning about the effects of deficiencies and are continually adding new elements to the list.

Several factors have been accelerating the exhaustion of the available supply of trace elements. These factors include weathering and leaching (enhanced by acid rain), stimulating increased yields by chemical fertilizer, the increasing purity of these chemicals, and the decreasing use of organic fertilizers. The United Nations Food and Agricultural Organization remarks: "Trace elements are not regularly applied to the soil by the use of common fertilizers. The removal from the soil has been going on for centuries without any systematic replacement."[259]; p.1

Undoubtedly, as time passes the situation worsens and human health along with it. "An honest set of figures on trace elements would show a lot of zeros on a part-per-million basis and damn chemical agriculture for the monstrous fraud it is."[522]; p.115. The importance of trace elements in human health has long been a major concern. A scientist comments on this, "The time may thus be fast approaching when evaluation of trace element concentrations will play a fundamental role in the diagnosis of illness and when manipulation of the concentrations may play an even greater role in its prevention."[811]

What makes the situation worse is that deficiencies in trace elements make the impact of heavy metals more of a threat. Iron, for example, is becoming less available in conventionally grown foods. When a person is low in iron the body will chemically react with lead or cadmium as if they were iron, and both lead and cadmium are toxic in small amounts. Both can severely interfere with mental and physiological functioning. Many similar relationships exist and mild deficiencies can help create a great deal of biochemical imbalance. Furthermore, most metals are important for regulating brain functions and vital organ performance.[712]

For example, schizophrenics have been known to have elevated levels of copper and low levels of zinc.[544] Schizophrenics have malfunctioning limbic system structures. Part of the limbic system, which communicates between the brain's two halves, the hippocampus, has "a high concentration of trace minerals," notes a scientist. Research tells us that metals, such as zinc, are important to nerve function (*neurotransmission in the synapses*).

A variety of neurobehavioral disorders may be simply related to abnormalities in trace metal availability and metabolism. Many physical diseases and disorders are known to have mineral imbalance.[577,907,986] A justified conclusion is that both chemical agriculture and the excess of heavy metals are a "time bomb" ready to explode leaving behind the "debris" of mentally and physically inferior humans, as the result of human actions against life.

16.6 Life-Centered Holism

A scientist expounds on our need for a new direction:

"In this last eventuality we have to anticipate that the continual attempts to rectify calamities caused by technology (after they have occurred) by means of new technological improvements will mean that more and more disastrous calamities are risked and that they will bring us to an ever more 'artificial' lifestyle and society. The only realistic solution would then be a return to a form of living and working that is closer to Nature and which again respects the natural order."[159]; p.145

We must forgive ourselves, who, like this scientist, often use the wrong words to describe our direction in this matter. It is not a "return" nor "going back" to a closer relationship with Nature. Rather it is moving forward—as in progress—to a time like *no* other, where we *help* Nature achieve its goal of maintaining life and its interrelationships. A goal, which once accomplished, will free us from any laborious tasks, because it is self-maintaining. We will have brought about what no other human society, even native peoples, has ever consummated. Then there will be none of the conventional stigmatism of a lack of "progress" and a "return" to former "ignorance." At last we will have broadened our perception to include what is essential to our own existence; a scope that includes loving the living world, as well as ourselves.

A scientist calling for "a holistic ecological food system" asserts:

"Since the beginning of human history, our relationship with the environment has changed from a position of subservience to one of dominance. If our species is to survive and individuals are to realize their full potential, we must develop a supportive, partnership role with the environment; otherwise, we could well return to the subservient position of our ancestors, through the inevitable development and multiplication of crises. Indeed, the harder that we try to satisfy our wants, the more unstable the system is likely to become and the sooner it will collapse."[559]; p.16

According to this scientist there are three major reasons we do not interact with the world holistically. Holism is best defined as maintaining the interrelationships of living things and the environment, which involves nurturing balanced ecosystems. Decision making should be dominated by a holistic perspective, but we ignore the root of the problem. "We tend not to view ourselves, or even our environment objectively; in fact, our view of reality is usually significantly distorted by the residual effects of our past experiences."[559]; p.17

To put this in terms which reflect our mental state, our dependent social structure leads people to be self-aggrandizing. To put it another way, collectively we humans blindly take to fulfill wants, rather than needs, or giving unselfishly in love. The second reason, which stems from the same dependence, is the specialization of scientific professionals. Specialization is obviously in direct contrast to holism. As a result, information outside one's chosen discipline is usually not available, and so, interrelationships go unnoticed. Again, stemming from the same source is the third reason, which is our misperception of being detached from Nature.[559] We have sought to be selfish, but unwittingly have not learned that the only way to do so is to act unselfishly, even altruistically, in the service of Nature—the organism Earth, FEM (this will become clear, if it is not already).

An objective view of Nature is the only way that we can fulfill humanity's total potential for blissful existence. Nature demonstrates that the most complex life-systems are also the most stable. For example, organic matter in soils maintains the maximum yield even when the conditions are not ideal—another example of the homeostasis of FEM. As discussed, organic matter increases soil organisms who in turn stabilize other soil factors. Nutrients are released gradually, trace minerals and nutrients are made more available, and plants are less vulnerable to attack. Soil structure and water-holding capacity are also improved and erosion is reduced or eliminated. Macronutrients are typically positively charged (*cations*) and micronutrients are typically negatively charged (*anions*). Soil organic matter makes both of these more available, and so, enhances electrical activity. An enhanced and stabilized electrical environment then contributes to the maintenance of the Earth's magnetic field and electrical environment.

Under the conditions created by increased soil life we see ideal conditions: "plants can be protected from pathogens."[554]; p.166. "Together, these organisms may be able to replace mineral fertilizers and chemical pesticides, thus lowering costs and reducing pollution and environmental hazards."[514]; p.169. "The biological activity induced by the organic matter has the potential to antagonize root diseases and supply different growth factors."[93]; p.9

The best news about our present ill-fated direction is that we merely need to stop our harmful practices and life can clean up the mess. Bacteria can transform heavy metals, such as cadmium, lead, zinc, mercury and others (*arsenic, palladium, thallium, tin, chromium, platinum, gold, free cations, etc.*), making them non-toxic.[254,1133] Water hyacinths are known to remove heavy metals, toxic organic compounds and pathogens from waste waters. They can also upgrade industrial toxic substances (*effluents*).[483,619] Though pesticides initially reduce the number of organisms, within months to a year they can rebound if no new spraying occurs.[868] The human body is completely capable of cleansing itself, if one employs water fasting and uses chemical-free, organically grown foods. All of the environmental hazards associated with the production of chemicals—mining, pollution, ozone depletion, toxic wastes and so forth—can be considerably reduced or eliminated if we, as consumers and participators, insist on the elimination of chemical agriculture.[48,146,598,619,1014]

Organic agriculture, which avoids the use of pesticides and chemicals fertilizers, and employs the use of composted soils is one such alternative. Detrimental effects caused by these chemicals are eliminated, including the improvement of soil health and the availability of trace elements.[118,159,449,560,790,1190,1213, 1220,1255] Higher yields and better quality are some of the beneficial effects, partly as the result of organic sulfur.[147,148,969] It can also help to prevent animal waste problems brought about by the domestication of livestock. As one scientist warns: "If we do not use these wastes and if we continue to abuse their handling they shall in the long run return to haunt our children."[218; p.193.] This is in addition to all the other benefits previously discussed for soil organic matter, such as reduced erosion, better quality crops, and improved mental and physical health for humans and livestock. However, there is even a better alternative.

CHAPTER 17

Forests: Our Friends Forever
(Silviculture and Ecosystems Approaches to Future Needs and Environmental Stability)

Even though organic agriculture is a better alternative than conventional agriculture it is still not the best of systems. When compared with the conditions created by wilderness there is greater erosion, less nutrient availability, reduced soil life, a thinner layer of topsoil, more energy expenditure, and so forth, with almost any type of farming itself (except forest farming). Like conventional agriculture, though to a far lesser degree, organic agriculture requires endless maintenance.

Establishing the interrelationships between plants, animals and the environment that typically occur in Nature, and are called ecosystems, are the only relatively maintenance-free and problem-free food-producing systems. Our lack of effort in establishing ecosystems and the problems created are commented upon by an ecologist: "Man's very existence is being threatened by his abysmal ignorance of what it takes to run a balanced ecosystem."[901]

A far superior food-producing system is referred to as silviculture, forest farming, permaculture or sustainable agriculture. The labels all basically refer to the same concept, which involves growing trees or other plants that are not harvested, and are usually grown in a forest setting. The trees that are grown produce food, fuel, fiber and fodder (livestock feed). For the most part it is a self-sustaining system that also enhances the environment, and increases and enhances wildlife.[587,706,1150,1176]

Nature herself seeks to establish forests more than any other ecosystem. All life systems go through stages of more and more complex and stable forms in a process known as succession. Forest is the most mature stage and is accompanied by an increase in the types of life forms (species diversity), which creates greater stability, both in the ecosystem and the environment. At times animals create other ecosystems by transforming the forest, such as beavers' dams producing meadows, and elephants' felling of trees bringing about grasslands, but through succession they again return to forest. In order for us to take our stewardship of the Earth seriously we must work with Nature, and therefore, forests should be our main natural resource, and our source for food, fuel, fiber and fodder.

17.1 Ideal Soils

Soil and plant growth are rendered ideal conditions in forests. Trees can retrieve nutrients from deep below the surface where other vegetation cannot reach. Descending down to about 100 meters (325 feet) below ground-level in some cases, trees can recover trace elements that have seeped beyond the reach of other plants. These elements then become part of the tree's leaves, and when these leaves are shed or eaten they fortify the upper humus-rich layer of topsoil with trace minerals that otherwise would have been lost.

Forest soils have more humus content than other soils, thereby producing humic substances that allow for greater absorption of minerals (*also high electrolyte concentrations*).[93, 254,1012,1055] For example, greater mineralization of nitrogen has been observed under pine (*conifer*) and mesquite (*leguminous*) trees.[218] Humus and compost stimulate fungi, which, for example, when in association with pine roots, stimulate root and shoot development. Pine seedlings under such conditions are noted to have better size, health, vigor and disease resistance.[104] Fungi are normally microscopic, but in forest soils strands (*of mycelia*) consist of long "ropes" or "mats" (*hyphae*).[93,514] In fact, a single, huge fungus has been discovered in a forest in Washington state that covered 600 hectares (1,500 acres) and weighs 10,000 kilograms (22,050 lbs.). This makes one wonder how different the global optical characteristics and electric potential would be if more fungus like this existed.

The greater levels of soil organic matter increase nutrient uptake, balance acidity and enzymes, soil texture is enhanced, water storage is increased, and plant pathogens are reduced. In addition, there are less root diseases and pest attacks, while there is also increased nutritional quality. The elimination of environmental hazards also come with the best soils (less erosion, pollution, etc.).[20,93,254,255,339,514,554,685,694,1016,1094]

The benefits are so widespread that it seems completely counterproductive to do anything but establish forest ecosystems. Forest soils are the best water filters, which would eliminate the major problem of unclean water. Forest can even allow for food production on rocky hillsides, and trees require little or no maintenance.[109] Summing up the conditions present in forest soils a scientist remarks: "Consequently, herbaceous vegetation and grasses were more luxurious under the trees."[775]; p.191. Healthier foods and improved environmental quality are certainly a necessary present and future direction.

Forests and the other two environments that Nature works to establish, pasture and meadow, produce superior soils. Grasses enhance soil structure and create topsoil. Pasture is known to increase soil organic matter within two years.[218] Under natural conditions, soils are characterized by the continual addition (synthesis and breakdown) of carbohydrates, which increase soil fertility and structure.[775] Materials that rot in the soil are far more active in supplying nutrients than either the compost heap, or especially, chemical fertilizers.[1055] Humic substances are produced more readily through time (increasing turnover time as a function of age), and this is why many ecosystems have a large reservoir of dead organic matter, especially forests.[1012] In addition, humus improves soil structure by increasing aeration, allowing rapid seed germination, improving mineral uptake, and producing some hormone-like activity.

Forests, as well as pastures and meadows, have a greater diversity of plant species than other habitats. Different species utilize and distribute varying amounts of trace minerals. In fact, sections of plants above ground, below ground, and leaves of upper and lower levels contain different amounts of trace minerals.[220] This would tend to distribute trace elements among the different species of animal who eat different parts of the plants. As these animals migrate they disburse nutrients and trace minerals in their waste or when they perish. Pasture, in particular, is frequented by herds of migrating animals who distribute nutrients far and wide. Gains in soil nitrogen, organic matter and trace minerals are equal to or greater than those lost by leaching, thereby enhancing the system permanently.[1055]

Soil studies confirm that forest conditions are the most ideal, and both pasture and meadow will again return to forest. Humic materials are higher under undisturbed (*anaerobic*) soil conditions.[1012] Certain types of fungi (*ectomycorrhiza*) are most common among forest trees.[514] Uniformly, these three habitats produce the most fertile soil, and the healthiest plants and animals, but ironically, are continually destroyed by humans.

Plant health can also be enhanced by the addition of biological controls that naturally occur in forests. For example, a native (*Tortricid*) moth (*Bactra verutana*) has been shown to be effective in controlling infestations of yellow nutsedge (*Cyperus eschlentus*) and purple nutsedge (*C. rotundus*). Control of a serious weed (Northern Jointvetch) in rice fields was accomplished by applying spores of a local (*endemic*) fungus.[1303] Plant diseases have been controlled by applying microorganisms that destroy plant diseases (*soilborne plant pathogens*).[514] In contrast, the use of chemical pesticides and herbicides destroy these beneficial, natural controls.

For example, the yields of Indonesian rice fields rose by one-fifth as the result of using Wolf Spiders to control pests. The Indonesian government saves at least $35 million each year that it would otherwise spend on pesticides. Farmers enjoy greater profits without the risk of losing their crop to a pesticide-resistant pest or losing their life or health to pesticide-triggered diseases.[636]

Many naturally occurring biological controls can eliminate the need for pesticides. For example, mites (*chrysomelbia lapidomerae eickwort*) that live on adult beetles, such as Colorado potato beetles, which also attack eggplants and tomatoes, control the populations of beetles. The mites can shorten the beetles' life span by as much as 50%, reduce the number of eggs laid, and hinder the beetles' migration. A Chinese wasp (*apanteles rubecula*) can stop a butterfly that eats broccoli, cabbage and Brussels sprouts. Numerous studies indicate that nitrogen-fixing (*rhizobia*) and free-living bacteria (*i.e., Azotobacter spp. and Azospirillum spp.*), fungi (*ecto-mycorrhizal, vesicular-arbuscular*), and biocontrol agents (*i.e., trichoderma spp.*) can do all that modern agriculture tries to do, but without the high cost, risk and environmental abuse.[51,52,56,591, 633,911] "Together, these organisms may be able to replace mineral fertilizers and chemical pesticides, thus lowering costs and reducing pollution and environmental hazards."[514]; p.169. In forests all of these organisms are readily available and naturally stabilizing.

17.2 Diversity and Stability

Once more the physical world gives us an example of how life's abundance and diversity brings about stability. This stability holds true from the microcosm—quantum, subatomic, atomic and molecular—to the macrocosm—organisms, communities, land-forms, and the Earth's electrical environment and magnetic field (and beyond, as will be demonstrated in Volume Two, and *In Defense of Nature—The History Nobody Told You About*). Nature continually seeks to establish through succession, the most successful wildlife ecosystem for manifesting this stability: the forest.

Forests produce the most humus-rich soils, which support a large population of microorganisms. These microorganisms have the ability to maintain community stability and integrity in a variable environment so much that it is termed homeostasis.[514] The presence of humic substances, soil organic matter, fungi, microorganisms, and the greater capacity for water storage establishes the conditions that yield a greater availability of nutrients and trace minerals. This creates an activity gradient of charged particles in the soil (ions)[254], and enhances the bioelectric capacity of plants.[1012] In turn, this generates more electrical activity in the Earth's electrical environment, and enhances the production of air ions and Earth currents (*telluric currents* and *geoelectricity*), which strengthens and stabilizes the geomagnetic field (*through electromagnetic induction*). Later discussions will clarify this understanding.

The enhanced uptake of nutrients and trace minerals increases plant growth, health and abundance, which causes animals to thrive, too. Also, negative ions are greatly enhanced producing all of the beneficial effects noted earlier (see section 2.1). Through time the healthier, more abundant wildlife contributes to even more and more stability at every level.

Humans have generally neglected a positive input when interacting with wildlife systems. We have even gone as far as to destroy forests, while producing the most unstable and simplest of life systems (commercial agriculture). This must change if we are to fully benefit from the impact of our participation in the physical world.

We can make the transition more easily than might be imagined. Bacterial[97,189] and fungal[97,514] cultures can be used to establish trees more quickly. One of the most useful transitional ways of using our present technology would be to employ computers for coordinating the translocation of entire forest ecosystems.[625] Many have expounded on the subject of how and why forest-farming, permaculture or silviculture—all wildlife systems with stands of trees—will meet our future needs with much less energy expenditure, and much more environmental stability.

Included in forest farming are such outcomes as putting an end to growing world hunger and improving the environment. This can be done by involving ourselves in environmental stewardship while fulfilling our needs.[587,706,1127,1150,1176,1227] When contemplating those goals that are essential to our future existence one scientist suggests: "Any methods that are developed or employed should be sustainable. Thus, they cannot be based on finite resources or on practices that degrade either humans or the support environment."[559]; p.19.

Planting a forest and nurturing it to produce food, fuel, fiber and fodder is sustainable and even self-sufficient, especially after humans have nurtured its most efficient functioning. In the long term it will have freed humans from the tasks involved in conventional farming and all the energy required. The environment will be enhanced, including the regeneration of the ozone layer (forests are major producers of ozone), and an end to serious erosion, dangerous chemicals, and any global warming or Greenhouse Effect (trees store Greenhouse gases). Food production in previously unproductive areas, such as, rocky hillsides would be possible. An end to the further dwindling of finite resources would take place, while bringing about the ecologically sound production of new natural resources. Even deserts can be reclaimed with forests.[1015] That is about as far away as we can get from degrading humans, particularly if we consider our actions in this regard as the altruistic care—love—of a forest wildlife community.

CHAPTER 18

We are Born Well and Made Sick (Health and Human Participation in the Physical World)

Our understanding of health can be traced back to two ideas that originated in classical thought. One was personified by the god, Asclepius, who depicted the role of the physician as the healer of the sick. The other was associated with the goddess, Hygenia, whose attributes illustrated the concept that health could be achieved by a rational way of life. One can find both beliefs in the Hippocratic writings, while they persist in present-day medical practice and thought. It has only been in the last few decades that we have learned that these concepts are very, very limited. In fact, they are, even in a very basic sense, inaccurate.

In this section we will take a look at examples of how our choices can effect our health. The effect is not simply individual choice, but collective choice that makes a impact far greater than might be imaged. Overall, there is a direct link between the diversity and nurturing of living things, and the molecular basis of health.

18.1 Nutrition and the Environment

The health of any human is determined essentially by what is consumed (food, water, drugs, etc.), and secondly by one's behavior, and third, the condition of the world around us. Fourth is the concepts discussed in the previous paragraph, but in most people's eyes those two ideas dominate the picture of health care today. The truth is that personal medical care is marginal, at best. Many of us believe we are victimized when sick and are then made well. However, it is far, far nearer the truth to say that we are well and are made sick by unhealthy food, water, air and thinking.

Nutrition's paramount importance can be clearly understood if two simple facts are contemplated: (1) On the average a human being is assumed to consist of some 600,000 billion cells, while approximately 50 million cells are formed as each second passes replacing as many cells which have died within that time.[806]; (2) Water is 95% of the total composition of the human body. When considering these facts it is easy to see the significance of what enters the body for nutritional purposes and bodily maintenance. If the food, water and air are unhealthy there is an excellent chance for our bodies and minds to be weakened in some sense or another.

Looking at the world around us discloses that poor quality food, water, and air exist all too often. Polluted water is everywhere, brought about by the impact of our lifestyles. Oil spills, chemicals, acid rain, garbage, and nuclear, sewage and factory wastes are the prime offenders. The World Health Organization says that 80% of the world's cases of disease are traceable back to unclean water. Compounding the problem is the fact that 30% to 40% of the world's food is irrigated with unclean water.

In addition, there are toxic effects created by the use of pesticides, herbicides and fungicides, which amount to billions of kilograms (a kg. is about 2.2 lbs.) *each* day, not to mention the cumulative effects. This represents the fact that we are literally removing mountains of minerals, along with the life in those areas, for the production of these substances each year. In return, these substances can be responsible for anything from mental and physical disease to birth defects and death. Often these substances mimic the effects of hormones in what has recently been called the "endocrine disruption hypothesis," with potentially lethal effects. For wildlife the effects extend to endangered and extinct species. Growing statistics on crime, mental illness, physical disease, and the variety of other problems that mark modern "progress" demonstrates the increasing burden of unwholesome food, water, air and thinking.

It all begins at the time of conception, and even earlier, considering the rate at which cells renew. "A mother's nutrition is the most important single environmental influence in the life of her unborn child and by means of the food she eats a mother can have the most profound and lasting effect on her child's development—by the simple act of improving her diet where improvement is necessary she can greatly influence the development of her child toward normal healthy growth."[375]; p.14

Believe it or not, the optimum health of a mother's womb is the single most basic step to an improved society. There would be, at least, three million less birth defects per year in the United States alone, if proper care were given to the nutritional status of pregnant women. Millions of other problems, such as disease, mental disorders, learning disabilities, criminal behavior, and so forth, may also begin in the womb. For those of us who are able to make it without any noticeable flaws, it can mean ranging anywhere from a beautiful genius to an unbecoming imbecile. Nutrition does not do it all, psychological development and genes determine many of the probabilities. However, improper diet in the pregnant woman can be the deciding factor in a host of problems for the human that is yet to be born, and thereby, the society it will grow up in.

The mother's body undergoes anatomical and physiological changes during pregnancy in order to create an ideal environment for the growth of the fetus. The most significant growth occurs between the twenty-fifth and fortieth weeks of pregnancy.[375] During that period the human fetus is most prone to nutritional deficiencies, especially when there is undernourishment. Any human is the *most* vulnerable to the environment, womb or otherwise, from before conception, in the parents' physiology and genetics, to one year old. This vulnerability does not stop after the first year of life, but its importance is less paramount.

In those countries without sufficient food supplies the developing infant suffers the most. If you will, imagine yourself in the place of such an unfortunate human being. The placenta, which plays an indispensable role in the transfer of nutrients and shielding toxic substances to you, the fetus, has fewer cells. Cell division within your developing body is drastically reduced too. The weight of your brain could be reduced by as much as two-thirds, if you manage to survive. With a reduced body weight, your brain suffers further starvation.

Two deficiency diseases, *Kwashiorkor* and *Marasmius*, have low levels of an essential protein synthesizing chemical (DNA) that "interprets" the genetic code in brain and body development. *Kwashiorkor* is characterized by stunted growth, a swollen body due to body fluid making up 5% to 20% of overall weight (*i.e., oedema*), diarrhea, reddish hair and skin, skin wounds, lesions, irritability, apathy, and predictably, a miserably sad look on the face. *Marasmius*, also a disease of malnutrition, is characterized by lower weight and height than usual, a shrunken appearance, an "old man's face," and an anxious look. Such diseases are a product of over-populated urban slums and ghettos. Diseases of civilization that reap about 100,000 lives *each day*, they are one of the products of human problems that stem from wildlife's destruction.[375]

While malnutrition is the worst of nutritional deficiencies, any one or number of missing nutrients can cause problems. Foods which are grown in natural wilderness conditions have greater levels of more available nutrients than conventional agriculture. This is especially true of trace minerals. Our participation in the physical world has an impact on great numbers of the participators—we humans.

When the expectant mother consumes a diet that leads to a deficiency in vital nutrients, her unborn child suffers the most. A deficiency in zinc, for example, can cause dwarfing in humans. Many dwarfs have been born in Iran and Egypt where their diet is predominately beans and grains (*high phytate*), which prevents the utilization of zinc. During pregnancy blood levels of zinc are known to decrease, thereby, deficiency disorders are more likely to arise in the fetus and the expectant mother.

If animal studies are a parallel, then she may experience difficulty with delivery, and her child is likely to be small with low chances for survival. A child born from a zinc deficient mother will have its ability to reproduce and learn diminished, along with behavioral problems throughout life. About 98% of the zinc-deficient test animals had infants with birth defects that ranged from club feet, cleft lip, and fused or missing fingers to small or missing eyes, and heart and lung abnormalities.[970] Likewise, it may be zinc deficiencies in alcoholic pregnant women that lead to the birth defects noted to occur in their offspring (fetal alcohol syndrome).[426,671] Many diseases—Down's Syndrome, tuberculosis, indolent ulcers, uremia, myocardial infarction, cystic fibrosis with growth retardation, malabsorption syndrome, cirrhosis of the liver, chronic alcoholism and others—display reduced zinc levels in the blood. In most of the cases individual choice (dietary preference) in combination with collective choice (deficient soils, acid rain, etc.) brings the unsuspected blow.

Imbalance in the environment brought about by collective choice can be the deciding factor that brings about a mineral deficiency in any pregnant woman. Needless to say, such deficiencies can be disastrous to the unborn child.[352,375,856] Here is another way in which we are reminded of both our participation and the need to work with Nature if we are to care for ourselves.

For example, iron deficiency is second only to malnutrition as a world health problem. Third-world and developing countries show a high incidence of iron deficiency anemia. Pregnant women in Latin America exhibited a 48% incidence of anemia, compared with 21% in non-pregnant women. In Africa there is a 15% to 50% incidence, and in Asia a 20% incidence, of anemia.[267,359,375,635,970,975,1278] The child born from an anemic mother will have a disease that effects all cells. Due to the lack of oxygen the child will be born small and anemic, leading to impaired physical and mental performance. Iron deficiencies can be created by poor dietary choice, iron deficient soil, missing soil organic matter, and imbalanced soil pH, including the effects of chemical agriculture and acid rain.

Similarly, the lack of any essential mineral is disastrous to the infant who will enter the world. A selenium deficiency can lead to various disorders and diseases. These include dandruff, increased skin aging, varicose veins, sterility in males, fetal death, and deterioration and calcification of muscle. In addition, selenium deficiency has been correlated with various cancers, and is synergistic with Vitamin E, thereby resulting in the effects of a Vitamin E deficiency (muscle disorders, ruptured red blood cells, wrinkled skin, slow healing, reduced cancer resistance, etc.). Manganese deficiency may result in disorders of the bone and cartilage, diabetes, sterility, retarded growth, and a lack of coordination and equilibrium (*ataxia*), to mention only a few. And the list goes on.

Each and every essential mineral, if deficient, has influences that can lead to numerous burdens for the new born. So far 30 essential minerals have been discovered and new ones are continually being added to the list. At this time it appears that there may be 50 or more. Any of these minerals can be deficient in soils that are imbalanced by human actions:

chemical agriculture, deficient soils, missing organic matter, imbalanced pH, and acid rain, among others.

Vitamin deficiencies in pregnant women can equally be a source for major dilemmas to the developing fetus and its survival. For example, vitamin A deficiency may affect the newborn with obstacles such as, mental and physical retardation, deformed eyes and skin layers (*epithelial metaplasia*) of the urinary, genital, respiratory and gastrointestinal tracts, as well as other imbalances. There is also the possibility of color blindness, anemia, apathy, skin disorders, susceptibility to infection, and altered chemical activity in the central nervous system.[262,352,712] Vitamin A can easily be destroyed, even by simply having one's food take a long time to get to market, and/or by sitting on the shelf or in the refrigerator for a significantly lengthy period.[712] Such conditions are typical of urban areas, especially during the winter or if the food is irradiated.

The major cause of impaired fetal growth, low birth weight and high death rates in newborns is nutritional. Multivitamin supplements have been shown to prevent neural-tube birth defects in women who have previously given birth to a child with such a defect.[831,833,1154] Infant cases of diseases, such as tetanus, jaundice, anemia, malaria and measles, can be prevented by better nutrition.[738] SIDS (*Sudden Infant Death Syndrome*) or crib death may be due to a marginal deficiency of a B-vitamin, called biotin[628], among other nutrients (and those factors that affect their availability). Low levels of trace minerals may be implicated in the high rate of birth defect-related deaths.[1100] Conventional agriculture, storage and food processing can easily create foods that can cause all of these problems if regularly consumed by the mother.

Infertility can also result from nutrient deficiencies. Infertile women may only suffer from a vitamin deficiency, and this has been shown in the case of vitamin B-6 (*pyridoxine*). The women in this study had PMS, and were infertile for anywhere from 18 months to 7 years. Yet, they were able to conceive during a course of high-dose B-6 therapy.[828]

Even if one's diet is properly balanced with vitamins and minerals, there are other hazards for the unborn child. One out of every fourteen babies has been born with birth defects, and there is an ever present potential for that figure to increase. The General Accounting Office published data that uncovered these facts: an estimated 20% of birth defects are genetic or hereditary, another 20% are caused during pregnancy (environmental), and the remaining 60% is an indistinguishable combination. This means that it may be possible to eliminate as much as 80% or more of birth defects if our interaction with the natural world was more wholesome and supportive. As populations grow worldwide and support ecologically unsound practices the inevitable surfaces: "The equilibrium is thus out of balance, and new mutants are being added to the populations faster than they can be eliminated."[402; p.143].

Food, one part of our environment that we nearly totally control, brings the burden of thousands of chemicals. For example, meat that is consumed in the United States has 143 drugs and pesticides used in connection with its production. Of these chemicals at least 42 are known to cause or are suspected of causing cancer, 20 cause birth defects, and 6 cause mutations. Compounding this possible threat is the fact that only 46 of the 143 drugs and pesticides are monitored, because no testing methods exist for 97 of them.

Chemically sprayed crops are also a major hazard to the fetus. The Environmental Protection Agency conducted a study which revealed that significant amounts of pesticides end up in the womb and the vital organs of fetuses. On the other side of the world, in Moldavia, the same was true.[507] Stillborn infants displayed high concentrations of pesticides, mainly DDT and its metabolite DDD, largely in fatty tissues, brain and kidney. What is so surprising is that DDT has not been used for decades. The total effect chemicals have is hard to determine, because together many are synergistic, making them stronger and more lethal in combination. Worst still is that incomplete testing is performed on newly developed chemicals, including pesticides and food additives.

There were 63,000 chemicals in commercial use in the United States alone in 1978. An estimated 1,000 new chemicals are added each year.[812] Once again, our actions perpetrated against other life, often claimed to be in the name of human survival, actually end up harming and destroying humans.

Even though a given human may have been lucky enough to have overcome these obstacles and was born normal, his or her vulnerability has not stopped. The first year of life is one of remarkable change. After being in a warm and fairly protected environment, provided with the essentials for growth, a child is then thrust into a world in which she or he must independently carry out functions on her or his own.

For the following twelve months, the child will probably triple in weight and increase height by 50%, which amounts to more growth than at any later age. To match this growth his or her stomach capacity will increase tenfold. The central nervous system becomes more stable as the child learns to coordinate body movements, develops psychologically and becomes increasingly social. Environmental contaminants, including pesticides, food additives and imbalanced foods and/or diet could lead to lifelong changes.

"With such rapid development in a short time, the importance of health care, nutritional adequacy, and psychosocial attention during the first year cannot be overemphasized. When growth is most rapid the organism is most vulnerable to any deficiencies in the environment. The first year is second only to the gestational period as a determinant of health throughout life."[116; pp.212-213]

In 1993, the New York Academy of Sciences conducted a study that showed pesticides in foods were too high for children. Yet, chemically grown foods also have less trace minerals, vitamins, fiber and usable protein, which could cause just about any problem. Quite simply, human actions that destroy wildlife (the effects of conventional agriculture) destroy humans.

18.2 <u>Breast-feeding</u>

One of the most profound influences on the life of a newborn infant is his or her mother's choice to breast or bottle feed. A child's nutritional needs are fully met by human milk for which there is *no* substitute. Cow's milk formulas have amino acids that are either too high or too low for human children, have more saturated fat, and are low in lactose.

Human milk has perfectly balanced levels of amino acids, fatty acids and minerals. More unsaturated fatty acids, which are particularly essential to the growth of the developing nervous system, are available in human milk. Lactose is very important in mineral and protein utilization (*chelated amino acids*). For example, breast milk enhances iron utilization (*chelation of lactoferrin*). When compared with cow's milk, there are higher levels of vitamins important to skin, body, brain and vital organ development (inositol, A, E, B-3, C, pantothenic acid, etc.).

This is especially true when one considers the fact that healthy infants consume more human milk in total than any substitute. The hormone, prolactin, is transferred from mother to child and enhances protein utilization. Also transferred is the hormone, oxytocin, which relieves anxiety, lessens negative environmental input, and enhances bonding throughout the family. Not only does human milk offer superior nutrition, but it also allows for more intimacy and touch, essential nutrients for body and brain development. This is clearly shown in studies which reveal that bottle-fed infants are more likely to fail to thrive, and at a rate of 1.5 to 2 times more often experience some form of malnutrition.[375,965]

Furthermore, nearly all studies have shown that infants which have *not* been breastfed are twice as likely to become sick or die. Breastfed infants are less likely to have infections, particularly of the respiratory and gastrointestinal tracts. Common infections of the bottle-fed include diarrhea, septicemia, otitis media, dehydration, pneumonia, cholera, sepsis, strep throat, oral thrush, salmonella and numerous viruses, among others. Unlike substitutes, immune, antibody, anti-allergic and detoxifying factors (*immunoglobulins, opsonization, chemotaxis, lymphocytes and macrophages*) are available to the infant as the result of breast milk only. High concentrations of lactose help produce a healthy intestinal environment (promotes proper pH and bacterial flora), prevents the growth of dangerous bacteria, and enhances brain development.[965] In composite, recent studies have shown that less disease and a longer life span are the lifelong benefits for those who were breastfed.

Formulas, in contrast, have repeatedly caused nutrient deficiencies and imbalance, because they lack the nutritional balance of human milk. Early in the century it was rickets, in the early fifties it was neonatal tetany, in the late fifties and sixties it was a vitamin B-6 deficiency. More recently, it was the disintegration of red-blood cells (*hemolysis*) due to a vitamin E deficiency, and the risks associated with high sodium.[965] In the 1980s and 1990s there has been a "bottle baby epidemic" in Asia, Africa and Latin America that caused tens of millions of infants to suffer from this syndrome. Many of these bottle-fed infants died or have had their physical and/or mental capacities lessened for life. Again, the connection is clear, many of the components in formulas are artificially derived from chemicals obtained and processed at the expense of wildlife (due to mining, factories, etc.), and obviously, humans.

18.3 Growing-up

Imagine yourself in the place of the growing infant. With a diet that provides the necessary nutrients your life is literary better. When you are six months old you sleep less during the day, and prefer being free to play rather than be confined to a cradle or being carried. At eight months you are spoken to more frequently, and are often praised by the members of your family. As you reach one-year/five-months you are venturing away from your mother frequently. On the whole you receive more attention from your mother and the family. You are touched, cleaned and washed more often, and protected more from physical injury, and receive more emotional contact. Everyone in the family is proud of you and rewards you with presents, toys and clothing. Curious, active and aggressive, you have a good deal of energy, are not sick often, and are a good size for your age.[743]

Next door there is a child who is about the same age, but the child's life is worse simply because the necessary nutrients are not available. When six months old she or he sleeps more during the day than you do, does not mind being confined to a cradle, and even prefers being carried. At eight months she or he is not spoken to as often, and rarely receives praise from family members. The mother is frequently clutched as the child experiences anxiety and fear. Even at one-year/five-months the child stays close and is often carried. Unlike yourself, she or he would cry often, rarely talks, very rarely plays, and is quiet more than half of the time. Basically, the child is reserved, timid, sick more often, insecure, mentally confused, and small for her or his age.[743]

Sound like a nutritionist's fairy tale? Well, these are actually the results of a study showing the effects of improper diet on the behavior and health of children. Scientists gave supplements to mothers and their children in a poor Mexican village and discovered these startling effects. One group was supplemented with vitamins, minerals and powdered milk, while the other group (the control) ate the usual diet of poor villagers.

In addition to what was reviewed above, the supplemented group was three times more physically active at the age of one, and six times more active at the age of two. In the supplemented group, mothers spoke, smiled and played with their child when it was six months old. It was not until two years old that the non-supplemented child received the same attention. The non-supplemented mother lacked initiative, and did not know how to play an active role with the child. She did not speak to nor educate the child actively for at least the first two years.

Fathers in Mexico do not normally give attention to their children and were not supplemented in this study. Meanwhile, the fathers of the supplemented children did give attention to the child. When the child reached its first year the fathers spoke, carried and played with their children. The supplemented child had somehow warranted more attention from the father. Certainly, part of the reason was that the child was more curious, active, assertive, sick 30% fewer days, larger in size, and had more energy than the non-supplemented child.[743]

Similar in many respects to the non-supplemented child is the child who grows up in a family that consumes processed foods regularly. Processing destroys B-vitamins, among other vitamins, making processed foods deficient in essential nutrients for both the brain and body. In the classroom a child from such a family could easily have defective vision. For example, a deficiency in vitamin B-6 alone could make a child unable to effectively see what the teacher writes on the blackboard. One study of nearly 1,000 school children indicated that B-complex vitamins can improve and prevent defects in vision.[321] Furthermore, B-vitamins are essential to a healthy state of mind, and central nervous system biochemistry. Vital to the entire body, including the brain, B-vitamins are necessary for metabolizing proteins, carbohydrates and toxic substances, as well.[712,1340,1349] Deficiencies of B-vitamins have been noted to cause anything from fatigue and failed muscles to insomnia and mental illness.

Processed foods, also referred to as fast foods, are destructive to wilderness and the environment. They require more raw materials and machinery than unprocessed, whole foods. Partly as a result of its life and environmental destructiveness, processed foods are nutritionally deficient, eventually eliminating the problem by increasing mental and physical disorders and diseases, and aging in those consuming it (see also section 2.6b).

18.4 Overcoming the Hazards

About 90% of all cancers are caused by environmental factors controlled by humans. Carcinogens, chemicals that initiate the cancer process, can range anywhere from livestock hormones and industrial waste to pesticides and food additives. Our environment is awash with carcinogens that enter our systems through unclean water, impure food and polluted air. Like most modern diseases, cancer continues to increase as our society becomes more industrialized, technological, and urbanized.[1207]

Conventional agriculture requires the use of extensive tracks of land, not only for cultivation, but also for the mining and industrialization that result from the methods employed. This undoubtedly causes the loss of a great deal of wilderness and the life associated with it, both initially and for subsequent generations. In fact, in the United States alone approximately *78 square kilometers (30 sq.mi.) each day* has its life eradicated, much of it due to the exploitation of natural resources. The chemicals used have frequently be shown to be cancer causing, and have various mental and physical debilitating effects.

Likewise, the loss of trace minerals due to repeated harvests can also cause cancer. For example, in Lin Shen, China one out of every four people were dying of esophagal cancer. Observations disclosed that some of the foods that were being consumed contained nitrosamines, but this cancer-causing substance could be blocked by Vitamin C. Sampling the soil, which had been repeatedly farmed, revealed that a trace mineral, molybdenum, was no longer present. As a result, plants produced less Vitamin C, causing a widespread deficiency, which made many susceptible to cancer. Human participation in removing life is always self-destructive and in this case had the wilderness remained the molybdenum would have also, the vitamin C would have been produced, and the cancer would have rarely existed, if at all.

A disease that causes hardening of the arteries, called arteriosclerosis, which is a factor in heart disease, also has its roots in diet. Fat deposits on the artery walls, particularly due to consuming red meats, are at the heart of the problem. Vitamin B-1 combats arteriosclerosis and has been utilized as part of the therapy against chronic heart failure.

Meanwhile, any toxic substance, such as pesticides, some additives, and livestock hormones, destroy vitamin B-1.

Other B-vitamins also play a role. Vitamin B-3 or niacin tends to depress blood cholesterol. For example, moderate decreases in blood cholesterol were accomplished by administering vitamin B-6 to patients with coronary heart disease. Vitamin B-12 decreases cholesterol, and this can be enhanced when another B-vitamin, choline, is administered simultaneously. This vitamin combination also regulates heart beat, reverting it back to normal (according to electrocardiograms), and reduces the frequency and severity of chest pains (*angina pectoris*).[1137] Likewise, Vitamin C reduced chest pains (*angina pectoris*), and produced a substantial decrease in cholesterol. As a treatment for patients with coronary heart disease vitamin C is recommended, and also prevents the recurrence of heart and brain blood-vessel clotting (*myocardial or cerebral infarction*).[1171,1172] In fact, so influential is Vitamin C that some researchers insist: "Vitamin C is the only factor involved in arteriosclerosis."[1172]

Other studies indicate Vitamin C is a very significant factor, but also demonstrate that choline deficiency, or an abnormality in its metabolism, is involved.[1137] Choline breaks down fat deposits.[712] All of these vitamins are destroyed by food processing, storage, agricultural chemicals and a polluted environment, demonstrating once again that factors under human control, which destroy living things, are the real culprit.

Aside from pollution, human-created imbalances in water can cause significant health hazards. A strong correlation exists between the hardness of water and the incidence of cardiovascular disease. The findings show a lower death rate with harder water.[531] Areas that have soft water have a higher incidence of cerebrovascular, (*ischemic*) heart and hypertensive diseases. Soft water has lower levels of calcium, magnesium, sodium, potassium, barium and sulfates, along with higher levels of copper and manganese. Metal pipes, such as copper, are more corrosive under the effects of soft water, and this may be a major factor for increasing the disease rate.[531] Hard water may also protect one against (*urinary*) kidney stones, because prevention depends in part on high levels of magnesium.[716,974] It has been noted that (*ischemic*) heart attacks occurred more frequently in areas where magnesium was 13% lower and calcium was 8% higher than normal. For sudden death, magnesium was 5% higher and calcium was 10% lower than the norm.[397]

Mineral imbalances are the result of mining, irrigation, agriculture, erosion, acid rain, industrialization, and the absence of deep-rooted vegetation, such as trees, and migrating animals (responsible for mineral distribution and balance). Furthermore, water softeners require extensive mining and industrialization, which destroys wildlife. Again, acts against wildlife also destroy humans.

Similar to the fundamental principles of life and survival, health and disease depend on a balance between ourselves and the environment.[119] This is especially true if the environment includes psychological and social balance in a context of harmony or disharmony with the natural world. These factors are all interrelated. When a society's harmony with the natural world is out of balance a large portion of the population becomes susceptible to widespread disease. This susceptibility to widespread disease restores harmony by eliminating the problem: those humans who are making the wrong choices.

18.5 Mental Health and the Environment

Similarly, survival and health are directly related to the proper functioning of the brain. The human brain is essentially a lacy network of interconnecting nerve cells embedded in a supporting mass of (*glial*) cells. In its entirety, the brain has a custard-like consistency.

Dangling from each cell's (*soma's*) wall are root-like fibers, called dendrites, which receive electrical signals from as many a several hundred other nerve cells. Signals are conducted along another fibrous extension of the cell (*axon*) that forms connections with the dendrites and the cell bodies (*soma*) of other nerve cells. Chemicals burst out of one cell that electrically charge nearby cells.

These chemical substances are divided into two groups known as neurotransmitters and neuromodulators. The neurotransmitters send the electrical signals of communication, while the neuromodulators maintain the efficiency of that communication. Both of these biochemicals require the necessary compliment of nutrients in order to be produced in the body.

Intelligence, learning and memory are all coordinated and interpreted by neurotransmitters that carry messages between neurons. However, they control more than one's ability to think. They have just made it possible to interpret what these words mean. Last night they put each of us to sleep or their imbalance prevented sleep. They can make one blissful or deeply depressed, as well as energetic or calm. Neurotransmitters control appetite, coordination of body movements, sex drive, and much more. A nutritious, balanced diet, complete with all the necessary vitamins, minerals and amino acids, is essential for producing the correct levels of both neurotransmitters and neuromodulators.

Mental retardation affects about 32 out of every 1,000 of the general population, and offers a illustration of this. Most people who are afflicted with this burden are typically unable to get by on their own and have to be taken care of. However, when treated with vitamin and mineral supplements mental health is improved. For example, retarded children have had their IQs raised an average of ten points in a mere four months.

When this therapy is administered over longer periods and early in life even greater responses occur. For example, one severely retarded seven-year old could not speak, had an IQ of 25 to 30 (average is 100) and was still in diapers. After supplementation he began talking, in a few weeks learned to read and write, and began to behave like a normal child. When he reached nine years old he read and wrote on an elementary school level, was moderately advanced in mathematics, and developed an IQ of 90. Even those with Down's Syndrome, which is more of a genetic disorder, show improvement.[533]

Many mentally disturbed people could be helped with supplements, or better yet, more available nutrients in foods. If such findings would be applied, it could mean the difference between socially functional and non-functional human beings. Collectively, we can make a difference.

All substances that enter the body are metabolized by the same B-vitamins that are essential to proper mental functioning. Vitamin B-1 stabilizes and strengthens the electrical field of the nerve (*nodal*) membrane. A deficiency of the vitamin causes an inactivation of the "feeding center" (*of the lateral hypothalamus*), gaps in the system that communicates between the brain's two halves (*lesions in the limbic system, especially the diencephalon*), and changes in the structure that controls respiration (*brain stem*). Vitamin B-1 also reduces oxygen consumption in the brain, and a deficiency decreases the production of certain neurotransmitters (*acetylcholine, serotonin and catecholamines*). Deficiencies in these neurotransmitters cause behavioral and physiological alterations that include mental depression, apathy, anxiety, irritability, sexual dysfunction, memory loss, depression, eating disorders and a schizophrenic-like psychosis (*Wernicke-Korsakoff Syndrome*). Vitamin B-1 is depleted by environmental pollutants, pesticides, fungicides, herbicides, sugar, food additives, drugs, stress, and it is also lost in cooking and food processing, and protein and carbohydrates are metabolized by B-1.

All B-vitamins have an effect. A vitamin B-6 deficiency can cause neuromuscular irritability, central nervous system depression, damage to nerves (*peripheral neuritis*), insomnia, abnormal brain-wave (EEG) activity, convulsions, and lowers neurotransmitters (*serotonin, norepinephrine and dopamine*). A niacin or B-3 deficiency may bring about apathy, depression, anxiety, dementia, memory loss, hyper-irritability, mania, delirium, emotional instability, apprehension, nerve degeneration, and a schizophrenic-like psychosis (*pellagra*). A vitamin B-12 deficiency causes abnormal brain-waves (EEG), irritability, depression, confusion, memory loss, delusions, hallucinations, paranoia and psychosis. It is not necessary to review each vitamin to understand each is essential to mental fitness.[319,763,912,1161,1170] All of these B-vitamins' availability depend on our interaction with the natural world, particularly the way in which we produce food, and generate pollution.

Other mental illnesses can be triggered by nutrient deficiencies, as illustrated by vitamin B-6 and tryptophan. One neurotransmitter, known as serotonin, regulates mood, sleep and the hunger drive, and low levels can cause depression. The amino acid, tryptophan, and vitamin B-6 (*pyridoxine*) are essential to serotonin's formation.

Studies have indicated that those with eating disorders and those with severe depression have low levels of either tryptophan or vitamin B-6.[1349] Furthermore, many women experience depression just prior to menstruation, which is a time when tryptophan metabolism is disturbed.[1340] Similarly, tryptophan metabolism disturbances occur in post-menopausal women who experience depression.[36] In one study, patients hospitalized for depression were given tryptophan and vitamin B-6 for one month at the end of which depression had decreased by 82%.[1183] Depression dropped in patients given only tryptophan, as well.[1349] Likewise, the B-vitamin, niacin (B-3), which helps in the utilization of tryptophan, also resulted in a 38% decrease in depression.[263] Tryptophan and B-6 can be of therapeutic value in most mental illness, including violent criminal behavior. Chemical agriculture and/or food processing make less tryptophan available (less chelated amino acids) and destroy vitamin B-6 (i.e., due to food processing, as well as toxin metabolism).

Depressed individuals may experience other imbalances. Low levels of calcium due to problems with absorption can cause depression (*lowers catecholamines*). A parallel exists with the postmenopausal woman or the elderly individual, both of which are prone to calcium deficiency and depression. The woman expecting a baby or one which has had a newborn may undergo a calcium drain through fetal development or breast-feeding, and experience depression as a result.[307] It has been noted that patients suffering from depression have significantly lower levels of magnesium than healthy people.[557] Magnesium is essential for nerve firing.[712] Many nutrient imbalances can trigger mental disorders, but like calcium and magnesium they are typically the result of human controlled environmental factors (chemical agriculture, food processing, pollution, soft water, etc.).

Minerals, which are often missing from processed and chemically grown foods, are also essential to mental health. Calcium is important to nerve impulse generation and the production of the neurotransmitters, known as catecholamines, which regulate mood, body movements, memory, and other psychobiological functions. Without enough calcium, convulsions and hyper-irritability can even occur. Copper also stimulates the production of one of these neurotransmitters (*dopamine*), and the coverings (*myelination*) of nerves. Iron increases the binding capacity of a neurotransmitter (serotonin) that regulates mood, sleep, memory, and other psychobiological functions. Iron also aids in bringing oxygen to the brain and is involved in the production of neurotransmitters (*catecholamines*). Surely it is unnecessary to review each mineral, because all have some influence.[215,712] The major reasons for the deficiencies are chemically grown foods (less chelated minerals), poor food-combining, food processing, insufficient fiber, and environmental pollution, as well as other factors under human control.

Consider as an example some of the observations made concerning schizophrenics. Studies of hospitalized schizophrenics have shown that many of the biochemical changes diagnosed as schizophrenia were simply the result of vitamin deficiencies.[665] Minimum daily requirements or recommended daily allowances for vitamins are not valid assessments for all people, because of the biological variability of the population, individual habits, and environmental insults. Simply what is sufficient for one individual may not be for another.[544] As one scientist indicated: "Researchers agree that inadequate nutrition can produce deleterious psychological effects."[311]; [p.81].

Large doses of vitamins, known as megavitamin therapy, or more complete approaches, called orthomolecular psychiatry, have proven to be effective in a wide variety of cases. This is especially true when the mental illness is detected early. Many analogous factors exist between schizophrenia and several rare degenerative neurological diseases caused by "slow viruses." The effects of the virus are not manifested until years after the initial infection, and there is no doubt that sufficient nutrient levels and proper diet enhances immunity, protecting one from viruses.[1230]

Hospitalized chronic schizophrenics are known to require more vitamin C than others in order to reach a normal concentration in their urine. Zinc can displace excess iron and copper, but schizophrenics are deficient in zinc, which partially leads to imbalanced brain-chemistry (*high dopamine*). In fact, zinc produces an "anti-anxiety effect" in schizophrenics, but not necessarily for those people who are considered normal. Therapy involves a combination of B-vitamins (particularly, B-3, folic acid, B-1, B-12 and pantothenic acid), zinc, manganese, calcium, and vitamins C and E. Often a discontinuation of this therapy causes a "sudden and dramatic" return to mental illness. Cerebral allergies, which can be caused by anything from food additives and pesticides to plastic containers, have also been known to cause symptoms. The best way to cure schizophrenia with nutrition is in its early stages, because if allowed to progress irreversible damage may occur.[170,544]

It was formerly believed that diet had little to do with mental functions. The discovery of how important diet can be came almost by accident, though some had long been suspicious of its role. Improper diet as a cause of disturbed mental functions became clear when a disease, known as pellagra, was understood to be the result of a niacin (B-3) deficiency. Symptoms resemble those of schizophrenia, involving perceptual distortions and thinking, and are part of the late stages of this deficiency illness. When niacin was added to the diet, pellagra with its schizophrenic-like symptoms receded completely.

Schizophrenia, in general, responds to niacin. Supplemented groups showed great improvement and less readmissions to the hospital than non-supplemented groups.[337,567] One of the pioneering researchers in this area comments: "There is no doubt that a major portion of schizophrenics recover on vitamin B-3. Any psychiatrist will find that 90% of his patients are well, the rest are improved, none will be worse."[564]

Since discoveries like these, many similar findings too numerous to mention were uncovered. For example, a vitamin B-1 (*thiamine*) deficiency also causes a schizophrenic-like disease, called the Wernicke-Korsakoff Syndrome.[504] Mental retardation is sometimes the result of inborn errors of folate metabolism, and low folate has been observed in 10% to 33% of psychiatric illnesses.[35] A learning disabled child's ability to learn is enhanced by the use of large doses of vitamins, mineral supplements, and the elimination of junk foods and additives.[299] Diet with the proper whole essentials has helped probationers stay within the law and out of jail.[170,1014] At least ten forms of mental retardation are treatable with large doses of vitamins.[337] Hyperactivity was controlled by low fat, low sugar and additive-free diets, and large doses of vitamins.[565] Substantial doses of vitamin B-6 may have some benefit in the treatment of childhood autism.[1033] It's no wonder improper diet is considered the "root of psychopathologies."[1045] Again, all of these social ills stem from an urbanized and industrialized social structure that imbalances the food, water and air, as well as creates inborn errors in metabolism (which is not solely genetic).

Optimum nutrition from conception throughout life pays off regardless of one's genetic endowment. The rewards yield optimum mental and physical growth, greater intelligence, fewer diseases and less disability from accidents, as well as less sickness and a greater capacity to withstand stress.[712,1302] More healthfulness, productivity and happiness in a longer life span with less suffering—as well as the social effort to produce environmentally friendly, wholesome foods, and eliminate human-created environmental imbalance—certainly justifies the little effort required to make the right choice and eat right.

18.6 Aging: The Role of Human Participation

Aging basically progresses in two phases. A slow degenerative phase extends roughly from age 30 to 60. Then somewhere around the age of 60 a quickened phase starts. The exact age at which these phases take place depends on the environmental stresses and imbalances one experiences, one of which is diet, that come with the passage of time.[618,1301] The more of these insults that take place the earlier and more rapid the phases progress, yielding a shorter life span, less youthful vigor, and more suffering.

Few people realize that the recent increase in life span is basically an illusion created by a greater survival rate at birth. Only marginally is the increase due to better medical care. Lately, research has revealed the process behind aging. Studies have disclosed: "A substantial body of data [is] suggesting that diet influences the rate of aging—at least it influences many if not all of the major components of this aging scale. Certain specific foods and nutrients are importantly involved."[235]; p.41. Once more, the evidence reveals that we are at the "controls" of our environment, and therefore, we also control aging, individually and collectively.

On the individual side of this control something as simple as overeating has a great impact. A person who eats excessive amounts of food shortens the life span, partly as the result of being prone to disease. Such a diet typically means the over consumption of fats, protein and refined carbohydrates. Furthermore, these foods usually lack a sufficient quantity of essential vitamins, minerals and high-quality protein, as well as fiber, which aids in nutrient utilization. The usual components of such a diet consist of foods like cake, white bread, potato chips and processed foods in general. Because these foods lack nutrients, one tends to overeat merely to meet some bodily needs.

At the heart of the problem are poor nutrient-availability in conventionally-grown foods, and poor food-combining, such as not periodically eliminating beans and grains (phytates), and certain foods (oxalates: coffee, spinach, peanuts, kale, rhubarb, beets, beet and turnip greens, etc.), which block nutrient absorption. This often results in obesity or excess weight, which is frequently accompanied by a lack of exercise.

Typically the outcome can be fatal. For example, a 45-year old man about 10 kilograms (25 lbs.) overweight has his life expectancy reduced by 25%! An overweight person is more likely to suffer from heart disease, hypertension, cancer, difficulty in breathing (*dyspnea*), diabetes, gall stones, gout, arteriosclerosis, and just about any disease.[1036,1037]

Another problem that arises is the possibility of too much of a certain vitamin or mineral and too little of others. Imbalances such as these are typical of disease states, and the result of conventionally grown and processed foods. This creates imbalances that offset bodily functions increasing the aging process, including one's susceptibility to disease and death.

In contrast, being underweight can actually increase one's life span.[980,1037] However, being undernourished can only lessen one's life span. Therefore, being underweight must be due to leanness. Semi-starvation is common in many elderly people due to insufficient financial resources to buy food, or a lack of interest in food or preparing it. Other barriers to elderly nutrition are poor teeth or a loss of the will to live. Some lose the will to live because they have suffered the loss of a loved one or believe no one needs them. With the rejection of food or the lack of nutrients due to inadequate diets, the elderly can experience mental dullness, senility, anemia, brittle bones, and accelerated skin aging.[1037,1352] Meanwhile, leanness, as the result of eating moderate amounts of nutrient-rich whole foods and exercise, increases longevity, lowers blood pressure and decreases the risk of cancer, heart disease and other diseases.[980,1037] Again it's the care of life in natural systems that produce the best foods, which also includes caring for one's own life.

The factors behind the process of aging actually begin at the time one is still in the womb.[1036,1037] What one's mother eats during pregnancy and bottle-feeding can lead to certain weaknesses that will remain with the child throughout life.[116] A person's parents may have an incomplete knowledge of nutrition, which affects that person during the most developmental stages of life, and this can carry over into adult life. If the diet they consume lacks the necessary protein (*i.e., chelated amino acids*), vitamins and minerals, physical and mental disorders or weaknesses can develop more easily. This is especially true if one continues the same inadequate diet. The life span of succeeding generations is in the hands of each of us who become parents and our knowledge of nutrition, including the way in which the food was grown and prepared, and the condition of the world around us.

Without that understanding, as today in general, the elderly population is plagued with poor mental functioning. Senility is often diagnosed when, in actuality, other problems exist.[433] Under the influence of environmental insults the aging brain is extremely sensitive.[144] Any drug, including pesticides, sugar, additives, nicotine, caffeine, and psychoactive prescription drugs, have been known to create the imbalance.[315] Drugs can produce depression-like symptoms, which include apathy, lethargy and sedation.[1069] Furthermore, in general practice depression is often mis-diagnosed as dementia, and this leads to administering prescriptions that offset biochemistry further.[1043] Such circumstances have actually led to invalidism and even death.[157]

Nutritional deficiencies are the most frequent problem, and involve vitamin C and the B-complex, especially B-1, B-3, B-6, B-12 and folate.[99,1042] A medical expert states: "The extensive degree of subclinical vitamin deficiencies may contribute to physical and mental dysfunction far more than previously realized."[99] Of those over 65 years of age, about 80% have heart conditions, diseased thyroids or weak kidneys.[99] These conditions are typical outcomes of nutritional imbalance throughout life.

Arteriosclerosis may actually change blood flow and oxygen consumption in the elderly brain, bringing about the degenerative changes labeled senile dementia.[1157] Summing up the findings one doctor comments: "Old people do not, in fact, become weak, frail, immobile or demented through any near-universal change coupled to chronological age."[282] All drugs, comprising prescriptions, pesticides, sugar, additives, nicotine, caffeine and others, also pollutants, destroy vitamins B-1 and C.[215] Both of these vitamins are used to produce the memory neurotransmitter (*acetylcholine*) that is deficient and poorly utilized in dementia.[140]

Common sense dictates the reality that the health of any living thing is directly related to the health of its surroundings. For example, we could never expect a healthy baby from a heroin addict's womb, nor a healthy plant from nutrient-deficient soil. Likewise, we could not expect a healthy society from unhealthy food and an imbalanced environment. The quality of our foods is directly related to our participation in the physical world, particularly our care of life in general. In fact, a recent study has shown that there is an increase in life span when a person is a caregiver for any living thing, regardless of their nutritional status.

CHAPTER 19

In Nature, Nurture Begets Nurture! *(The Optimum Love/Life Principle in Nutrition)*

It can be easily discerned from the previous sections' discussions that a food's nutritional quality is directly related to the abundance and diversity of life forms that are involved in interrelationships with it. Because the full scope of this understanding embraces our care and maintenance of other living things, especially in abundance and diversity, it will be referred to as the Optimum Love/Life Principle in nutrition. In this section a few examples that illustrate this principle will be examined, and again shows that we affect more than we assume.

19.1 <u>Nurture</u> <u>Nutrients</u>

Consider vitamins A and C as an illustration. Both vitamins immediately begin to decay or lose their potency when crops are harvested. Anything that delays a food from reaching a consumer, including transportation and storage will take their toll. Air, water, and especially light and heat destroy these vitamins so much, that little or none may be left when a food is finally eaten.

These vitamins are especially important for today's "modern" or "civilized" world. Both are essential for producing stress hormones and neurotransmitters. They help the body defend itself against toxic substances, because they are essential to the functioning of the liver, pancreas and adrenal glands. Both metabolize substances, such as, pesticides, herbicides, fungicides, and other substances that typically contaminate chemically grown foods. Protection from nitrites, hydrocarbons, nitrosamines, gaseous pollutants, dust, radiation and even noise are among the benefits of these vitamins. Vitamin C diminishes the negative effects of heavy metals like arsenic, cadmium, lead, chromium, mercury, titanium and others. Both vitamins A and C are essential to the maintenance of the entire body, helping to prevent anything from aging and cancer to the common cold and mental insufficiency.[215,712,864,1008,1141]

In spite of these benefits and their defense against environmental hazards, civilization creates conditions that destroy these vitamins. Civilization, particularly urbanized and industrialized, holds with it certain precedents. Any archaeologist will tell us that an earmark to a civilization's beginning is its agricultural fields. As populations become centralized and increase, foods must be brought from further and further away. With the civilizations of today this distance has grown even greater, often thousands of kilometers. Unlike previous civilizations, foods are often preserved by canning, refrigeration, freezing, dehydration and irradiation. All of these practices destroy vitamins A and C to the extent that even the so-called fresh items, like lettuce, are months old, and potatoes are years old in our supermarkets.

When the Optimum Love/Life Principle is considered it becomes evident why this is the case. Wilderness must be destroyed for the agricultural, urban and industrial centers. Producing and maintaining these centers, and the transportation that, for one, bring foods to the populous, also require resources that devastate wilderness. In return, civilization is relegated foods that are deficient in these vitamins making those responsible more imbalanced and prone to disease, eliminating the problem of devastated wilderness—those humans who made the choice to consume foods that destroy wilderness.

Agriculture, too, poses some problems in this respect. Due to chemical fertilizers, nitrates and nitrites find their way into water supplies and foods. These substances deplete vitamin A in the liver, and also prevent carotene from converting to vitamin A. Often foods are harvested unripened, in order to prevent damage in packaging and shipping, but this practice creates higher amounts of nitrites, and far less carotene than tree and vine ripened fruits and vegetables. Vitamin C blocks nitrites from being converted to the carcinogenic nitrosamines, but vitamin C is much less available due to poor harvesting and storage practices. Any toxic substance, including fluorides in water, pollution, nicotine, sugar, caffeine, alcohol, pesticides and so forth, destroys vitamin C. On this level alone the immune system, which requires vitamin C to function properly, is weakened, which could create anything from the common cold to cancer, and poor memory to mental illness.

Consider vitamin A, which is one of the necessary defenses against cancer. Vitamin A is extremely important to the body's purification efforts, because it acts as an enzyme in vital organ function (liver, pancreas and adrenals). Cancers of the stomach, lung, colon, nose, nose and throat (*nasopharynx*), skin, breast, bladder, vagina, uterus, cervix, trachea and bronchus are partially the result of vitamin-A deficiencies.[345,829,830,1067,1173]

A study of 16,000 men measured the levels of vitamin A (*retinols*) in blood. It was discovered that low levels of vitamin A increased one's chance of cancer by 2.2 times.[830] Zinc, which increases the effectiveness of vitamin A, also helps combat cancer.[1067] Another study of 8,278 men over a five-year period indicated that the incidence of lung cancer was 4.6 times higher for those with a low vitamin-A intake.[830] Conducted in Japan, a study of 122,261 men, whose ages were 40 and over, demonstrated that those with a low intake of vitamin A suffered more than twice as many deaths from prostate cancer.

Three-hundred-thirty patients with bronchial cancer were two-times more susceptible to death if they were born in winter than in summer. Fresher vegetables and dairy products with more vitamin A are available in summer. As a result, those infants born in winter had received less vitamin A, and as a result, were more likely to develop cancer later in life.

A study of 10,126 Americans ranging from 1 to 74 years of age was conducted by the National Center For Health Statistics. About 40% of the one to five-year age group, 57.4% of the 18- to 44-year age group, and 56.2% of those 60 or older were not getting sufficient quantities of the vitamin. Vitamin A begins to be destroyed as soon as something is harvested, deteriorates as food is stored, and is nearly totally absent in processed foods and conventionally grown crops. All of these factors are typically a part of urbanized and industrialized areas where cancer is the most prevalent.

Vitamin C illustrates other ways in which human participation increases our risk of cancer. Leukemia cells were reduced markedly when vitamin C was added to cultures.[921] Skin cancer (*melanoma*) underwent a 50% reduction in colony formation, number of cells, and survivability when vitamin C was introduced. Healthy cells were left unharmed.[176]

Nitrites, a chemical preservative recently banned due to its implication in cancer causation, can be blocked by vitamin C. The use of chemical fertilizers releases nitrites into our foods, air, and water making it an ever-present threat.[353] Overall, the situation increases requirements for vitamin C.

Vitamin C can aid in the prevention of any cancer whether caused by nitrites or not.[570,1342] For example, a decrease in the incidence of bladder cancer appears to be due to increased vitamin-C intake.[1313] Insufficient intake can lead to cancer, such as breast cancer, and cancer patients show an increased requirement for the vitamin.[832] Vitamin C begins to breakdown immediately following harvest and the longer it takes to get to the consumer the less there is of the vitamin in a given food (unless it is still alive). Also, light, heat, processing, microwaves and food irradiation destroy the vitamin.

Vitamin C is needed to metabolize drugs, including sugar, caffeine, alcohol, additives, pesticides and so forth, in addition to detoxifying other toxic substances, and metabolizing carbohydrates and protein. Considering these facts it is easy to understand why those most at risk for cancer live in urban/industrial environments, and consume processed, chemically-grown foods that are not fresh. Again, those who are most involved in a lifestyle that destroys life are the most likely to destroy themselves.

This illustration of the Optimum Love\Life Principle discloses the underlying factors. Urbanization, agriculture and industrialization, along with the eradication of life, create the conditions that lessen these two vitamins. Furthermore, this is merely an example, other vitamins and minerals are also destroyed or absent. As a result the immune system becomes less effective, which in turn leaves a given individual open to various disorders and diseases. For example, both of these vitamins are deficient in those who develop cancer. They are also importantly involved in brain chemistry with deficiencies leading to mental diseases and disorders. In the final analysis, the unloving act of destroying wildlife reflects back our attitude by eliminating those humans responsible. We are simply not really loving ourselves, because we are *not* separate from Nature.

19.2 Processed Foods

Another example of the Optimum Love/Life Principle is afforded by refined and processed foods. Factories, machinery, chemicals and energy are required to separate a whole food into its components. White sugar is far less nutritious than unrefined raw sugar cane. Raw, uncooked, unfiltered honey has more amino acids, minerals and vitamins than filtered, cooked honey. Whole grains and rice have more protein, minerals and B-vitamins than white flour and rice. Any processed food is far nutritionally inferior to fresh, whole foods. When we look at the bottom line, making processed and refined foods requires the destruction of wilderness, and as a result, these foods are deficient, which increases the likelihood of a myriad of human problems.

Consider a few of the details about some forms of food processing. Frozen vegetables are often treated with chelating agents that inactivate the already low levels of trace minerals, and usable protein of conventionally grown foods. Heat is also often used, which destroys many vitamins (vitamins B-1, B-2, C, A, etc.) and minerals, particularly in combination with food additives and sugar. French fries are immersed in a freezing chemical that is also used as a solvent in dry cleaning (*difluoridichloromethane*).[566] Some fumigants and solvents alone can destroy vitamins, minerals and amino acids often causing cancer, such as leukemia in factory workers.[592]

Every chemical additive carries upon it traces of the other chemicals used in its manufacture. When the additives are combined they can produce totally unsuspected compounds that are potentially dangerous.[566,592] Processing nearly always results in the loss of fiber, while low fiber is implicated in every serious, widespread disease in industrialized society.[190,207,697] B-vitamins are always eradicated in processing, even something as simple as making brown rice cakes from brown rice. Meanwhile, B-vitamins protect the body from heavy metal contamination, carcinogens, pollutants, ultraviolet radiation, and other environmental insults that can lead to disease.[215,712]

Processing always requires substances or apparatus that have contributed to the eradication of wildlife. As a result, processed foods have been linked to physical disease, mental disorders, criminal behavior, and even aging. Again, our lack of concern for the living world is reflected back on us, demonstrating that it is really an unconscious form of self-hate.

Numerous studies disclose the wide range of problems associated with the consumption of processed foods. Eliminating these foods from the diet of antisocial people has lead to a reduction or elimination of delinquency and violence. Additives have been correlated with learning disabilities in normal children, as well.[1314] Brain (*cerebral*) allergies due to additives can result in psychiatric disorders and criminal behavior.[821,948,994,1120] A 2,000% higher requirement for vitamins occurs when an individual consumes high levels of sugar and other refined carbohydrates.[1337,1338] Summing up the effects in one respect are those studies that show nutritional approaches that eliminate processed foods have been successfully used to rehabilitate criminals.[562,1013,1034,1080-1082]

Vitamin C can produce a stronger brain-wave (EEG) response, and increase overall mental ability, which was evident, for one, in better grades.[664,864,1034] Processed and refined foods are always low in vitamin C. Even when vitamin C is added as an ingredient the product stays on the shelf or in storage so long that it is no longer potent when consumed.

Levels of vitamins A and C are also substantially reduced in processed foods. A vitamin A deficiency can cause disorders ranging from weak eyesight, eye disorders, unhealthy tissues and organs, improper growth, weak bones, the inability to heal well, and reduced immunity to infection and disease.[712] Its no wonder that vitamin A deficiency increases infant mortality.[666] When vitamin C is not adequately available in diets impediments range from a slowed healing process, less energy, more colds, mental disorders, memory loss, and more viral and bacterial infections. The entire story begins to unfold when one understands that vitamin A is essential to vital organ functions (liver, pancreas and adrenals), and most of the brain's and body's chemistry is regulated by vitamin C.[712]

Processed foods typically contain food additives, which are known to cause or aggravate hyperactivity in many children.[1149,1199] Children with hyperactivity typically ingest 100 to 150 milligrams of additives per day. Meanwhile, avoiding additives is prescribed for overcoming hyperactivity.[1120,1199] The hyperactive child's attention span is poor, and the capacity to learn is reduced. Characteristics of overactivity, distractibility, impulsiveness, excitability, and sleep disturbance are common. Ranging anywhere between 3% to 20% of school children, hyperactivity's incidence typically affects one in four boys, and one in ten girls. Diets without additives were shown to cure a majority of the hyperactive children tested, even when researchers attempted to disprove that additives were the villain.[116]

Sugar, in particular, and also refined carbohydrates, such as white flour or starch, can be far more detrimental than is generally realized. Consuming substantial amounts of sugar or refined carbohydrates can cause the release of excessive amounts of insulin. Insulin is utilized to release the energy of nutrients in the blood, but if too much is present all the energy is "wiped clean" from the body. The result can be low blood-sugar possibly causing dizziness, headaches, angry outbursts, depression, moodiness, and other behavioral or physical outlets. Other possibilities, if the improper diet is continued, could be hyperinsulinism, hypoglycemia, and possibly, diabetes. The most common and proven result is hypoglycemia, which has been implicated in depression, chronic fatigue, nervous breakdowns, juvenile delinquency, childhood behavior disorders, alcoholism, drug addiction, inadequate sexual performance, criminal behavior, and even America's high divorce rate.

The body can tap energy reserves, but the brain cannot. Thereby, low blood-sugar effects mostly behavior and those body functions regulated by the brain. For example, millions of American children have learning disabilities that are linked to hypoglycemia. Learning disabilities include characteristics of a shortened attention span, emotional instability, and poor memory, as well as wild, erratic behavior. In fact, one study showed that nearly 8 out of 10 children afflicted with learning disability had hypoglycemia.[370,1149]

Sugar, like food additives, can cause many burdens in life. In school, a child with hypoglycemia could end up jumping out of his or her seat, talk out of turn, and even though the child has a high potential, he or she does not work up to capacity. At home, it is a restless, nightmarish sleep.

Neurotics, those who have comparatively mild behavior disorders, may only have diet as the triggering factor for such behavior. Many neurotics it seems are hypoglycemic. Of 220 neurotics tested in one study, 205 were noted to have hypoglycemia. Testing 700 mildly neurotic people disclosed that 600 had hypoglycemia. By correcting diet, behavioral problems, such as temper tantrums, claustrophobia, and suicidal attempts, can be eliminated. Correcting diet cured most of the behavior problems as they regained more complete mental health.[5]

Nearly 90% of all inmates in prison today have hypoglycemia, indicating one of the precipitating causes of crime.[487,1080-1083] A drop in blood sugar imbalances the levels of neurotransmitters (*especially serotonin, but also catecholamines and acetylcholine via depletion of vitamins B-1 and C*) that control sleep, mood, motivation, stamina, memory, sex, and learning. Often the results are hyperactivity or violent disruptive behavior.[1352] Improved diet has reduced the recurring crime of released prisoners, probationers, and juvenile offenders. Low blood-sugar can close off the higher brain (*cortex and temporal lobes*) with the consequential loss of the conscious knowledge of what is right and wrong.[1149] Considering that many of history's most violent personalities, such as, Napoleon, Mussolini, Hitler, Stalin and others, had hypoglycemia, we can begin to see the potential impact of such an unhealthy dietary choice.

In spite of this readily available knowledge, the average American consumes about 55 kilograms (120 lbs.) of sugar per year. In addition, there are greater quantities of refined carbohydrates, like white flour, which can also contribute to hypoglycemia. Many Americans are no longer consuming sugar, which indicates how some individuals eat greater quantities. Such is the case with the criminal hypoglycemic whose yearly intake of sugar is 155 to 275 kilograms (300 to 600 lbs.), which amounts to at least a half kilogram (1 lb.) per day or more. Worst of all is the long-term effects of consuming sugar, which includes a greater prevalence of diabetes in newborn infants (2 to 3 per 1,000), along with strong evidence that it causes brain damage and death.[299]

Sugar and refined carbohydrate foods have also been implicated as factors contributing to many diseases. "Sucrositis" is the label used to describe the various problems caused by sugar.[1071] For example, the now-defunct German Federal Republic, since 1948, had increased its intake of sugar, which was accompanied by a similar increase in diabetes, gout, hypertension, and coronary heart disease, and possibly contributed to the increase in hyperuricemia and gastrointestinal disorders. Refined carbohydrates, including sugar, are the cause and the disease is the effect. One scientist sums up the overall picture:

> "Whenever a community has been examined under conditions of its natural habitat, manner of life and traditional diet, diabetes has appeared to be virtually absent, as among Laplanders, purebred Eskimos, Australian aborigines, South African bushmen or Hottentots. When certain developing groups, such as Yemenite Jews, South African Bantu and New Zealand Maoris, all of whom ate large amounts of lightly processed carbohydrate foods, adopted Western patterns of living in towns, the prevalence of diabetes increased markedly. In the process of this change many food habits altered, but the largest dietary change involved the consumption of refined carbohydrate foods."[1249]; p.229

Processed foods are also deficient in fiber. Fiber is that portion of food which is not absorbed into the body, but contributes other necessary functions. Those functions include the slow release of nutrients into the body, the absorption of bile acids in the intestines, which leads to healthier intestines, and increased immunity and greater utilization of vitamins, minerals and protein. So vital is fiber that it has recently been designated as a vitamin.

Many diseases and disorders are the result of fiber-depleted foods. Among the diseases associated with a deficiency are constipation, appendicitis, diverticular disease, hiatus hernia, colonic cancer, polyps, hemorrhoids, dental cavities, obesity, diabetes (*mellitus*), heart disease, gallstones, and ulcers.[156,204-207,487,622,623,797,987,991,1250] These diseases are rare in undeveloped countries, but are widespread in industrialized urban areas where there is a prevalence of processed foods, among other things.

Processed foods contribute significantly to mental and physical diseases and disorders. Literally, a multitude of facts demonstrate the problems associated with such foods, but there is no need to labor the point. Underlying it all is that these "fast foods" require more raw materials, machinery and energy than whole foods. For example, the average American consumes 5.5 kilograms (2.5 lbs.) of additives alone per year. This amounts to about four million kilograms (660 million lbs.) per year in the US alone, not to mention exports. Again, the effects literally remove mountains of resources along with wildlife.

Demands of this sort require the wholesale destruction of wildlife through greatly increased mining and industrialization. Individual choices become a collective force that destroys life, which in turn destroys those humans most responsible: the laws of Nature preserve all life in general. Once again, we see evidence of a homeostatic mechanism on Earth that eliminates the disease—destroying wildlife (Earth-cells)—by getting rid of the disease-causing organisms: humans who make the wrong choices.

19.3 Food Irradiation

Another, newer form of food processing is food irradiation. Little is known about the changes that occur with radiation. When a food is irradiated, damage occurs that is not detectable with the unaided human senses. Even a small amount of radiation destroys nutritionally important compounds or creates potentially dangerous ones.

For example, an altered form of an amino acid (tyrosine) is produced in meat.[839] This amino acid is important to the production of neurotransmitters and hormones (i.e., catecholamines, etc.). These biochemicals are low in those who suffer depression, suicidal urges, lethargy, and other mental or physical disorders and diseases.

Irradiation also produces chemicals known as free radicals.[1316] Free radicals can cause cell damage and disease.[1030] They are also responsible for some aspects of aging.[85] Aside from this, heating and grinding also produce free radicals, therefore all of us get them just by cooking, but especially through the consumption of processed and irradiated food.[1316] Moreover, irradiation may conceal the fact that a food has already spoiled, but without eliminating any of the toxins produced by decay. The so-called "fresh" foods may be much older when consumed, with virtually no vitamins C and A left. The brief intense ionization generates unknown chemicals with changes taking place in the largest molecules, notably DNA, which is essential for maintaining life and reproduction.[1252]

A Food and Drug Administration law, known as the "Delaney Clause," forbids the use of any additive that is proven to be carcinogenic. Gamma rays, used in food irradiation, are undoubtedly cancer causing, but the industry made its way around the definition of additive.[1316] It's no wonder that many, including the major medical associations, have a general fear that irradiation poses a health hazard.[74,185,1252,1307]

The nutritional quality of irradiated foods is compromised. Vitamins C, B-1 and B-3 are reduced as though these foods had already been cooked, while vitamin E is affected more so, and folate levels are also reduced.[1252] Irradiation also affects amino acids, such as the essential lysine, tyrosine and isoleucine. Fats (lipids) are altered into a greater number of detectable by-products, including increases in the degradation products of cholesterol.[1316] This means that irradiated foods may pose a threat with regard to those diseases correlated with a high-fat intake and low levels of vitamins C and B-1, such as cancer, heart disease, arteriosclerosis, hypertension, and various mental disorders and diseases.

Studies with irradiated Valencia oranges, produced oranges that frequently showed rind disorders, ruptured oil glands, which lead to increased blemishing, and an increased softening of the outer texture. Both forms of vitamin C (ascorbic and citric acids) are reduced significantly, and the product is off-flavor.[878] As radiation dose increases an even more substantial decrease in vitamin C occurs. An increase of a natural gas, known as ethylene, which is used in the production of plastics and vinyls, also takes place.[373] There is no doubt that ethylene is unhealthy in the human body. In addition, a decrease in the total amount of digestible or soluble solids is noted, which means less of the food is utilized.[785,815]

Soybeans that were irradiated had less available protein *(decreased nitrogen solubility, and inhibition of trypsin and chemotrypsin inhibitor activity)*, inhibited fat breakdown *(inhibition of lipoxygenase)*, and less of an essential fatty acid (linolenic acid).[515] Increased chemical decomposition, as the result of radiation *(radiolysis)*, is continually observed in irradiated foods.[839,878,1216,1316]

Genetic mutation *(polyoloid cells)* was noted in the cells of bone marrow of those mice, rats and monkeys who consumed irradiated foods. Freshly irradiated wheat was fed to starving children with Kwashiorkor, and they too experienced similar mutation.[145,1174] This study was criticized for not storing the wheat, which is claimed to reduce the hazard.[1252] Meanwhile, there is ample evidence that foods are much more nutritious if they are not irradiated, nor stored.

Acquiring irradiation technology destroys wildlife. So do the storage facilities. Again, disregard for non-human life is disregard for human life.

19.4 Environmental Factors

Let us consider this Optimum Love/Life Principle from a more environmental perspective. Acid rain has recently captured the attention of every industrialized nation as it threatens to undermine the quality of life. The combustion of fuels and the use of nitrogen fertilizers, which emit nitrogen oxides and sulfur dioxide into the atmosphere, return as acid precipitation. As these gases combine with oxygen in the air they produce sulfuric and nitric acids, which then plummet to Earth in rain, snow, hail, mist and fog.

Given enough time, the acidity of lakes change to the point that fish and other aquatic life perish. At the end of 1979 some 170 Adirondack lakes had become fishless due to acid rain. Of the 85 lakes along the Boundary Waters Canoe Area stretching across the Minnesota-Ontario border, some two-thirds were on the brink of acidity where fish could not survive. The Great Lakes region had been receiving precipitation that was 5 to 40 times more acidic than what is considered normal for rain.[1320] Sweden is one of the hardest hit with over 4,000 lakes fishless, 9,000 lakes with threatened fish, and 18,000 acidified lakes. The acidity of waterways is only one aspect of the problem.

It seems that the normal pH of rain has changed worldwide. For example, in the Colorado section of the Continental Divide the pH of precipitation became considerably more acid (5.43 to 4.63 pH) in only three years.[494] The concern about acid rain has been and still is nation-wide and international.[38,942,1222]

Acid rain has been affecting Europe, Brazil, China, South Africa, and the Arctic, as well. There is no doubt that its origin is industrial to a significant extent, because soot and black graphite carbon are observed in the precipitation, which can only be created by fuel combustion. This is confirmed in samples from the Greenland ice sheet where both acidity and these particulates increase with the advent of industrialization.

Significant reductions in the growth and vitality of several tree species in the eastern United States and Europe have taken and are still taking place. This is more of an economic and human threat than the problem affecting waterways. In northern Vermont, for example, tree growth has decreased drastically since the 1950s. Red Spruce has had its numbers reduced by 80% since 1965. Sugar Maple has been reduced by 84% and Beech by 65%. More than one half of the Black Forest in western Germany has been seriously damaged, with an estimated loss of $1.2 billion per year. Tree loss can easily lead to an increase in the number of avalanches, floods, landslides, mudslides and crop losses.[968]

Trees derive their nutrition from elements such as calcium, magnesium and potassium, but acid precipitation adds hydrogen ions that displace these nutrients.[968] An increase in pH below 5.0 leaches out aluminum, which leads to stunted root growth and inhibits calcium uptake. Metal analysis of trees since 1950 has shown that aluminum uptake has been much greater. Heavy metals are also leached from the soil, these include lead, zinc, copper, cadmium, chromium, manganese and vanadium. Trees may be the only realistic answer to the carbon dioxide problem, whether the result is the Greenhouse Effect (global warming) or global cooling, as well as replenishing the ozone layer.

The relationship between human actions and their effects on the natural world is clearly evident in the case of acid rain. Soils become more acidic as the result of acid precipitation.[1329] This in turn affects the availability of nutrients to plants, either directly or due to its effects on humic substances.[254] The end result is deficient food crops, lost forests, less fish, and all humans and other species suffer the effects, while the environment becomes less and less stable.

Our waters become contaminated with heavy metals and other hazardous compounds, as the acid rain transports and leaches them into our groundwater.[1048] As previously noted, soils are becoming more and more laden with heavy metals.[971] Cadmium, for example, is a contributor to liver and heart damage, and hypertension, and is present in our water.[215,1048] Likewise, lead has been known to impair mental functions lessening performance and abilities, and causes psychological disorders.[215,272,1034,1116] Cadmium, too, is closely linked to learning disabilities.[215, 945] Investigations of ancient Peruvian native's bones disclosed that 500 times more lead can be found in the bones of present-day people.[403] All of this becomes worse when we consider that low-income households may not be consuming enough calcium, which protects one from lead.

The Environmental Protection Agency indicates that 90% of about 57 million tons of toxic waste, in 1980, was disposed of unsoundly, and this still continues. Acid rain will more readily bring this into our groundwater, while preventing nutrient uptake by plants. Moreover, a survey on the effects of heavy metals on the immune system's response indicates that most suppress immune defenses.[1236] Mineral imbalances, including low levels of trace elements and high levels of heavy metals, have been implicated in many diseases, including the top killers, cancer and heart disease.[432,451,678,840,967]

Considering the effects of acid rain on mineral uptake, let's examine just one mineral: selenium. Acid rain (i.e., sulfuric acid) produces an abundance of sulfates, which inhibit the uptake of selenium by plants. Furthermore, the more acidic a soil is the less selenium will be utilized by plants, or even remain in the region of the roots (selenium is leached).

Selenium is noted for inhibiting the formation of various cancers, as well as other diseases. An examination of cancer rates across the United States uncovered the fact that higher mortality occurred in the areas with low selenium. Selenium inhibits the onset and growth of cancerous tumors, especially in conjunction with its "helper" (synergist) vitamin E.[603,613]

Low selenium has been tied to increases in cancer. For example, in the continental United States cancers of the digestive and respiratory tracts, as well as cancer of the breast, are the highest in low selenium areas.[162,302] Likewise, New England has the highest rate of colorectal cancer, and has selenium-deficient soil. In contrast, the lowest cancer rate in the United States is in the region where there are soils with high selenium; Seneca County, New York.[621] On the other side of the world, in China, low selenium areas have high lung-cancer mortality.[266] Data from seventeen countries unveiled the fact that when selenium levels in the blood rise, breast cancer rates fall. Inhabitants of the Philippines, Japan, Taiwan, Thailand, Puerto Rico and Costa Rica have three times higher levels of selenium in their blood than do the people of Europe and the United States. Cancer rates are two to five times higher in Europe and the United States. Direct laboratory evidence and regional studies show selenium's cancer prevention capability is genuine.[162,266,302,603,614]

Cancer is not the only disease affected either. A regional heart disease in the People's Republic of China, Keshan's Disease (*endemic cardiomyopathy*), is associated with low selenium levels in cereals and in hair analysis. The disease occurs most frequently in areas with low selenium.[253] Cystic fibrosis, Legionnaire's Disease and disorders of the pancreas (*atrophy*) can be caused by low levels.[201,887] Selenium is involved in the reactions in plants that produce vitamin C, and therefore, all the problems associated with a Vitamin-C deficiency are present. Selenium deficiencies have been noted in muscular dystrophy, white muscle disease, pancreatic fibrosis, liver necrosis and neonatal jaundice, as well.[712] As acid rain continues to be a problem, it is obvious that numerous humans are perishing as a result.

Many less obvious symptoms result from selenium deficiency. Selenium can prevent ozone irritation to the respiratory tract, and chromosomal damage, limiting birth defects. Protection against radiation, pesticides, herbicides and other chemicals, possibly cancerous, are attributed to its biological effects.[215,281] Heavy metals, such as lead, cadmium and mercury prevent the absorption and metabolism of selenium, while selenium's presence prevents heavy metal toxicity. Selenium is utilized in the liver, testicle, muscle, kidney, lung, brain and ovary, which indicates that a deficiency can cause physical, mental and/or reproductive disorders and diseases. Compounding the problem is the fact that food processing decreases selenium, as well as vitamins C and E, and niacin, which work with selenium (*i.e., they are synergists*).[712] Aside from being important to the elasticity of the skin and blood vessels, lessening wrinkles and Blood-vessel diseases, selenium has also been shown to work against the aging process itself.[662]

Also contributing to the problem is the fact that conventional agriculture's repeated harvests ship the selenium off to markets and never replace the selenium in the soil. Acid rain, produced by generating energy for industry and using chemical nitrogen fertilizers to grow crops, prevents the uptake of selenium into plants. Undoubtedly, the whole situation produces greater rates of cancer and other diseases, such as cerebrovascular disease, and also premature aging. The created mineral deficiencies are also apparent in the observation that migrating people who take on the diet of a new region also acquire the Cancer-types of the area, though their native region did not have those cancers.[484] One is certainly living a perilous existence if one lives in an area with acid rain, and low selenium levels in areas of crop production, and is eating processed, chemically grown foods, while surrounded by urbanization and industrialization. This is merely an example involving one mineral and holds true for all trace minerals.

The Optimum Love/Life Principle is again conspicuous. Urbanization, industrialization and agriculture destroy wildlife in order to be established and maintained. All also contribute to the production of acid rain, which then produces all of the aforementioned hazards. One such hazard is low levels of selenium in soils where crops are produced increasing cancers, heart disease, aging and numerous diseases and disorders. Human participation in destroying life ends up destroying humans, partly as the result of deficient foods and acid rain. Again, a more holistic and Life-centered food production is an essential aspect of caring for ourselves, as well as our planet and the biosphere.

19.5 Cultured Foods

Another illustration of the Optimum Love/Life Principle in nutrition is cultured foods. Foods such as these have edible bacteria or fungi—life—added to them. These cultures invariably improve the nutritional quality of those foods in which they are found.

For example, milk increases cholesterol levels, while yogurt, a cultured milk product, lowers cholesterol. Yogurt also has more B-vitamins than milk. The bacterial culture (acidophilus) used to make yogurt enhances the intestinal environment, nutrient absorption and immunity.

Soybeans are not as valuable a food as the cultured product tempeh. Higher levels of essential amino acids and more B-vitamins than either soybeans or tofu develop in the fermentation process. Tempeh even has B-12, which is not found in other vegetable sources. The fermentation process both increases the nutritive and decreases the antinutritive components. A decrease of biochemicals that inhibit protein and mineral utilization also takes place (*i.e., tannins and phytic acid*). Increased digestibility follows from these changes. It has components that have been shown to be anticarcinogenic, analgesic, antiviral, antioxidant, antifungal, and enhance immunity and lower cholesterol. The changes enhance brain chemistry because of the increases in amino acids and vitamins essential to the formation of brain chemicals, known as neurotransmitters. All cultured foods, which add life, are superior to the same food when it is not cultured. Again, the presence of life makes foods more wholesome.

19.6 Livestock Production and Health

Undoubtedly, one aspect of love is freedom. Being allowed the freedom to be what one is, without inhibitions, is another way of describing it. To be granted the right to live a long, fulfilling, prosperous life with the full potential to give and receive are, without a doubt, attributes of love. Most basically, freedom is being given the right to experience life in its fullest capacities.

Most livestock produced today do not have the luxury of being free. Typically they are confined to small spaces or cages, given a minimum of light to prevent movement, and are fed or injected with hormones.

All of these conditions speed growth, most of which is fat. Certainly, animals raised under such conditions are not being allowed the right to live a long, fulfilling, prosperous life in its fullest. They are not being loved.

If the Optimum Love/Life Principle holds true, some things about human meat consumption should reflect the principle. Data from 37 countries indicate that there is a direct relationship between cancer and the consumption of animal fat, but not vegetable fat. Ovarian, breast, skin, pancreatic, prostate and colorectal cancers head the list.[229,1350]

There is less fiber in a diet that includes meat, especially because of its fat content. The vast majority of meats come from an animal whose flesh consists of 53.8% fat.[1350] A diet that includes high amounts of fiber takes 33 hours to be eliminated by the body. In contrast, a diet that is low in fiber, such as one that includes significant portions of meat, can take as much as 83 hours.[156,190,697] The extra time causes bile salts to breakdown into substances that irritate the intestines. As a result, the potential for causing cancer and other diseases is greatly increased. In fact, meat is correlated with colorectal cancer, which is the second most prevalent cancer.

The increased bile acids also destroy the (acidophilus) bacteria that help in nutrient absorption, thereby allowing a yeast (candida) to multiply and block nutrients. Problems of this nature (candidiasis) are typical in diseases like cancer and heart disease. Moreover, low-fiber diets contribute to constipation, appendicitis, diverticular disease, hiatus hernia, colonic cancer, polyps, hemorrhoids, tooth decay, ulcers, obesity, diabetes (*mellitus*), (*ischemic*) heart disease, arteriosclerosis, gall stones, osteoporosis and numerous others.

Livestock are usually fed crops that were grown on trace-mineral-deficient soil along with the use of chemicals, and therefore, should also reflect these human-created imbalances. Pesticides, toxins and poisons tend to increase as one goes up the food-chain. For example, carcinogens increase in concentration from plants to animals that consume plants, and from plant eaters or vegetarians to meat eaters. Compounding the problem are the 42 drugs and chemicals used in the production of livestock that are known or suspected carcinogens.

Other problems arise from meat's contamination by bacteria. For example, contaminated chicken accounts for about 48% of gastroenteric disease and many of the salmonella poisonings.[77] Irradiation of meats also poses a threat by increasing free radicals, which can increase aging and the chances for developing any disease. Meanwhile, the story only begins with an increased potential for disease and a shortened life span.

Those who consume meat from supermarkets are also feasting on some dangerous chemicals. As discussed in a previous section, the United States government investigative agency for congress, the General Accounting Office, examined meat and found that there are 143 drugs and pesticides connected with the production of livestock. Of these, no less than 42 are known to cause or are suspected of causing cancer, 20 cause birth defects, and six cause genetic mutation. The USDA monitors only 46 of these chemicals leaving a total of 97 chemicals unmonitored.

Furthermore, about 13% of all meat goes to market with illegally high pesticide residues. Compounding this further, too little funding results in short cuts when making inspections. In 1977, for example, 3,100 cases of high pesticide residues were reported, but only 37% were investigated. Pesticides alone are a serious health threat, but any toxic chemical contamination is an undetectable potential health risk, too.[72,178,519,1038,1196,1269,1324]

Meanwhile, there are other potential dangers with other chemicals, as well. A synthetic estrogen called DES (*diethylstilbestrol*), was mixed with cattle feed in order to speed-up the conversion of feed to protein. DES was eventually proven to cause cancer in many animals, including humans. Humans were also noted to acquire reproductive and sexual disorders in those who where born from a mother that consumed meat with DES.[139,525,655,1186,1187] Due to its cancer causing potential, law required that DES be removed from feed 48 hours prior to slaughter in order to prevent residues from remaining in the meat. However, the USDA inspection and sampling was so inadequate that less than 500 of the 30 million cattle slaughtered annually were examined for DES. The USDA records show that DES was detected in one out of every 200 cattle tested. Compounding the problem is the fact that the techniques used in testing cannot detect residues less than two parts per billion. In other words, all the cattle may have had DES residues. DES was eventually identified as a drinking water contaminant, because it seeped from feed lots into the groundwater. Finally, the whole situation had gotten so out of control DES had to be banned.[402]

In spite of the federal ban, it was discovered that a widespread disregard for the law was taking place. No one can be sure that DES is not being used somewhere this very minute. Some new chemical may have already been produced that is just as hazardous or worse, but it will not be known as such for years or even decades.

Most of us never consider the possible dangers posed for those who slaughter or raise livestock. Yet, the danger is clear: "Most and probably all of the distinctive diseases of civilization transferred to human populations from animal herds."[826; p.51]. In fact, humans share many diseases with domesticated animals. Diseases that are shared with animals total 26 with poultry, 42 with pigs, 46 with sheep and goats, and 50 with cattle.[590] One such disease, a flu transferred from the pig, known as the Swine Flu, took twenty million or more human lives worldwide in 1918 and 1919. Livestock, especially if fed with chemically grown feed, can and have produced famine.[913,1101] By purchasing commercially grown meat, we are asking fellow humans to slaughter or raise animals that might give them disease, ignite an epidemic, or bring about famine.

Consider the contrast afforded by the free-ranging, additive free, organically fed animals. Because they are free-ranging or fed organically grown feeds they are healthier, affording more trace minerals and less contaminants, such as pesticides and hormones. Being healthier they are much less likely to get disease and transfer it to humans.

Less fat makes consumption of their meat less likely to cause those diseases associated with animal protein and fat. For example, free-ranging beef has 23.7% fat compared to the 53.8% fat of conventionally raised animals. That is, conventionally raised beef has nearly two-and-a-half times more fat.

In addition, free-ranging animals preserve a fairly natural setting, possibly even wilderness itself, preventing all those problems associated with conventional agriculture. These problems include global threats, such as soil erosion, nitrogen loss producing acid rain and ozone depletion, chemical hazards, and endangering wildlife. Furthermore, there is not all that extra farmland necessary to produce feed crops, thereby eliminating pesticide use, mining, factories, fuel and the environmental problems created. In short, loving livestock animals brings about better nutrition and a better world, or conversely not loving them literally kills us.

In fact, the recent FDA food pyramid reflects this. At the base, which are those foods that should be consumed the most, there are vegetables, fruits and grains. A one moves towards the peak, where there are the foods that should be eaten the least, we find those foods that require more energy expenditure to produce; fats, oils, and meats. This pyramid was formulated from a myriad of studies that showed what is healthy, and what is not.

19.7 Urbanization and Industrialization

Certainly true freedom for living things involves the preservation of wilderness. Only then will the conditions for the fullest development of any creature, particularly humans, become available. The relationships between air, water and soil would not be disturbed yielding higher quality foods and environmental stability (including the Earth's electrical environment, ozone, soil, succession and nutrient availability, among others).

If we harvest only ripe fruits—fruits meaning the widest definition of seed-containing fruits and vegetables that do not destroy the plant when harvested—there will be more available nutrients, trace minerals, usable protein and environmental stability. If tree and bush crops are made locally available, eliminating transportation and storage, far superior quantities of vitamins A and C, and usable protein, along with the elimination of harmful nitrogen compounds will also exist. A diet consisting of such ripe "ovaries" of vegetables and fruits is also more alkaline and fibrous. A fibrous-alkaline diet enhances mineral absorption and increases immunity. The Optimum Love/Life Principle is 100% effective in all situations we humans control.

As a contrast, consider the antithesis of wilderness: industrialization and urbanization. Diverticular disease, appendicitis, colorectal cancer, (*adenomatous*) polyps, ulcerative colitis, varicose veins, deep-vein clotting (*thrombosis*), pulmonary obstruction (*embolism*), heart disease, hemorrhoids, and (*hiatus*) hernias are among the common diseases of industrialized society.[190,204-207,622,697] The trend reflects the effects of increasing industrialization in the past five decades, because the evidence suggests that most of these diseases were rare even in the western world before the present century, and that the prevalence of each has greatly increased along with industrialization and urbanization.[204]

Other diseases of industrialized society are obesity, diabetes (*mellitus*), heart disease, cancer and gall stones.[622,697] A pathologist makes an observation that reflects the Optimum Love/Life Principle: "There are several diseases which are rare in rural areas of under-developed countries, in particular rural Africa, but which are common in industrialized countries and becoming more common in those inhabitants of rural areas who become urbanized."[622; p.162]

Even infectious diseases show a similar relationship: "Only in communities of several thousand persons, where encounters with others attain sufficient frequency to allow infection to spread unceasingly from one individual to another, can such diseases persist. These communities are what we call civilized; large, complexly organized, densely populated, and without exception directed and dominated by cities."[826; p.50]

Reminiscent of industrialized and urbanized society's conventional food production, and the diseases shared between livestock and humans, the same scientist comments: "Human success meant larger numbers of fewer plants and animals: an improved feeding ground, therefore, for parasites able to flourish by invading a single species."[826; p.53]

When one confronts the cancer issue the evidence is there too. After collecting a mass of data one finds themselves confronting the chemical industry, the most life-destructive component of urbanization and industrialization.[43,50,178,191,274,296,388,419,519,530,804,990,1038,1196,1269,1324]

Among the substances responsible for causing cancer we find manufactured goods that require the destruction of life: synthetic clothes, construction materials, photographs, televisions, plastics, fuels, fluorescent lights, pesticides, medical equipment, drugs and so on. In addition, those jobs that involve the use of chemicals show an increased rate of cancer.[388] Moreover, the combined effects of carcinogens are more lethal.[1207]

Amusingly one author has attempted to implicate "natural" substances as cancer-causing. However, the examples cited are processed or refined products, or require an absurd situation, like somehow suspending oneself above a forest in the case of ozone.[388] It is certainly no lie that one's chances of contracting cancer increase when there are life-destructive factors involved in the production of a given substance.

Industrialization and urbanization are directly responsible for cancer, either as the result of producing carcinogens and/or creating a certain state of mind. Minerals in unnaturally high concentrations are the outcome of mining, erosion or an imbalanced life system. Some foods are reported to cause cancer, but they are grown with chemical fertilizers, pesticides and an imbalanced environment. Fungal contaminants, such as aflatoxins, are present only because the food was harvested and stored. Oxygen itself could be considered cancerous, because it creates superoxide radicals in the body. Meanwhile, foods rich in vitamins A, C, E and selenium have antioxidant effects, and as noted, it is urbanization, industrialization, and conventional agriculture that limit the availability of these nutrients.

Ultraviolet (UV) radiation is only a problem because of ozone depletion and open spaces, both of which are the result of urbanization and industrialization. Forest canopies protect one from direct exposure to UV, and forests produce ozone, which blocks UV radiation. Furthermore, significant levels of a B-vitamin, PABA, and vitamin C are a natural sun screen.

Those substances naturally found in some foods are isolated compounds and are typically from conventionally grown foods, which are less complex. They have been consumed by humans for decades, centuries or even millennia, and pose no real threat to human welfare. All the other so-called "natural" sources of cancer are explainable in terms of human interference with the natural order.

Looking at the geographic distribution of cancer discloses the actual culprit. Breast cancer, for example, is a major cause of death among women of industrialized countries, such as the United States, Canada and Europe. In most countries breast cancer is known to have increased along with industrialization.[229] Britain's Imperial Cancer Research Fund, examined the rate of breast cancer around the world and discovered that the highest rates were in those countries with the highest Gross National Product (GNP).[353] GNP is a direct measure of a country's level of industrialization, and the exploitation of wilderness for resources and the waste generated. Likewise, the more economically developed a society is the greater is its incidence of colon cancer.[1350] Those most responsible for Nature's exploitation are among those who suffer the disease more often. "In general, it has been shown that in countries with a low rate of colon cancer, the disease is found more commonly in people of a higher economic level."[1050]; p.58. All cancers are more prevalent in industrialized and urbanized regions.

Other studies on cancer clarify this relationship. A geographic analysis of cancer mortality in the United States, from 1950 to 1969, disclosed excess rates of cancer among males in the 139 counties where the chemical industry is most highly concentrated.[576] Thousands of chemicals are in current use in the plastics industry alone and unknown combinations will likely prove to be cancerous.[1284] Most cancers have a latency period of 15 to 40 years, indicating "that much of the cancer from recent industrial development is not yet observable [and] a steep increase in the human cancer rate from suspect chemicals may soon occur."[419]; p.2. Nearly all cancers are known to be due to an imbalanced environment.[1233] And this is apparent in that, migrating people, both nationally and internationally, acquire the predominant cancer of the area they move to, and lose the predominant cancer of the area that they came from.[419] Cancer also involves factors of certain dietary choices and psychological states that are more likely to occur in industrialized and/or urbanized areas.

Consider PCBs as an example. PCBs (*polychlorinated biphenyls*), used in the production of vinyls and plastics, are carcinogenic, and increase as much as four-fold from mother to infant.[684] Moreover, some studies have revealed that those mothers who consumed fish contaminated with PCBs had children with IQs that were an average of six points lower. PCBs have been found in the most remote spots on Earth (i.e., Arctic, Antarctic, etc.).

Even specific components of industrialized society have a relationship to disease and environmental problems. Human-made lakes, for example, pose numerous health problems.[1178] Large dams increase the incidence of malaria, viruses and other diseases (*schistosomiasis, filariasis, leptospirosis, etc.*). Nutrient imbalances, erosion, earthquakes, and degraded water quality are also created. Wildlife habitats, particularly for fish and aquatic plants, are eliminated or severely altered. The over population of certain insects results, as well. Terrible consequences also befall native peoples.[602]

Sewage from industrialized nations ends up in the oceans. In 1987, mussels poisoned a number of Canadians, while hundreds of

Bottlenose Dolphins and Humpback Whales were discovered dead along North Americas' coast. It may well have been toxins dumped in the ocean.[788] Many of the recent sicknesses experienced by those who consumed shellfish are attributable to dumping sludge or excessive levels of heavy metals offset by human interference. Shellfish (mussels, clams, crabs, lobsters, shrimp, etc.) are filter-feeders and scavengers consuming particles suspended in the water or on the ocean floor. Coastal-water fish consume plankton grown on the pollution generated by industrial-urban centers. As a result, these particular sea creatures have been known to pose threats to human welfare when consumed.

Consider the psychological effects. When psychiatric emergency room visits were compared with the level of two common pollutants (NO_2 and CO), it disclosed more mental illness when pollution was greater.[1189] Pollution also affects the health of plants, soil nutrient availability, and pest attacks on plants.[735,847,932,1102,1310] This will effect the health of humans and livestock that consume these crops. Examples of such phenomena are literally endless.

Once again the world at large displays the characteristics of the living Earth. Our lack of care for life in general ends up stopping us. FEM's collective nature operates like an organism's homeostatic regulatory mechanism against a disease. We must heed the lessons and begin to work with Nature if we are to experience the true ascent of humanity.

Let's not get the wrong idea here. It's not that we should acquire an anti-establishment state of mind, and terrorize the industries. Rather, we need to be involved in a life-centered holism in our consumer activities for our own good. The industries will follow, as they want to satisfy our consumer needs.

CHAPTER 20

Can You See the Light?
(The Effects of Ocular Light Perception on Humans and Animals)

An essential nutrient for physical and mental health is full-spectrum light. The Sun emits full-spectrum light, but it is often filtered by window panes, and nearly all indoor lighting is limited in spectrum. Light enters the eye activating photoreceptor cells that stimulate the pineal gland. This gland then excites the release of hormones by the hypothalamus, which in turn activates the pituitary gland's hormones. The pituitary gland is the main activator of the entire hormone system regulating one's entire body chemistry. Without full-spectrum light, only some of the photoreceptor cells are activated and a person's biochemistry can become imbalanced.

20.1 Biological Effects of Full-Spectrum Light

Studies undertaken on blind people illustrate what the absence of light can do to an individual's biochemistry. As could be suspected, there is a pituitary of reduced size. This smaller pituitary lowers blood-sugar (*serum glucose*), protein, sodium, chloride, inorganic phosphate and various hormones (*cortisol, STH, E, NE, thyroxine, LH, etc.*).

Low levels of cortisol alone can have a profound effect. Cortisol regulates protein and carbohydrate metabolism, and is a stress hormone that is secreted during infections, colds, injuries, extremes in temperature, and other stressful experiences. A deficiency leads to a loss of appetite, weakness, low blood-sugar, water retention, and shock under stress.

Blind people are known to mature sexually and physically much later than those with eyesight, disclosing another effect of full-spectrum light. Physical abnormalities of the skeleton are typical. Those hormones that are lowered are known to control vital functions.

Consider some of these hormones. Adrenalin, now known as epinephrine, is involved in the general maintenance of tissues, heart function, blood flow, metabolism, kidney function, and the functional movement (*motility*) of the lungs, intestines and sex organs. Furthermore, it excites the central nervous system, stimulates the release of other hormones (*ACTH and cortisol*), and regulates the balance of at least ten hormones.

Another hormone is noradrenaline, now known as norepinephrine, which aids in nerve impulses, increases blood pressure, decreases kidney function, and aids in the functional movements (*motility*) of the lungs, gastrointestinal tract and sex organs. Thyroxine, one of the most important hormones, regulates body metabolism, temperature, growth development, reproduction and the nervous system. In addition, it controls the metabolism of cholesterol, other fats (*lipids*) and trace minerals (*chelation*).

Other hormones are also involved. Human growth hormone (*HGH or Somatotropin; STH*) regulates general tissue control, insulin and overall metabolism (protein, fat, carbohydrates, water and minerals). Hence, it is essential to general body growth and is involved in slowing the aging process, because it controls nitrogen balance and promotes protein (and RNA) synthesis. Oxytocin, promotes a sense of well being between intimate humans, whether lovers or mother and child, which would make "love hormone" a rather fitting common name. It also promotes uterine contractions and milk ejection, and increases blood sugar, and urinary sodium and potassium. The love hormone also balances fourteen other hormones.[712] Each of these imbalances can be attributed to the poorly activated pineal gland as the result of the absence of full-spectrum light due to blindness.[571]

Light has a major impact on physical and mental biochemistry. Full-spectrum light stimulates hormones that effect the sex glands, adrenal glands, liver, thyroid and pituitary. Reduced liver metabolism alone inhibits the metabolism of toxic substances, protein and carbohydrates. Low-spectrum light results in less hemoglobin, white blood cells, and other disease-combating (*eosinophilic*) cells. In addition, elevated levels of cholesterol and free fatty acids are also observed under the influences of low-spectrum light.[571] All of these effects have a profound impact on an individual's mental and physical states.

Full-spectrum light is essential to the proper metabolism of calcium, partly due to it stimulating the production of vitamin D.[327] In turn, this effects the production of a neurotransmitter called dopamine, which regulates body movement, sexual functions and mental activity, including memory consolidation and mood.[476] Artificial light also reduces muscle strength, partially as a result of reducing dopamine and calcium.[626] Schizophrenics are known to blink more, increasing the levels of dopamine, which is at the heart of the problem.[648] Depressives tend to receive too little full-spectrum light, lowering dopamine and another neurotransmitter (*serotonin via melatonin*), which can cause depression. Depression triggered by insufficient full-spectrum light is called Seasonal Affective Disorder, which is a sensitivity to the shortened days of winter or staying inside too often during the summer.[120]

Mental and physical well-being are affected by the availability of essential amino acids (*tryptophan, cystine, etc.*) regulated by light.[571] One of these amino acids (*tryptophan*) is important in the production of the neurotransmitter known as serotonin, which controls mood and sleep. Light is known to regulate the level of serotonin, by stimulating its by-product (*melatonin*).[120,759,889] The by-product (*melatonin*), in turn, has an effect on behavior and sexual maturation (puberty) and function.[759,1026,1294] Full-spectrum light is so superior for maintaining a balanced mental and physical state that it is known to increase the life span.[881,1225]

20.2 Artificial vs Natural Light

A scientist comments on the distinction between artificial and natural light: "In contrast to natural light, artificial light has an unusual and hence unphysiological stimulatory effect; the greater the intensity of light involved, the greater the effect."[918]; p.96. Hence, the brighter the artificial light is the more it offsets biochemistry.

The behavioral effect of limited-spectrum lighting was clearly illustrated in one situation. Initially the behavior of first graders under the influence of the limited spectrum of a standard cool-white, fluorescent-lighting was observed. The first-graders displayed irritability, lapses of attention, nervous fatigue and hyperactive behavior. Students could be seen leaping from their seats, fidgeting excessively, swinging their arms, and were unable to pay attention to their teachers much of the time. Afterward, full-spectrum lighting was installed and a dramatic change was observed in the children's behavior within a week. Teacher's reported an overall improvement in class performance, which included much less nervousness and far more interest in their work. The experiment was conducted by the Environmental Health and Light Research Institute in four Sarasota, Florida classrooms.[918]

In a more dramatic premise a scientist notes: "Above all, in the case of children who attend schools without windows, we must be prepared for the eventual appearance of pathological consequences."[918]; p.91. One might wonder how different the world would be today had this been known long ago.

Often children will return home almost immediately after school and watch television with its limited-spectrum light. At first, no explanation could be given for the symptoms that were experienced by thirty children after usual tests were conducted for infectious and childhood diseases. Even diet disclosed no factor for which the illnesses could be attributed. The symptoms of nervousness, continuous fatigue, headaches, loss of sleep and vomiting seemed to have had no cause. Finally, a common factor surfaced, each child had been watching television an average of six hours per day during the week, and six to ten hours per day on the weekends. The American Academy of Pediatrics prescribed abstinence from television, and the symptoms vanished in two to three weeks in twelve cases, and five to six weeks in eighteen cases when television was cut to two hours daily.[33] The limited spectrum of television had offset these children's biochemistry.

Other studies have uncovered additional imbalances caused by limited-spectrum lighting. Work places that do not have full-spectrum lighting are noted to have a greater incidence of illness and lower productivity. Women that are exposed to artificial lighting may have their endocrine system off-balanced enough to cause irregular menstrual cycles. A possible outcome is that Premenstrual Syndrome (PMS), which has been implicated in violent crimes, could be triggered by the absence of full-spectrum light.

Likewise, when any criminal or mentally-ill individual is observed it becomes apparent that these individuals typically spend daylight hours indoors. When they do venture outdoors it typically occurs at night for the sake of being socially isolated. The absence of full-spectrum light has been demonstrated to be a factor in mental illness and many criminals are mentally ill. Certainly, the health of average individuals, and more so, the rehabilitation of mentally or physically ill people must include unfiltered natural light.

Cancer causation is also partially related to limited-spectrum lighting. A direct relationship between light and the effects of certain cancer-causing chemicals was discovered. Illuminated by the white light of a tungsten lamp were cultured cells exposed to each of eight chemicals (*polycyclic hydrocarbons*). Of these, the five which are known carcinogens reacted the most, rather than the three that are not carcinogens.[1207] A clear association between the photodynamic activity of limited-spectrum light and the development of cancer was uncovered.

The facts reveal a clear contrast between the conditions present in Nature and those that are artificial. Natural full-spectrum light enhances health, while limited spectrum light is unhealthy. Again, this is evidence of a homeostatic mechanism that through disease eliminates those who are not working with life and the organism Earth, FEM.

20.3 Urbanization and Industrialization

Each of the disorders just examined, and numerous others that partially stem from the absence of full-spectrum light, are more likely to occur in urbanized and industrialized societies. Air pollution is known to alter the light environment by changing the intensity and wavelength of light. The Smithsonian Institute, for example, reported a loss of 14% in the overall intensity of sunlight during 60 years in the District of Columbia. Likewise, at the relatively high elevation of Mount Wilson Observatory, California (compared to L.A. itself) scientists reported a loss of 10% of the overall intensity, along with a 26% reduction in the ultraviolet portion of the spectrum within 50 years.[918] It is very much worse today. Typical air pollutants, such as carbon dioxide and particulates (dust, etc.), particularly absorb the ultraviolet portion of the spectrum.

Aerosols, from spray cans, and nitrous oxide from power plants, nitrogen fertilizers and disturbed ecosystems, destroy the protective ozone layer. As a result, the dangerous forms of cosmic radiation (x-rays, gamma rays, etc.) enter our environment causing, among other things, skin cancer and mutation. Recently, along with ozone depletion, Chile reported blinded livestock and unusual sunburns in children as the result of ozone depletion.

All chemicals and minerals have a maximum wavelength absorption. Therefore, imbalances in the light spectrum combine with nutritional imbalances or biochemical impurities by allowing the absorption of more of certain types of light and less of others. Furthermore, an imbalanced light spectrum can activate certain hormones more than others creating biochemical imbalance that can lead to nutritional imbalance (in attempts to compensate). The likelihood of developing cancer or some other physical or mental disorder is thereby increased.

An example of this effect involved experiments with mice (*C3H strain*) which have a high susceptibility for developing spontaneous tumors. Under pink-fluorescent light spontaneous tumors developed and within 7.5 months the mice perished. Under other types of light with an increasingly wider spectrum there was a progressive increase in life span (up to an average of 16.1 months).[918] In contrast, with humans full-spectrum light has been therapeutically shown to combat cancer and even symptoms of AIDS, among other disorders and diseases.

Too much direct sunlight can create damaging effects, as well. The skin loses its elasticity and the immune system falters. Excess exposure can produce free radicals, which increase aging and the chances of becoming diseased. Increased requirements for vitamins A, E, C and the B-complex occur (particularly paraminobenzoic acid; PABA). These vitamins are essential for maintaining mental and physical balance. As a result, one may experience premature aging, disease and toxic poisoning (heavy metals, chemicals, etc.), among other risks. This is more of a problem in urbanized and industrialized society where extensive areas are cleared and the most damaging influences on the ozone layer take place.

Again, when all the facts are examined Nature seeks to establish the ideal situation. Indirect natural sunlight, as occurs under the canopy of a forest, reduces blood pressure, sugar levels and cholesterol. Natural light can help prevent and is often used therapeutically to relieve asthma, allergies, osteoporosis, psoriasis, depression, sexual dysfunction, and various physical and mental diseases. The metabolism of oxygen in human cells is improved and stress is reduced. Above all, the immune system and general biochemical activities are improved and enhanced. The formation of hemoglobin and white blood cells is increased. The "love" hormone (oxytocin) is stimulated by full-spectrum light. Thereby, sexual intimacy is intensified, more pleasurable and stress reducing in the outdoors. The ideal place for indirect sunlight is under the protective canopy of a forested wilderness.

Light is a nutrient and the indirect full-spectrum light in the forested wilderness is health "food." A scientist comments on its importance: "The results also show that natural light, 'the forgotten parameter,' is a basic element of vital processes. The extraterrestrial light emitted by the Sun acts as an energetic transmitter via the energetic portion of the optic pathway for all living things. Entering the eye, it stimulated the hormonal glands and metabolic processes to act together harmoniously."[571]; p.93. Again, inner harmony is achieved by a harmonious relationship with Nature, which is also harmony with the organism Earth.

CHAPTER 21

Quiet! Please.
(The Psychobiological Effects of Noise)

Like light, sounds have effects on physical and psychological health. They can bring imbalance when the sound is noisy. Natural sounds, in contrast, can relieve imbalances and bring about health. Natural sound recordings have been made to promote health for this reason. The impact of noise is greater than most people imagine.

21.1 <u>Physiological</u> <u>Response</u>

Our modern world with its technological "progress" brings about an ever-increasing clamor. In fact, the noise level increases approximately one decibel per year. Nerves always react to noise even while we are asleep, and whether or not we realize it. Arousing the entire body, noise impulses move from the lower brain down the spine to nerves in the heart, veins, arteries, stomach, lungs, pancreas, and muscles. These responses involve the involuntary (*autonomic*) nervous system, producing increased heart rate, blood pressure, respiration, and stress hormone secretion (*corticoids*).[269,703] Sudden, unexpected bursts of impulsive or steady-state noise cause changes in the heart and circulatory (cardiovascular) system, blood pressure and volume, hormone levels, digestive system function (*gastrointestinal motility*), breathing, and other neurophysiological activities.[509] While these responses are not necessarily injurious (pathological) they are similar to stress reactions, which can cause mental and physical disorders or diseases if prolonged.

Under the influence of continued stress from noise, one will eventually suffer some physical or mental (physiological) damage, or dysfunction in the circulatory (cardiovascular), digestive (gastrointestinal), hormone (endocrine), and/or nervous (*neurological glandular*) systems.[509] For example, school children in noisy areas have higher blood pressure. High-noise areas, which affect brain waves (*desynchronized EEG*) and reflexes (*hyperflexia*), are noted for a higher incidence of ear problems, insomnia and other sleep disturbances, headaches, irritability, nervousness, depression, swollen ankles, skin problems, and burns, cuts or other minor accidents. Also, noise can cause an irregular heart beat (*cardiac dysrhythmia*), and a higher prevalence of heart disorders and hypertension are observed in noisy areas.

Though not conclusively proven, some evidence indicates noise contributes to premature death and abnormal births. The problem is that noise cannot be singled out as a sole factor, but there is little doubt that it amplifies the total level of stress, which can contribute to these problems. Furthermore, noisy areas have more premature deaths and abnormal births. Anxious people are most susceptible and respond more with greater blood pressure, heart rate, respiration rate and skeletal muscle tension, and lower electrical activity of the skin.

Typical of the psychological basis of disease, repression and denial, or those who suffer in silence, are the defenses used by those who tend to show more of a physiological response to noise. In other words, ignoring the noise makes its negative effects worse. Summing up the entire physical effects of noise are the facts that any physical problem requires a calm peaceful environment to recover, and industrial noise contributes to most disorders and diseases.[509,703]

21.2 Psychological Response

Mental problems also arise from noise stress. A greater number of mental hospital admissions occur in areas with the greatest aircraft noise.[509] Likewise, more admissions to mental hospitals and more neurotic disorders were observed around noisy airports.[269] One of the most primary effects of noise is the development of hearing loss, impairment or handicap, even deafness.[703] Paraphrenic patients, a disorder similar to schizophrenia with hallucinations and delusions, show a higher prevalence of hearing loss. Several studies have disclosed a relationship between hearing loss and psychiatric disorders, with hearing loss occurring four times more often in mental patients.

Annoying noise can cause irritability, headaches, and depression. Responses are similar to dependence, otherwise referred to as learned helplessness, as well as aggression.[278,631] Continuous low-level noise is adapted to, which may indicate a desensitizing of the sensory cortex, an area disturbed in the most severe mental problems. Once again, the whole situation is summed up in the fact that stress from noise can lead to some mental health problems in noisy urban and industrial areas, and recovery requires a calm peaceful environment.[509,703,1347]

21.3 Performance

A combination of the mental and physical effects of noise can be seen in studies involving noise's influences on performance. Both the onset and offset exposure to noise lead to a noticeable deterioration in performance. Responses were more extreme and judgments were uncertain when work environments were noisy. Less information can be processed and errors increase.

Noise shifts resources towards one activity and away from others; usually towards the activity that is already performed well. In total, the performance of tasks overall was reduced.[631,703,1347] Reduced scholastic averages, less quality production and a decreased overall workload, among other facets of productivity, take place in noisy industrial and urban areas.

21.4 Interpersonal Relationships

As anyone might surmise, human relationships also suffer from the effects of noise. Decreased attraction to others occurs under the influence of noise. Decreased responses to those who need help also takes place both during and after exposure to noise. Relationships are oversimplified and perceptions of complex social relationships are distorted. In fact, others are frequently judged impulsively, and judgments tend to be more harsh or unfair. Aggressive acts are more likely to occur, as well.[278] Noise makes communication more difficult, but not only because it interferes with hearing.[1307,1347] These effects are in addition to those discussed above about noises' negative impacts on mental and physical functions, as well as performance.

A few examples of settings illustrate the effects of noise. A comparison of two villages demonstrated that the village with increasing aircraft noise showed a corresponding increase in drug use.[269,1347] Traffic noise caused more disagreement relative to argument, more tension, more uncertainty, and more irrelevant dramatizing in communication.

Likewise, those residents that lived in areas of light traffic had three times more friends and two times more acquaintances than those in heavy traffic areas. Those in the noisy, heavy-traffic streets said it was a lonely place to live, while the quietest streets were perceived of as friendly and sociable areas. The average length of residence was highest (16.3 years) in light traffic, while in moderate (9.2 years) and heavy (8.0 years) traffic areas residence was substantially shorter.[278]

A noisy neighborhood of Boston was noted for its residents being arrested more often, and were more likely to be truant or absent from school. They were also less likely to take care of the entry ways to their houses and businesses.[278] Noise alone has a profound effect on society, including interpersonal relationships and individual stability.

21.5 Wildlife

Animals are also affected by noise. Changes are noted in brain waves (*EEG*), heart function, respiration, digestion, growth, hormone levels, and reproduction. Even the migration patterns of wildlife were altered.[421] As urban and industrial areas grow, even the few areas left in their natural state will be affected by noise, especially by aircraft. Few of us realize that this restricts the already limited area of natural habitats increasing the likelihood of species endangerment or extinction.

21.6 Nature's Symphony

In contrast to the noise of the urban-industrial scene, Nature produces musical sounds, even the Earth itself emits a hum.[32] These sounds are harmonious with internal rhythms and can alleviate many of the symptoms caused by noise stress. Modern studies demonstrate that pleasant sounds keep blood pressure low, relax tense muscles, and promote rest, sleep and self-awareness. They also relieve troubled moods, and readjust disturbed organic patterns in the body, repairing damage done by other forms of stress. This is why numerous recordings produced for relaxation employ natural sounds, such as ocean waves or bird songs.

A study of an African tribe living in a noise-free society revealed better health, including less heart (cardiovascular) disease, and better hearing in older people than in urban-industrial societies.[269,1347] Our participation in the physical world is self-destructive, because noise levels rise as the result of actions that in the final assessment are destructive of other life. Again, we find a homeostatic mechanism in the biosphere of FEM that seeks to eliminate what is destructive to life.

CHAPTER 22

Which Way to the New Creature?
(Molecular Evolution
and the New Earth Model)

For more than a century we have contemplated the origin and extinction of species within the framework of evolution. Darwin claimed that evolution was the result of natural selection, a term coined to describe what is also known as "survival of the fittest." Meanwhile, nearly a century of genetic research has disclosed that the ultimate source of variability is at the genetic or chromosomal level. Mutation in the chromosome or its components—DNA, RNA, proteins, etc.—is the primary source of producing new species. Survival of the fittest merely plays a minor, at best, secondary role. Evolutionary transitions rely primarily on mutation at the molecular level, not with competition for a given niche in the changing environment, as has been hypothesized. Evidence proves that it is mutants that are fit to survive that lead to new species.

22.1 The Molecular Basis of Evolution

Hereditary information is carried in DNA within the chromosomes of a cell's nucleus. Chromosomes are known to have increased lengthwise as evolution progressed from the simplest to the more complex organisms (*prokaryotes to eukaryotes*). The simplest mechanism for accomplishing this lengthening involves the rearrangement of preexisting identical, small DNA sequences into longer groups.[186,563,760]

The most important characteristic of the DNA segment forming the gene is its position within the structural unit, which defines its function.[358,747,748,760] Mutation or other transformations (*base substitution, transposition, rearrangement, etc.*) can cause small or widespread physical (*morphological*) changes in an organism.[186,357,358,563,638,747,748,760,1051] In fact, modifications in DNA can be utilized to demonstrate a species relatedness to other species when similar patterns are observed.[186] "Molecular clock" is a term used to express the periodic or episodic changes that have occurred in many different species simultaneously throughout time. One level at which the molecular clock is manifested is in the sequences of DNA.[357,638]

RNA reads the genetic code of DNA which cranks out the protein that make up the developing organism. The evolution of RNA played a very important role in the earliest history of life. For one, genetic variation can be introduced as the result of mutation and RNA makes the mutation workable (*RNA-catalyzed recombination*).[355,356,573,637] Protein synthesis involves coding information from DNA, but this requires three types of RNA (*i.e., messenger, transfer and ribosomal*).[760] Interrelated checks and balances between DNA, RNA and protein determine what direction a given mutation will take in an organism.

One of the major evolutionary controls is that a given protein only recognizes a given type of DNA or RNA, and thereby, contributes to determining the type of hereditary information present. A protein's recognition of DNA directs the frequency and type of mutation.[760] Studies of the molecular clock reveal episodes in protein evolution that occur at a nearly constant rate throughout various types of organisms (*i.e., lineages*).[468,469]

When we combine all of the components mentioned above and include genes, we then have the structures known as chromosomes. A vast arsenal of remarkable molecular devices endow the chromosome with properties that ensure its own survival. The chromosome can bypass any assault on its integrity by repairing, reconstructing, substituting, improving and innovating its own molecular environment. Chromosomes can dispose of whole regions, whole chromosomes or sets of chromosomes. Genes function according to their position in the chromosome. The same gene sequence in a different location on the chromosome can alter the gene's action, leading genetic pathways into new functional alleys, and hence, new creatures.[760]

When genes are broken into pieces the reordering can facilitate evolutionary transformation.[356,466,467] The splicing of gene pieces does not have to be 100% effective, because molecular devices in the chromosome repair it, making the altered chromosome functional.[466] The genes of higher organisms (*eukaryotes*), such a mammals, are not continuous but broken into coded sequences (*exons*) and non-coded sequences (*introns*).[356,467] Evolution can then be described as the "shuffling" of these coded sequences into new positions, which produces new species.[356,466,467,1118] Because chromosomes have built-in "rules of conduct" that escape natural selection, natural selection is an explanation made in the face of our ignorance about these molecular mechanisms.[563,760] The understanding that molecular mechanisms, not natural selection, are responsible for evolution will be clarified subsequently and is now generally accepted by scientists.

Recent research has uncovered the molecular mechanisms accountable for evolutionary transmutations, but not how they took place throughout the history of life. The position of genetic information within the chromosome effects the order (*including polarity*), structure and function of genetic information, leading to transformations at all levels (*i.e., organism phenotype, chromosome phenotype, RNA transcription, and DNA replication*). Distinctions are made between the two strands of a DNA duplex and different, but similar, (*homologous*) chromosomes. The same is true of different DNA within the same cell, DNA segments, chromosomes, segments of the same chromosome, and sets of chromosomes. Gene expressions, mutations and rearrangements occur as the result of the controlling elements within the chromosome (*also episomes and transposons*).[356]

Rearrangements of the chromosome, and changing coded and non-coded sequences (*exon shuffling*) can cause sudden evolutionary transition.[356,466,467,1118] Examining a great variety of protein from different back-boned animals (vertebrates) disclosed that shuffling is a major determinate in evolutionary transitions.[356] There are episodes in evolution when DNA underwent bursts of substitutions followed by long periods of no substitution.[53,356,466] The major unanswered question is what caused the episodes of shuffling to take place.

22.2 FEM, Ionizing Radiation and Evolution

The reason that this question is unresolved is that the new model of the Earth, the Field-dynamical Earth Model—FEM—is not yet known. The Fields are particle accelerators which routinely produce X-rays and gamma-rays. Hydrogen fusion by-products include neutrons, helium nuclei or alpha particles, and high-energy protons and electrons. Collectively, x-rays, gamma-rays and fusion by-products are referred to as ionizing radiation. Ionizing radiation in turn ionizes the atmosphere producing electrostatic fields, microwaves, hydrocarbons, and changes the acidity of precipitation. Furthermore, during such times the geomagnetic field experiences reduced strength, and drifts to new locations or reverses (i.e., polar wandering and reversals). As will be discussed, the process of evolution involves mutation caused by ionizing radiation, electrostatic fields, magnetic fluctuations, microwaves, alkalinity, and hydrocarbons (in the order of their greatest effect).

Ionizing radiation, in the form of x-rays, gamma-rays and neutrons, is the most effective mutagen, as has been seen by its effects on

DNA.[596,1074,1202,1203,1235,1260] Damage to DNA is physical, chemical, and biological, leading to modifications in genetic information.[208,596,1135] Amino acid (*purine and pyrimidine*) bases are stacked in a parallel fashion in the core of the double spiral or helix of the DNA. One of these amino acids (*pyrimidine*) is more sensitive to radiation damage, leading to DNA breakage or degradation.[295,1265] Often one or both strands of the DNA helix break leading to cross-links between DNA molecules, chemically active sites (*ring openings in purine types*), or liberation of a base (*N-glycoside bond breakage*).[295,736,1218] Double strand breaks can cause the loss of the original genetic information more than single strand breaks.[736] Slippage during DNA replication occurs when the strands "mis-pair" in relation to their original coding.[691] Also, randomly produced unstable chemicals, known as free radicals, in DNA can undergo a series of reactions ending in a stable radiation product with stable damage centers that confer new genetic expression.[499]

DNA can be directly or indirectly affected by ionizing radiation, causing strand breaks and cross-links, transposing the sequence of genetic information.[295,309] Induced currents can greatly increase the charge separation in molecules, producing DNA that is more reactive.[944] Fast electrons captured by DNA, for example, can alter base functions, producing stable chemically active substances (radicals).[499] The electrical or electrostatic potential of the DNA helix is large and attracts chemically active substances (*counterions*) effectively.[596,760,1265]

Factors that enhance the sensitivity of the DNA include its phase state, dehydration, temperature, oxygen, and base substitution (*halogenated DNA-base analogs*). Oxygen enhances radiation damage because oxygen reacts with hydrogen, hydroxide and electrons (*this is a radiobiological law*).[579,686] The full effects of radiation on DNA mutation are still being studied.

Regardless of how much, it is well known that ionizing radiation's effects on DNA alter the genetic information. DNA conserves and replicates genetic information into a usable form, repairs genetic and molecular damage or errors, and transmits genetic information to the developing offspring.[760] DNA repair undergoes metabolic disturbances due to radiation damage, and is therefore likely to remain altered.[516]

The vertical stacking of bases makes DNA stable.[1265] However, information can be shuffled, changing genetic expression, including the functional pathways of RNA and protein synthesis.[295] For example, rejoining DNA pieces together (*transposition*) creates new genetic information. Shuffled information that interacts with enzymes or proteins is known to have radically altered plant development.[92,420,487-489] Furthermore, protein sequences only accept certain types of DNA and contribute to the type of DNA present in the chromosome.[237,760] This genetic change is true of any organism, because the most important characteristic of a DNA segment forming a gene is its position within the structural unit.

The genetic effects of ionizing radiation in offspring are due to DNA transformations in the reproductive cells of the irradiated parent. This transformation is of fundamental importance, because DNA double-strand breaks can lead to all of the different biological end-points observed in the fossil record.[237,314] The expression of genetic information by DNA occurs as a result of RNA and protein synthesis. As a consequence, the radiation induced structural alteration of DNA molecules will alter genetic expression, producing new species.

Other molecular regulatory phenomena may weaken or amplify radiation's effect on DNA.[308,314,516] The effects of radiation on DNA occur in body cells (*mitosis*) and reproductive cells (*meiotic drive*), bringing about uniform changes in a group of the same organisms, hence extinction and speciation take place.[246,308] "Imperfect" corrections of DNA damage produce mutations that can lead to evolution and extinction. Episodes in evolution parallel those of DNA, which shows bursts of base substitutions (*exon shuffling*) followed by long periods of no substitutions.[355,356,466-469] Collectively, the evidence reveals the fact that periods of increased radiation can be responsible for the mass extinctions and speciations seen in the fossil record.

RNA also undergoes direct and indirect mutations from ionizing radiation. One indirect effect involves alterations in DNA that exert modifications on RNA. Expression of the genetic information encoded in DNA involves the synthesis of RNA and protein. Alterations may modify the genetic information of DNA, leading to different bases in (*messenger*) RNA. This is in addition to the sensitivity of RNA (*polymerase*) itself, the effects of which include a decrease and occasionally an increase in RNA synthesis.[516] Like DNA, RNA demonstrates a history of (*exon*) shuffling that could be the result of episodes of intensified levels of ionizing radiation.[324]

Protein also undergoes radiation induced transmutations. Recognition of protein interactions with DNA or RNA entails the electronic features of the molecules and their (*steric*) arrangement. Molecules with this type of molecular recognition include the major cellular components known as cytochrome C, haem protein, other proteins, and the enzymes of DNA and RNA (*nucleases*).[760] The electronic and energetic aspects of the molecules correspond to their physical and chemical properties. One characteristic involves the electron donor or acceptor properties, which bring about a protein's potential for being affected by ionizing radiation. Because ionizing radiation carries charges it effects the electronic and energetic properties of the molecules. This charge shift affects protein bonding, and protein bonding is part of the stability of the genetic information.[1265]

Both the built-in (*intrinsic*) energy states of proteins and direct tunneling of electrons through impurity centers are involved in electron transfer. This appears to be particularly true for impurity centers with iron or an element with the same degree of reaction (*valence*).[944] For example, a component of the cell, called the mitochondria, regulates the electrochemical gradient of the cell, especially protons. Within the mitochondria is cytochrome, which is involved in electron transport and utilizes iron-sulfur proteins. Examination of the evolutionary rates of mitochondria should therefore be more rapid if ionizing radiation is offsetting the electrochemistry of the cell (by causing internal ionization). Observations indicate mitochondrial DNA evolves at a rate that is five to ten times faster than nuclear DNA in various types of organisms. Such a finding indicates that mitochondrial and nuclear heredity (*genomes*) evolve independently while under similar environmental conditions.[1275]

Furthermore, experimentation discloses that irradiated chromosomes (*i.e., chromatin*) yield a transfer of electrons from the protein to the DNA. As a result, damage to the protein is transferred to the DNA.[314] As could be predicted from the effects of episodes of ionizing radiation, two unrelated approaches, protein and DNA sequencing, support each other in terms of the existence of the molecular clock.[638] "Distinct subcellular compartments within an organism are under different controls and evolutionary constraints, which are not uniform and evolve differently in different taxa."[1275] Taxa are the different categorizations of organisms. Ionizing radiation is known to effect different cellular components and different organisms differently, and this is what we find in molecular evolution.

Other evidence confirms this understanding. Ionizing radiation causes stable highly active chemicals (*radicals*) that become localized on the protein (*pyrimidine*) bases of DNA and RNA.[499] This alteration becomes more important when we consider that an acceleration of protein (*amino acid*) substitution takes place during gene duplication in some cases. Again, iron-containing structures, or similarly reactive (*valence*) elements containing protein, are more affected, such as hemoglobin genes.[905]

Proteins are put together from mosaics of simpler structures alternating between coded (*exons*) and non-coded (*introns*) components, as well. Non-coded components (*introns*) serve to assemble genes for interaction with proteins that have repeated structures. Furthermore, their arrangements are not random. Protein's non-coded components are the late products of evolution, and have been inserted into the gene or omitted changing the genetic information.[467] Ionizing radiation is known to readily affect protein. In fact, ionizing radiation is employed to produce plant mutations with higher protein levels.[3,4,594,721,794,860,922,923,1155,1206]

Chromosomes, altered by ionizing radiation, contain the genes and other components discussed above, and are most important in the transmission of genetic information. The function of genes depends on their position within the chromosome.[760] Ionizing radiation results in the exchange of information through recombination, crossing-over or mutation, yielding new genetic information.[91,237,1074,1235] Other modifications include a "rewiring" of the code (*transcription*) and duplication of genetic information (*replication*), as well as alterations in DNA, RNA and protein.[1300] The alterations produced are particularly influential if they occur in the reproductive cells of an organism.[246,1235] A scientist asserts: "Ionizing radiation has turned out to be a powerful tool for changing the genetic material."[1235]

Numerous examples illustrate ionizing radiation's potential for mutation, which can bring about new species and extinction (*dominant lethal mutation*). Cultured mammal cells, for example, undergo transformations in chromosome structure and stable metamorphosis (*morphological and phenotypic*).[196,303,503,656,1073,1177,1277,1293] In mammals, irradiation leads to whole body and germ cell mutations, and skeletal abnormalities.[246,335,389,390, 841,1056,1057,1074,1235]

Plant mutation has been studied extensively, and changes in only small sections of the chromosome, called alleles, can lead to a whole new group of species, known as a genus. Alleles are in the thousands on each chromosome, but only tens or hundreds need change at a few locations on the chromosome (*loci*) to bring about a whole new genus.[92] New species have been brought about by using ionizing radiation in crop improvement, where allele mutation is noted.[354] In some cases there is evidence of diverse genetic modification (*tetraploid induction and cytochimeras*), and a high frequency of mutations.[649] Most plants studied yield mutants that have improved protein-content, structure, and resistance to disease or pests, and better adaptability to varied environmental conditions.[3,4,333,594,649,794,922,923,1155,1206] One scientist studying the effects of radiation on plant communities in the wild commented with surprise that the alterations mimicked the "patterns apparently related to the evolution of life."[1346]

Another possible way to speed evolution is if a virus infects an organism and swaps genes with it.[49,745] Viruses are very susceptible to transformation induced by radiation making this possibility even more likely. No scientist would deny that ionizing radiation is capable of bringing about the patterns we see in evolution, and has often been suggested to be the reason for widespread speciation and extinction.[1054,1068,1346] However, the idea has been scoffed at because no reasonable source for the ionizing radiation is known without FEM.

Studies of atomic bomb survivors and the possible consequences of a nuclear war reveal a similar scenario. Increased body (*somatic*) cell mutations were observed in atomic bomb survivors, and involve gene loss and chromosome alterations (*chromosome missegration, miotic recombination leading to homozygosity, chromosome aberrations, and GPA expression-loss variants*).[719] Offspring of atomic bomb survivors had a greater incidence of reduced brain and head size, mental and growth retardation, and cataracts. Survivors have a higher incidence of cancer, especially leukemia, and exhibit more chromosome abnormalities than the average individual.[284,886] Survivors of the atomic bombing of Hiroshima and Nagasaki are known to have experienced damage to proteins as the result of radiation. In fact, protein damage is even being utilized to assess their degree of exposure.[284,886]

Ionizing radiation impairs immune functions in both humans and animals.[732] Such consequences could easily cause extinction or mutation (especially in the case of viral infection leading to gene swapping). Studies of the effects of a nuclear war, which produces ionizing radiation, indicate that there would be species extinction and genetic (*and ontogenetic*) abnormalities.[392,1074] Such a scenario could be said to reflect evolutionary history, and this is particularly conspicuous when we consider that the effects of a nuclear exchange were supplemented by our studies of the dinosaur extinctions—meanwhile, ionizing radiation is only part of the story.

22.3 FEM and Other Sources of Genetic Mutation

Other effects on genetic information come in the form of the indirect effects of ionizing radiation, which include radio-frequency and electrostatic fields. Electron transport and atomic and molecular recognition, undoubtedly influenced by electrostatic charges and fields, may have laws that make even genetics and molecular biology secondary. Recognition of protein by DNA or RNA is electrostatic, and includes structures such as cytochrome, haem proteins, other proteins, and components of DNA and RNA (*nucleases*).[760,1265] Moreover, it is recognized that these structures have been transformed throughout evolutionary history.

Alterations in DNA synthesis are known to take place under the influence of electrical or electrostatic fields, such as those present in the Earth's electrical environment.[751,893] DNA mobility and the pattern of chromosome separation become transformed in relation to the orientation of an electrical field.[265] All biological molecules have magnetic properties (*ferromagnetic, antiferromagnetic, paramagnetic and diamagnetic*) that allow them to be shifted by electrostatic and magnetic forces. Such is the case with a component of RNA (*ribonucleotide reductase*) whose iron center is magnetically coupled (*antiferromagnetically*). A similar relationship is noted for cytochrome (*oxidase*), as well as biological molecules with certain (*transition*) metals that are found in haem and other proteins.[943,1063,1131,1185,1286] In fact, certain techniques (*nuclear magnetic resonance*) utilize this understanding to isolate particular biological materials in research.[943,1077,1131,1185,1286]

These electrical properties are one of the reasons that there is increased DNA, RNA and protein synthesis under the influence of an electric field.[482,561,825] Electrical forces communicate with living cells in such a fashion that their genetic expression is altered.[128,561,825] This is why electrostatic fields have been shown to bring about alterations in chromosomes.[803] Electric fields (*ELF*) break, fragment and cause the loss of part of the chromosome (*deletion*). They can bring about an uneven distribution (*aneuploidy*) or more than two sets (*polyploidy*) of chromosomes.[316] Radio frequency fields can also alter chromosomes.[552] The by-products of hydrogen fusion build-up in the magnetosphere producing electrostatic fields and lightning, which produces radio frequency fields (*whistlers, etc.*), that can alter genetic expression.[86,1058,1059,1123,1124,1283]

Another source for genetic mutation is the reversal or reduced strength of the Earth's magnetic field. Again, it is the electronic and magnetic features of biological molecules that make them susceptible to the influences from such forces. DNA becomes aligned perpendicular to a magnetic field when using the axis of the DNA helix as a reference point.[177,751] One scientist studying the possible role of magnetic reversals on DNA comments: "This would constitute a new type of mutational force and perhaps could be used to explain, in a rather direct fashion, the interrupted speciation accompanying geomagnetic reversals."[752]

DNA is not the only genetic material affected by magnetic fields. Also affected are RNA[597,1063] and protein.[127,197,943,1063,1286] Magnetic fields have been shown to stimulate the formation of mutations.[113,365] And reduced magnetic field strengths have produced profound irreversible body mutations in higher animals.[212,365,520,1267] The return of full strength of the Earth's magnetic field enhances the production of reproductive cells (*shortens the mitotic cycle*), which would intensify the spread of any mutation.[1235]

New genes are created repeatedly by recombining basic components (*domains*) throughout the course of evolution.[905] If ionizing radiation breaks up the genetic material, then magnetic reversals, and reduced strength should reorder or shuffle the (*exon*) materials (*i.e., due to properties such as ferromagnetism, etc.*). Magnetic reversals and/or sudden drops in intensity are correlated to extinctions and speciation, which verifies that this relationship is, at least, very probable.[150,305,306,364,365,471,534, 546,547,651,652,661,707,751,773,903,919,1035,1054,1096,1139,1198,1257,1262]

Other factors that enhance mutation, and could be produced by ionizing radiation, include microwaves, hydrocarbons and acid rain (*nitric acid*). Microwaves can disrupt chromosome-associated proteins and can cause chromosomes to uncoil due to a certain property of DNA (i.e., its oscillation resonance).[1131] Hydrocarbons or any aromatic molecules influence the functions of DNA and RNA by binding to them, leading to inactivation or mutation.[1265] In fact, any molecule or molecular fragment containing hydrogen can alter DNA or RNA function.[171,595,596] Changes in acidity have an effect on DNA (*cytosine monohydrate*) increasing its formation (*crystal growth*) in acid conditions, and leading to strand breakage under basic (*alkaline*) conditions.[295,894] The fossil and stratigraphic records disclose that a few or all of these influences were present during the mass extinction and mass speciation events.

22.4 FEM's Role in Evolution

The series of events at times of extinctions and speciation follow a certain pattern. First ionizing radiation breaks chromosomes and their constituents: DNA, RNA and protein. Then electrostatic effects alter active sites, and due to electronic and magnetic properties of the molecules (*i.e., ferromagnetism, etc.*), reorders the genetic material. A weakened or reversed geomagnetic field shuffles the material more so. Finally, microwaves, hydrocarbons, acid rain and other chemical fluctuations produced by ionizing radiation add the last of the transitions.

The end result is organisms that have transformed so completely that they are categorized as new organisms—speciation—or organisms that could not survive the mutation—extinction. The scenario is more on the order of extinction leading to speciation, because mass extinctions are followed by mass speciations, and the number of species has grown with time.

Much in the fossil record confirms this evolutionary scheme. Ionizing radiation is evident in the geologic and fossil records as irradiated minerals, such as iridium, tektites and microtektites, bones that are radioactive, mummified fossils, abrupt shifts in the levels of elements known as isotopes, and selective extinctions. The huge deposits of fossilized bones that make phosphate rock deposits is staggering, and they are often radioactive. The conditions under which these fossil bones were deposited do not exist today, as they appear to have been cut off from both sea and air, and no sedimentation took place as they were laid down. The chemical process that transformed the bones into phosphate is unknown and could have involved ionizing radiation, especially since the deposits are radioactive.

Microwaves and ionizing radiation dry vegetation, and the decay of neutrons produces lightning that could ignite wildfires, which are also recorded throughout the geologic record during these times. Everything that takes place has parallels with a nuclear war scenario, even shocked minerals, hydrocarbons and acid rain. Cycles in geological events are accompanied by cycles in the molecular clock (they might better be referred to as episodic). Mass extinctions are always followed by a blossoming of new species. Often the Earth's magnetic field drops in intensity and/or reverses or wanders. These observations will be addressed to some extent in Tome Three with particular reference to the time of the dinosaurs' extinction (*i.e.,* K/T Boundary event).

The scientific community is looking for new interpretations of the fossil record. "Conventional dogma is being questioned and in some cases discarded. We are seeing a change from dominantly gradualist interpretations of natural phenomena to those that emphasize chaotic events."[1001]; p.12

Many scientists have suggested that ionizing radiation was responsible for mass extinctions. For example, a supernova, asteroid or comet, and a super solar flare have been proposed as sources for ionizing radiation bringing an end to the reign of the dinosaur. Some have asserted that the repeated events of extinction and blossoming of new species is the result of ionizing radiation.[1068] Many events show dramatic fluctuations in elements (*isotopes*), such as oxygen and carbon.[160,653] Some events have been associated with iridium, an irradiated mineral.[270,352,659,663,823,914,919,958,1136,1223,1330] Wildfires have occurred repeatedly throughout geological history.[294] And abrupt climate transitions and extinction events have taken place throughout Earth history.[312] All of these observations could be expected of FEM, while no conventional theory explains all of the facts.

The fireball of a nuclear exchange is composed of ionizing radiation (largely gamma-rays and neutrons), and reflects much of what is found in the fossil record. There would be the extinction of large fraction of the Earth's animals, plants and microorganisms, including the mass extinction of plankton, genetic mutation, nutrient dumping, climatic cooling, wildfires, hydrocarbons, acid rain, soot, and shocked and/or irradiated minerals.[392,542,883] In fact, the nuclear winter scenario was partly conceived by considering the effects of the fireball created by an asteroid impact, which has been theorized to have caused the mass extinctions that accompanied the dinosaurs' demise.[663] However, many facts contradict the impact theory, as will be discussed in Tome Three.

Marine mass extinctions also occur in cycles. Two scientists comment: "Perhaps most importantly, it indicates that mass extinctions are not independent events, but rather are dependent on some single ultimate cause that recurs at regular intervals."[1109]

Mass extinctions reveal that ionizing radiation is responsible and its source is the new model of the Earth: FEM. Marine extinctions offer the most compelling evidence for cycles of mass extinctions. Throughout a long stretch of Earth history (*Permian to Pleistocene*) it is those marine organisms with calcium, a high atomic-number element sensitive to ionizing radiation, that display the most obvious cycles. A family is the category below an order comprised of several genera and numerous species. Of the 1,800 families within that time interval, 970 have become extinct. These organisms consist of shellfish (*ammonoids, echinoderms, brachiopods, gastropods and bivalves*), fish (vertebrates) and sponges.[1109]

Many reptiles and reptile-like animals (*amniote*) mass extinctions correspond to the marine events, but not all.[137] Most of these creatures were very large animals, such as the dinosaurs, which due to their size would absorb lethal amounts of radiation. Furthermore, those that were reptiles were more sensitive to radiation than other land animals, as can be noted in how reptiles bask in the Sun or hide in the shade to regulate body temperature.

Meanwhile, mammals internally regulate body temperature, and also develop in the protection of a placenta in the womb. As a result, mammal extinctions do not appear to be as cyclic or as severe. Furthermore, they reveal only the slenderest coincidence with the marine extinctions.

Other facts confirm that this scenario, involving FEM, occurs in evolution. As could be predicted from ionizing radiation, there are some evolutionary correlations between North America and Europe even when they are geographically separated.[1078] Plants are much less vulnerable to radiation than animals (by a factor of ten), and compared to animals, plants are relatively invulnerable to mass extinctions.[681]

Overall, it is the shallow water or surface dwelling, high-calcium marine creatures that show the most mass extinctions and in cycles. Those in deep water are the least affected. This relationship is apparent in what are referred to as "living fossils," which have not changed since they first appeared.

One is *Lingula,* a shellfish (*brachiopod*), that has been uncovered from very ancient rock (*Ordovician*). Its shell is not as abundant in calcium as most other shellfish (*i.e., it is a horny phosphate*). They lived in mudflats and shoals with a few specimens known from deeper water. Mudflats and shoals tend to have very silty waters, which would buffer the ionizing radiation.

Other "living fossil" shellfish include *Cranis, Neopilina* and *Perotrochus,* some older than *Lingula,* that either always lived in the deep sea or moved into the deep sea. Another is the Horseshoe Crab, *Limulus,* which walks along the ocean bottom. This is also true of the oldest marine vertebrates, such as the shark (*Devonian*), which never needs to surface, and inhabits the deep ocean. Others are the turtle and the crocodile (*Triassic*) who laid their eggs in mud.[506,760] Water, particularly deep ocean water and mud, are exceptionally good protection against radiation.

Studies at the molecular level reveal much of the same evidence that indicates the effects of ionizing radiation. Regardless of the great variety of organisms on Earth, most, when studied at the molecular level, show striking similarities.[760] Genetic material has been broken, shuffled and spliced, producing the great variety of organisms we witness today and throughout the fossil record.[186,338,356,358,466,467,760] For example, there is a "uniform average rate of DNA evolution" across a wide range of very different (*phylogenetic*) groups.[1130] Just like the cycles or episodes in evolution, there are episodes of widespread transformations at the genetic level.[357,468,469,638,747,748,890,1179,1180,1333]

Like the effects of ionizing radiation on different organisms, the rates of evolution are not constant, but vary significantly from one type of organism to another (*from lineage to lineage*). Ionizing radiation produces electrostatic forces, and a component of the cell that is more affected by electrostatic fields (and magnetic fields), known as the mitochondria, displays a faster rate of protein evolution (*silent substitutions are greater than DNA*).[469] Comparing DNA is an effective means by which genetically related organisms, such as birds, reveal their evolutionary relatedness (*phylogeny*).[1130]

Genetic codes are the fundamental units for determining whether an organism survives and reproduces.[331,749,760,1335] A given species' ability to adapt is forged by recombination, DNA alterations, mutation, and/or gene fusion.[231,760,1146] Radiation induced mutation is suggested by the observation that various species transform at the same time on a very broad scale, and may comprise most changes, which occur in cycles of *co*evolution.[1270] This is supported further by the fact that mass extinctions occur in cycles and steps, and are not random.[612,773,1003,1096,1109,1333]

Extinction and speciation are better understood at the molecular level where ionizing radiation has its greatest influence. Structural rearrangements cause the elimination of the chromosome segments that are involved and produce lethal effects (*i.e., translocations, inversions, etc.*), hence, extinction. Likewise, rearrangements in the chromosome can bring about an entirely new species while the original organism becomes "extinct."[760] It has long been understood that it is small transpositions at the genetic level that cause extinction and new species.[480,490,492]

Uniform rates at the genetic level bring about the patterns observed in the fossil record.[479] In mammals, for example, there is a striking correlation between adaptive changes, and an increase in the number and decrease in size of chromosomes (both by approximately a factor of two).[800] Structures are repeated disclosing the fact that proteins and genes are mosaics of simpler structures, called minigenes.[106,467]

Evolution seeks new solutions without destroying the old at the genetic level. Silent splicing areas (*introns*) are frozen remnants of history, and possible sites of future evolution.[466,467] As a result, there are multigene families of organisms which have groups of genes with similar sequences (*homology*) and related overlapping functions. Or, supergene families that possess a broader group of genes containing one or more functional units within proteins, and DNA-coded segments, called domains, that are of common origin. Several important domains are used repeatedly in evolution, and new genes are created repeatedly by recombining basic domains.[905] Again, this is more evidence of homeostasis in the biosphere.

Studies at this level clearly disclose that mutations are not random nor is natural selection a primary influence, debunking the long held tenets of Darwinism. Meanwhile, FEM, entailing periodic enhanced levels of ionizing radiation, having both direct and indirect effects, and magnetic field fluctuations, explains every detail; even the geographic locations of enhanced extinction and speciation as uncovered by paleo-biogeography.

CHAPTER 23

Here, There and Everywhere
(Mass Extinctions, Speciation and Paleobiogeography)

Studies of mass extinctions have uncovered facts that are predictable when considering the possibility that there are events with increased levels of ionizing radiation. Mass extinctions are not random and are not related to an organism's ability to survive. Unlike other (*background*) times of extinction, mass extinctions are abrupt at the family level.

Those species which have a broader geographic dispersal and few species within a group (*low speciation rates*) have extinction resistance during mass extinctions. Typically, it is those marine organisms that are in the open-ocean or deep-ocean (*benthic*), such as certain shellfish (*ammonites*), that survive or are the last holdouts at mass extinction boundaries. However, during all large mass extinctions both surface (*planktonic*) and some deep-water (*benthic*) marine organisms become extinct. This is sufficient to rule out any cosmic origin of the ionizing radiation, such as a supernova.[612] Meanwhile, this does not rule out FEM with the ionizing radiation being released through Fields in the oceans below the protective effects of the atmosphere and magnetosphere. Studies do disclose very disturbed oceans and atmosphere at times of mass extinctions.

Without FEM, mass extinctions remain enigmatic: "In spite of the obvious importance of extinction, and in spite of the fact that hundreds of thousands of extinctions in the geologic past have been documented, we know surprisingly little about the process itself and surprisingly little about its actual role in evolution."[1001]; p.2

23.1 <u>Many</u> <u>Win</u>, <u>Some</u> <u>Lose</u>

Other observations confirm the reality that mass extinctions are due to episodes of ionizing radiation. The shallow-water reef-building communities are so disrupted that it takes the longest for them to reappear in the fossil record.[612] Reefs are very high in calcium, which absorbs a great deal of radiation, and they do not have outer tissues or deep water to protect them. Moreover, reefs are typically located in the Field areas as can be seen both in the past and in the present.

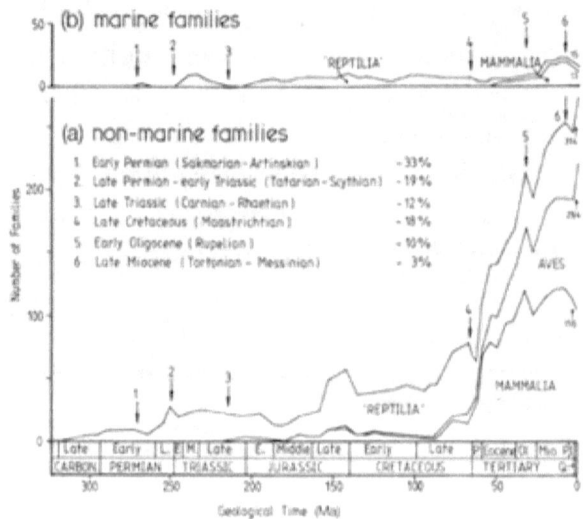

Figure 2.2a. Extinction and Speciation of Marine and Non-Marine Families. Note how this graph displays the close coincidence of extinctions and speciations. In fact, they are almost a mirror image of each

other with the exception of the fact that the speciation rate that follows is greater than that of extinctions. The standing diversity increases with time, punctuated by mass extinctions (numbered 1-6). The curves for reptiles, mammals and birds (Reptilia, Mammalia, and Aves) are shown separately [from reference 137]. Note how the marine families suffered much less extinction than the land animals. This is because water is a buffer against radiation. Also mammals, due to a placenta (and also many lived in burrows), had the lowest extinction levels. These facts could be predicted from the effects of ionizing radiation, but *not* from natural selection. [permission granted by The Systematics Association (London)]

The severity of reef extinction is not the case with other groups of organisms. Mass extinctions of four-footed land animals (*tetrapods*) are not as great as marine extinctions, nor are the extinctions due to a decrease in fitness.[137] Reptiles (*amniota*) are radiation sensitive, and show the cycles or episodes of the marine extinctions. Meanwhile, mammals display only the slenderest coincidence with those cycles, and those with more protection, because of a placenta, are less vulnerable than those with pouches, known as the marsupials.[1078] Radiation sensitivity is greatest in reptiles followed by marsupials and finally placental mammals, and the extinction record displays the same trend. During extinctions there is the mass disappearance of many types (*taxa*), then a few remain that later become extinct, and finally, totally new types emerge (see Figure 2.2).[824]

Plants offer further evidence of such events, and have been studied extensively under the influence of ionizing radiation. Plant breeding using ionizing radiation has been going on for decades. A few of the most obvious alterations are increased protein, greater yields, and resistance to disease.[3,4,594,794,860,922,923,1155,1206] There is even better adaptability to varied environmental conditions after radiation exposure.[794] In most studies, plants showed superior height, weight and reproduction.[333] Under high doses there is a great deal of mutation (*e.g., tetraploidy, cytochimeras, etc.*).[649] Overall, plants are much less susceptible to ionizing radiation than animals, by a factor of ten.[1346] The fossil record shows that plants are the least susceptible to mass extinction, again reflecting their sensitivity to ionizing radiation.

The fossil record discloses that, compared to animals, plants are much less vulnerable to mass extinctions and do not show the cycles or episodes of extinction.[173,681] When high levels of plant extinction do occur, it is at those times when great mass extinctions occur in animals (*Late Devonian, Permo-Triassic and Late Cretaceous*).[681] Dry spores under oxygen-deficient (*anoxic*) conditions are very resistant to radiation.[686] Typically such conditions occur during mass extinction episodes, at times due to wildfires and dramatic climate fluctuations. Spore-bearing plants have lasted the longest in the evolutionary history of plants.[489]

During one of the great mass extinctions (*Permo-Triassic*) a pine (*conifer*) embryo was discovered in North America. This fossil disclosed that a significant delay occurred between fertilization and seed germination in the earliest pines (*conifers*). Today it is the most modern seed-plants that undergo a delay in germination, and dormancy is more common in flowering plants.[799] Meanwhile, ionizing radiation is well known to delay seed germination.[1115] Furthermore, this pine embryo displayed no larval or fungal damage, and ionizing radiation is employed today for this very reason to preserve food, as it destroys organisms in the food.

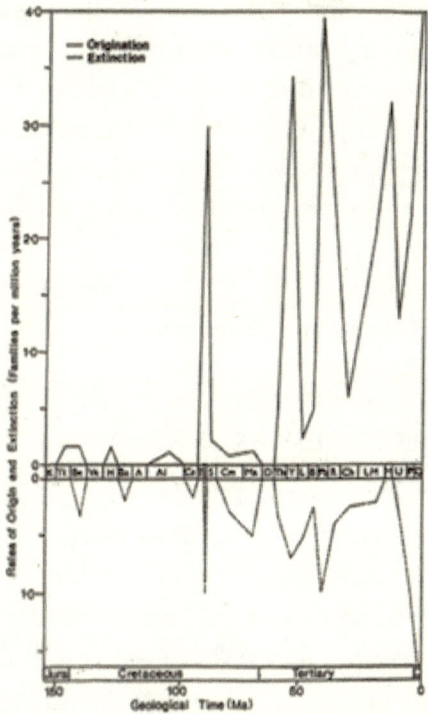

Figure 2.2b. Extinction and Speciation Through Geologic Time. Here we see speciation (origination) at the top of the graph and extinction at the bottom (inverted). Again, it is quite clear that the speciation rate is much greater than the extinction rate. This could be expected if ionizing radiation lead to extinction of the original animal which then diversified as a result of genetic mutation. [permission granted by The Systematics Association (London)][137,1261]

Developmental processes are similar at the genetic level in flowering plants. Then, unique development accounts for the evolution of plant diversity.[487,488] A scientist studying the effects of ionizing radiation on plant communities comments that the effects follow "predictable patterns apparently related to the evolution of life."[1346]

Another indication of the effects of ionizing radiation is the appearance of new species. Four-footed land animals (*tetrapods*) experience family extinction and speciation in unison and in cycles, and are not the result of changes in fitness.[137] Even plants display a clear cyclic relationship of first and last appearances.[173] Calcium-bearing, surface dwelling shellfish, such as foraminifera, are so frequently replaced by new species after extinction that they are used to date rocks (*i.e., they are index fossils*).[396] The history of life clearly shows a blossoming of new species after extinctions.[1068] Every fact could clearly be predicted from ionizing radiation's effects.

23.2 Coevolution and Parallel Evolution

Further support comes in the form of what is called parallel evolution. Australian pouched mammals, called marsupials, parallel the evolution of placental mammals on other continents. The "surprising" aspect of this fact is that Australia has been isolated from the other continents since the mass extinction that made the last of the dinosaurs (*Late Cretaceous*). In addition, they evolved independently in very geographically isolated habitats. Yet, many species of marsupials have become very similar (*in phenotype*) to those of the placental mammals. Examples that are very similar to mammals include: Marsupial Moles, Marsupial Mice, Banded Anteaters, Tasmanian Wolves, Koala Bears, and Flying Phalangers or Gliders that resemble the Flying Squirrel. Though they evolved geographically separated with different plant species to eat, and other soil and climatic conditions, their evolution proceeded along essentially the same line as did mammals on other continents. This is against one of the tenets of Darwinism, which states that organisms which are geographically isolated should evolve in very different ways (*due to extrinsic barriers or extrinsic isolating mechanisms*). Random mutation and natural selection could not possibly produce such specific parallel evolutionary transformations.[760]

A specialist on molecular genetics discusses this: "Instead one needs to look for other agents which can much more easily account for the phenomenon, namely, molecular determination conveyed by molecular recognition in the form of the organization of the chromosome and of the other cellular pathways."[760]; p.1039. In other words, global events including ionizing radiation and the other mechanisms altering genetic information breaking and rearranging the chromosome in geographically isolated mammals and marsupials could account for the observations.

Further evidence is available in what is referred to as coevolution. Coevolution involves totally different organisms that evolve in ways which reinforce each others' success. Fungi coevolved with plants and animals, and plants and animals coevolved in such a way that none would have survived otherwise.[465,495,955] Transformations in plants are matched by transformations in other organisms. For example, the first flowering plants were matched by the first pollinating insects.[630] In fact, various species change at the same time on a very broad scale and may comprise most of evolution, which occurs in cycles of coevolution.[1088,1270] Such phenomena could not be explained by random mutation and natural selection, because had they evolved independently they would not have survived.

Most organisms, particularly the ones most vulnerable to ionizing radiation, undergo extinction only to be replaced by new species. With the exception of living fossils, it seems to be the lot of all organisms to become extinct. In some cases there is a kind of relay of species, a new one comes in to occupy the vacant niche left by the extinction of a predecessor. In particular groups there is a definite correlation between a high rate of speciation and a high extinction rate. Any paleontologist will readily confirm this and that ionizing radiation could be responsible if the source were not extraterrestrial (*with the exception of a bolide*). In fact, such an observation could be predicted if events (including ionizing radiation) were breaking chromosomes producing extinction, and by rearranging them create new species. This is most clearly evident in shellfish[580,612,925,1004,1005,1109], but is also apparent in marine animals, reptiles, mammals and birds[137,1261], as well as plants, but to a much lesser extent[681]; all of which again reflects radiation sensitivity.

23.3 Geographic Segregation

As discussed, greater ocean depths shield organisms from ionizing radiation, and a relationship with FEM is obvious when considering this. The newer, more advanced species, genera and families are nearest the surface, while increasing depths reveal the more primitive groups. "Along the lower continental slopes, from about 2,000 to 3,000 [meters], there is a striking representation of archaic organisms."[183]; p.14. At such depths there is the still-symmetrical Hermit Crab (*pylochelidae*), a "living fossil" known as the Segmented Mollusc (*neopilina*), and many other ancient organisms (*stalked crinoids, archaeogastropods, paleoconchs and protobranch pelecypods, primitive enchinoderms, an ancient suborder of isopods and pogonophorans*). Many had originated on the shelf, a shallow water area, but could only be found at greater depths in more recent times (*i.e, Paleozoic and Mesozoic*).[183,184] The ancient types only survived under the protection of deep water, which shields radiation.

One type of shellfish (*ammonites*) is among a few of the long-lived taxa and tend to survive or be the last holdouts at times of mass extinction.[612] These shellfish tend to be mostly from the open ocean or the ocean bottom (*benthic*). Likewise, a certain type of jawless fish (*agnatha*) became extinct in shallow waters (*Detritus Feeders or Mud Grubbers*), but survived by becoming bottom dwelling types (*Benthonic Detrivore Placoderms*), and parasites on other fish (*Antiarchs*).[521] Similarly, the shallow-water reef-communities are typically the most devastated ocean life at times of mass extinction.[213,612]

Considering the position of the Fields, which release most of the ionizing radiation along the 30° to 40° latitudes, a latitude segregation should also be apparent. Fossils of marine mammals (*Cetaceans*), such as whales and dolphins, certain (*Hermatypic*) corals, and shellfish (*bivalve molluscs and benthic foraminifera*) disclose that the newer, more advanced species are found in the tropics, while the higher the latitude—the further from the Field latitudes—the more ancient species are found. This Field-latitude segregation is also true for both open ocean (*pelagic*) and bottom-dwelling (*benthic*) invertebrates, fish, reptiles and mammals.[183,184,1079]

Latitude constraints are common to all extinctions for all types of organisms. This is especially true of shallow-water marine creatures, particularly reef communities, where low latitude types are more severely affected than those with polar and worldwide distribution (*cosmopolitan*).[612] The facts support what could be expected from FEM.

The positions of the Fields would tend to produce more extinctions, more species and more diversity in the tropical to temperate regions. The study of the geographic distribution of life, biogeography and paleobiogeography, tells that tale. Some of the earliest plant distributions (*Early Devonian and Carboniferous*) are mostly between the 30° latitudes. When considered in terms of shifting continents, coals, evaporites, easterly and westerly winds, low and high pressure systems, and rainfall during these times demonstrates a latitudinal bias that conforms to the Field regions.[1109,1010] This holds true for all time periods.[240,245,512,518,588,710,959,1041,1145,1234,1319]

In fact, this geographic restriction has been used to find the location of the ancient pole, and for reconstructing plate tectonics.[184,1181] Furthermore, a number of land separations in continental drift are contradicted by the geographic distribution of species, if we assume that natural selection and random mutation are at work.[184] However, with ionizing radiation a parent stock of organisms can undergo parallel evolution even if they are separated by geographic boundaries. Those species that have a broader geographic distribution have extinction resistance during mass extinction.[612] This extinction resistance exists because they are also geographically located in, at least, some regions not as severely affected by ionizing radiation.

Another confirmation of the influence of the Fields in biogeography are what is called centers of origin. These centers of origin are labeled as such because they display the earliest known fossils of a given organism, and also greater diversity than surrounding regions. Both polar regions, the Arctic and Antarctic, are two such centers. The other Field regions are represented by centers in the southeastern United States, West Indies and Central America (North Atlantic Field), and Australia and New Zealand (East and West Australian Fields). Other centers of origin are southeastern Africa and Madagascar (South African Field), North Africa (Mediterranean and Persian Gulf Fields), Eastern Europe, Ethiopia, Turkestan and other countries around the Persian Gulf (Persian Gulf Field), Brazil (Brazilian Field), and Southeastern Asia, Indonesia and the Orient (Japanese Field).[183]

Of all these regions it is the region comprising Southeastern Asia, Indonesia and the Orient that is the most active center. Interestingly, this region has been affected the least by plate tectonics, because it has not moved much for quite some time (*Late Jurassic/Early Cretaceous*).[1327] Among those creatures which had originated in the area are (*Hermatypic*) corals, seagrasses, mangroves, salamanders (*order Caudata, family Hynobildae*), certain beetles (*family Chauliognathidae*), wasps (*family Vespidea*), earwigs, pit vipers, rattlesnakes, flowering plants (*Angiosperms*), plants with two seed leaves (*Dicotyledons*), many groups of shallow-water marine animals and plants, and rats, mice, voles and lemmings (*family Muridae*).[183] The reason this evolutionary center has been so active is that the Japanese Field has been affecting the area due to inconsequential "continental drift" so that ionizing radiation more often affected the region, resulting in more frequent genetic mutations.

An observation that is particularly interesting in this respect involves the newer, more specialized species. For example, (*Hermatypic*) corals are younger and more advanced in the center where diversity is also higher. Conversely, the further away from the center the more primitive the species and less diversity there is. This not only holds true for the Japanese Field, but also the South African Field for age, and the Persian Gulf Field for diversity.

Other species display similar relationships with respect to the Japanese Field. These include Shore Fishes (*family Gobiesocidae*), Demsel Fishes (*Pomacentridae*), Blennies (*Blenniidae*), Tongue Soles (*Cynoglossidae*), marine shellfish (*Gastropods; Strombidae and Cypraeidae*), freshwater snails (*Pomatiopsidae*), Turtles (*Emydine*), Flat Bugs (*Hemiptera, Aradidae*), lizards (*Agamidae and Scincidae*), Coral Snakes and Cobras (*Elapidae*), Tree Squirrels (*Sciurinae*), primates (*Infraorder and Lorisidae*), flowering plants (*Angiosperms*), and a host of others (*echinoderms, Comatulid Crinoids, Ostariophyan Fish, nematodes, Water Striders, Stoneflies, ant fauna, mosquitos, modern bird fauna, Flying Lemurs, Tree Shrews, pines, firs, larches, hemlocks and so forth*).[183] More of the recent species and a greater diversity of species are in the region of the Japanese Field center of origin.

As discussed, those times when ionizing radiation is released by a Field or Fields it often spews iridium or tektites. The most recent and one of the largest tektite strewn fields effectively pinpoints this area. This tektite strewn field is known as the Australasian strewn field.[906] It stretches from the Japanese Field to both the East and West Australian Fields. The area encompassed includes not only Southeastern Asia, Indonesia and the Orient, but also Australia, New Zealand, the East Indies and New Guinea (see Figure 2.3).

Figure 2.3 Tektite Strewn Fields. These are all of the presently known tektite strewn fields. The small circles are the general locations of the Fields of the Field-dynamical Earth Model. The strewn field that nearly stretches around the globe is from the time when the dinosaurs became extinct, known as the Cretaceous/Tertiary Boundary (this event will be discussed in Tome Three). The strewn field along the First entry should be "southeastern North America" is the North American tektite strewn field. The Australasian tektite strewn field is the egg-shaped area at the lower right of the figure. This strewn field clearly covers an area that embraces the Japanese, East Australian, West Australian and South African Fields (Fields marked by circles) [after reference 906].

These other areas are also important centers for developing the newer species. For example, the center for tropical seagrass is between Malaysia and northern Australia. Water Striders (*family Gerridae*) are found in Southeastern Asia and Australia. Skinks-lizards (*Scincidae*), and Cobras and Coral Snakes (*Elapidae*) are most diverse in both Australia and Asia. Pines, firs, larches and hemlocks of the most recent species are concentrated in East and Southeast Asia, Australia and New Zealand. Parrots (*Psittacidae*) greatest diversity and pigeons (*Columbidae*) richest development of families are in the Australian region. The Duckbilled Platypus and Spiny Anteaters, which are intermediate between reptile and mammal, can only be found in Australia. In the midst of this strewn field, and as will be discussed in Tome Three, at the confluence of these three Fields is Wallace's Line, an imaginary line used to describe the east-west separation of distinct (*endemic*) types of plant life.[184,1327]

Other Field regions are also represented by centers of origin and finds of tektites. For example, located in the North Atlantic Field area is the North American strewn field.[906] This Field and the strewn field are near the southeastern United States, which is the center of distribution for plants and animals of North America.[184,672] In fact, for amphibians and reptiles, diversity is greatest in a concentric arrangement around the Field area.[672]

Another confirmation that these Fields release radiation involves observations of thermoluminescence or bioluminescence. In Tome One it was discussed that these Field regions were noted to have luminescence in the ocean and were often associated with earthquakes. A close link exists between environmental radioactivity and luminescence.[880] Alpha and beta particles form as decay products from neutrons and can create luminescence.[1357] The Fields release the by-products of hydrogen fusion one of which is neutrons, and this creates the luminous seas. These same regions of luminescence are also centers of origin.

23.4 Evolutionary Enigmas

The understanding that the Fields release ionizing radiation, and that this causes what has been observed in evolution, explains enigmas which present theories find it hard, if not impossible, to reconcile with. Iguanidae lizards, for example, are found on Madagascar (*two genera*), near the South African Field, and also thousands of kilometers away in the Southwest Pacific (*one genus*) near the East Australian Field. This is interpreted by present theory as an African origin for the lizards with subsequent extinction in Africa. However, no fossil evidence has been found to support this contention. Its occurrence in the Southwest Pacific is said to be an example of long distance rafting by ocean currents. That is, at least two lizards, or one "pregnant" lizard, are purported to have ridden some floating vegetation or earth for thousands of kilometers. In contrast, neither of these explanations is required if an ancestral lizard is irradiated as a result of their proximity to the Fields, and thereby, undergo the same genetic mutation while geographically isolated.

A good example of how inadequate present explanations are is afforded by aquatic earthworms. Their greatest diversity (genera and species) is in the tropics of the Americas, and therefore, it is considered the aquatic earthworms' center of origin. Meanwhile, it is also found around the North Pole in the Arctic (*Holarctic*), and in South Africa and Madagascar. It is utter nonsense to consider that the aquatic earthworm somehow traversed the entire Atlantic Ocean and in two different directions, one towards the Arctic and the other toward South Africa. Yet, again, each of these regions is near a Field, and the same genetic mutation of a common ancestor is a far more realistic explanation. Similarly, the explanation for parallel evolution in widely separated geographic regions and the origin of species in polar regions, some of which are present-day tropical species (e.g., alligators, etc.), can be better understood on this basis.

A scientist who studies ancient and present-day geographic distributions of living things, expounds on an unsolved problem: "What we really need to find out is why the evolutionary process that goes on in centers produces species that are dominant in terms of their ability to displace other, older species and to become widespread."[183]; p.72

Again, it is the genetic mutation of the older species that gives rise to the newer species. It is not that they are "dominant" and "displace" the older species, as would be assumed in the Darwinian scenario of competition and natural selection. Hence, we could predict that the older species would be replaced by the younger species in the region of enhanced ionizing radiation. Species are not produced by competition in terms of natural selection, but by genetic mutation on a broad scale brought on by ionizing radiation, and other factors (electrostatic fields, magnetic reversals, pulsed radio-frequency fields, microwaves, hydrocarbons and alkalinity) produced by the Fields, and the particles which flow within them (neutrons, gamma-rays, X-rays and energetic electrons).

In those times when these evolutionary events take place there are also alterations in the physical environment that bare the signature of ionizing radiation. Reduced geomagnetic intensity, and polar wandering and/or reversal typically occur. There is also evidence of strong magnetic activity around the 30° to 40° latitudes between reversals, which has been referred to as a transitional field (see for discussion). This transitional field is the Field(s) spewing out ionizing radiation as the North-South dipole is weakened. Many of the events that take place occur in cycles or episodes of similar duration.[18,135,150,305,306,471,534,546,547,651,652,707,919]

Evidence of highly disturbed oceans include sedimentary records of changes in sea level, fluctuations in salinity and isotopes, and deposits of black shale, all of which occur in each episode. Likewise, there is a highly disturbed atmosphere, resulting in severe temperature drops leading to temperature minimums or glaciation, shifts in precipitation, large-scale winds and storms, pressure modifications, and fluctuations in carbon, hydrocarbons and isotopes. Climatic transitions and evolution go hand in hand.

As could be suspected from fluctuations in ionizing radiation, isotopes in both the ocean and the atmosphere undergo transitions in their abundance (^{10}Be, ^{26}Al, $^{87}Sr/^{86}Sr$, $^{18}O/^{16}O$, etc.). Seafloor spreading, continental displacement (tectonic episodes), volcanic eruptions and so forth often accompany the other transformations. Records frequently show extensive areas of missing sediments (*hiatuses*) on submarine ridges and rises, and occasionally, even the deep abyssal floors, which reflects strong, deep ocean currents at such times. Iridium, tektites or microtektites occasionally occur at unusually high levels in the sediments laid down at these times. Cratering occasionally takes place, but not of the impact types (*mostly cryptoexplosions or geoblemes, not volcanic nor impact*; see discussion in Tome Three). All of these phenomena tend to occur in cycles together or nearly so, though it is rare to find a complete record for a single event with all of these phenomena occurring at the same time.[18,135,312,707,773,919,1003,1088,1096,1109]

During these episodes mass extinctions take place, and eventually new species emerge. First there is the mass disappearance of many types (*taxa*), then a few remain that also eventually become extinct, and finally, totally new types emerge.[822] One paleontologist discusses these "radiations" or blossomings: "Many of the radiations of the geologic record are as spectacular as the mass extinctions, although they have not attracted as much attention."[1002]; p.9. Likewise, the last and first appearances of plants and animals tend to coincide.[137,173,612]

Another paleontologist describes this phenomena in four-footed animals (*tetrapods*): "The total origination rates generally track the total extinction rates quite closely."[137]; p.278. The selective nature of the extinctions involve types (*taxa*) that are big-bodied, tropical, have few species, and are land or terrestrial types; a predictable scenario for FEM. Furthermore, origination and speciation track each other so closely because extinction is the result of mutation, which leads to origination and speciation.

In the marine realm, it is the surface dwelling plankton, reef communities and immobile, deep-water (*sessile benthic*) shellfish that are most affected.[18,612] The immobile deep-water shellfish are those that typically inhabit the waters in the Field areas. Meanwhile, the least affected are the mobile deep-water (*benthic*) types.[18]

Increased diversity occurs through time, and speciation follows each extinction event.[137,173,612,822,1002] Meanwhile, in spite of this overall increase of creatures, the number of four-footed animals (*tetrapods*) per classification (*taxon rates*) decreases toward the present.[137] Latitude variations are common to all mass extinctions with tropical and temperate types more affected than polar or worldwide (*cosmopolitan*) types.[612] These facts argue in favor of ionizing radiation and the associated factors causing the extinction of old types, and bringing about an equal number or more numerous new types through mutation. The new types are so totally different they must be put into new classifications, hence a decrease in the number of members for each classification (*per taxon rates*) and an overall increase in diversity towards the present. These facts and the latitude restrictions confirm that the phenomena associated with FEM is the unknown process behind evolution, including extinction and speciation.

23.5 Mass Extinction Events

The following events will serve as examples of what accompanies mass extinctions. These same events are followed by a blossoming of mass speciation, as well. An activated FEM during times of mass extinctions could produce episodes of plate movement, climatic shifts, highly disturbed oceans, wildfires, isotope fluctuations, shocked and irradiated minerals, hydrocarbons and acid precipitation. The following discussion will not exhaust the subject, because information is still being gathered, and what follows is only meant to be a brief review.

The Late Devonian was one such major event in Earth history. In this period the Earth's rotation slowed down from an earlier 400-day year, which was revealed in coral growth rings. The mechanism behind changes in the Earth's rotation or length of day will be discussed in Tome Five of Volume Two (Chapter 18), and entails solar-FEM linkages.

Late Devonian events produced oxygen-depleted water, a sign of a highly ionized ocean, which caused the extinction of many species of shallow, bottom-dwelling marine creatures. Another sign of highly ionized oceans is that sea level fell, as salt, calcium carbonate, and organic carbon were precipitated in greater quantities.[572] Geologic evidence tells us that there were highly disturbed oceans, including "tidal waves" or tsunamis, and a highly disturbed atmosphere.[18,572] Isotope data indicates sulfate, carbonate, strontium, and other isotopes were offset.[572] Cool water began to flow into equatorial regions and glaciers formed.[822,824]

The Late Devonian was a time when fine clay, gas, coal and petroleum were deposited in great quantities. This scenario indicates that there was the possibility of hydrocarbon precipitation, erosion and wildfires, too. Cratering and an iridium layer are also evident.[823,958]

Barren landscapes were colonized by plants, insects and vertebrates, particularly reptiles. Newly evolved plants took over as the extinction of older species commenced.[681] Extinctions hit the high-calcium, radiation-sensitive primitive corals and sponges, as well as some shellfish. The first egg that enables an embryo to develop on dry land emerges, and ushers in a totally new and major family assemblage.[137]

Another major event is marked by the boundary between the Permian and Triassic. No evidence of a geomagnetic field reversal exists at the boundary itself.[919] The continents shifted and buckled dramatically so much that many areas which had been submerged by inland seas had become dry land. In fact, one tenth of the salt of our modern oceans was stored in sedimentary deposits that were laid down at the close of the Permian.[18,612] Climate was so disturbed that warm weather shifted to a cold climate that produced glaciers.[312] Predictably, evidence of unusual chemical levels (*geochemical anomalies*) is plentiful, including an excess of iridium (at least in the Orient).[572,919,1351]

The Permian/Triassic Boundary was an another time of mass extinction. It was one of the five most pronounced marine mass extinctions in Earth history, with estimates ranging from 77% to 97% of the families becoming extinct.[1000,1001,1109] Crinoids, corals and certain shellfish (*brachiopods and cephalopods*) had lost twenty or more families. Classes such as *rostroconchs, trilobites* and *blastoids* disappeared totally.[580,1109] All forty genera of large *fusulinid foraminifera* disappeared and *bryozoans* lost eighteen families.[18] Latitude differences are evident, as for example with the shellfish, *articulate brachiopods*, whose family extinction reached 75% in the tropics compared to 56% elsewhere.[612] Again, it was the organisms with high levels of calcium that experienced the greatest extinctions.

Amphibians (*labyrinthodont*), fish (*anapsids*) and mammal-like reptiles (*synapsidia*) had dominated the landscape from the end of the Devonian. In the beginning of the Triassic they had given rise to the early diapsids, dinosaurs, pterosaurs and mammal-like reptiles (*synapsidia*), whose diversity had peaked in the late Permian only to dwindle at this time.[137] All of these creatures were large and would, therefore, absorb more radiation, yielding more extinction and mutation. The more radiation sensitive organisms, amphibians and reptiles, lost 75% to 80% of their families.[18,137] The Permian/Triassic Boundary marked the end of those life forms that had dominated the Earth since the Late Devonian.

What followed was the beginning of a new era, called the Mesozoic, which comprised many new life forms. New types of *diapsida*, who were the ancestors of lizards, snakes, crocodiles and dinosaurs, emerged. Also, new types of *synapsida*, the mammal-like reptiles that gave rise to mammals towards the end of the Triassic, appeared on the scene. This was an era that would bring about the dinosaur explosion, as well as the pleisosaurs, icthyosaurs, pterosaurs, mosasaurs, and other huge creatures.[137] New orders and families of plants emerged, as many earlier types became extinct.[681] As could be expected, new types of shellfish, fish and corals also came forth. The Mesozoic was the beginning of the reign of the dinosaur, and one of three major explosive transitions in the history of life.

Another time of major transition is marked by the Cretaceous/Tertiary Boundary at the close of the Mesozoic. At this time the Earth's magnetic field underwent a reversal.[919] Massively destabilized oceans and atmosphere brought extensive erosion, huge storms, a drastic temperature drop, and offset trace element and carbon isotopes. A rapidly spreading ridge shoved the continents apart, buckled land forms, pushed mountains skyward, created tsunamis (tidal waves), and drastically increased sea level. Volcanic eruptions were some of the most extensive in Earth history. Ionizing radiation brought acid rain, hydrocarbons, black shale deposits and wildfires worldwide. Iridium, shocked minerals and unusual abundances of chemicals (*geochemical anomalies*) could also be found worldwide.[18,312,456,572, 919] A more detailed discussion of this truly catastrophic period can be found in Tome Three, and was not due to an impact(s) as has been claimed by some.[18,1224]

With regard to all life forms, no mass extinction was as great as this time in Earth history. Every animal 25 kilograms (100 lbs.) or more in weight, with the exception of those protected by water or mud, became extinct.[18,271,456] Here was the end of the dinosaurs, plesiosaurs, mosasaurs, pterosaurs, icthyosaurs, and numerous others. Thirty-six families of the ancestors of reptiles, mammals and birds (*Amniote*) died out. Pouched mammals, the marsupials, also decreased substantially.[137,456] Shellfish (*bivalves, gastropods, foraminifera, etc.*) and corals experienced extensive, widespread extinction. In fact, this was one of five of the most pronounced marine extinctions on record.[456,1001] The end of the Cretaceous was a truly catastrophic episode in Earth history.

Latitude restrictions are clearly apparent in this episode. Shellfish, for example, showed that 80% of tropical and 9% of other regions' gastropods, and 86% of tropical and 17% of other regions' bivalves, became extinct.[612] Plant extinctions were more pronounced between the 30° to 60° North Latitudes than elsewhere.[681] Those areas where the devastated vegetation is noted, and overlaid by a layer of fern spores, are in western North America and eastern Asia.[1224] Both of these areas were located near the present-day North Atlantic and Japanese Fields during the Cretaceous. In western North America there is latitude restriction at about the 40° North Latitude[1254], and cycads disappeared below the 40° latitudes.[681] For the most part, extinctions were highest in mid-latitudes and to a lesser extent at the Poles, and more pronounced in the Northern Hemisphere.

New life forms ushered in a new major era in Earth history, the Tertiary. Mammals, birds and smaller reptiles underwent an explosion of new species. A drastic increase in the number of families of four-footed animals (*tetrapod*) brought "modern" groups of frogs, salamanders, lizards, turtles, birds, crocodiles and mammals.[137,183,271] Almost all of the present-day insect families and many genera became well established.[183] The number of new families and orders of plants increased dramatically.[681] Some new species of plants only reached into the beginning of the next (*Early Paleocene*) period, and then became extinct; obviously mutants that could not survive. There was also an increase in leaf size and the diversity of deciduous trees in mid-latitudes. In fact, there was at least a three-fold increase in certain plants (*dicotyledon*) families in mid-latitudes.[173] All of the surviving life-forms became dramatically more diverse, particularly placental mammals in an era, the Tertiary, known as the Age of Mammals. This was the greatest increase in diversity in Earth history, just as the extinction catastrophe was the worst.

The Eocene/Oligocene Boundary is the last of the examples to be presented here. The Earth's magnetic field again experienced a reversal.[919] Sea level changed as ridges expanded lifting continental mass above water.[18] Mountain ranges pushed skyward in Europe and an inland sea withdrew from the Eurasian land mass making it a single continent. Isotopes were again offset, disclosing that the oceans and atmosphere were highly disturbed.[572] An abrupt cooling of the climate took place, which caused waters to cool and some ice formed.[18,312] Cratering, shocked minerals and iridium are also in evidence.[18,919] In fact, this is the period when the North American tektite strewn field was laid down.

Extinctions were not as dramatic as other periods, but still quite extensive. Shellfish (*marine microplankton, dinoflagelates, coccolithophorids, silicoflagellates, ebridians and planktonic foraminifera*) again underwent major declines in species and generic diversity.[1109] Twenty-eight families of four-footed animals (*tetrapods*) became extinct.[137] In North America, fifteen families of mammals died out and in Europe nine families disappeared.[1078] Again latitude restrictions dominate the extinctions.[612] In fact, North America experienced more extinctions, as well as new species, and is the area in which the tektite strewn field is noted; obviously it was a place that experienced more ionizing radiation.

Once again new species emerged. Mammals underwent another radiation of new types, such as the "modern" deer. Whales and dolphins (*cetaceans*) began to adapt to cool waters. Flatbugs (*hemiptera and aradidae*) evolved into modern types.[183] Flowering plants developed flowering in response to changes in daylight (*photoperiodicity*), and began widespread dormancy. More new species appear in North America and the Caribbean than elsewhere, and it is these areas that are covered with tektites.

These extinction-speciation events are only some examples of numerous others which are either less well known or not as dramatic. Certain horizons and boundaries are definitely sudden events. Yet, each that is known to be a sudden event was suspected previously, and so, detailed, high resolution studies were undertaken as a result. Simply looking at the number of types (*taxa*) of organisms against time does not necessarily disclose the suddenness.[822] This is because diversity increases through time, and often new species out number those that became extinct. Again, this is proof of ionizing radiation and FEM, and more detailed studies may disclose that additional sudden events have occurred.

One thing that must be noted is that through evolution and mass extinctions more complex systems emerge. Diversity is not lost as new and more numerous species emerge. Through time the biosphere has become more and more complex, which amounts to more stability, and at the subatomic and molecular levels chiral media are created. This scenario allows for greater energy flow.

CHAPTER 24

The Enigma of Life
(Origins of the Proto-cell and Complex Organisms)

One of the biggest mysteries in science is the origin of the first cell, called the proto-cell. Self-assembly processes in the proto-cell are the most important prerequisite for life, even more so than the formation of genetic material. If we are to consider that all life evolved from a single cell, then self-assembly had to appear first.[137,760,1343] Yet, even the probability of producing a protein from the earliest physical conditions was "essentially zero."[1343] Meanwhile, the oldest rocks that provide evidence of life demonstrate that cellular life had already become well established.[137,637,760] How did it get here?

24.1 The First Cell?

Many of the scenarios conjured to explain this apparent enigma have failed to produce viable results. Amino acids have been produced from what was believed to have been the gases in the primitive atmosphere (H_2, CH_4, NH_3 and H_2O), and an electrical discharge simulating lightning (i.e., the Urey-Miller experiment). However, few people realize that some of the chemicals that were used were derived from biological sources.

Any inorganic chemicals present on the early Earth would have become contaminated by sulfur gases. Meanwhile, the first conditions probably eliminated some gases (CH_4 and NH_3), because they decompose in the primitive light conditions, or time would progressively imbalance them (CO and CO_2 over CH_4, and N_2 over NH_3). The prebiotic atmosphere is not known and could have been strongly reducing, mildly reducing or non-reducing. Reducing conditions produce less amino acids in terms of both yield and variety. Many compounds necessary to life would have to be protected in a micro-environment, because no free oxygen existed to manifest the ozone layer, which would protect the cell from cosmic radiation. Oxygen was not in its free state until oxygen-producing photosynthetic bacteria and algae emerged.

Theories claim that self-replicating clays, with genetic information stored as a distribution of charges, and replicated by ion interactions between existing and newly-formed surface layers, were responsible. Others claim organic compounds other than nucleic acids, or minerals and organic compounds began the first cell. However, none has provided any substantiating evidence from either fossils or ancient rocks, nor experimental research.[637]

DNA and RNA are essential to life, but also appear impossible to form in pre-life conditions. The synthesis of the most important amino acids (*i.e., pyrimidines*) is itself problematic. Yet, the greatest mystery is not synthesis, but finding a substance or mechanism for attaching them to nucleic acids (*i.e., pyrimidine nucleosides*).

RNA was recently understood to have two functions (*enzymatic and catalytic*), and therefore, seemed to be a good candidate for life's origin (*due to both genotypic and phenotypic functions*). The objections to RNA beginning life are among two categories: (1) it is relatively inactive biochemically (*inept*), and (2) it is not a plausible pre-life molecule, because little was available on the primitive Earth. An article giving a detailed account of RNA's possible role concludes: "The most reasonable interpretation is that life did not start with RNA."[637]

In the case of DNA, like RNA, significant events in evolution had to have occurred before DNA was formed.[760] One such event appears to be the formation of the ozone layer, because DNA is damaged by ultraviolet light, and degraded by "cosmic" or ionizing radiation.[308,1061] Essentially the same conditions apply to RNA. Yet, life capable of photosynthesis was a necessary prerequisite for producing the ozone layer. Also, a micro-environment capable of both protection and allowing photosynthesis appears completely untenable. Meanwhile, all life requires RNA and DNA to supply genetic information, and in order to interpret genetic information.

The chances that life could originate from a pre-life environment are virtually non-existent. Most scientists examining the problem admit that it is unsolved problem. This is particularly so, because life's origin requires so many conditions that most scientists admittedly say were essentially impossible to achieve.[308,310,329,338,582,637,760,858,1061,1257,1343,1361]

For example, a Nobel Prize winning biochemist, and co-discoverer of the structure of DNA explains: "An honest man, armed with all the knowledge available to us now, could only state that in some sense, the origin of life appears at the moment to be almost a miracle, so many are the conditions which would have had to have been satisfied to get it going."[310]; p.88

A team of scientists completely aware of the problems of life originating on Earth attempt to explain that life began elsewhere and was brought to Earth by meteorites or comets. Yet, this only brings the whole unsolved problem elsewhere, as their own comments on protein enzymes alone indicate, because a single enzyme is one major feat alone.

"By itself, this small probability could not be faced, because one must contemplate not just a single shot at obtaining the enzyme, but a very large number of trials such as are supposed to have occurred in an organic soup early in the history of the

Earth. The trouble is that there are about two thousand enzymes, and the chance of obtaining them all in a random trial is only one part in $(10^{20})^{2000} = 10^{40,000}$, an outrageously small probability that could not be faced even if the whole Universe consisted of organic soup."[582]; p.24

Although these scientists admit they had been atheists all their lives, the high degree of order and complexity of life in face of the chances against it brings them to this conclusion: "Once we see, however, that the probability of life originating at random is so utterly minuscule as to make it absurd, it becomes sensible to think that the favorable properties of physics, on which life depends, are in every respect deliberate. It is, therefore, almost inevitable that our own measure of intelligence must reflect higher intelligences…even to the limit of God."[582]; p.141 & 144

Even more astounding is the endless breakup and reforming of the organic soup's complex molecules, leading to a small virus: $10^{2,000,000}$ to 1! [329] That is, the simplest organism forming by chance is ten followed by two million zeros to one!

What makes this even harder to reconcile is that there is not the long period of time we often envision for things to evolve. The first fossils of life indicate that about 75% of evolution had already been completed. An evolutionary scientist comments on this: "We are left with very little time between the development of suitable conditions for life on the Earth's surface and the origin of life. Life apparently arose about as soon as the Earth became cool enough to support it."[492] Again, the evidence discloses the organism-like characteristics of the Earth, where life—the Earth cell's components—come into existence early in the Earth's life history.

24.2 The Origin of Species

What complicates things even further is arriving at an acceptable theory that can explain the origin of all the species which have existed.

Charles Darwin offered his theory of Natural Selection which claimed that small alterations due to a changing environment and competition brought about the various species. Today, many philosophers of science and scientists question the validity of Darwinism, or its recent counterpart, Neo-Darwinism, as a scientific theory.[326,350,563,760,904,1195] Darwin, in his famous work, *The Origin of Species*, states the shortcoming of his own theory: "The geological record is extremely imperfect and this fact will to a large extent explain why we do not find intermediate varieties, connecting together all the extinct and existing forms of life by the finest graduated steps. He who rejects these views on the nature of the geological record, will rightfully reject my whole theory."[326]; pp.341-2

After more than 130 years of deciphering and exploring the geological record, Darwin's beliefs remain without evidence. Of the hundreds of billions of fossils uncovered we have found no intermediate varieties. In fact, the gaps between major groups of organisms have been growing for the most part undeniably wider and wider.[888] Furthermore, not only does evolution occur in spurts without "missing links," but it has little to do with an organism's ability to survive, and it is not random. "Natural selection still gives a distinct flavor to evolutionary studies, but it loses its place as the all-powerful creative force"[563], comment a group of scientists.

The misuse of chance and randomness has been referred to as the "folly of probability." Yet, science is based on cause and effect, which translates into the ability to predict an outcome. Prediction is the result of a determining factor or factors, not of randomness nor chance. "Selection appears to be impotent in producing major changes in the face of the molecular determination locked in the chromosomes and the cell."[760]; p.147. Many mutations have no effect on genetic expression, such as "silent" mutations and pseudo-genes. Other genes are not functional due to repression by a primary gene or genes that can alter function and substitute for other activity. Position effects, rearrangements and DNA repair mechanisms correct molecular mistakes. In short: "The molecular mechanisms which are an integral part of the organization of the chromosomes actually hinder the process of selection."[760]; p.151.

Most, if not all, of what we know about life and its history argues against such a simplistic view of the origin of species, as is suggested by Darwin. Bacteria, one of the simplest life forms, and humans have identical biochemical pathways. A protozoan, a single-celled microscopic organism, and a human have the same cell organelles. Abrupt explosions of new species are followed by long periods with no change. Often these new species have no ancestors at all, such as the first appearances of invertebrates and vertebrates. Recent experimentation has disclosed that the quantum level is not random nor chaotic, so species' origins could not be due to random quantum processes either.[163-165,325,329,543,904,966,1125]

In fact, the fossil record proves it could not be random. Molecules such as DNA are in a continual process of rapid exchange of energy (*quanta*) with their surroundings. Yet, the non-random features of living things have produced an ever developing hierarchy of higher orders of structure and function. Evolutionary history constitutes an internally ordered process that is committed to certain general lines of development.[329,543, 904] In a book titled the *Cosmic Blueprint* the author comments on shuffling:

> "A highly ordered sequence of cards will almost certainly become less ordered as a result of shuffling. In the same way, one would suppose that random mutations in biology would tend to degrade, rather than enhance, the complex and intricate adaptedness of organisms. This is indeed the case, as direct experiment has shown: most mutations are harmful. Yet it is still asserted that random 'gene shuffling' is responsible for the emergence of eyes, ears, brains, and all the other marvelous paraphernalia of living things. How can this be? Intuitively one feels that shuffling can lead to chaos, not order."[329]; p.109

Gaps punctuated by sudden explosions of new species demonstrate that the Universe works towards more complex and ordered systems that are not random. Darwin himself admitted he had absolutely no hard empirical evidence of any major evolutionary transition. In his letter to a friend, for example, he states: "Imagination must fill up the very wide blanks."[325] No one has ever found a missing link, and imagination is about all that can fill the blanks. Even the first bird, archaeopteryx, is a full-fledged bird when it emerges, and was as capable of flight as are modern birds.[338] In fact, archaeopteryx was thought to be the direct line between reptiles and birds, but it is now known to have been preceded by modern bird fossils. An enigma to Darwin, as well as modern evolutionists, was the sudden emergence of new species.

One such episode was the sudden appearance of life in the earliest fossil bearing rocks in what is often called the "Cambrian Explosion." In these rocks one can find a multitude of highly complex creatures with no ancestors, but they represent every major group (*phylum*) of animals. Here was the first fossil rocks and 75% of evolution had already taken place.[1195]

One formation deposited at this time, the Burgess Shale, gave paleontologists a shock when groups of animals (*invertebrate phyla*) were found that defied all efforts to link them with known groups, and "whose existence had not even been suspected."[867] The transition to the first mammal or mammal-like reptile is still an enigma. Such phenomenon is an almost universal picture of the fossil record, as the author of the book, *Evolution: A Theory in Crisis* states: "And as far as the individual defining characteristics are concerned, one could continue citing almost ad infinitum complex defining characteristics of particular classes or organisms which are without analogy or precedent in any other part of the Living Kingdom and are not led up to in any way through a series of transitional structures."[338; p.107]

Specialized organs also make sudden appearances in the fossil record. Consider how the eye with its optic nerve, pupil and lens appears fully functional in a number of Cambrian creatures. Eyes are found on the squid, fish and trilobites at that time. No evidence can be found that the eye gradually formed from, say, a light sensitive spot on the skin. Trilobite species alone had such distinctly different eyes that they would have had to evolve eyes separately some 30 or 40 times.[1195] A scientist asks an important question regarding this enigma: "What are the chances that just the right sequence of purely random mutations would occur in the limited time available so that the end product happened to be a successfully functioning eye?"[329]; p.111. A similar question could be asked about any complex structure, such as the ear or the brain. They could not develop slowly bit by bit, because there is no advantage to having an ear that cannot hear, an eye that cannot see, or a brain that cannot perceive.

Consider the enormous number of random mutations required to bring about the human brain and its complexity. What are the chances of such a structure originating out of pure randomness? A scientist comments on the human brain:

> "In terms of complexity, an individual cell is nothing when compared with a system like the mammalian brain. The human brain consists of about ten thousand million nerve cells. Each nerve cell puts out somewhere in the region of between ten thousand and one hundred thousand connecting fibres by which it makes contact with other nerve cells in the brain. Altogether the total number of connections in the brain approaches 10^{15} or a thousand million million. Numbers in the order of 10^{15} are of course completely beyond comprehension. Imagine an area about half the size of the USA (one million square miles) covered in a forest of trees containing ten thousand trees per square mile. If each tree contained one hundred thousand leaves the total number of leaves in the forest would be 10^{15}, equivalent to the number of connections in the human brain!"[38]; p.330

Other concepts used in evolutionary theory also make random mutation and adaptation questionable in terms of selective advantage. Altruism, giving up one's life or chance to mate for the good of the species, contradicts selective advantage. After all, it stands to reason that the more a given species is involved in mating the more likely will the species successfully compete for its place in the environment. Greater numbers of individuals would allow for greater success. Yet, there are numerous examples of altruism.

Examples of altruism are found among mammals, birds, and insect communities. For example, there is the White-Fronted African Bee Eaters who postpone opportunities to breed in order to aid others raising a brood.[405] Some bees give up their lives to protect the hive, but bees have been one of the most successful insects since the origin of flowering plants. Aggressive behavior might be considered a selective advantage, but it is not a built-in behavior (*not endogenous*). It involves a complex set of variables that depends on various features of the physical and social environments. For the most part animals will not initiate a fight if their needs are met or no threat exists.[593] Behavior like this was always thought to be genetic and evolved from competition. It seems we have gone too far in trying to explain everything in the terms of the dogma of natural selection.

A philosopher of science warns us of not being trapped into this mode of thinking:

> "The stability achieved, the semblance of absolute truth is nothing but the result of an absolute conformism. For how can we possibly test, or improve upon, the truth of a theory if it is built in such a manner that any conceivable event can be described, and explained, in terms of its principles? The only way of investigating such all-embracing principles is to compare them with a different set of equally all-embracing principles, but this way has been excluded from the very beginning.

The myth is therefore of no objective relevance, it continues to exist solely as the result of the effort of the community of believers and of their leaders, be these now priests or Nobel prize winners. Its 'success' is entirely man made."[417]; p.179

This is no secret even to those who write on the subject.[809,1053] Such as the foreword in a book titled *Macroevolution*, "We might wonder whether the doctrine of evolution would qualify as anything more than an outrageous hypothesis."[1179]; p.2.

24.3 A Different Timescale

By witnessing the effects of ionizing radiation and the failure of Darwinism or Neo-Darwinism, we are left with other considerations. The effects of ionizing radiation will also be addressed in Tome Three. Among the observations presented are calcium-dependent mutations, mummified dinosaurs, and radioactive bones, among other fossil evidence. The presence of iridium, and the process of mass extinction and speciation, including its geographic restrictions argue in favor of episodes of increased ionizing radiation. There have been some observations of increased radiation during a magnetic reversal, and at other times.[711,714,981] These observations bring into question our dating methods, which are based on a fairly constant influx of ionizing radiation from space, and the protective effects of the ozone layer, geomagnetic field and magnetosphere.

Half-lives of elements are utilized in dating the accepted evolutionary scheme, but decay rates or half-lives are *not* constant. The inherent limitations of the early instruments that were used led to erroneous conclusions. Experimentation has disclosed that variations in decay rates are produced by changes in pressure, temperature, chemical state, electric potential, and the stress of single molecular layers. All of these phenomena were observed to have changed, as noted in the fossil and geological records, during times of mass extinctions. Ionizing radiation, electrostatic fields, microwaves, a reduced and/or reversed magnetic field, and pulsed radio-frequency fields also greatly increase decay rates. Without all these influences it is still apparent that "half lives are not constant."[367]; p.68.

The catastrophic nature of these events has obscured the fact that they have taken place by the overturning, mixing and deposition of sediments, shifting inland seas, increased precipitation and so forth. In the overall picture the events and evolution itself have taken place on a much shorter timescale than has been envisioned. A more detailed discussion of this can be found in Tome Three.

Numerous observations support this view. For example, sediments laid down have been shown to have taken one percent of the time normally allotted to them in terms of radiometric dating.[1060] The geologic record is not complete, there are gaps where spans of time are missing, and no where on Earth have we found a completely continuous record.[888,1195,1264]

A look at the recovery of plant life after the Late Cretaceous catastrophe (*K/T Boundary*) shows how off the dating system might be. That record discloses: "an ecological trauma followed by a steady recovery that mimics normal serial succession, but over a time scale of perhaps 1.5 million years."[173]; p.21. Normally succession takes about 150 years.[696,1086] It seems quite ridiculous to consider succession, the gradual changes from ferns to forest, taking so long, yet our dating system forces us to such a conclusion.

Making predetermined theories and then applying them is a short route to dogma. We must look at the facts, like so many philosophers of science have stated, as if there were no theory, as fresh new insights to a possibly overlooked truth. Radiometric dating is based on the assumption that decay rates are constant, and that there are no unique events causing increased radiation (a closed system vs what is actually an open system), both of which are evidently false theoretical "visions."

A scientist comments on the problem of being taught to see things a certain way:

> "The literature in the field illustrates forcefully that our perspectives, even as scientists, are unavoidably constrained by our disciplines. I agree with Kant that the human mind is organized in ways that restrict our perceptions. In addition, I think that our training in a specific discipline (as well as the proclivities that led each of us to select a particular discipline) imposes on each of us an additional mind-set, which we take as self-evident."[909; p.243]

We need to open our minds to the fact that the Universe is fundamentally biological and that life is a physical law that goes beyond other physical laws.[582] This is difficult for scientists, in general, to accept, because it means that the Universe is teleological or purposeful, and science has focused on trying to explain everything as random events. In fact, life itself even brings us to the following question.

CHAPTER 25

Do Things Really Always Fall Apart, Decay and Become Chaotic?
(Is Entropy a Mutable Law?)

Entropy is the scientific term used to describe the amount of randomness or disorder in processes and systems. In the physical sciences the concept is central to the descriptions of heat-transfer properties, or the thermodynamics of molecules, heat engines, and the Universe as a whole. Also known as the Second Law of Thermodynamics, entropy implies that all processes operate at less than 100% efficiency when energy is transferred or utilized. In the classical sense, entropy is a measure of that portion of energy unable to produce useful work (or the unavailable energy).

Another aspect of entropy is an object's or system's tendency to move toward greater disorder, decay or chaos as time passes. For example, a house will eventually fall to pieces if not maintained. Along the "arrow of time" this process is said to be irreversible. That is, the house will not spontaneously reassemble itself.

Statistical thermodynamics states that entropy is a measure of microscopic disorder or the uncertainty associated with the microscopic state. In information theory it is a measure of uncertainty or lack of information. In short, entropy states that no process in Nature involving chemistry or physics will occur spontaneously without some loss of energy and order.[1029,1032,1113, 1237,1298] However, many facts call these assumptions into question as a universally applicable law, especially with regard to biological systems.

25.1 The Second "Law"?

Entropy is said to be the "premier law," "supreme metaphysical law," or "ruling paradigm" over the next period in the history of science.[1032] That is, every hypothesis or theory must be tested in terms of its adherence to the Second Law or the theorist is in jeopardy of being labeled a crackpot by the scientific community. However, in spite of the insistence that entropy is the ruling law no one has proven it. In a recent article a scientist comments: "The fact remains that the irreversibility of entropy—the H-theorem—has not been proved, and, until it is, we cannot be certain of the Second Law's reliability."[1126]

Some scientists indicate that entropy is a human-measured or anthropometric concept. A physicist comments on its uncertainty: "Even at the purely phenomenological level, entropy is an anthropometric concept. For it is a property, not of the physical system, but of the particular experiments you or I choose to perform on it."[624]

It might be better to say that it is the entropy of the data or that it is relative to the observer's data.[336] Another physicist tells us that entropy is a subjective viewpoint which only measures "our degree of ignorance as to the true unknown microstate."[624] Furthermore, entropy is a statistical law and is, therefore, true only on the average.[336]

What makes it even more uncertain is that it involves the necessity of making a record of an irreversible change. With subatomic or quantum particles this requires stopping a particle at some point to make a record, which brings about the uncertainty of what that particle would have done if it were not stopped.[329]

Regardless of this ambiguity, physicists and chemists still verify or quantify any discipline in a way that relates to entropy. Furthermore, a number of observations contradict the Second Law of Thermodynamics. Self-gravitating systems become hotter (i.e., gain energy). Gravity inducing structure, energy and organization runs counter to the Second Law's greater disorder, decay with time, and energy loss in work. Meanwhile, the structuring tendency of self-gravitating systems is a fundamental principle of Nature. In fact, the Universe itself is progressively becoming more self-organizing, and *not* more disordered and chaotic.[329]

Another contradiction to the entropic phenomena of disorder and randomness increasing through time is life itself. The evolution of life displays a record of increasing order and complexity. More and more species exist the closer we reach the present. The more specialized structures, such as the primate brain, show up late in the fossil record. If everything is supposed to become more disordered, how did life become more ordered and complex?

"The emergence of life is compatible with thermodynamic principles but not required by them. As we have seen..., this is to be expected from the nature of thermodynamics, which deals with the limits of what is possible rather than with a determination of what will necessarily occur. We still must answer the question, what led to this emergence of order out of chaos?"[909]; p.255

Those scientists who write on the subject of life's ability to overcome the typical traits of entropy are often not open minded about their approach. Consider this statement written in a book titled, *Entropy, Information and Evolution*, "Attacking the autonomy-of-biology citadel is therefore an implicit theme of this volume."[1328]; p.140. The point of making such a statement or taking such a stance is the firm belief that entropy is an immutable law that everything must adhere to, because it is predetermined to be a "universal law," but is it?

25.2 <u>Life's</u> <u>Non-Chaotic</u>, <u>Ordered</u> <u>Complexity</u>

Meanwhile, the biological world brings entropy, as a universal law, into serious question. Life's energy flux does not conform to classical thermodynamics.[1086] The evolution of life on Earth shows decreasing randomness, and increasing order and complexity, which is the opposite of entropy. This is especially true if we consider coevolution. One author comments on life's strong organizing principles: "Nature is unusually efficient at conquering its own Second Law of Thermodynamics and bringing about organized complexity."[329]; p.151. Quite some time ago one scientist even coined the term "negentropic," while presenting a great deal of evidence that life overcame the Second Law.[1090] Biology integrates chemistry and physics at a "higher hierarchial level" than either chemistry or physics could alone, therefore thermodynamics must accommodate biology, not the other way around.[629]

For quite some time, philosophers and scientists have argued that the laws of physics as presently conceived are wholly inadequate to explain complex organized systems, such as life. Biology needs to be better understood in its own right.[329,909] The new set of rules or paradigm must emphasize the collective, cooperative and organizational aspects of Nature with a perspective that concentrates on synthesis and holism, rather than the reduction or analysis of its finer physical and chemical makeup. The unifying of the sciences requires a broader definition of law than has been entertained in physics, and in which quantum mechanics is only a limiting force, not an all-powerful one. One scientist writes:

> "Quantum mechanics refers to the results of measurements on collections of identical systems, i.e., systems which belong to homogenous classes. It has nothing to say, at least in its usual formulation, about regularities in inhomogeneous classes. But in biology one is interested in regularities in different but similar organisms, i.e., inhomogeneous classes.

Quantum mechanics places no restriction on the existence of regularities of that sort. Therefore we are free to discover new, additional principles which refer to members of such classes."[329]; p.180

Another scientist makes a statement, in an article titled "Observations on Evolution," that reflects the underlying problem: "In the final analysis, my critique of them and others who are trying to reduce evolution to an entropic or informational phenomenon is not that they are too bold. Rather, I think biologists are being insufficiently bold. Instead of trying to make evolution a consequence of thermodynamics and/or information theory, they need to develop it more fully in its own right."[909]; p.260. Such is the purpose of this volume.

Consider photosynthesis on this level. Photosynthesis is nearly 100% quantum efficient.[341] In fact, it is so efficient that studies have been conducted on synthesizing photosynthesis for use as an energy source.[755,862] Unlike the chemically synthesized photosynthetic reactions, plants change to maintain their high quantum yield. For example, an environment with low light intensity reduces the photosynthetic rate, but dry matter yields are less affected (proportionally) by leaf expansion and larger photosynthetic unit size, both of which are changes that increase the quantum yield.[92] Certain molecules (*porphyrin, quinone and carotenoid*) shift in such a way that they conserve energy and reduce entropy.[132,755,862] The full scope of the photosynthetic world can be appreciated when one considers that all living things derive their energy from photosynthesis, and more human energy sources are the product of past photosynthesis (i.e., fossil fuel, wood, etc.).

An important molecule, the porphyrin-ring structure, in photosynthesis and respiration has remained stable throughout the history of life. It is essential that it did, because it is a key component that has basic but diverse functions in the cell. Two essential functions provide the energy used in maintenance and renewal: one is photosynthesis and the other is respiration. The remarkable aspect is that the porphyrin ring has been preserved as essentially the same structure in all its derivatives (algal chlorophyll, bacteriochlorophyll, haem and vitamin B-12), and has been a key factor in cellular evolution.[670] How can increasing disorder and randomness, entropy, produce such essential stability throughout all the trials and tribulations ingrained in the history of life?

A number of characteristics of the living world run counter to entropy. Life originates spontaneously, unlike most other physical things in the world, which if left alone tend to disintegrate. If an organism becomes extinct a new species that is more complex takes its place. If a natural system is perturbed entropy increases, but the system reacts by returning to a state of least entropy. Both evolution and succession progress towards more order and complexity, and least entropy and randomness. Observations make it obvious that a species interacting within the boundaries of an ecosystem assumes a state of least entropy production. The non-living components of Nature express higher entropy and less order than any living cell, and this is the problem with the Second Law, it is based on the non-living world.[329,589,859,1308]

Ever since the middle of the Fourth Century BC, and the competition between Aristotle and Democritus, scientists and philosophers have attempted to explain away the order in life. Most have tried to reduce it to the mechanistic or atomistic level, but have continually failed to explain the order. A geneticist reveals the inadequacy of molecular physics as it pertains to life:

"All attempts to establish a mechanistic interpretation were frustrated by the following facts: (a) the inadequacy of physical laws to explain biological finalism; (b) the crudeness of physical schemes for such fine and complex phenomena as the biological ones; (c) the failure of 'reductionism' to realize that at each level of integration occurring in biological systems new qualities arise which need new exploratory principles that are unknown (and unnecessary) in physics."[859]; p.3

Two recent attempts, dissipative structures and information theory, have also been inadequate. Life as dissipative structures tries to explain the increase in order of life by increasing entropy in the environment. By releasing mass and energy into the environment dissipative structures have the capacity for self-organization. Entropy is locally circumvented for an increase globally.[629,1331]

Meanwhile, dissipative structures can evolve in two directions: least or highest entropy. It does not explain why genetic systems should produce more order, just that it is possible.[280] Much discloses that this dissipative-structure approach is not capable of explaining the order. For one, life and the environment have evolved in a mutually reinforcing way that has brought stability to both (Gaia Hypothesis).[778] Earlier in this volume we witnessed how life reduces entropy in its surroundings, not the opposite. The presence of life produces more balanced and stable soils, which produces more nutritious and healthy foods. This in turn lessens mental and physical diseases and disorders, including social disorder and antisocial behavior. There is a more balanced environment with life, which was examined in terms of air ions, ecosystems and even the stability of the Earth's magnetic field and electrical environment. The following chapters, Volume Two, and *In Defense of Nature—The History Nobody Told You About* that life produces even more stability, which can even be observed in human history and the Universe as a whole. Aside from our lack of information on human participation (*and the Uncertainty Principle, and EPR Paradox*), a scientist comments on our lack of knowledge: "We do not know nearly enough about organisms as chemical systems to be able to tell whether their functioning as dissipative structures has anything to do with the creation or even maintenance of inherited order."[280]; p.231

Information theory or communication theory has been applied to explain the order in biological systems, but also fails. Information theory is a branch of probability theory that has been provided to measure the flow of information from an information source to a destination. In the evolution of biological order the gene is said to be the source of information, which through mutation and natural selection produces new species, which is the destination. A scientist comments on its failure: "Communications theory allows us to make qualitative and (in principle) quantitative measurements of biological order, but it cannot explain the origin of the order it measures."[280]; p.233

Changes in the environment could more easily produce disorder in the information (genetic heredity), yet we find always increasing order. Consider a photocopy being copied, and then the copy being copied, and so on, being continually recopied. Surely its information becomes less and less clear with each copy. A certain sequence of cards that is ordered and shuffled will certainly more likely become disordered than more ordered. Yet, evidence discloses that genes are shuffled and always lead to more order. The reason is clear: "Living beings prevail at a number of levels of organization above the molecular one. The existence and the reproduction of these levels imply a state of improbability, a kind of 'information' which cannot be dealt with by the current theory of information."[779]; p.92

Evolutionary history conflicts with the Second Law in a number of ways. The number of species, structural complexity, diversity of species, and acceleration of energy flow have all increased with time.[629,1086] The increase in complexity is measured in informational, structural, energetic, and flow component terms.[1086] Diversity or more species (*heterogeneity*) creates stability. Conversely, simplicity or less species (*homogeneity*) can be lethal. At the microscopic level many interactions are essential to the stability of complex organic molecules that depend on the more complex (*noncovalent*) chemical bonds.[629] Cooperative (symmetrical) interaction is essential for coevolution, which is abundantly displayed in the fossil record.[629,1328]

Stressed systems display an increase in energy or material cycling as a fundamental property of living systems. Stress causes the system to revert to a lower state of efficiency, then to a more efficient state afterward.[1086] Increased amino acid cycling occurs in stressed biological communities, and this cycling involves the basic, essential amino acids the most. It is these amino acids that are the most crucial to forming the nucleic acids and proteins essential to life.[1086,1328] This is one of the reasons why we see times of stress—mass extinctions—followed by more stability and complexity—blossoming of speciation (mass origination), and this is true throughout geologic time. Likewise, this is one of the reasons why throughout human history wildlife systems have continually reestablished themselves (see *In Defense of Nature—The History Nobody Told You About*).

Disturbed biological systems also show other responses that ensure least entropy production. Population dynamics display a constant size and frequency distribution following disturbance. After the displacement there is a return to former density and size without fluctuation. This occurs in all types of organisms and is internally mediated, because it can take place with single species.[629] Again the tendency is towards stability, order, structure and diversity, not decay, randomness, disorder and breakdown.

Diversity also increases energy flow by reducing internal resistance (*and damping*). Reduced resistance (*damping necessitates finer-tuning*) represents specialization, while diversity overcomes resource limitations. In the overall picture this allows the tracking of energy fluctuations more closely making energy acquisition in a fluctuating system more available.[629] All of these observations are in contrast to the increasing disorder and breakdown of entropy, and cannot be explained by information theory nor dissipative structures.

Natural selection, which is usually the most pervasive explanation for evolution, also fails to account for these observations. Adaptation to a changing environment can just as easily or more easily produce disorder.[280,329] A scientist examining evolution on thermodynamic grounds concludes: "The extreme functional reductionist, who maintains that all biological order can be explained by adaptive value, is wrong." [280]; p.235. Another makes a broader statement: "The theory of evolution as an entropic phenomenon…seems to be conceptually flawed in ways that make it untenable."[909]; p.259. As we witnessed earlier, evolution is not random with regard to mass extinction and speciation (*origination*).

As we will delve into further, it is known that quantum processes, the most fundamental level of the physical world, are not random either.[143,428,966] The molecules of genetics, such as DNA, RNA and protein, are in continual process of rapid exchange of quanta of energy with their surroundings. As a result, the current laws of quantum theory may be leading to seriously wrong inferences, because they do not take into account biological systems at a fundamental level. A scientist expresses concern about this: "If quantum processes are not random, then the whole basis of Neo-Darwinism is undermined."[32]; p.156. Recent experiments disclosed that quantum processes are *not* random.[143,428,966]

There even exists the possibility that evolution is a reversible process, and entropy is a history of unique changes or irreversible processes. Throughout the fossil record, particularly around the periods of mass extinctions, there are spans when particular organisms (species, clades, genera, families, etc.) disappear for a length of time, and then reappear again later.[137,612] This is called the "Lazarus Effect," after the biblical character who died and was resurrected.

A striking example is the time when virtually no insect fossils can be found for most of the entire Cretaceous period. After the end of the Cretaceous and the demise of the dinosaur, insect fossils return in full force along with a striking increase in flowering plants. This has been described as anything from extremely "bad luck" to an incomplete fossil record. But, could it be that evolution has locally and periodically undergone a reversal? Are there times when conditions cause genetic material to revert to lost codes? These questions may seem utterly ridiculous to present-day paleontologists and biologists. However, they do need to be asked if we are to be an open and scientific society, because we are assuming all systems show irreversible processes by insisting entropy is a universal law, and therefore, do not see it when it is observable.

Furthermore, forest plant communities exposed to ionizing radiation do display reversible processes related to the evolution of life.[346] So, experimentation has shown it is possible, but we do need to explore the possibilities more. The answers will clarify the nature of life: "Entropy is that which increases in irreversible processes. If a function does not have this property, it is not an entropy."[1328]; p.144. In this regard a certain quote deserves repeating:

> "The literature in the field illustrates forcefully that our perspectives, even as scientists, are unavoidably constrained by our disciplines. I agree with Kant that the human mind is organized in ways that restrict our perceptions. In addition, I think that our training in a specific discipline (as well as the proclivities that led each of us to select a particular discipline) imposes on each of us an additional mind-set, which we then take as self-evident."[909]; p.243

Succession, the progressive process of a habitat changing from weed to grass land to forest communities, also displays characteristics that are not typical of entropy in the physical world. The short-lived pioneer species are replaced by long-lived species ensuring energy is reduced and more evenly (symmetrically) distributed as biological mass or biomass accumulates. As with evolutionary history, the number of species, diversity of species, structural complexity, and acceleration of energy flow increase with time.

The increase in energy flow is relative to the biomass, so that in the course of succession there is a decline in production (i.e., overall reproduction) relative to biomass (production to biomass ratio). Thereby, flow rates accelerate and decelerate simultaneously, allowing more efficient energy utilization (energy equipartioning). This acceleration is time-varying, which is a more efficient acceleration (*i.e., Faraday's Law applies*).

This allows a state of near equilibrium (homeostasis) that permits interrelationships (*interspecific organization*) to develop towards a common goal that eliminates potentially lethal fluctuations (*excessive heterogeneity or homogeneity*). Even a partial decoupling of age and size brings about a size uniformity within the group in such a way that could not be predicted by individual behavior.[629] Such increasing complexity, order, energy utilization, and structure display a tendency toward least entropy.

An interesting distribution of diversity displays geographic restrictions that relate to the new model of the Earth (FEM).[629] Diversity decreases with increasing latitude, which translates into more energy flow along the equator, and less the more one moves towards the poles (*i.e., a bioelectric/bioelectrostatic time-varying particle accelerator*). The more diversity there is, the more continuous variation there is (i.e., it is time varying). These variations in energy flow and its time-varying characteristics are employed to produce particle accelerators. As we will see in Volume Two, other celestial objects also display similar energy flow distributions.

25.3 <u>Quantum</u> <u>Mechanics</u> <u>and</u> <u>Life</u>

At the most fundamental level life is encoded quantum mechanically, and interacts with quanta in the environment. All the important processes of molecular biology are quantum in nature, such as the stability of genetic information. In order to fully understand life or reality we must also comprehend the quantum level. So the question arises is disorder, randomness or chaos a property of the quantum state?

Generally speaking, chaos is a type of randomness that appears in certain physical systems and is "built-in" (*intrinsic*) to the system, rather than caused by external influences. Scientists have found chaos in a number of physical systems, from something as large as the orbit of Pluto to something as small as a pair of atoms. However, does it occur at the quantum level?

The answer shocked most physicists, because the random behavior of chaos, that is frequently observed in classical (Newtonian) physics, does not exist at the quantum level. Quantum systems behave peculiarly when their classical counterparts are chaotic, but there is no evidence of real chaos at the quantum level. This lack of chaos at the quantum level has been demonstrated in a number of experiments.

Another surprise to physicists is that when the quantum level is offset by classical chaos the reaction is *not* irreversible. At the quantum level another property of entropy does not exist. A quantum system will return to its origin behavior. What quantum systems do is "mimic" chaos that is present in classical systems.[143,428,966] One physicist comments on the observations at the quantum level: "For, given that quantum mechanics is universally accepted as our most fundamental and all-inclusive description of Nature, the undeniable existence of chaos in Nature clearly implies that chaos must also occur in quantum mechanics. And yet the evidence exposed...indicates that it does not!"[428] Another describes this finding's implications: "Theory predicts that...quantum mechanics will eventually suppress classical chaos."[143] The question arises, how did the classical systems become chaotic?

Chaotic systems have the property of being essentially unpredictable. However, this may simply be our ignorance of what is causing the chaos. It could simply involve the progressive weakening of the properties embodied in the initial conditions.[336] However, the evidence reveals a more orderly situation than merely a disintegration of initial conditions.

Along with entropy there is clearly another "arrow of time" that is equally fundamental, if not more so, but it is "mysterious." The Universe is also progressing with a steady growth of structure, organization and complexity. This "optimistic arrow" has been somewhat ignored by scientists for the more theorized "pessimistic arrow" of entropy.[329] Yet, there is no doubt that complexity, structure and order is manifest, and life is one of its main goals. A molecular biologist describes this: "Carbon is the key element of all organic molecules mainly due to its ability to build chains. Carbon, like all other elements, is supposed to be derived from hydrogen, and hydrogen is built of protons and electrons. Thus the conclusion seems to be inescapable. That life has no beginning, it is inherent to the structure of the Universe."[670]; p.1047

25.4 A New Physical Law

The self-organization and complexity in Nature reveals that there is a wholly new law above the known laws of physics. We are on the verge of discovering ways of thinking about Nature that depart radically from traditional science. However, it is not in conflict with previous observations, but involves a reinterpretation of those observations. It is not the lowest levels, not the quanta and their fields and particles that unify the sciences, but a hierarchy of organization and complexity based on life itself. Life cannot be deduced by studying the thermodynamics of physical systems, because biological organization and biological phenomenon cannot be explained by even the levels of biology, much less physics and chemistry.

The practices employed for examining the quantum level are flawed with inaccuracies that can easily lead us to illusionary conclusions. More than half a century ago it was realized that one cannot measure both the position and momentum of a particle simultaneously. For example, an electron cannot have a precise location and motion at one and the same time. Whenever one quantity is more strictly measured the other becomes less precise. Furthermore, the process of measurement itself disturbs the system being measured. This is known as the Uncertainty or Indeterminacy Principle.[163,166,329,551]

Other complications arise when observing the quantum world. When a quantum system is not being observed it evolves deterministically (*i.e., according to the Schrodinger Equation*). Yet, when the system is being inspected by an observer the system's behavior changes. If a measurement is made, a quantum system evolves differently than when no one is looking. What it all comes down to is the theory of measurement may not be a meaningful concept.[329,1101]

Another confounding effect is that the quantum world is non-local (*EPR Paradox*). In one respect, the apparatus we construct to measure a quantum system changes the behavior of that system. To put this another way, we are not just observers of the physical world, but participators that change what goes on.

In another sense, the fate of any given particle is completely linked to the fate of the cosmos as a whole, because in reality it is interwoven with the rest of the Universe. A form of inseparability exists in the sense that there is a connection between events for which any form of cause and effect is not known. Because we do not take into account our participation, our conclusions about the physical world may simply be an illusion. A scientist comments:

> "The two complementary aspects of reality reveal themselves not only in acts of observation and measurement, but also in other actions that we might be involved in. A free will—if it exists—acts in one direction only: toward the future, otherwise we would merely be passive spectators as in classical physics, rather than actors as is admitted by quantum physics. Indeed, both aspects of reality are essential for us to be regarded as active participants rather than mere spectators. We

may therefore distinguish two sides, or aspects of the act of measurement: An 'instructive' and a 'destructive' side. The instructive (or constructive) side increases one's information, while the destructive side produces an uncontrollable change in the values of other, complementary, observables. It decreases information and produces entropy."[1011]; pp.263 & 266

Here is the heart of the problems with our perceptions of entropy when applied without discretion to components of the physical world. We destroy life to make the apparatus that is used to measure the quantum world. At another level we witness life, including its past evolutionary history (*retrodiction*), as capable of overcoming entropy in at least some respects. Our participation—the physical world without a continuous, maximum abundance of life—produces uncontrollable changes in "other complementary observables." This is what produces much of what we define as entropy. Our destruction of life prevents the operation of life's ability to stabilize the physical world and its ability to reduce, or even overcome, entropy.

Because of the different kind of perceptual and interpretative processes characteristic of life, living organisms can possess knowledge that is more detailed in certain respects than is the knowledge designated by the quantum theory. Higher discrimination and selectivity in a process that makes contact with Nature as a whole is characteristic of living things only. This ability can be compared to a process that makes contact with individual atoms—relative to one that makes contact with the macroscopic aspects of a system only—thereby creating order at the quantum level by responding to fluctuations in the system as a whole. One part of the biosphere "knows" what is happening to the other part. Moreover, just as the eye developed to perceive objects at a distance, so too can the Earth's life system as a whole react at a distance to perturbations of the Earth as a planet and those objects reacting with the Earth in the Universe.

Furthermore, life's ability in this sense would also extend back to biological phenomena and make it more efficient, and thereby, the possibility of ending entropy could exist. It is not merely the so-called fundamental particles of modern physics that are "really real," but rather the higher level laws imposed by biological phenomena. We must seriously consider the non-local aspects of quantum physics in biological phenomena as part of the cosmos and treat it as a coherent whole. To put that in the form of a question which will be addressed: Is entropy a product of human participation and does it exist in a physical world that has the most abundant life system possible?

Entropy is a theory of probability and "probabilistically defined concepts are flawed because the properties of the system and those of the observer can become confused."[1086]; p.113. To put that another way: "A system may not be separable from distant features of its environment and may be non-locally connected to other systems that are quite far away from it."[165]; p.3. The bottom line is that the concept of measurement is always rooted in the classical world of familiar experience. We see disorder and decay never questioning its true origin, and insist that it is a universal, immutable law.

All processes at lower levels of hierarchy are controlled by and act in conformity to the laws at higher levels. Some have referred to this as "downward causation," and "the strong cosmological principle," among others. It involves a single universal law governing processes that produce order, and order is generated by several hierarchically linked processes. At the top of this hierarchy is life, which is unusually efficient at conquering its own Second Law of Thermodynamics, and bringing about organized complexity and stability.[329,726]

Recent studies of quantum chaos suggest this hierarchy and the role of human participation in causing disorder. A scientist reviewing the experiments describes the unexpected:

> "What the quantum system does is 'mimic chaos.' The classical system remained spread out, but the quantum system

returned to its starting point. Although the quantum system had mimicked the classical system and looked chaotic, it was not random and had not lost track of where it had been. In this view, quantum systems mimic chaos, but they do not have the randomness or irreversibility that mark classical chaotic systems."[966]

What this means is that quantum systems do not have the randomness, disorder and irreversibility of entropy unless they are disturbed by a chaotic classical system, such as human participation in the physical world. If humans introduce chaos into the system "quantum mechanics will eventually suppress classical chaos."[143] And this statement better be taken seriously, because all the instability, including 'natural' disasters and growing world tensions, is the quantum system suppressing our creation of chaos and entropy.

Consider this from another point of view which relates to coupled oscillators. The idea of coupled oscillators was discovered by a physicist who invented the pendulum clock, and found that if two clocks were mounted in a common support they tick in unison. Such cooperative vibrations are very familiar in physics, and occur in systems like crystal lattices where each atom acts as a tiny oscillator bringing about organized collective motion. Biological compounds are composed of a crystal structure in lattices.[29] Human activities that destroy life inhibit the formation of coupled oscillations in the biosphere as a whole. Furthermore, organisms are also resonators, which amplifies the energy flow more so than mere coupled oscillators.[629] What this comes down to is a world filled with a maximum abundance of life is a *totally* different physical cosmos, not just the immediate environment.

Studies of energy flow in life systems both in the present or future (prediction) and past (retrodiction) disclose this hierarchy of life.[629,1101] The greater the variety of (*heterotrophic*) interactions, the greater the energy flow. The acceleration of energy increases up to the point of maximum diversity and then declines when a dominant species, such as humans (*homogeneity*), takes over. Evolutionary history reveals the ultimate ascendancy of a trend towards increasing diversity and energy flow, and more energy flow means less entropy.[629]

25.5 Ethical Responsibilities

Energy availability for life does not only come from sunlight, but moisture, nutrient availability and ambient temperature as well. As reviewed earlier, moisture and nutrient availability are at their best in organic soils in life-abundant situations. Temperature is also more stable, particularly when compared to the cleared fields and heat islands that are at their worst in agricultural-urban settings. Energy flow requires source to sink cycles, which are often interrupted by mining, fuel burning, and habitat destruction, among others.

Meanwhile, it is making energy equally available (*homeokinesis and energy equipartitioning*) that allows life's "charged atomisms…temporarily to evade the rule of the Second Law."[629]; p.94. This higher energy flow demands that the system be recharged with energy input—something humans rarely do—and if maintained globally could do more than temporarily evade the Second Law. Energy flow is greatest with the greatest diversity, which is lowest when the environment is frequently disturbed (*an inverse relationship between environment fluctuation and diversity*). The least dissipation of a system requires a "static" (*quasi-*) equilibrium where energy input and output are essentially equal. Stability depends on complex relationships and multiple interactions, and only life can give these meaning.

The linkages between succession and the principles of thermodynamics are nowhere more evident than in the succession of plants and animals recovering from human-made disturbances. The so-called "natural" disturbances are typically human-created instabilities that offset the physical world (see *In Defense of Nature—The History Nobody*

Told You About). The most stable system minimizes irreversible decay, and irreversibility depends on the microstate-macrostate relationship.

Furthermore, entropy is a measure of the indeterminability of the microstate, which humans participate in. Throughout evolution, life has been finding ways to overcome resource limitations; life is ecological and holistic. Humanity has been developing less diversity (*homogeneity*), unavailable energy and instability, in contrast to evolutionary history, which displays increasing diversity (*heterogeneity*), energy flow, complexity and stability.[629]

Human-centered interaction (*homogeneity*), without regard for other species, will become lethal. Our interaction must bring about a near equilibrium, one which reinforces the homeostatic capability of the system. In terms of our discussion in Tome One, humans must take part in the overall workings of the organism Earth. In fact, this is why the species with the highest energy flow will be the most successful, which also requires conservation (taking part in the source to sink cycles). Self-organization in the face of environmental fluctuations necessitates complexity, which involves increased, as well as smoother and more ordered, energy flow. Helping to maintain life in its greatest diversity is essential for human survival. A scientist discusses this at the end of a paper titled, "The Thermodynamic Origin of Ecosystems":

> "Many of the foregoing aspects of the living world are most clearly illuminated by man's present condition and his relationship with the rest of the biological world, for we are a species in transition, imposing changes of great magnitude on the biosystem. Action based on simple, asymmetrical notions of causality has led to much undesirable ecological change. Highly asymmetrical interaction may bring immense immediate reward, but it also induces potentially lethal fluctuations. If we wish to maintain the world in a near-steady state then we must think in terms of symmetrical interaction and cyclic processes. Only in

this way will it be possible to regain the fragile quasi-equilibrium on which all existence ultimately depends.

If we wish to reestablish a near-symmetrical biological world it will be essential to reconnect ethics with the sciences. Ethics depend on judgement, and judgement and the biological sciences become intimately associated for there are no fixed points of reference in the nonequilibrium world; life is a compromise. Ethics may be defined as the maintenance of near-symmetry through the conscious limitation of the degree of asymmetry that we impose on the rest of the world, both with respect to our conspecifics and with respect to all other species. Social interaction is the thread from which the fabric of moral character is woven. The foregoing of immediate self-interest with the aim of future general good is thus an essential ingredient of near-symmetry maintenance at all levels of organization. If this is not recognized then the system will itself eliminate the inequalities, without regard to ethical or 'humanitarian' precept."629; pp.100-101

This quote may be clarified by comprehending some of the terminology. "Asymmetrical" interaction involves those actions that involve human interest in terms of self-aggrandizement. "Symmetrical" interaction involves those actions that concern the maintenance of all species and the environment. Economically speaking, we must make life, not lifeless things, paramount. Ethically speaking we must love and nurture life, not kill it nor destroy what enhances life's abundance. If not the disorder or entropy will grow, and bring back the order and stability of diversity by eliminating those humans who create the instability. This human-created instability is something we have repeated examples of throughout human history and in the present (see earlier sections of this chapter, and *In Defense of Nature—The History Nobody*

Told You About). Quantum processes will suppress the chaotic classical systems that we are a part of.

A trend in life systems works towards increasing diversity, which indicates that the rate of energy flow increases through an ecosystem. "This may be interpreted as the inherent tendency to return to near-instantaneous dissipation of energy with the emission of photons."[629]; p.90. A photon is a quantum of radiant light energy. The electromagnetic field between particles involves a force of electrons and other charged particles in terms of force carriers called photons.[1197] Life, in terms of diversity, actually enhances the flow of light through the system (*Fermat's Principle of Least Time*). In contrast, individual species tend to increase the time-delay and tie-up nutrients making them unavailable for other species. Light travels at speeds that are relative to diversity (space-time relationships). The greater diversity of life at near equilibrium can reach a state of energy flow without resistance (superconductivity; see earlier chapters and Tome One), leading to a relativistic relationship that could transcend some laws of physics, such as entropy. This may be hard to reconcile for some physicists. However, biological systems can show phenomena not accounted for by purely physical theories.[329,1331]

Part of the problem of studying life systems physically involves the Uncertainty Principle, as this quote from a scientist studying entropy indicates. He defines a niche as the "totality of thermodynamic boundaries of a living system."[1086]; p.122. "Boundaries" is the key word here, because he has, so to speak, stopped the particle making its momentum uncertain. Similarly, another states: "It is evident from observation that biological species interacting within the boundaries of an ecosystem express the tendency to assume a state of least entropy production."[629]; p.82. Again, "boundaries" make the observation uncertain, yet extending the boundaries to include an ecosystem brings about "least entropy production."

The question arises, what probability would exist if the boundaries included the entire biosphere (or beyond) with an abundance of life? Could entropy be reduced or even cease to exist as a "law"? This uncertainty is in addition to the confounding effect of our participation in the biosphere, and not taking that into account too. The biosphere has never been fully nurtured by intelligent beings, such as us, who could perfect its functioning.

The very physicists that made some of the most important contributions to quantum mechanics had reservations about its application to explain biological phenomena. The originator of the Uncertainty Principle, Heisenburg, suggested that unified theory would be governed by broader natural laws with quantum mechanics only being a limiting force. Hidden variables in quantum mechanisms were made more conspicuous by a physicist, David Bohm, who questioned the true randomness of quantum events in biological phenomenon. One of the first physicists to study atomic theory in quantum mechanics and received a Nobel prize for his atomic theory, Niels Bohr, indicated that biological systems were a completely different way of looking at Nature or the physical world. Another Nobel prize winner for the Theory of Relativity, Einstein, and two others, Podolsky and Rosen, certainly did not leave out biology or human participation in their theory of non-local phenomena (*EPR Paradox*). In the final analysis we will find that there are higher biotic laws that through non-local effects, including human participation, hold the key to the Grand Unifying Theory—the GUT—of the Universe.

25.6 Collective Mind

This becomes even more obvious when the human mind is considered. Mind is predominant over the physical world.[532,574, 617,1046,1047,1166-1169] Mind is not brain and not limited in ways implied by the physical brain.[574,1166-1169] The mind can alter brain structures and nerve function, which amounts to the power of mind over brain matter.[1166-1169] Experimental evidence of psychokinesis, the mind moving external physical devices, discloses that goal-directed mental activity can produce measurable changes in the operation of external objects.[179,532,574,617,1046,1047, 1332] Without consciously knowing it, each of us are influencing the environment and the Universe—reality itself.[317,617,1046,1047]

Ordinary consciousness is only the most minute fraction of the total activity of the mind, which has the ability to overcome time and space (*it is spatially and temporally extended*).[532] A mind can, for example, see a picture or scene experienced by another person who is a hundred or more kilometers away.[574,979] Most of us are familiar with people who have been able to correctly envision the future or the past. Numerous studies disclose that minds are joined in unseen ways.[532,639,640,1121,1122] In a book published by the American Association for the Advancement of Science, titled *The Role of Consciousness in the Physical World*, a scientist comments: "The perception of separateness of an individual from other persons (or from the Universe) is an illusion."[532; p.118.]

We have not recognized this full extent of human participation in our scientific perspectives, and therefore, only deal with probabilities of the preconceived sort. And many scientists are more skeptical of collective mind issues, it seems, than most other topics. Yet, suggestion and expectation influence perception far more than has been assumed.[532]

A good example of this misguided perception was a study conducted to examine the effects of experimenter bias on the outcome of an experiment. Some albino rats perform better than others if the experimenter is led to believe (falsely) that the rats belong to a strain selected for intelligence. The conclusion of those conducting this study is that "research assistants who do not know what outcome is desired" should perform the experiments.[1044] This is of course not the answer, because the researcher may communicate his or her intentions to the assistants in the same fashion as the experimenter did with the rats.

This is far from the only example, and though there are people opposed to the idea, there is no doubt that minds are joined.[532,639,640,979,1121,1122] Furthermore, a person or people in general "know" (unconsciously) how to hide information that is not desired or threatening (the psychological defense of repression).[532] What this comes down to is that, "the reality perceived tends to be a consequence of the reality believed, as well as the other way around. [And as a result] individuals and whole cultures are susceptible to self-deception."[532]; p.117. We see what we want to see and see what we are told to see, and thereby, create what is seen, but *is it reality*? This is not to undermine the direction of science, but we cannot assume a "law" is immutable when it has not been proven to always exist, and in fact, has been should to not exist under certain conditions. No one doubts the Second Law of Thermodynamics, but no one has proven it yet, especially with regard to biological systems.

Progress in science or scientific discovery depends on one's selection of "pertinent" data and one's instruction during learning. However, we must probe into the possible weaknesses of theories and attempt to refute them.[964] Studies are selected to fit existing beliefs, which gives short-term success, but according to the history of science, long-term failure. Meanwhile, thought experiments, the roots of new science, often rely on observational (empirical) data that is well-known and generally accepted. "How then, relying exclusively upon familiar data, can a thought experiment lead to new knowledge or to a new understanding of Nature?"[705]; p.7. An extremely valid question that is generally ignored.

Many physicists state that there is a crisis in the physical sciences, and crises end when imagination weaves a new fabric of laws, theories and concepts (i.e., new paradigms). Such a fabric must assimilate the incongruous experience and most or all of the previous assimilated experience.[705] However, new theories are tested by the old, and thereby, are judged as to their adequacy.[1115] Yet, theories may prejudice observations, while objectivity rests on criticism, critical discussion and the critical examination of the evidence.[964] In order for us to acquire all of the evidence we need to enhance life, but we are doing the opposite.[391,1134] Then again, if we continue on our present course, the "pessimistic arrow" of entropy will follow and confirm our pessimistic view of the physical world at our own expense, as it has done so many times before.

TOME THREE

Prehistoric Recollections
Geological and Paleontological Investigations of Earth History with Special Reference to the Cretaceous/Tertiary Boundary

"The advocates for terrestrial causes for extinctions may seem to be party poopers, but the facts seem to lie on their side." Charles Officer and Jake Page (*Tales of the Earth*, N.Y., Oxford University Press, c1993, p.146)

Forces of a restless, primitive Earth had crushed together all of the land-mass, creating a super continent called Pangea. One enormous warm-water ocean encompassed Pangea, around which a great diversity of sea creatures prospered. As time passed the Earth's occasionally turbulent surface buckled and broke the super continent, bringing about shallow inland seas that favored a greater variety of life forms.

Ocean life abounded. Varied and numerous, shellfish (*Molluscs*), such as clams and oysters (*Bivalved Pelecypods*), flourished so much that they made up entire immense reefs. Other ocean creatures included coiled single-chambered (*Gastropod*) and many-chambered (*Cephalopod*) shellfish. Prospering in other niches were shrimp, crabs, crayfish and lobsters (*Arthropods*), mingling among sea urchins, crinoids and sea cucumbers (*Echinoderms*). Colonies of small shelled animals (*Bryozoans*) that attached themselves to underwater surfaces had been fashioned into shrub-like and fan-like arrangements. Predictably, fish, particularly bony fish, were fruitful with the ample supply of water and reefs. As swamps were becoming less common-place, amphibians, similar to modern frogs, toads, salamanders and newts, started to decline. In total, it could be said that the changing face of the Earth was complimented by an equally revolutionary shift in life-forms.[133,180,274,444,704,709]

Upon the broken Pangea, with its sea-divided lands, various plant and animal species were thriving. At first, fertile forests of tree ferns, seed ferns, scale trees and seed-bearing trees (*Gymnosperms*), such as pines, covered that massive continent. Scouring rushes—grasslike marsh plants—were so tall that today they could easily be mistaken for trees. Later, as Pangea separated into a number of continents, true ferns and cycads, which were palm trees with fern-frond-like leaves, dominated temperate regions when today they only inhabit the tropics. The picture of plant life portrayed a dense growth of ferns and more or less pure stands of cycads with occasional Sequoia Redwood pines towering high above the general level of vegetation.[133,180,274,704,709]

As time progressed, plants with flowers and bearing seeds encased in fruits or pods (*Angiosperms*) emerged. Oaks, maples, buttercups, sagebrush, peas and violets were of the type that have two-seed leaves and net-veined leaves (*Dicotyledons*). A forest of mixed pines (*Conifers*) and flowering plants (*Angiosperms*), with ferns occupying a subordinate position and the cycads rapidly waning, was the picture of transition. Pollinating winged insects that included bees, butterflies, beetles, wasps and flies had arisen along with the emergence of flowering plants. Birds, and mammals the size of rats or mice, flourished on the new plant species' food offerings. The Mesozoic, often referred to as the Age of Reptiles, was an era when reptiles and warm-blooded reptile-like creatures dominated the air, land and sea. Among them were giant sea animals (*Plesiosaurs, Icthyosaurs and Mosasaurs*) and the dinosaurs.[133,180, 195,274,704,709] So the Earth appeared, rendering a scene unlike any that would follow.

As time progressed, Pangea was further torn into fragments that would eventually become today's continents. Occasionally the ocean would fill low-lying areas, forming inland seas, while mountains pushed skyward elsewhere. From one moment to the next the Earth's landmass lent the impression of a clay pot as it is dropped into shallow water and fractures into pieces.

CHAPTER 26

The Land of the Dinosaur
(The Cretaceous Earth)

26.1 A Diverse and Plentiful Earth

North America was the first fractured fragment to separate from the huge Pangea. North America, along with Europe, was severed, forming a sea that today the Mediterranean, and that part of the Atlantic just outside of its mouth (Straits of Gibraltar), seem to mimic (*Tethys Sea*). At that point in time it appeared as a large continental island with the tallest mountains, whose worn down remnants are today's Appalachians on the East Coast. Not far outside the Straits of Gibraltar it rested in what today would be considered the Mid-Atlantic. Only a shallow sea separated North America from the remaining continental mass.

Western North America was merely a group of volcanic islands divided from Eastern North America by a vast inland sea that stretched the full north-south length of the continent. Further west one could see the endless horizon of an enormous ocean which was about eighty percent of all our present-day oceans put together. By traversing this truly vast ocean one could reach all of the remaining continental mass, which then could be considered a single landmass divided by long shallow seas.[567,657]

Climatic conditions at that time were much more uniform than those which had gone before or those which would follow. Moisture was plentiful and plant life grew everywhere with a remarkable uniformity and abundance. Aside from the growth rings of some trees, there is a total lack of evidence that indicates well-defined seasons, at least in low mid-latitudes.[147,397,795] Possibly, only something similar to spring's blossoming and summer's abundant harvests existed.

The facts make one wonder if cold or snow ever entered the picture of a planet with ice-free poles. Subtropical forms reached as far as the 70° latitudes and monstrous trees grew thousands of kilometers closer to the poles than those of today. The boundary between seasonal and nonseasonal plant life had been shifted 15° poleward. An abundance of ferns, pines (*Gymnosperms*) and flowering plants (*Angiosperms*) grew in Alaska, Greenland, Spitsbergen, Siberia and Antarctica, signifying the existence of warm temperatures in present-day polar latitudes.[177,282] In high northern latitudes there is rich fossil evidence of now extinct Polar Broad-leaved, Deciduous Forests.[699,794]

Today, polar ocean surface temperatures range from -1.7° Celsius (18.4° F) to 1.0° Celsius (33.6° F).[272,624] However, during this last period of the Mesozoic, the Cretaceous, temperatures ranged from 5.0° Celsius (37° F) to 19.0° Celsius (62.4° F), or in other words, it was never cold enough to freeze fresh water, let alone salt water.[45,646] This remarkable global warming cannot be explained by the repositioning of the continents (*plate movements are insufficient*).[177] Furthermore, it contradicts the pole-to-equator temperature gradient, and requires a deep water source from both the poles and the subtropics.[646] The warmest in the entire history of the Earth, the Cretaceous climate is so unusual it defies explanation by the best scientific minds equipped to unravel such mysteries.

Those scientists that examine ancient climates, paleoclimatologists, describe the Mesozoic climate: "There is a strong consensus that the Mesozoic Era was appreciably more equable than the [present]

(*Quaternary*), with temperatures characteristic of the tropics extending into mid-latitudes, and polar regions experiencing temperate conditions."[282]; p.434 (brackets, parenthesis & italics added). "No evidence of glaciation occurs anywhere in the geologic record"[277]; including "no polar ice caps."[282]; p.433

Fossil evidence promulgates that the Earth was warmer, without winter, had very little change from season to season, and experienced abundant growth throughout the year. Dinosaur footprints were discovered at the 77° North Latitude (10° north of the Arctic Circle), a location that today would produce temperatures typically ranging from 20° below to 5° Celsius (-64° to 40° F). Breadfruits, magnolias, laurels, and Sequoia Redwoods, which cannot stand freezing temperatures, thrived in western Greenland at 70° North Latitude.[316,420,421] At 57° North Latitude, fossil shells from North America, Western Europe, and Russia reveal ocean water was 15° Celsius (59° F) warmer than today.[178,316] Plant remains from Alaska indicate that it was an unbelievably rich and lush growth. In fact, no place on Earth today could compare to it.[577,700]

The crowded fossil remains of dinosaurs in Alberta, Canada, suggest large herds, communal behavior and migration.[154] In fact, a large dinosaur trackway appears to have stretched the entire length of North America.[454] Dinosaur fossils were found in Australia, which in the Cretaceous was attached to Antarctica, and was as far south as 85° South Latitude.[615] Dinosaur fossils were also found in northern Alaska, which was near the North Pole during the Cretaceous.[93] In the Cretaceous, the 30° to 70° North Latitudes were like the present day's 5° to 32° North Latitudes.[316]

Arctic fossils really gave us quite a surprise. Alligators, land turtles and even Flying Lemurs were uncovered, demonstrating that its climate was mostly tropical. The first species of *Perissodactyls*, the ancestor of the horse, camel and rhinoceros, were found there. Plants, such as the elm, walnut, redwood, birch, sycamore, maple, sumac, fern and horsetail rush, were first witnessed in the Arctic, though today they only exist thousands of kilometers away from the pole. Leafy temperate forest plants were located well into the present Arctic region.[316] In contrast to the Darwinian perspective that most creatures originated in the Tropics, it is now known that many were preceded, for long stretches of time, by those that originated in the Arctic.[317] Imagine, the Arctic was riddled with swamps and marshlands, and had no well-defined seasons with a climate comparable to San Diego or Marseilles!

In recent geologic history, the Earth has been tilted 23.5° in relation to the plane of its orbit. For this reason, regions from the Arctic Circle to the pole experience two to six months of darkness centered around winter. The Northern Slope of Alaska was only about 250 kilometers (400 miles) from the ancient pole during the Cretaceous, yet plants were growing in the region that today we would only find thousands of kilometers away from the pole. Even if the climate was warm in the Arctic Circle, these plants would have required sunlight to grow. But today an area 250 kilometers (400 miles) from the pole would experience months of darkness. It was not until the end of the Cretaceous that evidence for well-defined seasons is again apparent in the fossil record. The entire planet having been almost entirely ice-free suggests that the Earth may not have been tilted as much on its axis during the Cretaceous.[178,317] In fact, some scientists believe that the Earth's axis of rotation was nearly perpendicular to the Sun during this period, so that the polar regions received light almost year-round.

The shifting of land masses does not explain these observations as was once thought. A prominent paleontologist states: "Thus, the once-mild climate on the now-frigid islands cannot be explained by the theory of continental drift. The North Pole itself must have been warm once."[317] Present iron-nickel core theories of the Earth cannot explain the reinstatement of the Chandler Wobble. Yet, here is something far more dramatic than the reinstatement of the Wobble: the disappearance and return of well-defined seasons brought about by changes in the tilt of the Earth on its axis!

Meanwhile, a sharp, sudden and short-lived climatic deterioration is well-documented for the end of the Cretaceous. Oxygen isotope studies of singled-celled chalk-forming organisms (*planktonic foraminifera*) confirm the temperature drop was real, rapid and sharp. Deep-sea cores display a remarkable temperature difference across the boundary that marks the transition to the following geologic period. This boundary is referred to as the Cretaceous (K)/Tertiary (T) Boundary, or K/T Boundary.

The temperature decrease just prior to the end of the Cretaceous was like that of the ice age, which took place in the Pleistocene. As this comment indicates, the transition at the end of the Cretaceous "seems to be as severe as the temperature changes documented between glacial and interglacial stages of the Pleistocene."[463] Studying ancient soils (*paleosols*) in the period following the Cretaceous (*Paleocene*) disclosed that it was a time with greater rainfall and cooler temperatures.[446] A number of other sets of data indicate a long-term cooling.[31,458,713,775,793] The period which followed has been linked to a progressive climatic deterioration; the beginning of a general cooling of the Earth's climate.

With the exception of Antarctica, only a brief period of glacial cover occurred around the pole, according to geological dating, about 300 million years ago, in the Permian, and other times extending further back. But in the entire Mesozoic, which includes the Cretaceous, only a brief glaciation took place on Antarctica before the Cretaceous, in the Jurassic. From that point on, little evidence of glacial cover exists until the most recent geologic time (Pleistocene), which included an ice age. More recently, data indicate that glacial cover may have briefly existed on Antarctica in a period following the Cretaceous (the Eocene).[470,471,499,500,668,716]

26.2 A Global Metamorphosis

The exceptional aspects of the Age of Reptiles only begin with its unusual climate and environment. Many other unique events occur at the end of the Cretaceous. At this time, worldwide changes take place; among them are mass extinctions, sea level fluctuations, widespread seafloor spreading, continental displacement, mountain building and volcanic eruptions, as well as major alterations in climate.

North America's facelift was one of the most major transitions, as rapid spreading of the Mid-Atlantic Ridge took place.[301,459,582,583] Mid-continent, just east of the Rockies, covering the Plains region from the Gulf Coast to the Arctic and from Minnesota to western Wyoming, was an inland sea, known as the Western Interior Seaway.[379,567,657] When this sea first swept the continent, its mineral-rich waters had toppled and solidified the trees that make up the Petrified Forest. After a long time of depositing many layers of sediment, this huge inland sea was drained as North America buckled under the strain of the rapidly spreading ridge. Left behind were the sedimentary rocks that would become the Grand Canyon, Painted Desert, Bryce Canyon, Mesa Verde, and numerous other unique places stretching the full north-south length of North America.

Immense environmental disturbances overwhelmed the planet. Aside from the displacing of the Western Interior Seaway, extensive sedimentary evidence indicates vast, powerful storms (*erosion/bypass events*), and drastic temperature changes. Massive destabilization not only took place in the atmosphere, but in the oceans and seaways as well. Masses of water moved rapidly in and out, while overturning and mixing (*anoxic and dysaerobic events, migration of oxygen minimum zones, oxygenation, and rapid ventilation and stratification*). Seawater became fresh water (*desalination events*), and rapid water temperature and chemical (trace element and carbon cycle) changes occurred. Often thick, coarse rock fragments can be found in those (*conglomerate and sandstone*) rocks that had formed, attesting to the vigorous erosion of the highlands. Within a very short time, fragments of pre-existing rocks were deposited in the Western Interior Seaway, forming a (*clastic*) wedge of large rock fragments that today reaches a depth of 25 meters (82 feet). Masses of sand and mud were swept into thick layers (*mass flow sedimentation*). All of these events and more were associated with rapid continental movements that pushed North America with such fury that the Southeast Coast dropped beneath the waves to remain there until this very day. Likewise, volcanic mounts east of New England submerged to thenceforth become part of the Atlantic seafloor.

On the West Coast were recurring gargantuan earthquakes and widespread volcanic activity as the Rocky Mountains, Sierra Nevada, and Cascade Range (Cordilleras) were shoved skyward. Along the Pacific Coast, thousands of kilometers of the continent's edge are believed to have been pushed under the Earth's surface to be melted in processes known as subduction and convergence. All of these stresses had also brought some of the seafloor above the waves. Having previously been a group of islands, the western section was now part of the mainland as the Western Interior Seaway, which had separated them, swept back into the ocean.

Explosive volcanic eruptions altered the landscape of Western Canada, Montana and the Mexican border. Other volcanoes were widely scattered and volcanic vents spewed thick clouds of ash that covered much of Texas, Arkansas, Louisiana and Mexico. The geologic record displays missing sedimentary layers both on land and the ocean floor.[293,379,527,800] As this period closed, North America, like the remainder of a very turbulent and restless Earth, along with the life upon it, would never be anything like it had been before.

Evidence that shifting seas and stupendous storms were an active force shaping the landscape can be found worldwide. Inland seas were everywhere leaving abundant sedimentary deposits on all of the continents. At times these seas covered one third of our present land area long enough to have left behind extensive evidence up to the close of the Cretaceous. Abrupt increases of the radioactive counterparts of the same element, but with different masses, known as isotopes, ($^{87}Sr/^{86}Sr$ *analysis*) resulted from "greatly increased continental weathering."[463] A picture such as this is testimony of an extremely stormy, very unsettled planet.

Much of what occurred was the result of sudden and rapid seafloor spreading. Scientists at the Lamont-Doherty Geological Observatory examined the ages of sediments, and the width of seabed magnetic stripes obtained from deep-sea drilling. From this research it was deduced that a sudden burst of seafloor spreading had taken place. Such a pulse, it was noted, would have raised the high-water mark of all the shorelines in the world. Mid-Atlantic Ridge expansion, apparent at this time, would have boosted sea level, at the very least, 300 meters (about 1,000 feet).[301,459,582,583] Yet, ridge expansion was taking place worldwide, as for example, also in the Indian Ocean[177,525], and displacing Africa, Antarctica and Australia.[80,111,521] Meanwhile, sea level was 300 meters higher than today solely due to the spreading Mid-Atlantic Ridge.[292,301,582,583] An uplift of the Pacific (*due to the heating of the lithospheric plates*) is also said to have increased sea level.[47]

The development of ocean ridges, at times of continental displacement, typically results in worldwide flooding (*transgressions*).[753] Throughout Earth history, no other time had as much continental displacement and ridge expansion as that of the Cretaceous period. No polar ice caps, due to the warm climate, would have also raised sea level. As a result, no other time experienced as much flooding (*transgressions*), and the frequent occurrence of missing sedimentary layers may have been the result of shifting inland seas, and the tsunamis, more commonly known as tidal waves, caused by a rapidly spreading ridge system. Afterward, as the ridge expansion ceased, and a cooler climate set in, there was a sea-level fall starting just above the boundary.[582,755,756]

Violent earthquakes and widespread volcanic activity were other by-products of the wholesale continental breakup. Landmasses were rended and buckled (*structural warping*), while elsewhere they split and shifted position (*considerable faulting*). Everywhere deep erosion cut at surfaces (*mass flow sedimentation, clastic wedges, erosion/bypass events, conglomerates, sandstones and increased weathering*). Many vast expanses sunk beneath the ocean, while sea bottoms brought their sandy faces above the waves. Withdrawing seas left behind swamps, flood plains and tidal flats, either accompanied or caused by mountain building. The poles traded north for south and wandered to new locations.

The continents pushed far and wide. Both Americas were driven westward with the opening of the Atlantic Ocean. Mountain ranges were propelled skyward from Alaska to South America's Tierra del Fuego. Africa was rended and rotated. Eurasia crinkled and cracked. Greenland and Antarctica were shoved towards the poles. Australia and New Zealand were propelled to isolated positions. India was pushed into mid-ocean by the rapidly spreading ridge of the developing Indian Ocean. All of today's continents, once a part of the one vast continental mass, Panagea, had fully separated, moving to new positions not far from their present-day locations. Geologic evidence indicates that most of the original continental fragmentation was spent by the end of the Cretaceous, and several landmasses had arrived near their present geographical positions.[3,133,148,180,195,201,379,473,567,657, 704,708,738,806]

The events of the Late Cretaceous put an end to the dinosaur, and its close was the most catastrophic of any era in the history of life on Earth. More than one half of all life-forms on Earth became extinct. In some places 75% to 90% of the species vanished from the scene never to be witnessed in the fossil record again. Along with the close of the Cretaceous there came the widespread extinction of large marine creatures (*Plesiosaurs, Mosasaurs and Icthyosaurs*), ancestors of the Chambered Nautilus (*Ammonites*), (*Scleractinian and Hermatypic*) corals, flying reptiles (*Pterosaurs*), shellfish (*Bivalves, Inoceramids,*

Rudists, Gastropods and Echnoids), chalk-forming creatures (*Planktonic Foraminifera, Coccolithophorids, Beleminites, large Benthic Foraminifera and Radiolaria*), and the dinosaurs.[232,631] Here was the end of one of the most successful life systems in all of Earth history. What was it that overwhelmed the Earth?

CHAPTER 27

The Sounds of Silence (Cretaceous/Tertiary Boundary Event Theories)

Deciphering the hieroglyphics of Nature's past, fossils and rock formations, has never been a simple task. Most scientists are quite certain that the change was catastrophic. Very few believe that the metamorphosis was slow and gradual. Interestingly, the theory of catastrophism is more than a century and a half old.[157] However, geologists had lost sight of the theory when Darwin and Lamarck proposed their theories of slow, gradual change.

The prevailing theory for about a century had been that of slow, gradual transition, referred to as gradualism.[160,461] As a result, some mainstream scientists have found it difficult to alter lifelong opinions and work. Nearly every geologist and paleontologist in pursuit of their degrees has had to learn the theory of gradualism. Compounding the problem, they had also been taught that catastrophism went the way of the dinosaur, into extinction, with the emergence of evolutionary theory. However, along with new, advanced methodology (*micropaleontology, high resolution stratigraphy, isotope analysis, etc.*) geology has had to face the rigors of change. Now the school of gradualism has rapidly lost its enrollment, as article after article espouses catastrophe. If we listen closely, with our eyes and mind, we can "hear" the sounds of silence speak of that catastrophic past.

27.1 Boundary-Event Theories

Many theories, both gradual and catastrophic, have been proposed to explain the extinctions, but have lost support from the scientific establishment. One theory states that the dinosaurs were poisoned by new toxic (*alkaloid*) plant species. However, the diversity of flowering plants through the latter part of the period (Late Cretaceous) was accompanied by a simultaneous increase in the diversity of plant-eating dinosaurs (*Ceratopsidae and Hadrosauridae*) that show no late decline.[541,628, 631] Furthermore, the fossil record shows that they survived the (*palynofloral*) change at the end of the period (*Cretaceous-Paleocene Boundary*).[631] Therefore, the dinosaurs could not have become extinct as the result of toxic plants.[631,720] Furthermore, this theory offers no explanation whatsoever for the marine extinctions, which were the most rapid and severe.

Inland (*epicontinental*) seas receded worldwide at the end of the Cretaceous (*Late Cretaceous through the Cenozoic*).[283] The regressive trend, it has been argued, resulted in major environmental deterioration that profoundly affected life.[137,375,535] The brief, profound drop of sea level, amounting to 100 meters (328 feet), would have produced global degradation.[360] Short-term worldwide sea-level trends can be hidden by a region's change (regional tectonics), and therefore, more profound fluctuations could have taken place.[78,360] Meanwhile, the end of the period (*Maestrichtian*) was preceded by an earlier (*Campanian*) lowering of sea level that did not produce any extinctions of exceptional severity.[360] Another, much later event (*Miocene-Pliocene*) also caused no major problems for life (*no faunal discontinuity*).[53,56,776] Furthermore, there were advancing seas for those lands bordering the South Atlantic, yet the extinctions took place there also.[612] Evidence is lacking its convincing element when considering that the lowering of sea level, at the end of the Cretaceous (*Maestrichtian-Danian regression*), could produce large-scale extinctions, especially worldwide.[376,720]

A strong warming trend that was accompanied by various Earthly (terrestrial) stresses embody other theories. A greenhouse warming trend took place that brought temperatures past a biologically critical limit, thereby leading to mass extinction.[494,495] Large reptiles are particularly vulnerable to heat stress, and an increase of temperature could lead to their extermination as a group. Likewise, the dinosaurs could have succumbed to the long-term cooling that came at the end of the Cretaceous.[41] However, evidence demonstrates that not all dinosaurs were true reptiles, but were warm-blooded or hot-blooded mammal-like creatures.[168,564,734] Therefore, they would have survived. Furthermore, temperature data indicate a rise long before the extinctions (Early Upper Cretaceous).[631]

A combination of various terrestrial stresses, including lower sea level (*marine regression*), global temperature fluctuations, and increased volcanic activity led to at least some of the extinctions.[2] Contradicting this scenario is the fact that these events occurred over a greater time span, and equally dramatic changes at other times did not coexist with mass extinctions to such an extreme.[631] In particular, these theories fail to explain the selective character of the extinctions, the particular species involved, and the geographic regions affected.

27.2 The Impact Theory

Suddenly a boulder the size of a small mountain pierces the atmosphere, igniting a huge fireball that scorches the sky. Intense blistering heat from the blast ignites rapidly spreading, global wildfires. The huge object smashes into the ocean near a continent, ejecting massive amounts of water, dust and debris into the air. Within a short time the dust encircles the globe, creating a thick, black darkness. Every last bit of vegetation withers, followed by widespread famine, and thereby, mass extinctions. With the Sun's rays blocked by the dust, an extreme cold, something like a nuclear winter, comes with the passage of time. Eventually the dust falls, but much of the water remains aloft, causing a greenhouse effect that heats the Earth by as much as 10° Celsius (48° F). Acid rain plummets to the ground as the energy released in the atmosphere combines nitrogen and oxygen to form nitric acid. So goes the theory that a comet or an asteroid, or a number of either, hit the Earth at the close of the Cretaceous.[5-11,166,333,395]

Embodied within this widely accepted theory is the idea that the heat stress caused by the fireball, or the eventual greenhouse effect, would lead to the extinction of the heat-sensitive dinosaurs.[6,166] However, the dinosaurs were not so heat sensitive, because some were warm- or hot-blooded, mammal-like creatures. Notwithstanding, the isotope record shows a rise in temperature long before the dinosaurs' demise.

Supporting evidence for the theory comes in the form of an element, iridium, which is rare on the Earth's surface, but fairly common out in space and in the Earth's interior. Iridium was deposited worldwide around the close of the Cretaceous.[9,10,562] Also found worldwide was shocked quartz that is claimed to be the result of an impact[72,73,400], and carbon soot, which suggests global wildfires.[761,791] In addition, another mineral (*stishovite*) that is typical of impacts was discovered.[487] On the basis of these findings the theory of an impact or multiple impacts is the most widely, but not universally, accepted theory.

It is not a fully accepted theory because there are a number of problems with the impact theory, particularly some of the physical evidence. One of the first problems to be noted was the absence of a significant-sized crater or craters capable of producing what is observed. A few craters of the wrong age existed, and were not anywhere near the right size, reaching only a maximum of 65 kilometers (40 miles), not the required 200 kilometers (125 miles) necessary for the observed effects.[125,129,283,592] In a time period ranging from long before to long after the dinosaurs (*Phanerozoic*), the largest is the Popigai Crater, which is still not large enough (90-100 km.; 56-62 mi.), much too young (Mid-Tertiary), and did not cause mass extinctions.[479] One crater was found in Canada of nearly the right age, but is far too small, meaning a number of other craters would be required.[401] Another crater was discovered in Manson, Iowa, of the right age, but it is also too small and may not be an impact structure (*it is often claimed to be a sub-surface volcanic-like explosion*).[432] Furthermore, magnetic (*paleomagnetic core*) studies indicate that the central uplift had no evidence of the reversed polarity typical of the other boundary sites, which means that it does not match the proposed time frame.[124] The double Siberian craters of Kara and Ust-Kara where thought to be the impact site for a number of reasons, including the main crater's size (about 60 km), but are now known to be much older than the boundary.[423] Many craters extend back in time long before the dinosaurs (*Phanerozoic*), so it is very unlikely that the crater(s) lie hidden under sediment or were destroyed by erosion.[283] When scientists finally found what they believed to be the right crater, they ignored a number of facts that also make it questionable.

Moreover, the very high levels of what is believed to be an extraterrestrial component (*about 100% for the basal layer of Woodside Creek, New Zealand*) suggests that it was either a comet, meteorite swarm or an asteroid.[434,436-438,653] Some studies indicate that the marine and continental clays require at least two different types of projectiles with different compositions.[460] An extremely implausible scenario.

At first no one crater or number of craters the right size could be found on either the land or the ocean floor, including beneath the ice of Antarctica.[54] The original explanation for the lack of a suitable crater(s) was that it disappeared beneath the Earth's surface (*subducted*) in the Pacific with the opening of the Atlantic Ocean.[8] The evidence pointed to a probable impact in the Pacific. However, such reasoning was found to be no longer tenable, because the most likely region of impact, if it did occur, would have been the Atlantic, near the Yucatan Peninsula, or

Caribbean Ocean, which were being formed and uplifted at the time.[76,125,321,352,353,670]

Now the two proposed impact sites are said to be near Cuba or the Yucatan Peninsula (Chicxulub).[76,319] A Caribbean site seems highly unlikely, because it was part of the Pacific Plate during most of the Cretaceous.[15,101,187,188,260] During the Cretaceous, a convergence of the southern and northern margins of the Caribbean was taking place. The Caribbean ocean floor moved east with respect to North and South America, making the Colombian Basin, Venezuelan Basin and the Caribbean ocean floor between one to two kilometers (about 2 to 3 miles) shallower than expected for the Late Cretaceous (*according to thermal subsidence*).[101,187,188] Volcanic eruptions in the region, typical of island arcs and oceanic plateaus, created the uplift.[187,188] An impact would be a highly unlikely cause for bringing about this type of volcanic activity. It is a scene that is very hard to reconcile with the effects of an impactor, which should have deepened the district, not uplifted it and causing volcanic activity.

The Earth's attempts to subduct the thick crust of the Caribbean were unsuccessful for this reason (buoyancy).[101,187,188] Western Colombia lies within sedimentary (*accretionary prism*) structures mostly of the Late Cretaceous and Early Tertiary times. The formation of these structures, along with the Gulf of Mexico, which Yucatan borders, and the Caribbean, is related to a major jump in ridge expansion in the Atlantic.[47,101] The recent analysis of a well coring (well log No. 6) at Yucatan shows a volcanic sequence not characteristic of an impact, but typical of a tectonic or volcanic origin. Furthermore, it was discovered that the crater is smaller and extends more deeply than thought, and it is a multi-ringed crater with a central peak. Multi-ringed craters with a central peak, like those on Venus, are thought to be explosive structures and related to the core, not impacts. As will be discussed in this chapter, and in Volume Two, Chapters 2 and 3, some cratering is the result of explosive ejection, not impact.

The high-pressure (*metamorphic*) event that affected the Caribbean/South American plate margin occurred in the Late Cretaceous *prior* to the K/T Boundary (*Coniacian to Campanian*).[47,187,188] Around the time of the K/T Boundary the Caribbean Plate changed its relative motion from northeast to east and began to be "shoved" (*underthrusted*) into and under the South American Plate. However, the shape of the Caribbean Plate has not changed greatly since its creation, except when Cuba was severed from the Caribbean Plate and sutured on to the North American Plate, the Yucatan Basin began a small spreading center, and the Caribbean Plate moved east.[478] Any consuming of the Caribbean ocean floor took place prior to the K/T Boundary.[187,188,478]

As a consequence, any evidence of an impact crater(s) should still be visible. In fact, the boulders and coarse sediment (*deep- and shallow-water clastic, epiclastic and moderate to intense deformation*) discovered at Cuba are more likely to be evidence of its severance from the Caribbean Plate at the end of the Cretaceous. The Caribbean Plate encompasses the proposed sites of Yucatan, Cuba and the Colombian Basin. This scenario, including the island volcanism, occurred mostly *prior* to the K/T Boundary and during *uplift*.

The change in the Caribbean Plate boundaries during the Cretaceous was the conversion of subduction to "slipped" (*transform*) boundaries. Important changes in tectonic and depositional styles of both the Caribbean Plate and the Colombia Basin took place before the K/T Boundary.[187,188,440,478] Seismic sections across the southern margin of the Caribbean Plate reveal a structure related to convergence with the South American Plate, but no signs in the regional tectonics show that an impact occurred.[440] The uplift (*regressive phase*) of Venezuela, Colombia, the Caribbean, and the Western Interior Seaway, and the Coastal Plains of North America, and probably South America took place from the Late Cretaceous into the following period, the Tertiary.[236,669] The evidence indicates that the tracts of possible impact sites were being uplifted, not depressed by an impact. Furthermore, tektite-like glass preserved at the boundary at Beloc, Haiti, suggests continental crust material.[680]

Meanwhile, the type of shocked quartz found argues against an ocean impact or that it even originated by an impact. Sediments in the deep ocean are generally not quartz (*calcareous or argillaceous*), with only occasional layers of (*turbiditic*) quartz sand. Most of this sand is not completely consolidated into rock, making it very difficult to imagine how extreme shock deformation could occur, which is required to produce the shocked quartz. In fact, most of the expelled rock would have been quartz-free basalt.[283] Furthermore, water is incompressible, and if the object were to target the ocean it would shatter into millions of pieces as soon as it began to penetrate.[202] The Yucatan and Gulf of Mexico, the proposed impact region, were under the sea at the end of the Cretaceous. Furthermore, the quartz is of a different variety than known impacts, as a specialist with the Geological Survey of Canada tells us: "I'm not convinced that it's the same stuff you get from an impact."[401] Observations such as these are ignored.

The lack of material at the boundary, which indicates higher shock pressures, contrasts with the wide range of shock pressures associated with typical impact-generated debris.[74,407] The quartz associated with the boundary is typical of high-temperature forms, such as those from volcanic sources[109,568] or tectonic stresses.[4] Furthermore, at Gubbio, Italy, shocked minerals (*shock mosaicism and lamellae*) occur over a four-meter interval that is bisected by the boundary. That is, the shocked minerals cover a much longer time span than the boundary itself.

Meanwhile, shock waves by high-temperature internal explosions have only recently been seriously considered in any context.[108,620] It has not even been pondered as a possible explanation for the physical evidence. The problem is that if you look to support a hypothesis, then you interpret the evidence in favor of what you are looking for. As will be discussed, it is high-temperature internal explosions that produced the shocked minerals, not an impact.

Iridium is considered the most important evidence of an extraterrestrial object, yet it too displays numerous problems for the impact theory. The clays at the end of the Cretaceous (K/T boundary) containing iridium are not the same ages everywhere. For example, at Gubbio, Italy, successive iridium peaks indicate a necessity for five impacts.[149,530] Similar extended zones have been found in the Pacific, Atlantic, Denmark, Spain, France, Germany and New Zealand, which indicates multiple dynamic events not possible from an impact.[79,109,200,374,621,622,745] This makes the problem of the missing craters worse, because this would require multiple impacts, and therefore, multiple craters.

The iridium at Raton Basin, New Mexico, was not deposited at the right time, according to data of ancient magnetic pole positions (*paleomagnetism*). The Raton Basin iridium was deposited during "normal" polarity, not the reversed polarity of the other sites containing iridium.[19] Many irregularities in iridium abundances occur worldwide (by orders of magnitude), when for the most part the greatest abundance should be near the impact site(s). It has been argued that these variations are due to different rates of sedimentation.[6] However, such a vague possibility remains to be verified, and is an ad hoc assumption in order to support a deficient theory.

The Hess Rise in the mid-Pacific has the highest iridium abundance, and thereby, suggests that the impact site was on a now subducted portion of the Pacific oceanic crust.[7] Meanwhile, other evidence suggests that the most likely possibility for a proposed impact is the Caribbean or the Atlantic near Yucatan, about 10,000 kilometers (6,200 miles) away.[76,322] In contrast, the large percentage of shocked quartz found at nonmarine sites, combined with other (*granitic and metamorphic*) associations, strongly suggests a continental impact.[125]

The unusually high concentration of iridium at the boundary is always associated with biologically altered material (*kerogen, etc.*). This suggests that it is the result of the rapid burial of a multitude of dead creatures, rather than an indication of the cause of the extinctions.[125,362,653,654,792] Moreover, the (*noble*) metal enrichments are associated with the decomposition of organic matter (*and a redox boundary*).[430,653,654] At a number of marine sites there are peaks in the concentration of organic carbon and/or coal deposition that occur along with the iridium.[125,361,385,386,653,731,741] The relative abundances of certain (*platinum group*) elements below the boundary clays at Stevns Klint and in New Zealand imply a continuous source, unlike an impact.[125] Also unlike an impact, Caravaca, Spain sources and a Deep-Sea Drilling Project hole in the central north Pacific (*hole 465A*) show that the other metals (*i.e., siderophiles: AU, Co, Ni, and Cr*) are strongly concentrated in a (*basal*) layer, while the iridium is largely outside of this layer, when they should be concentrated together.

Other later events contrast the proposed K/T Boundary impact scenario. A more recent and extensive iridium layer (Australasian tektites) accompanied only a minor upheaval (*stratigraphic discontinuity*).[749] Another event, in the late Eocene, displays evidence for three or four impacts, but no mass extinction occurred. Meanwhile, the Cretaceous event was possibly the most devastating catastrophe in Earth history.

Much of the evidence supports a terrestrially, rather than extraterrestrially, triggered event. The iridium is typical of Earthly types with compositions resembling the sea floor (*similar to manganese nodules*).[125,294] The abundances of (*noble*) metals are more consistent with earthly sources than with meteoritic materials.[125,228,320,436,460] Non-meteoritic trace elements (Sb, As and Zn) have been discovered at eleven boundary sites. The abundances cannot be reconciled with any wholly satisfactory conventional source, and ocean water most likely contributed some of the elements.[240]

While both the marine and nonmarine boundary clays exhibit anomalous, that is, "unusual," enrichments of certain (*noble and chalcophile*) elements, and certain differences argue against their common origin (conventionally speaking).[125] Two elements typical of meteoritic material (*Co and Ni*) show no significant correlation with iridium at Raton Basin. Furthermore, different elements (*Ti, Cr and Sc*) in enhanced concentrations at Raton Basin appear more like earthly types (*basalt or volcanic ash*).[240]

Other findings are also inconsistent with an impact scenario. Some, but not all, surface and deep sea marine sites in other sections indicate a strong correlation of meteoritic-like material with the iridium.[435,687,688,689-692,694] Another element (*Ti*) shows a strong positive correlation with iridium in the continental sites, but not the marine sites.[435,588] The abundances of iridium (Ir) and nickel (Ni) produce ratios that fall on either side of the meteoritic types (*chondritic*) to such a degree that there appears to be two components (Ir-rich/Ni-poor and Ni-rich/Ir-poor). Meanwhile, the ratios between copper and iridium in the clays from Europe and New Zealand fall within the range of earthly types (*terrestrial mantle values*).[125] All in all, the similarity of the major element abundances for the boundary clays and the enclosing sediment, and the contrasting dissimilarities of the major element abundances between sites, is evidence of a local origin. Supporting such a deduction are the similarities of quartz grains with those types that are associated with being transported by water (not atmospheric).[653] The conclusion is conspicuous: "These observations tend to contradict the consideration of the marine boundary clays as mostly either impact or volcanic fallout."[125]

Other facts bring about the same interpretation. Spherical forms, called spherules, in the boundary clays do not directly relate to the proposed impact event (*i.e., they are products of authigenic, diagenic and organic processes*).[92,290,353,354,585,652] Other (*sanidine*) microspherules do not conform with a high temperature impact (*sanidine is authigenic*).[166] Certain (*magnesioferrite*) crystals from the boundary could be derived from exposed and uplifted mid-ocean ridges (*mantle*).[72] Likewise, (*sanidine*) spherules occur over a longer period of time at Gubbio, Italy, with an enhanced concentration in the boundary. The prolonged occurrence indicates a volcanic or ridge source with enhanced activity at the boundary.[530]

The impact scenario does not explain a number of other observations, and suggest no single catastrophe. One is the permanent draining of the inland seas that took place at the end of the Cretaceous. Likewise, the rapid seafloor spreading and mountain building are unlike what an impact would cause. There was also a pulse of calcite dissolution in shallow marine waters that has no link to an impact. Such a pulse of calcite dissolution can be explained by the effects of ionizing radiation. The iridium and other trace elements are vastly different in sites only a few hundred kilometers apart, when they should be relatively similar. Moreover, the extinctions of some organisms takes place above and below—before and after—the boundary, and many that "should" have been affected remained unscathed.

Microtektites and tektites (also spherules) contain high levels of iridium. They are claimed to originate from either impact on Earth or on the Moon, with melting upon (re)entry into the atmosphere, causing teardrop- and round-shaped objects. However, there are flaws with both theories that even the originators of the theories admit, as will be discussed in sections 28.2 and 28.3, and Volume Two, section 2.2. At this point it is important to realize that they provide most of the evidence for iridium in the boundary.

The magnetic spherules present have been compared to microtektites.[512] Meanwhile, their iron content, much higher than any rock that could have been impacted, is unlike microtektites that would originate from an impact.[55] The high magnesium content (*of spinels*) is consistent with formation under high oxygen (*fugacity*), while tektites and microtektites are formed under low oxygen partial pressure.[89,123,125] Furthermore, the iridium content of the magnetic spherules is not in the range of (*iron or chondritic*) meteorites or tektites. Also, the nickel/iridium ratio (*of magnesioferrite*) is much higher than typical (*chondritic*) meteorites, but the ratio is like that of earthly materials (*crust and mantle*).[125] All in all, numerous studies question the impact origin of the (*platinum-group*) metals, which were the original bit of evidence that suggested impact.[125,736,757,781]

A relatively abrupt, but long-lasting climatic change is indicated by the data. It seems unlikely that a prolonged climatic change could result from a brief catastrophic event, such as that envisioned for an impact. The climatic system should have responded much faster to such an abrupt, short-lived disturbance.[446] Moreover, oxygen isotopes in the Weddell Sea (Antarctica) sediments imply that *before* the K-T Boundary the Earth's climate cooled suddenly.

The cause of the continental rejuvenation and increased erosion at the boundary appears less possible with an impact, as well. An impact is theoretically less able to initiate widespread tectonic activity. Furthermore, generating a general alteration in sea level and erosional processes indicates a major geodynamic disruption not possible from an impact. Likewise, another overlooked fact is that the impact scenario does not explain the retreat of shallow seas from the continents. Therefore, these and other facts indicate that the event was a major intra-terrestrial, rather than extraterrestrial, disruption.[212,283]

Many of the extinctions are also inconsistent with the impact theory. The mass extinctions at the end of the Cretaceous were not instantaneous and were selective.[212,283] This selectivity will be discussed subsequently. No permanent change for the animal kingdom has been linked to the specific time frame of the proposed event.[628,631] Various sea creatures (*Ammonites, Inoceramids and Belemnites*) and land animals (*terrestrial vertebrates*) show a gradual decline before the close of the Cretaceous and the extinctions occur at different times.[22-24,86,170, 171,176,212,279,283,342,396,404,642,760,787] One scientist comments on this: "Obviously, all three extinctions cannot be related with the iridium anomaly."[787; p.404.]

If the Yucatan/Caribbean region were the site of the impact then the greatest mass extinctions should have been in the southeastern North

America, Central America and northeastern South America, but are not. In fact, the greatest fossil graveyard for dinosaurs is probably in the Gobi Desert, in Southeast Asia, on the other side of the Earth. The types of organisms that became extinct and their geographic distribution reveal the actual nature of the event, which does not support the impact theory.

A huge cloud of dust shrouding the Earth after the impact is a scenario that is also contradicted by the facts. The dust cloud would have caused land plants to die out first (particularly considering the fireball as well), and mostly along the equator in tropical regions. Plant extinctions, more profound at mid-latitudes, also do not match the picture of a global event caused by an impact.[202,283,404,448,693,787] Tropical species have very short dormancy periods for their seeds and roots (tubers), and therefore, could not survive the long wait for the dust to settle, nor could they survive the cold.[315] Meanwhile, the fossil record demonstrates that many plants died out after the dinosaurs, and that those along the equator were affected the least. In fact, many tropical species remained untouched, while certain temperate regions had extinction rates as high as 90%.[202] Single-celled marine organisms that grow with light (*photosynthetic nannoplankton*) had their final extinction well into the following period (Tertiary).[693] In addition, the coexistence of characteristic Cretaceous and Tertiary plankton occurs beyond the boundary for inland marine sections.[31,385,386]

Other selective extinctions, or the lack of expected extinctions, also contrast the impact scenario. Tropical insects should have been seriously depleted if it were an impact, but instead survive into the Tertiary.[782] The dust-cloud scenario is too drastic to account for the marked selectivity of the extinctions with a high proportion of land and ocean groups surviving with little or no change.[212,283] The only possible way this theory could be maintained is if an already devastated biosphere was finished off by an impact, and then we are still left with a lack of plant extinctions, especially with regard to tropical species.

A scientist reviews the status of the theory, which still holds true today:

> "Even after nearly a decade of intense study of all aspects of the K/T boundary phenomena, a clear consensus about the fundamental details of the proposed impact event is lacking. These disagreements reflect conflicting lines of evidence regarding the composition and number of the impacting bodies, and the general character (marine or continental) of the target area(s)."[125]; p.111

27.3 The Volcanic Theory

The impact theory has been debated with its contender: a scenario of huge volcanic eruptions. Much like the impact theory the event could have brought about drastic weather changes that included a darkened sky, climatic cooling and acid rain.[140,141,142, 169,182,543,544,546,590,712] Both iridium and shocked quartz can also be caused by explosive volcanic eruptions.[108,118,140,142, 143] The hot lava is also purported to cause wildfires.[140,142,143,545]

India, near the end of the Cretaceous, had vast lava flows (*flood basalts*), known as the Deccan Traps. These lava flows covered a territory about the size of France. The Deccan Traps were the outcome of the largest volcanic catastrophe since the beginning of the Age of Reptiles.[125]

A high intensity of volcanic activity also took place in the Pacific Basin, northeast Asia, the Western Interior of the United States, and Peru. Meanwhile, by the end of the Cretaceous, volcanic activity may have already been spent. Eruptions within Britain (*British Igneous Tertiary Province*), Baffin Island, Canada, and northern and western Greenland began shortly after or at the end of the Cretaceous. Sea floor spreading in the Coral Sea near Australia, possibly in the Arctic, and the Ninety East and Walvis oceanic ridges in high southern latitudes took place around the time of the Cretaceous-Tertiary transition.[26,80,111,125,207,657]

In a number of points the volcanic theory is superior to the impact hypothesis. There are a number of iridium peaks, the iridium appears not to be like others of known extraterrestrial origin, the eruptions covered more time than a single impact, and it may account for some of the selective extinctions. In addition, the duration of the Deccan Traps' activity is similar in length to the irregular distribution of iridium and shocked quartz at the boundary marine sites.[149,336]

However, the volcanic theory is also flawed. It is very difficult to imagine lava flows in India, which was beginning to be pushed away from the other continents in major plate tectonic shifts, being responsible for global wildfires.[6] These wildfires seem even more difficult to reconcile when it is understood that this was a time for the wholesale breakup of continents attended by shifting inland seas and sea level.

Basalt lava flows can produce high iridium, but none has been found in the Deccan Traps themselves.[60] It is difficult to imagine a single location sending iridium worldwide, as well. Furthermore, the type of iridium is not typical of a volcanic eruption on land. Shocked quartz can occur with violent eruptions, but then the grains are so large they could not possibly be lofted high enough to be distributed worldwide. For instance, the Toba eruption, which was 400 times more powerful than Krakatau, the most powerful of recent times, lofted grains only 500 kilometers (310 miles). Volcanic eruptions cannot explain the sudden, sharp boundaries of deposited layers and extinctions (*biostratigraphic datum marking the Maestrichtian-Danian Boundary*).[631] Recent evidence indicates that wind-blown quartz can travel as far as 10,000 kilometers (620 miles)[61] and extreme pressures can be generated.[612] These data indicate that the Deccan Traps could produce only some of the Northern Hemisphere observations. The tremendous climate, temperature and ocean perturbations, as well as wildfires and selective extinctions, cannot be accounted for by volcanic eruptions.[379]

At least three other times of extinctions cannot be connected with any recognized similar type of eruption (*flood basalts*). For example, during one period (*Oligocene*), very violent eruptions, stretching from Colorado to Nevada, had little effect on the nearby animal populations, and no known mass extinctions occurred. All in all, the evidence displays characteristics of both an impact(s) and volcanic eruption(s), but does not conform to either alone, or in combination.

27.4 The Role of FEM

The most basic reason for the deficiencies of these theories is that the new Earth model, FEM, is not yet known to scientists. They are only considering the facts based on a solid iron-nickel core model of the Earth. The present model includes the assumption that the biosphere is protected from radiation due to the effects of the magnetic field, magnetosphere and atmosphere. As a result, theorists do not recognize the other possible mechanisms that could explain the evidence.

However, unlike the other theories, the effects of FEM account for *all* of the observations. For one, the Fields interact with the Earth's surface, bringing into being the mid-ocean ridges, rifts and so on (plate boundaries), and are the mechanism behind plate tectonics and plate motion. This was discussed briefly in Tome One, and will be addressed more in depth in Chapter 29. Polar reversals become as easy to understand as reversing the direction of the coil on an electromagnet (*reversed polarity of the Field, not core-mantle coupling offsetting the dynamo; see Volume Two, Chapter 29*). The mid-latitude Fields release great masses of ionized particles, which results in first producing shocked quartz, dust and spherules, and immediately following, strewn iridium. These features appear globally because the Fields are worldwide. The ionizing radiation (*by-products of hydrogen fusion*) that is released by the Fields causes huge climate and ocean disturbances, offsets element balances (carbon, oxygen and other isotopes), starts wildfires (microwaves drying vegetation and lightning igniting them) and precipitates carbon (soot) and hydrocarbons, produces acid rain, and has selective effects on extinction and brings into being new species (genetic mutation).

27.5 <u>Evidence</u> of <u>Radiation</u>

As a result of various observations, and consistent with FEM, a number of theories have been proposed that included the effects of ionizing radiation. One theory claimed that a nearby exploding star, a supernova, was the culprit.[445,635,742] The presence of iridium had made extraterrestrial objects very suspect.

For about two weeks a supernova would spew out intense energy in the form of strong radiation. If the supernova was within the Earth's vicinity, it could have killed the dinosaurs by irradiating them, and by adversely affecting climate. Evidence of such an event could be observed today by the presence of a gaseous nebula nearby in outer space, but only more recent and historically recognized nebula are observed. Furthermore, plutonium does not co-exist in high enough quantities with the iridium as it should if it were in fact a supernova. All in all, no ambiguous nor compelling evidence exists for the occurrence of a supernova during the Cretaceous.[213]

Similarly, a super solar flare has been proposed as the source of radiation.[626] However, this has been discredited because of the protective effects of the Earth's atmosphere, magnetic field and magnetosphere. Another theory suggests that the Earth's magnetic field underwent a reversal, yielding an increase in radiation. A reversal would not be enough to cause the extinctions due to the protective effects of the atmosphere and magnetosphere. If a reversal occurred in combination with a super solar flare (*solar proton event*), depleting the protective ozone layer, it begins to look tantalizing.[506,605,606] Yet, such a scenario does not account for the selective nature of the extinctions, especially with regard to tropical plants.[376,377] The super solar flare theory also fails to explain the increased levels of iridium, sea level fluctuations and the plate tectonics of the time, as well as the timing of the various phenomena (i.e., without FEM).

Meanwhile, the effects of radiation are clearly manifested in many fossil remains. Trachodon, a dinosaur discovered in a Kansas rock formation in 1908, can now be found in the American Museum of Natural History in New York. Mummified, its body remains fairly intact, shows no signs of predation, and the impression of its skin shows perfect detail, lacking significant signs of decomposition.[563] Likewise, proteins were found in the bones of one of the largest dinosaurs known, Seismosaurus or "ground-shaker," and are also the oldest preserved protein.[275] Considering the possible effects of radiation, one must wonder if Seismosaurus is such a large dinosaur because of radiation-induced mutation. Other amino acids were also found in Late Cretaceous fossils.[565]

Both finds are seriously questioned by scientists because protein is normally incapable of surviving such a long time. However, irradiated amino acids could be preserved, but are not even being considered. Today there are meats and other foods preserved by irradiating them, because the radiation destroys decomposers.

The Polish-Mongolian Paleontological Expedition discovered a number of interesting finds. Upper Cretaceous dinosaur and turtle bones from Nemegt Valley were unearthed that exhibited high radioactivity. A scientist comments on them by comparing these fossils with those of the period following the Cretaceous, known as the Paleocene: "It is, however, interesting that the radioactivity of the bones from Tsagan Hushu in Nemegt Valley is approximately seven times higher in Cretaceous specimens than in the younger Paleocene bones collected in the same area. This indicates that not only the locality of deposition of the bones but also the age of fossils influences their levels of radioactivity."[359]

In August of 1971, this expedition, while in the Gobi Desert, uncovered the skeletons of two dinosaurs, Veliciropator and Protoceratops, still locked in mortal combat.[168] Something very sudden had to overcome them or obviously both would have defended themselves against a common threat. Furthermore, both had their backs arched as if something had severely cramped their back muscles. Radiation affects high atomic-number elements more. The most abundant high atomic-number elements in biological organisms are calcium and potassium. These two elements are utilized in muscle function, and ionizing radiation may have caused extreme cramping of their muscles, arching their backs. Some have claimed that the tendons tighten after death producing this arching. Regardless, something had to overwhelm these dinosaurs very quickly in order for them to still be locked in combat.

A number of fossils show high levels of radioactivity, especially if made up of bones rather than some other material (calcium is a high atomic-number element).[50,162,175,359] Fish bones can be considered an exception due to the radiation-buffering effects of water (*not necessarily uranium-rich percolating groundwater and fine-grained sediments as proposed*).[50,162] Likewise, uranium-bearing bones may be due to an irradiated environment, which the animals lived in before their final extinction. This conclusion is suggested when one considers that those dinosaurs which show greater diversity—skull mutation—(*Ceratopsians and Hadrosaurs*) and surface-dwelling, air-breathing marine creatures (*Plesiosaurs and Mosasaurs*) dominate the samples of uranium-bearing bones.[50]

A number of dinosaur embryos have been discovered.[329,540] It is very difficult to imagine the preservation of such delicate specimens under normal conditions, but it could be expected of irradiated specimens. Though not a popular theory, it has already been suggested that the presence of high concentrations of radioactive minerals in fossil bones of Late Cretaceous age were the result of a simultaneous increase in background radiation.[645] Fluctuations in the deposition of calcium carbonate (chalk), calcite and carbon dioxide (carbonates), and in the levels of isotopes are also indicative of ionizing radiation.[138] Likewise, there are a large number of carbonatite deposits around the world at the time of the boundary. The origin of carbonatites—transformed sediments of calcium carbonate rocks (calcite and dolomite)—is obscure, and could be the result of irradiated reefs and corals. The selectivity of the extinctions at the end of the Cretaceous also shows the effects of radiation released by FEM and will be discussed later.

27.6 Global Destabilization

Because the Fields control plate tectonics there should also be a correlation between seafloor spreading, sea level fluctuations and extinctions. The K/T boundary corresponds to a short term maximum in sea level (*eustatic transgressions*), which supports the idea that the extinctions and a pulse of seafloor spreading took place at the same time.[125,292,651] Further support for this correlation is that other times of mass extinction are associated with sea level changes, as well.[102]

If the Fields were situated as they are today, in the oceans, we would expect evidence of highly disturbed oceans in the late Cretaceous. Studies indicate rapid water stratification, ventilation and temperature changes. Dark sediments (*laminated, organic-rich shales*) were deposited on a base of high organic carbon that is known to have originated from the open ocean (*pelagic*). A number of different sets of data indicate massive destabilization of the ocean to an unusual magnitude.[379] Rapid fluctuations in isotopic marine deposits reflect catastrophic mixing of previously stagnant oceans.[87,335] Both marine and continental sections demonstrate that the extinctions were often accompanied by changes in sedimentation that display abrupt coarsening (*clastic wedge*).[630,631] Much of what is recorded at the boundary could be due to a transition from a Cretaceous equator-dominated to a Tertiary polar-dominated ocean circulation system.[517,713] Mid-latitudes were affected the most, and there was an abrupt lowering of the surface salinity of the world's oceans.[376,377] Lowered salinity occurred because ionizing radiation, which is a chemical catalyst (*Na, Cl and NaCl have high ionization potentials, especially in aqueous solutions*), was released through the Fields in the oceans. All of these observations support what would be expected of FEM.

The Fields control weather and the evidence demonstrates massive changes there too. Giant storms left behind the signs of dramatically increased erosion and runoff.[201] Likewise, the sedimentary, biological and geochemical data suggest massive destabilization of the climate.[379] The end-Cretaceous has even been compared to what would follow an instantaneous discharge of more than the world's nuclear weapons stockpile.[531,539,743,791] Both the Fields and nuclear weapons release a similar type of ionizing radiation into the atmosphere. In fact, it was the scientific study of the K/T Boundary that led to the nuclear winter scenario, which essentially has the same climatic effects.[46,371,531, 539,743,791] Intense, sustained ionization events, whether nuclear, cosmic, solar or FEM, cause a global weather impact.[608]

Nitric oxide is produced as a by-product of ionization, and it depletes the ozone layer, allowing solar and cosmic radiation to enter with additional mutagenic potential.[46,153,155,238,303,363] Furthermore, the nitric oxide would undergo chemical reactions that produce nitric acid, resulting in acid rain. At the close of the Cretaceous the evidence indicates a sudden and sharp drop in temperature, as well as acid rain, which could be predicted from the effects of FEM.

Clay mineral associations reveal a "major geodynamic disruption" characterized by tectonic instability and sea-level changes. These events occurred prior to the boundary and the iridium excess. On the Walvis Ridge an abundance of a mineral (*illite*) indicates active uplift and erosion. The Maud Rise and Walvis Ridge display sediments associated with volcanic activity (*smectite*). A number of sections show deeply weathered soils indicative of a highly destabilized climate (*detrital kaolinite*).[619]

Due to ionizing radiation we could also expect elements normally suspended in the atmosphere to be affected. A major carbon-isotope event (*depleted* ^{12}C *and enhanced* ^{13}C) occurred at the boundary, followed by a negative excursion at the very earliest of the next period (*early Paleocene*).[68,69,377,504] Radiation increases the production of carbon isotopes in both the environment and biological materials, and this took place.[184,799] Moreover, a shift to lighter organic carbon isotopes was found in marine sediments from the south Atlantic[677], Israel[465], Tunisia[388], and the north Pacific.[804,805] Carbon isotope shifts such as these suggest biological productivity was depressed following the boundary.[504]

Along with the extinction and sedimentary (*biostratigraphic and stratigraphic*) layers there are ash falls, soot and variations in the production of organic carbon.[68,69,377,791] It has been suggested that this carbon fluctuation was the result of wildfires caused by an asteroid's fireball or by India's volcanics, but neither is capable of producing a global distribution.[12,25,791] Reentering ejecta would produce heat radiation, but at the lower limit for igniting solid wood. However, it could dry the vegetation and lightning could then start the wildfire.[498] Some sites do not have the evidence expected from a global fire, and have hydrocarbon (petroleum-like) characteristics. They display a distribution that reflects FEM; for instance, high levels are noted in New Zealand (East Australian Field) and Denmark (North Polar Field), not Italy (not near any Field).[761] Furthermore, carbon studies contradict the hypothesis that the carbon originated from a giant forest fire and are not typical of extraterrestrial material, but are like earthly types.[291]

Atmospheric carbon and hydrocarbon precipitation, and wildfires occur after the Earth is affected by ionizing radiation, as for example during a nuclear explosion.[531,547] Hydrogen and carbon combine into substances similar to petroleum, called hydrocarbons, which could ignite, producing other constituents of the K/T-boundary layer (spherules, soot and wildfires).[126,531,539,547,743,791] Derivation of the soot from fuel burning could not be ruled out by isotopic and other data.[75,125,241,273,761] The particle size distribution of the soot is, again, similar to a nuclear winter smoke cloud.[531,744,791] Specific hydrocarbon compounds (*polyaromatic and polycyclic aromatic hydrocarbons*) in the soot are the only indication that it is strictly wildfires (organic source), but the evidence is inconclusive.[744,791] However, these hydrocarbons, too, can be produced by an irradiated atmosphere (*due to the polymerization of methane in the atmosphere*).[531]

There is 300 times more charcoal above the boundary than in it, indicating a prolonged production. Increased charcoal production can result from either an increase or decrease in atmospheric oxygen (O_2), due to either increased ignition or less efficient combustion.[125,310] An ionized atmosphere would result in less atmospheric oxygen (O_2). The correlation of carbon with noble metals and other constituents of the boundary may simply be due to fuel combustion (coal and/or hydrocarbons) and plant ash.[267,339,522,587] Therefore, an impact(s) and/or volcanic eruption(s) are not necessary even for these constituents.

An event that first dries the vegetation is necessary for the global extent of the wildfires. A very effective agent for drying vegetation is microwaves, which could be produced by ionizing radiation reacting with the atmosphere and lightning (*whistlers*). These conditions would be present as the result of the dynamics of FEM. An unusual negative excursion of oxygen isotopes (^{18}O) occurs at, or shortly after, the boundary, which could be produced by ionizing radiation.[376, 377] All of these observed effects and their global distribution are readily explainable by FEM with its worldwide distribution of ionizing radiation. Further conformation of this is the fact that the composition of the iridium in the boundary is more like the manganese nodules of a seabed.[125,294] Every fact compels us to accept FEM as the Earth model, while the other theories are inconsistent with the full range of facts.

Even polar wander, apparent in the Pacific Plate from the mid-Cretaceous to early Tertiary, suggests FEM. The path indicates an uneven, rapid polar wander with a sharp bend occurring at the end of the Cretaceous (*Normal Polarity Superchron*). This observation can be explained by "time-varying non-dipole geomagnetic fields" which vary significantly with time.[639] The Fields are time-varying, which would resolve the polar wander record, because polar wander sometimes occurs very rapidly (see Volume Two, Chapter 19 for other discussions).

Mass extinctions and evolution with their nonrandom patterns and "missing links" can easily be understood, as well. Mass extinctions are now known to be very selective, and are not related to an organism's ability to survive.[355] Furthermore, it is the organisms proven to be very capable of adaptation, which have been around the longest (older families), that typically disappear at such times.[84]

Meanwhile, it has long been recognized that small changes at the genetic level can cause extinctions and produce new species.[256] One theorist called them "hopeful monsters," because many of the same species could have similar mutants and still be capable of reproduction.[256,265, 266] More than three-quarters of a century in genetic research has shown that the most consistent and efficient means of mutation producing new species or subspecies is at the genetic level. Of all the possible ways available for producing genetic mutations (including chromosome aberrations, crossing-over and recombining, and DNA and RNA mutation; see Chapter 22) none is more pervasive, uniform and perpetuating than the effects produced by ionizing radiation.[91,110, 262-264,344,346-348,441,645,647] Chromosomes are known to break and then recombine, producing a totally new organism, while the original organism becomes "extinct."[110,643,645]

A scientist looking into the process of evolution indicates genetic changes that alter development are responsible for the transitions observed: "Indeed, if we do not invoke discontinuous change by small alteration in rates of development, I do not see how most major evolutionary transitions can be accomplished at all."[266; p.30]. This is why patterns of mutation are not random and are reflected in what has been called "evolutionary divergence," or diversity in types of species. It is uniform rates of change at the genetic level that bring about the patterns observed.[255] Ionizing radiation is ideal for this very thing.

Animal tissues that contain elements with a fairly high atomic number, such as calcium, absorb more radiation than do soft tissues composed of less dense material. As a result, extinctions should show a relationship with organisms composed of greater amounts of the denser elements. Calcium-bearing (*calcareous*) plankton, shellfish, corals, open nest egg-laying animals, and animals with large skeletons would be prime candidates for mass extinction. Testimony of this relationship is witnessed throughout the K/T boundary region of the fossil record, as well as other times of mass extinction.[330,447,637,645,725,728]

Cretaceous marine extinctions offer a compelling example of the effects of ionizing radiation. Calcium-bearing, floating and drifting aquatic organisms (*calcareous plankton: Foraminifera and Coccolithophorids*) were the most deeply affected with widespread mass extinctions.[128,283,362,376,379,385,386,546,628,631,693,787] In fact, so widespread were they that this was the first compelling evidence that led geologists to admit a catastrophe had occurred.

Larger cells are affected more by radiation, and even among these aquatic organisms, the smaller ones (*Coccolithophorids or Nannoplankton*) became extinct later.[285,379,546,693] A section from Late Cretaceous (*Maestrichitian*) deposits in Alabama containing these fossils was examined by scientists who concluded that they were affected by an increase of radiation.[482] Other marine organisms that became extinct were calcium-bearing shellfish and reef-builders (*Ammonites, Inoceramids, Beleminites, Rudists, Gastropods, Echnoids, Radiolaria, Crinoids, Brachiopods, and Hermatypic and Scleractinian corals*).[283,289,379,504,546,787] Of these organisms it was the larger, surface types that experienced extinctions, often before the final episode (Late Cretaceous).[128,283, 355,377,379,391,488,546,631,633,787] The reason that surface-dwelling types were more readily affected is that water, especially salty water, is a good neutralizer or buffer against radiation. As a result, the deeper water (*benthic*) organisms were not affected much and if they were it was because they were large, calcium-bearing and immobile, bottom dwellers (*sessile benthic*).[128,283,355,377,379,391,488,546, 631,633,787] These deep-water types were typically those that could be found in the region of a Field where the ocean and sediment were severely disturbed.

Shallow water types were the first to become extinct with a peak (*in calcite dissolution*) at the time of the iridium.[546] For instance, Ammonites, relatives of the Chambered Nautilus, assume distorted shapes by uncurling and twisting their otherwise spiraling shells in late formations. Genera is a group of a given organism that is comprised of many species, and of the 34 genera of shellfish, Ammonites, none survived.[631] Likewise, it was the organisms with a narrow range of adaptability to environmental change (*stenotypic*) that were more deeply affected than those with a wide range and a global distribution (*eurytopic*).[379]

Bony fish are more readily affected, showing more extinctions than other fish (from 185 to 39 genera in marine types).[631] Changes in deep ocean circulation (*causing fluctuations in basinal carbonate compensation surfaces prior to and following the K/T boundary*) have been postulated as a cause of marine extinctions, but no real "genetic" link between the two exists.[631,733] Also, fluctuations in salinity and oxygen (*oxic conditions*) can be ruled out, because certain organisms (*Bryozoans and specialized Crinoids*) were at a peak in diversity at the time of their mass extinction.[64,65]

Examining dinosaur egg-shells, by measuring their thickness in successive layers of rock, disclosed certain trends that reflect what would be expected of FEM. Older rock layers show an acceptable shell thickness (2.5 mm. or 0.1 in.), but near the end of the Cretaceous the shell thickness reached a drastic fragility (1.0 mm. or 0.04 in.). In one rock face, near Corbieres, France, eight eggs were so poorly shelled that the embryos could not have utilized enough calcium to form their skeletons. It has been theorized that these thin shells resulted from the passage of unusually severe atmospheric cold fronts[730] or by the emergence of toxic substances in the environment.[204] Some shells are noted to have shells within shells and indicate an offset hormone system and a sharp increase in mutation.[205,412] All of these observations can be linked with the direct effects of ionizing radiation.[645]

Further evidence of calcium-dependent mutation is seen in the late changes that took place in dinosaur skulls. An exaggeration of the nasal plume in the duckbilled dinosaur (*Parasaurolophus*) and a bizarre solid-bone domed skull of others (*Pachycephalosaurus and Saurolophus*) make their appearance at the close of the Cretaceous. The bony frill behind the skull of another dinosaur (*Triceratops*) underwent abnormal development. In fact, each formation at the close of the Cretaceous houses a totally distinctive skeleton of dinosaurs related to this one (*Ceratopsians*) that cannot be found earlier.[168] The Beaked or Duckbilled Dinosaur (*Hadrosaurs*) had a skull which bore an ornate crest rising from the crown of its head like a top hat. Some had huge bony crests that are either solid or may have contained hollow nasal passages. So well preserved were these fossils that descriptions of stomach contents, including the composition of a last meal, were possible. In fact, the Duckbilled Dinosaur (*Hadrosaur*), the "mummy" fossil discussed earlier, and others (*Ceratopsians*) are among the samples of uranium-bearing and radioactive bones.[50,308,563] In North America there is the accelerated evolution of another dinosaur (*Kristosaurus*) that became flat-headed. So prevalent were these new species that one could easily wonder if they were merely mutants, while some scientists saw this divergence as evidence of racial senility and a possible cause of the dinosaurs demise.

More body surface area, or a greater cell or chromosome size, permits more radiation to be absorbed.[746,747] Every land animal (*terrestrial vertebrate*) with a body weight of 25 kilograms (55 lbs.) or more became extinct (with the exception of some which will be discussed).[584,630,631] One of the leading authorities on dinosaurs comments on this: "The dinosaurian extinctions would seem to be in conformity with an increase in background radiation."[630]; p.175. This is why dinosaurs and large marine surface feeding animals (*Plesiosaurs, Mosasaurs and Ichthyosaurs*) had no general decline prior to a fairly abrupt extinction (*at Maestrichitian-Danian Boundary*).[633] The young did not survive because the offspring of irradiated parents are experimentally known to be the most affected.[110] Skeletal mutations, such as the abnormal skulls observed, are typical of the dominant mutations induced in experiments with ionizing radiation.[192,193,417]

Gastroliths, the stones used by dinosaurs to aid digestion, can be found scattered about in rock layers, but not in the animals' stomach region had they merely collapsed and perished. Radiation causes a sickness which leads to vomiting, and this is the evidence of that sickness. The presence of ionizing radiation explains all of the observations.

Severe climatic deterioration can be ruled out as a cause of these highly selective extinctions. Animals very sensitive to cold survived the boundary event. More elaborate explanations than either climatic deterioration or impact are therefore necessary to accurately interpret the evidence.[95,96]

Extinctions of land plants also indicate what could be predicted from the effects of radiation. A gradual change for land plants took place, not a catastrophic mass extinction. Flowering land plants (*Angiosperms*) reach a peak in the number of species at or near the end of the Cretaceous (K/T boundary), and the change at that time is relatively minor. Plants are far less susceptible to short-term radiation (by a factor of ten compared to animals), and are even less sensitive to long-term (chronic) exposure.[698,797,798] This is also why cycles of extinctions in plants are much less defined than for other organisms.[419] Seeds are even more resistant than the parent plants, especially if protected by soil, low cover or soot.[315,533,631,645,797,798] In fact, the earliest fossil record of seed dormancy is an exceptionally well-preserved specimen with no signs of insect damage.[466] Typical of irradiated plants, these fossils offer evidence for the presence of ionizing radiation.

A brief dominance of ferns after the iridium layer is restricted to western North America and eastern Asia.[152,559,641,741] Ferns' shoots (i.e., meristems) are at or below the ground surface, and may therefore be partially or completely shielded. Furthermore, spores, smaller than most seeds, would absorb less radiation, and the fossil record shows that seed-ferns became extinct while spore types did not. Pines are the most sensitive to radiation and many species of pine became extinct. The larger plants and those with larger chromosomes, such as broad-leaved forests, were affected the most, again reflecting the known effects of radiation.[698,797,798]

Although many types (*taxa*) of land plants became extinct, the total change was much less dramatic than the other organisms experiencing mass extinctions (*marine phytoplankton, dinosaurs, etc.*).[231,278,313,358,404,448, 546,684,739-741] Fossil evidence shows that plant life was devastated, and then returns with only some genera and species becoming extinguished.[448,559,739-741] Radiation is known to cause the inhibition of seed germination, but does not prevent it.[91,346-348,533] In contrast, all of the evidence exhibits a scenario that was not the type of catastrophe one could attribute to asteroid or comet impacts, or volcanic eruptions.[96,238,355,377,448,488,631,634,739-741]

The boundary in North America suggests a sudden and traumatic vegetation disturbance, and a profound and long-lasting climate change. The selective extinction of broad-leaved evergreen species transpired.[699] Broad-leaved evergreen species are the most radiation sensitive of all plants.[186,698,797,798] Typical of FEM, there was a long-term restructuring of vegetation that is expressed most strongly at the mid and low northern latitudes.[699] Aberrant angiosperm pollen grains occur several meters above the boundary, suggesting a prolonged period of stress.[217,699] Again, this suggests the type of response noted in studies of the effects of radiation on plants.[698,797,798] A different sequence of events occurred in western Canada between the 60° and 75° North Latitudes (the confluence of the mid-latitude and polar fields) that cannot be explained by the impact scenario.[721] The expansion of the range of broad-leaved deciduous forests in the following (*Paleocene*) period is said to be due to increased genetic diversity and the development of dormancy, which would have increased the survival rate. A minor vegetational disturbance in the Southern Hemisphere argues strongly against a simultaneous global conflagration of the impact scenario.[699] The volcanic scenario is also inadequate for explaining the evidence. A scientist reviewing the evidence of plant extinctions comes to this conclusion: "Existing scenarios are clearly inadequate to explain all the patterns of regional extinction, survivorship and ecological restructuring that took place at the end of the Cretaceous."[699]

Other extinctions show the selective nature of ionizing radiation. Freshwater amphibians, such as crocodiles and turtles, show almost no change even when in excess of the 25 kilogram (55 lbs.) weight.[96,343] This is because water is a good buffer against radiation, and mud, which they lay their eggs in, is particularly so. Mammals with placentas, which are protective, show much less extinction than those with pouches that the fetus must climb into in order to fully develop (marsupials).[24,128,546,631,759] Of the freshwater organisms it is calcium-bearing bony fish and fish with cartilage (with calcium-salts in association) that show some extinctions.[631] No other theory has accounted nor can account for such selectivity in the extinctions. A scientist comments:

> "The high selectivity of terminal Cretaceous extinctions implies that, whatever the nature of the Cretaceous-Tertiary boundary events, their effects on the various groups of living beings were not uniform. This is obviously not in good agreement with hypotheses which involve tremendously violent and devastating events; such catastrophes would not have been very selective in their consequences on the living communities. It thus seems that more complex factors than mere body size or

resistance to cold are involved, and that the very variety of the groups that survived calls for a more elaborate explanation."[96]

The remainder of observations account for the lingering effects of radiation on a much shorter time scale than hypothesized. Foods that are irradiated would persist beyond the effects of initial radiation doses.[596] This would lead first to initial effects followed by secondary effects. Organisms most susceptible to the primary presence of radiation would metabolize the radiation, becoming extinct (calcium-bearing photosynthetic plankton), or they would pass the radiation on (plants). This is why we see a progressive wave of extinction from primary consumers to large consumers (*lower to higher trophic levels*).[630,631] Likewise, primary producers are the least affected by radiation, followed by those that consume plants (herbivores), and finally by those that consume animals (carnivores). First, radioactive materials end up in basic food sources and are then transferred through food, air and water to animals.[230] This is the reason for the observed steps in mass extinctions and the progressive wave.[378,379] Also resolved is the fact that there is no randomness to the extinctions as occurs at other (background) times when there are no mass extinctions.[84,355] Furthermore, the sets of extinctions do not occur at the same time.[182,280,544,546] The largest animals, as for example the dinosaurs, undergo both initial effects and latent effects, leading to the present debate that their extinction was gradual or instantaneous. The fossil evidence is also geographically segregated in a way that would be predicted of FEM, as the extinctions occurred more along the mid-latitudes.[24,452,546,618,631,632,684,759]

Here is the mechanism that is responsible for bringing about new species, as well. Ionizing radiation released by the Earth includes the by-products of hydrogen fusion: protons and neutrons (electrons are utilized in the Fields). Gamma rays and X-rays are routinely produced by particle accelerators, or the Fields in the case of FEM, because the Fields are particle accelerators. All of these particles are the most effective mutagens known for both plants and animals.[91,110,192,263,344,346-348,647,722-724]

The primary results of more than three-quarters of a century of research is that the gene is the physical basis of heredity and function, and that mutation is the ultimate source of variability. Ionizing radiation not only causes genes to mutate, but also breaks chromosomes, causing them to recombine in a different order or split them. When two chromosomes break in the same nucleus it sometimes happens that the broken ends join together and the order of the genetic information is changed. Some organisms carrying such genetic coding may be capable of giving rise to an adult individual whose germ cells lack the normal chromosomal complement, and incapable of full development, it perishes. For this reason, we see a greater diversity of species near the end of the Cretaceous before the final extinction. Other groups irradiated with similar doses and undergoing chromosome recombination could lead to new organisms capable of reproducing, hence new species (eventually new genera and families). This is the process behind what has been called evolution, and mechanisms at the molecular level were discussed in Chapter 22.

Other conditions operating during times of mass extinction and the origins of new species have genetic influence. The Fields emit pulsed radio frequencies, especially during times of increased activity. Pulsed radio-frequency fields have also been shown to alter chromosomes.[307] Ionization of the atmosphere would definitely produce static electricity, producing electrostatic fields. Chromosome breaks and translocations are generated by electrostatic fields.[475] Furthermore, there is evidence that a reduction in strength, or a reversal, of the Earth's magnetic field is capable of influencing the restructuring of genetic material.[185] Those studying paleomagnetism find that the geomagnetic field strength is about 25% of normal during reversals, which often occur during times of mass extinction.[501] So the process behind evolution involves ionizing radiation, and radio-frequency and electrostatic fields, as well as reduced strength and/or reversal of the Earth's magnetic field (more detail was given in Chapter 22).

Most scientists admit that the evolutionary theory proposed by Darwin and maintained by geology for about a century is deficient when examining the fossil record. In fact, Darwin made three mistakes that have influenced perceptions, not only in geology and biology, but in our society as a worldwide whole. Those mistakes were that he: (1) denied mass extinctions by claiming they were merely the result of an imperfect geological record; (2) assumed that species diversity tends to increase (*exponentially*) with time, much like one reproducing pair would gain more and more relatives; and (3) that natural selection (competition) or "survival of the fittest" was the major cause of extinction. The fossil record indicates that these three assumptions are *seriously* in error, yet they have controlled much of our scientific and social thinking for at least a century.[334] Hence, there has been a call for a new evolutionary synthesis.[334,705]

Fossils, however, display the type of record that could be predicted from sudden genetic mutation. Such is the case with flowering plants, called Angiosperms, who make a sudden appearance in the Cretaceous. A specialist studying these plants remarks: "The ancestral group that gave rise to Angiosperms has not yet been identified in the fossil record, and no living Angiosperm points to such an ancestral alliance. In addition, the record has shed almost no light on relations between taxa at ordinal and family level."[36; p. 230]

Darwin had thought that the sudden emergence of higher plants was the most extraordinary event to take place in the Plant Kingdom. Though the above quote was made more than three decades ago, today it remains an essentially unsolved problem after a century-long look at the fossil record.[36,37,39,223,224,337,338,709] Today the best theories are argued from hypothetical linkages, and at best there is only a very superficial resemblance.[39, 181,223,224,425-427]

The first fossils were noted in Brazil, then Israel, and worldwide near the end of the Cretaceous.[35,37,39, 88,223,224,709] Yet, during that time, sea level was high and no land bridges existed, while many continents were separated by inland seas or oceans. If we are not to totally dismiss the fossil record of inland seas, and the history of plate tectonics, we are left with the enigma of worldwide distribution, as well as origins. Genetic mutation caused by similar levels of ionizing radiation (and other factors) occurring worldwide with some original parent stock already worldwide, mutated into a new worldwide species, appears to be the only possible answer. Probably it was seed-fern fronds mutating into Angiosperm-like leaves[27-29,115], and maybe primitive leaves and pollen undergoing parallel evolution.[39,181] Molecular evidence points to a possible (*Permo-Triassic*) origin at an extinction boundary.[476] Certain characteristics of angiosperms (*stratified meristems*) promote long-term retention of many categories of (*somatic*) mutation.[416] Further implicating FEM is the fact that the first prominence of angiosperms was around the mid-latitudes.[35,37,145]

This is also why we observe that some species (*Aquilapollenites and Proteacidites*) became extinct at the end of the Cretaceous.[211,448,702,741] However, the extinctions are not the catastrophic type expected from impact theory or volcanics.[448,741] They are complex and demonstrate unique occurrences even within the same continent. According to FEM we should see distinct changes between western Canada and the western United States that divide along the 40° latitude, and this has been demonstrated![448,538,741]

In light of this scenario, consider this statement made by a specialist studying the effect of radiation on plants: "Ionizing radiation is generally thought to have played a very minor role among the selective processes of evolution. It is somewhat surprising therefore that the effects of radiation on natural communities follow predictable patterns apparently related to the evolution of life."[797] Developmental processes are reasonably similar at the genetic level in all flowering plants

(*Angiosperm ontogeny*).[262-264] First, similar development (*ontogeny*) occurs, then unique development. A scientist well versed in radiation-induced mutations in plants says that such a process can "account for the evolution of plant diversity."[263]; p.15. Ionizing radiation remains the most successful way of breeding new plants.[34,91,262-264,346-348,647,697]

This is the main reason we witness an upsurge in species diversity in the following (Tertiary) period when many of the modern families and genera appear. The end of the Cretaceous (*Maestrichtian*) is dominated by genera that occur in recent plants, but earlier it is dominated by extinct genera.[418] At the end of the Cretaceous (K/T boundary) we observe a mass mortality of plants followed by a succession that leads to a new dominant type (*Gymnosperms*).[448,741] Studying the end of the Cretaceous (K/T boundary) in New Mexico and Colorado shows some Laurels (*three-lobed members*) drastically reduced, and then they disappear.[404,793-796] In North Dakota and Montana a forest of broad-leaved trees and shrubs was replaced by another very different forest. Only 10% of 32 new types of classifications, known as taxa, survived and 30 taxa appear to replace them at the end of the Cretaceous (K/T boundary). The scientist studying these changes comments: "Each of these forests would be completely different. You wouldn't recognize the new from seeing the old."[404] Such an effect is by no means restricted to plants.

When other fossil evidence is examined, similar patterns are observed. A rapid diversification of most surviving organisms transpires in the beginning of the next period.[48,49,645] The K/T boundary is the label used for the geological and palentological remains of the event that took place between the Cretaceous (K) and the following period, the Tertiary (T). Within a narrow interval of the K/T Boundary in some marine sections is a peculiar transitional form of calcium-bearing plankton (*foraminifera*).[70] Freshwater amphibians and reptiles, along with land (terrestrial) snails and mammals increase in the number of genera.[631] In fact, so much did mammals begin to prosper that the following period (Tertiary) is often referred to as the Age of Mammals.

Such is the case throughout the fossil record.[683] A statement by one scientist, in a book titled *Evolution: A Theory in Crisis*, tells the tale:

> "It is one of the most striking features of the fossil record that most new kinds of organisms appear abruptly. The fossils have not only failed to yield the host of transitional forms demanded by evolution theory, but because nearly all extinct species and groups revealed by paleontology are quite distinct and isolated as they burst into the record, then the number of hypothetical connecting links to join its diverse branches is necessarily greatly increased."[165]; pp.164-166

Of course, what is being referred to is evolutionary theory as proposed by Darwin, and maintained by biology and geology for over a century. However, the situation is beginning to change as the labels indicate: NeoDarwinism, molecular evolution, and molecular biology.

Genetic mutation, as caused by ionizing radiation, requires gaps between different types of organisms. That fact is also why the development of an organism from an embryo to a complete organism (*ontogeny*) reflects what might be considered evolutionary steps. Encoded in the genes are all the previous mutations created by periods of ionizing radiation throughout its evolutionary history (*phylogeny*).[265,266] This is also why evolutionary changes occur in steps and cycles or are periodic, and are not random.[82,84,214,216,342,355,377,449,456,457,483, 520,574,593-595,597-603,645,659,662-665,711,761]

27.7 The Actual Event

In the overall scene we can say that life's abundance had already been greatly diminished, leading to a life crisis on Earth (*loss of biogeoelectrostatic properties*). As a result, the intensity of the Earth's magnetic field became drastically reduced. Then a solar flare of huge proportions was hurled towards the Earth.

Depending on the direction (polarity) of the Sun's Interplanetary Magnetic Field, or IMF, the Earth's magnetic field either reverses (IMF of opposite polarity), or wanders to a different location (same polarity). This will be discussed in Volume Two, Chapter 19. During the event that is marked by the K/T boundary the pole reversed.

Particle flow along the other Fields of the Earth induced ridge expansion (*electrostatic repulsion, etc.*), causing the continents to rend, buckle and drift (tectonic episodes), contributing to changes in sea level (*eustatic episodes and ridge expansion*), while volcanoes burst forth in titanic eruptions (*including mantle plumes*). Iridium (terrestrial iridium, microtektites or tektites) is ejected as the by-products of hydrogen fusion are released. The Fields shifting also leads to explosive cratering, spewing out shocked minerals (*quartz or feldspar*). Consequentially, colossal whirlpools are churned in the oceans, elements in the environment become altered (*isotope fluctuations*), and freezing cold moves in (ionization causing a vacuum allowing colder upper atmospheric layers to move in). Shallow-water, calcium-bearing organisms (*index fossils*), and large or otherwise radiation-vulnerable organisms became extinct, particularly away from the tropics, as the Fields are situated on the equatorial bulge pointing away from the tropics. Other creatures emerge from the mutations created by the altered genetic material that had resulted from the effects of ionizing radiation and the other factors present (i.e., geomagnetic reversal, electrostatic fields, pulsed radio frequency fields, microwaves, hydrocarbons, and acid precipitation). Life not only managed to survive, but new species would rapidly diversify, achieving greater stability in the environment than ever before.

Such is the scene that can be predicted from an understanding of FEM. In fact, such events should be periodic or cyclic (time-varying), and capable of explaining the fossil and geological records. Scientists have discovered that events do occur in cycles or episodes, and the geologic and fossil records of the events reveal that they took place in the order just described.

CHAPTER 28

Deja Vu
(The Cyclic Coevolution of Life and Geological Processes)

Cycles of events in the fossil and geological records are a new idea in the geosciences. Some, but not all, events have been correlated to steps in mass extinctions, and dramatic disturbances in the elements (*stable isotopes*), oceans and atmosphere.[379,502] Mass extinctions, climate, polar reversals, sea level fluctuations, plate tectonics episodes, and volcanic eruptions also occur in cycles or episodes.[284,300,425,456,659] Furthermore, such events often take place when cratering occurs, which may be cyclic, or at least, clustered around other events.[6,11,398,456,594,595]

Such cycles vary in length and cause claim various scientists. The existence of these cycles has neither been fully proved nor disproved, but are very possible according to statistical analysis.[399,600,601] However, those who deny the existence of these cycles must prove that they do not exist, because statistical analysis strongly suggests that they do exist.[493]

28.1 Geologic Cycles

Many proposals have been made to account for these cycles. One theory claims that they result from repeated impacts of comets or asteroids.[11,130,163,449,477,594,595,658,658,783,784] The various suggested ways in which cycles of impacts could occur include a "death star" (*Nemesis*), which is claimed to be a solar companion whose orbit destabilizes comets that impact Earth.[130,163,677,783] Yet, a solar companion seems impossible, because it would require a highly improbable orbit.[737] Another theory claims that it is a presently unknown tenth planet that causes comets to be dislodged from their orbit (*Oort Cloud*).[477,784] However, this would require either an extremely large planet or a drastically eccentric orbit far from the rest of the Solar System. Other hypotheses state that as the Earth moves through the galaxy it passes through interstellar clouds, which then perturb comets[594,595], or cause cosmic ray intensities to increase leading to mass extinctions.[658] Meanwhile, a group of astronomers insist that comet showers are not produced with sufficient frequency nor intensity by known bodies, whether molecular clouds or stars, to account for cycles in mass extinctions.[43] No known extraterrestrial mechanism can account for these theories or the cause of the cycles.[283,456,519]

Only one possible impact layer has been associated with a potential impact site.[261] The proposed impact site is within a volcanic (*caldera*) complex, and therefore, may not be an impact site at all.[146] Furthermore, evidence connecting the largest craters with extinctions is lacking, as well.[270]

One of the first to present the idea of cycles with enough statistical accuracy to make most scientists take notice, utters a comment that reflects his training as a scientist: "I'm entirely stuck for a mechanism for causing such a period, but I suspect that the forcing agent would not be earthbound."[449] Scientists have been trained to think of everything in terms of matter, mass and gravity, especially with regard to cycles. Therefore, they look for some object(s) affecting the Earth, not realizing it is the Earth itself in a much more dynamic state than hypothesized.

Another scientist reviewing the problems of astronomical theories with respect to the evidence concludes: "Rejection of the extraterrestrial option leaves us, of course, with the problem of putting forward a geologically plausible ultimate."[283] The dynamics of FEM are that geologic ultimate.

Likewise, a group of scientists reviewing cycles in mass extinctions, volcanic activity, polar reversals, sea level shifts and climate make the type of comment that implicates FEM: "We have argued that the correlations are due to an internal, rather than external, cause."[456]; p.10. The existence of cycles does imply a single ultimate cause.[449,456,490] The single ultimate cause is FEM and its solar linkage.

When we examine all of the possibilities of our present understanding of the Earth, such as magnetic reversals, they are inadequate for explaining all of the observations. Claims exist that similar cycles can be found in reversals and polar wandering as for the other cycles.[349,450,457, 532,574,598,711,715] The evidence, however, is not compelling with regard to the synchronicity of the various phenomena.[520] However, this may be due to the use of a closed dating system, which would skew dates.

A sudden drop in the intensity of the Earth's magnetic field precedes the reversal, which takes place prior to changes on the Earth's surface.[456,659] Reversals peak near times of mass extinctions, climate change, iridium deposition, cratering, plate tectonic episodes, volcanic eruptions, shocked minerals, and sea level shifts.[18,66,189,220,249,457,483, 520,574,605,606,715,762-764] However, reversals occur before other changes, which brings the inevitable conclusion: "This would, in fact, indicate an internal, rather than external, event that has subsequent effects on the Earth's surface."[456] Such a correlation completely discredits the idea that an object(s) impacted the Earth and caused the reversal.[456]

Those times of proposed impacts often show no evidence of a reversal, or they occur prior to the deposition of iridium, which is claimed to originate from the impact.[243,250] A reversal is not capable of itself to cause mass extinctions, because it increases radiation by only a few percent.[590] In addition, sea core isotope analysis indicates greater amounts of radiation than could be caused by a reversal.[318] Examples exist, such as at the end of the Cretaceous, when events take place both before and after a reversal.[393,439,546,631,715,749] Therefore, it is impossible for all of the effects to be due to magnetic reversals or impacts alone.

Major events resulting from plate tectonics also occur in cycles.[233,413,456,467,484,594,672,763] Volcanic eruptions from deep within the Earth (*mantle plumes*)[233], and mineral deposits occur within the same time frame.[456,764] Also following the same patterns is the solidification of molten rock (*magma*) deep within the Earth, and its crystallization (*plutonism*).[413,671] Accompanying all of these phenomena are sea level and evolutionary changes, as well as temperature minimums, possibly leading to glaciers and glacial advance.[38,40,526,644,706,787]

As could be predicted from the new model of the Earth, both the Age of Reptiles and Mammals (Mesozoic and Tertiary) show histories of inland sea formation (*downwarping*). Sediment accumulations occur on many widely separated margins, but they are nevertheless broadly similar, indicating a global event.[394] One of the most significant correlations is between marine extinctions and seafloor spreading rates, which infers earthbound processes controlling evolutionary pulsations.[456,614] Likewise, a strong correlation exists between both major and minor extinctions in the marine realm, and sea level changes.[284,330,447,732] As with other proposed single causes, plate tectonics cannot explain all of the transformations, especially non-marine extinctions, climate shifts and magnetic reversals.

28.2 Tektite Episodes

The formation of iridium, shocked quartz and cratering occur together in phase at times, and alone at other times, during these cycles. What are considered extraterrestrial iridium layers have been found in phase with the ends of only three major geological periods: Permian[30,403,665,719,801], Cretaceous[5,7,71,72,229,245,251,252,379, 460,512,580,665] and Eocene.[8,30,229,251,379,387] Meanwhile, the one at the Permian-Triassic Boundary has not been confirmed.[127,403,788] One (*Devonian*) period's layer was originally believed to be produced by impact, but was recently shown to be concentrated by biological processes instead.[341,496,497] The so-called terrestrial iridium has been discovered at the end of another four periods (*Precambrian, Ordovician, Frasnian and Mississippian*).[403] Iridium spikes occur along with biological extinctions, spherules, shocked quartz, soot, and tektites and microtektites, and other (siderophile and noble) metal spikes.

Iridium often comes in the form of glassy particles which have the appearance of being melted after entering the Earth's atmosphere and are called tektites. At least two iridium layers, known as tektite strewn fields, correlate with polar reversals (Austrailasian[249], and Ivory Coast[189]), though there may be more strewn fields in phase.[242,246] Meanwhile, one of these (Ivory Coast) has been shown, through high-resolution analysis, to occur prior to a reversal.[655] The source of iridium is thought to be the result of meteor impacts, and some cratering is in phase with extinctions and some iridium layers.[6,11,456,594,595] However, the majority of polar reversals are known to occur prior to cratering.[456] Furthermore, no extraterrestrial iridium exists for at least one (*Ordovician-Silurian*) and possibly two (*Permian-Triassic*) mass extinctions, and many other extinctions of lesser severity.[127,243,403,560,788]

Typical of Field locations, the various types of tektites display chemical characteristics that indicate a sedimentary origin.[132,156,243,671] Tektites can also originate from explosions and explosive volcanics.[370,537] An iridium layer in the Antarctic Basin (*Late Pliocene*) lacks terrestrial materials from an impact, indicating that if it were the result of an impactor, it fragmented before reaching the ocean floor.[433] First appearances of new species and extinctions coincide with only two of the iridium layers.[242,247,379] Cratering is not cyclic or periodic, but clustered around the boundaries between geologic periods.[6,268,379,749] There is no known crater for one of the four iridium layers, the youngest and largest (Australasian Tektites) whose crater should be the easiest to find.[6,18,163,379] Meanwhile, a map of tektite strewn fields effectively points out the Japanese, and East and West Australian Fields for the Australasian field (see Figure 2.3).

The end of the Cretaceous has been identified with craters that are insufficient for the amount deposited, and the extent of the iridium layer.[401] At least three of the craters produced in other periods and associated with iridium layers are questionable as far as resulting from an impact is concerned (i.e., Manson, Bosumptwi and Ries craters are likely to be geoblemes).[18,423,659] One of these craters (Ries; i.e., Czechoslovakian tektites) shows a polar reversal immediately following its formation.[18,243,655] Most cratering and strewn fields occur after a reversal.[456] However, the North American microtektite layer does not appear to be associated with a reversal.

All four tektite events are associated with major cooling events of the Earth's climate (*North American in Upper Eocene, Czechoslovakian in Middle Miocene, and Ivory Coast and Australasian in the Pleistocene*). This suggests that something dynamically related to the Earth as a whole is involved that also has significant effects on climate. The times around, but not directly associated with, the North American and Czechoslovakian tektites appear to be times of mass extinctions. Some minor changes in the biota are associated with the Ivory Coast and Australasian tektite events. Unlike the Cretaceous-Tertiary boundary, none of the microtektite layers are clearly associated with mass extinctions, iridium anomalies, or "impact" debris (shocked minerals; except one North American microtektite site off New Jersey).[243,280,284,382,536]

28.3 The Origin of Tektites

The origin of tektites has not been satisfactorily explained, but the facts clearly implicate FEM. Iridium is a siderophile or "iron lover," meaning it combines readily with iron. Most iridium that naturally occurs on Earth is theorized to be deep within the interior. As a result, theories embrace impacts on the Earth[243,727] or on the Moon[113,549-553], or an explosive volcanic-like origin (geobleme).[243]

Meanwhile, the scientists who originated two of these theories admit that there are major problems with their own theories. The facts indicate that the theory that tektites are caused by impact on the Earth is inconsistent with the laws of physics, and that origin by lunar impact contradicts the tektites' chemical composition.[552] However, the facts are in accord with the hydrogen fusion model of the Earth (FEM).

An expert on tektites comments on the speeds required to eject the material either from the Earth or Moon: "It appears that such velocities are essentially unattainable without the use of light gases such as helium or hydrogen."[552] In fact, this scientist even calls for a revolution in our understanding of planets in order to account for the tektites. He calls for a planetary model that includes hydrogen or helium (*other than radiogenic degassing*).[552] FEM involves hydrogen fusion which produces helium. The worldwide distribution of iridium at the end of the Cretaceous can only be produced by explosive, hot gases.[554] More will be discussed about tektites in Volume Two, Chapter 2.

Some craters display a central basement surrounded by a collar of uplifted or overturned sediments and volcanics. Also a positive concentric (*Bouger*) gravity anomaly is coincident with the central core. Meanwhile, impacts are typically negative. Therefore, the positive value signifies multiple dynamic events.[109] Looking at the distribution of some of the known iridium layers or tektite strewn fields, they point to the Field locations (see Figure 2.3). The samples obtained from the end of the Cretaceous, though relatively worldwide, appear to be restricted to mid-latitudes and the pole (found near the North Pole; the South Pole not known), and away from the equatorial region. That is, the iridium is found around the latitudes of the Fields.

In addition, amino acids typical of comet dust were found at the Cretaceous-Tertiary boundary, but were not destroyed as would have occurred if an impact took place.[806] The amino acids were distributed over and below the boundary, but not within the boundary itself.[808]

Such facts can only be reconciled with an earthly source, or the amino acids would have been destroyed and within the boundary.

Other facts argue in favor of the Fields being responsible for the formation of tektites. The major element, trace element and isotopic compositions, suggest that the parent materials appear like a terrestrial sedimentary deposit.[44,243,244,299,422] A radioisotope (^{10}Be), formed by ionizing radiation, was found in the Australasian and Ivory Coast tektites, and is consistent with a sedimentary origin.[243,575] Only the Ivory Coast strewn field has been clearly linked with Bosumtwi crater, and the Czechoslovakian tektites are loosely associated with Ries crater. However, there is a question of where the source material came from, and whether these craters originated by impact. No craters have been found to account for either the Australasian or North American tektite strewn fields; yet, they are the two most recent.

Furthermore, the widespread geographic distribution of tektites is difficult to explain when considering an impact.[243] Aerodynamic studies of the (*ablated*) forms indicate that tektites could not have formed on the Earth's surface, but had to travel within the atmosphere.[113] Typical of FEM, glass beads similar to tektites have been produced at nuclear bomb sites.[243,248] The Fields explosively ejecting material that travels through the atmosphere producing tektites corresponds with *all* of the facts.

Black shales, sediments often laid down during these cycles, are the result of hydrocarbons being precipitated out of the atmosphere and oceans due to ionizing radiation released by the Fields. Black shales often contain iridium and other heavy metals (*Cd, Ag, Zn, Sb, Mo, V, Cu, etc.*), which are stable sulfides or are strongly associated with organic matter or both.[5,7,30,72,73,94,228, 245,251,460,512,580] As discussed in the Tome One, the formation of a conducting ocean-floor surface in the Field areas involves sulfides, organic matter and heavy metals. Encouraging the idea that iridium (tektites, microtektites, and terrestrial iridium), shocked quartz, and cratering result from the dynamics of FEM are observations made at sites bombed by nuclear weapons: small spheres of glass and metal, similar to the way in which iridium or tektites are found, as well as shocked quartz and cratering.[136,243,248,678,708]

One of the main problems of present-day perspectives is that shocked minerals are most often thought to be produced by impacts.[620] A specialist comments on the underlying theoretical problem involved with explosive events: "Despite much speculation no mechanism to explain the generation of shock waves within the Earth has been developed to a point that permits critical evaluation and prediction."[222] Now there is a mechanism with FEM: the Field's explosive release of fusion by-products.

28.4 <u>Other</u> <u>Cyclic</u> <u>Processes</u>

Another phenomenon that occurs in cycles is changes in sea level.[159,281,292,456,489,493,594,595,665,750-752] Sea level fluctuations are not merely local but worldwide, and often involve the withdrawal of inland seas (*regressions*) from the continents.[177,456,750-752,762-764] They are noted to occur along with plate tectonic episodes, climate changes, volcanics and polar reversals.[456,520,659, 787] Many episodes involve a highly perturbed ocean, which is evident in major fluctuations in salinity, and oxygen and carbon cycles.[379] Some boundaries with evidence of a highly disturbed ocean (*i.e., Eocene-Oligocene, etc.*) cannot be attributed to ridge expansion in the oceans (major tectonic events), reversals, or impact cratering.[177,666]

Two specialists, examining ocean level fluctuations of the past, make a comment that indicates the need for a new understanding of the Earth: "However, if these short-term events are, indeed, real and involve sea level changes of the order of a few hundred meters, the major problem of a mechanism must be solved."[177]; p.191. Another scientist presenting evidence for a correlation between sea level changes and marine extinctions makes a similar statement: "Much uncertainty surrounds the frequency, rate and extent of the underlying causes of sea level change."[284]

Large-scale variations in climate are also known to occur in cycles.[40,179,219,220,489,491,520,663, 665,787] For one, severe drops in temperature have been correlated with polar reversals.[220,356,428,659] In addition, sea level, plate tectonic, abrupt extinction, volcanic and black shale events occur in cycles of similar lengths as those noted for temperature.[40,179,219,220,489,520,644,663,665,787] Large-scale temperature shifts indicate a highly disturbed atmosphere.[379] Again, the effects are like those of nuclear explosions: lower temperatures and pressure, rain, snow, large-scale winds, and carbon and hydrocarbon precipitation.[429,531,605,606,624,625,742]

The end of the Cretaceous is known to have had carbon and (*polyaromatic*) hydrocarbon deposits, large-scale storms, and a drastic temperature drop.[379,761,791] Either temperature minimums or glaciations shortly follow the other events.[379,644,645] One scientist examining ancient glaciations comments on their unusual character: "[Glaciations] occurred under physical geographic conditions that were highly unfavorable to their development."[645]; p.1272. They formed in regions of very warm climate along coasts where red rocks or other warm water formations were deposited (*i.e., kaolinite sandstones, dolomites, limestones and phytolites*).[645] All of these rock formations, both presently and in the past, form in the areas of the Fields along mid-latitudes.[277] The Fields release the highly ionized by-products of hydrogen fusion, creating a vacuum which the much colder upper atmosphere flows into. This causes drastic temperature drops that, with conditions permitting, produce glaciers or even ice ages.

For the same reason, observations indicate dramatic fluctuations in what are usually stable elements (isotopes). For instance, at the end of the Cretaceous, carbon precipitation reached ten thousand times its normal level.[6] A number of elements (^{10}Be, ^{26}Al, $^{87}Sr/^{86}Sr$ and $^{18}O/^{16}O$) are known to fluctuate along with other phenomena in cycles.[102,318,379,463,519,665] Looking at one element ($^{87}Sr/^{86}Sr$) in relation to geologic time, scientists comment: "The configuration of the curve appears to be strongly influenced by the history of plate interactions and seafloor spreading."[102] Variations in this element are then known to occur at times of shifts in plate movements, which is in accord with the effects of particle flow along the Field-lines of FEM. Black shale events are also cyclic and associated with petroleum or hydrocarbon deposits.[665,803] Worldwide conflagrations are also evident throughout geologic history.[648,649] Likewise, nuclear explosions produce hydrocarbons, precipitate carbon and offset stable isotopes.[429,547,645] The synthesis of these facts compels us to accept FEM's dynamics as responsible.

In this understanding there is a possible resolution to the so-called carbon problem that stretches throughout geological time. There are tremendous amounts of carbon deposited in coal and oil (hydrocarbon) beds, and carbonate rocks. So much is this the case that the Earth's surface is anomalously enriched in carbon. There are often enormous deposits of petroleum ranging from 16 to 1,050 billion barrels, and the deposits often formed in areas without the sedimentary beds that are "supposed" to be in association. Moreover, the presence of oil or bitumen in the cavities of fossils that exceeds what could have been produced by the mass of the animal alone have been uncovered a number of times.[138]

Likewise, coal deposits are extremely thick (often 30 meters or more), and often cover great expanses (thousands of square kilometers), both of which are not consistent with their formation by peat bogs (the accepted theory). If an entire full-grown beech tree forest were utilized in a such a coal seam, it would only be about two centimeters thick. A seam of 10 meters thick would require 500 forests. Yet, for example, in Morwell, Victoria, Australia the seams can be 345 meters (800 feet) thick. According to theory, the carbon deposits that eventually turn to coal are supposed to undergo compression under subsequent rock formations. Yet some have yielded uncompressed trees, and many are at the surface and are strip mined. Moreover, the coal is unusually pure leaving only 0.5% to 3.0% ash when burned. If it were the result of plant debris then there should be more ash from the trace minerals utilized by plants. Putting the whole thing into perspective, there are no coal seams being formed today in any peat bog, swamp or anywhere else.[138] Meanwhile, irradiation of the atmosphere could precipitate carbon from gases in the atmosphere (e.g., CO, CO_2, etc.).

Likewise, coal displays a number of other observations that do not conform with the proposed peat-bog origin of coal. Coal is often found without plant fossils or the vegetative structures of the peat bog. Often there are marine fossils found in association with coal, again unlike the peat bog scenario. Coal fossils that are found are replacement fossils, where a mineral has flowed into the space left behind by a completely dissolved plant. Moreover, plant fossils with great structural detail are filled with coal, making the fossil about 90% carbon when originally it was only a few percent. Yet a soup of carbon makes up the rest of the coal. Fusain, often referred to as mother-of-coal, has a woody structure that has been drastically altered by "some agency" (irradiation). Moreover, there is excess methane in coal, which is more compatible with a source of irradiation and polymerization.[138]

There are quite a number of observations of oil and other hydrocarbons that can also be explained by the irradiation of the atmosphere. If oil had an organic origin then there should be odd numbers of atoms per molecule (i.e., 27, 29, 31, etc.). Yet, the older oils show little or no odd-carbon predominance. Since they are much deeper than the younger oils, the temperatures and pressures could have prevented bacteria from altering the oil into the organic-like odd-carbon. There is also less optical activity in these older oils, and bacteria can alter material into optically active material. The older oils have no bacteria, but the younger oils do.[138]

Some coal (bitumen) and oils are radioactive. Heavy hydrocarbons are often found in association with radioactive minerals, suggesting that they formed from lighter fractions that have been polymerized by alpha bombardment (a product of hydrogen fusion) and the absorption of energy. Petroleum is unusually saturated with hydrogen, when biological materials would be deficient in hydrogen. Nearly all oils have live, active bacteria, which would give it all its biological properties regardless of how it formed to begin with. Oil and coal are claimed to have completely different origins. Yet chemically they have strong similarities.[138]

Oil can be produced by polymerization (i.e., irradiation) of methane and carbon in the atmosphere. Oil is claimed to be marine in origin, but there are deposits, such as in China, that are completely isolated from marine rocks. Furthermore, oil is supposed to take a million or more years to form, but oil was found in the most recent (a few thousand years old) sediments in the Gulf of Mexico (near the North Atlantic Field). Oil is strongly associated with tectonic features, especially plate boundaries, and as will be discussed (Chapter 29) plate tectonics is partly controlled by electrostatic forces (a source of polymerization). Furthermore, many areas with petroleum and methane also have large quantities of helium, which is impossible to explain with a biological origin.[138] Yet, helium is a by-product of hydrogen fusion.

The geographic locations of the largest oil and coal deposits, and other facts, also suggest their origin is due to irradiation and polymerization. Coal and oil are found associated together in areas such as, Alaska (North Polar Field), and Iran, Saudi Arabia, and the Ural Mountains (Persian Gulf Field). Other large oil and coal deposits are found in Venezuela and Columbia (near a plate boundary), and Pennsylvania and the Appalachian Mountains (near the North Atlantic Field). The Middle East is very rich in hydrocarbons, covering the mountains of southeast Turkey, the Tigris Valley, the mountains of Iran and Iraq, the Persian Gulf, and the flat plains of Saudi Arabia. All these areas have little in common, and all have different geologic ages and different rock formations. What they do have in common is being enormously rich in oil and gas, and they all surround the Persian Gulf Field. Furthermore, there is no biological debris in these deposits. Oil deposits have also turned up on the deep-ocean floor without source rocks. These areas include the Gulf of Mexico (near the North Atlantic Field), the Shatsky Rise (near the Japanese Field), and the western Mediterranean (near the Mediterranean Field). All of the giant oil fields have none of the source rocks of the proposed scenario.[138] As will be discussed in sections of Tome Five in Volume Two, and *In Defense of Nature—The History Nobody Told You About*, there are historical sources that describe the rain of carbon and hydrocarbons, in some cases accompanied by a fire-filled sky.

28.5 Cycles of Extinction

The idea of Earth history occurring in cycles had originated as the result of examining the record of extinctions. Thirty-nine stratigraphic stages (*from the Permian to the Tertiary*) disclose that one of the highest species or taxonomic classifications, families, had lost more than half of the original families to extinction (of 1,800 families, 970 became extinct).[665] Many of the mass extinctions involve calcium-bearing organisms who undergo steps in extinction.[342,379,382-384,703,770] As discussed previously, extinctions are correlated to cycles of polar reversals, cratering, tectonic events and other phenomena.[6,11,38,242,246,342,379,381, 456,520,594,599,602,659,665] Many of these extinctions had occurred abruptly.[218,242,246,489,493,534,556,594,595,602,645,656,661,664,665,687,688,690,694,762-765] One mass extinction involved the loss of 96% of all marine species (*Permian/Triassic Boundary*).[597] Though once a popular explanation, the closing of the ocean basins is not sufficient to explain such a widespread extinction.[218,706] Another mass extinction (*Mid to Late Carboniferous*) displays an abrupt event for shelled marine creatures (*invertebrates*) on the Russian Platform that cannot be attributed to any ecological or physical transformation. Furthermore, the scientist studying this region came to the conclusion that it was the result of a worldwide increase in ionizing radiation.[351]

After studying extinctions throughout the history of life, a scientist who attributes the events to radiation comments on a fact not explained by other theories: "It should be noted only that the sharp changes in organic life at stratigraphic boundaries are recorded almost synchronously in the most varied groups of fauna and flora."[645]; p.1276. Numerous studies disclose that extinctions are followed by the blossoming of new species[447,572,573,586,637,643-645,707,728] (see Chapters 22 through 24). Thus, observations always indicate a worldwide influence affecting all plants and animals (flora and fauna), which is not what could be expected of regional changes in habitat, gradual adaptation, and competition. Instead, the observations fit a scenario of genetic mutation induced by ionizing radiation and the other phenomena produced by a highly activated Field-dynamical Earth Model.

28.6 Cretaceous Extinctions and FEM

Mass extinctions display the type of geographic distribution attributed to the positions of the Fields. The Fields are located near coasts, and it is those organisms in the Coastal Plains Provinces that are the least capable of survival, regardless of whether there were many species or a few.[355] For the entire Age of Reptiles (Mesozoic), fossil finds of dinosaurs have been uncovered, for the most part, between the 45° latitudes and around the North Pole, with the vast majority between 30° and 40° latitudes.[734] While this has much to do with present-day population distributions in those regions, it cannot be attributed solely to a selection effect. Drastic temperature drops are also known to occur in the Field regions, occasionally producing glaciations.[643-645] In every respect, if ancient (*paleo-*) geographic data alone are considered, FEM is essential to the complete interpretation of the evidence (see section 22.3).

The latter part of the Cretaceous offers a good example of latitude restrictions generated by FEM. The most striking feature of the European Late Cretaceous fauna is the dissimilarity between the northern and southern regions. Northern types have been recognized in England, Northern France, Germany, the former Soviet Union and Scandinavia, while the southern types were throughout the Mediterranean Province. This segregation holds true not only for Europe but also in the Asiatic, African and North American regions.[58,99]

Other observations disclose this geographic segregation. The irradiated bones discovered in the Gobi Desert were discovered around the latitudes of the Fields, and the area is one of the richest provinces for dinosaur fossil finds.[14,359] Extinction rates in general were greatest in the mid-latitudes, and least near the equator, particularly for plants.[313-316] The mid-latitude Fields are situated above the equatorial bulge and point away from the equator. Especially heavy losses took place in North Alaska, Canada and Northeastern Siberia, all of which surrounded the Cretaceous Pole.[32,695] In Western North America, geographically located just opposite of today's East Coast during the Cretaceous, extinctions reached 70% to 80%. The North Atlantic Field was situated near Western North America in the Cretaceous, hence the richest fossil region in North America exists there. When the distribution and size of shocked mineral grains is considered, it is the North American western interior that displays the most abundant distribution and maximum size.[125,432] Unlike the global distribution of marine iridium, the nonmarine iridium has only been found in the western interior of North America.[125] Again, these facts contradict an impact in the Gulf of Mexico region. The northern mid-latitudes and the western Northern Hemisphere were more severely affected, which is especially notable for plants.[314,699] Slower, more gradual extinctions ensued on Antarctica, and extinctions were less in the higher southern latitudes and near the equator.[401,810]

The evidence suggests that a huge burst of solar activity took place and entered the South Pole (GMF-IMF interactions), and subsequent hydrogen fusion by-products were released through the remaining Fields, particularly the North Atlantic Field. As a result, the Fields along the 30° to 40° latitudes, pointing away from the equatorial bulge, released radiation, causing extinctions to peak in the Temperate and North Polar regions, while more gradual extinctions occurred in the Antarctic region.

A number of facts indicate that the North Atlantic Field was activated the most. For one, there is the higher abundance and size of the shocked grains in North America. Also, North America was the continent that moved the most in plate tectonic activity at the close of the Cretaceous. That is, more particle flow occurred along Field lines whose descending limbs control plate motion. Plate tectonics will be discussed subsequently. Underwater earthquakes, such as those caused by sudden Mid-Atlantic ridge expansion, generate tsunamis or "tidal waves."[20,323] As could be predicted by this model, evidence shows that a tsunami struck southeastern North America, as studies in Texas disclosed.[83] The lack of any tsunami-like deposit in the shallower Braggs, Alabama section indicates that an impact-generated tsunami is out of the question.[125,369] Small iridium anomalies for sites from the Cretaceous North American Western Interior Seaway and Northeastern Atlantic have been interpreted as due to volcanic emanations from a very active mid-Atlantic ridge.[125,239,561] Also a sequence (*of Mg-rich semctite bearing hentonites*) is believed to be derived from volcanic ash at the boundary in northwestern Europe.[570] The (*transgressive*) sea advances onto land in the Cretaceous were closely allied to increased oceanic volcanic activity.[651] Furthermore, extinctions were more severe on the North American continent, especially along the 30° to 40° latitudes.

28.7 Cycles of Species Diversification

The appearance of new species is also cyclic and takes place a short while after or coincides with the extinctions.[216,351,489,493, 645,656,694] One scientist expounds:

> "Almost all the biological revolutions have, in fact, been characterized not only by mass extinctions, but also by the synchronous appearance and subsequent vigorous expansion of other groups of organisms. The latest biological and ecological research shows that, generally speaking, the importance of the habitat (in the narrow sense) has been greatly exaggerated, and that genetic factors of various kinds have been far more significant in the evolution of organic life.
>
> On the basis of the fact that the biological revolutions have occurred synchronously over the entire planet within comparatively short segments of geologic history, and moreover, have involved all or almost all groups of organisms, there is reason to think that there is one principle cause of such phenomena. It seems to the author that this cause may be sporadic intensifications of hard radiation, subjecting organisms to sharp increases in the rates of mutations—sudden, abrupt, random changes in the genes or chromosomes that transmit their heredity."[645; pp.1279-1280]

The idea of radiation-induced mutations as the cause of extinctions and new species is not a new one.[57,296,309,424,607,645,650,675,676,729] Furthermore, the idea continually reappears because the evidence continually suggests that radiation is at work. The major problem and ultimate discredit of these theories is the proposal that some form of cosmic radiation is responsible, when the source is the Earth itself.

Longstanding problems of evolution can be easily solved with an understanding of the new model of the Earth. Biological or evolutionary data often seem to contradict data on plate tectonics.[90,623] For example, according to the history of plate tectonics, North America and Europe were to have separated around the middle of the Age of Reptiles (*Jurassic*). However, paleontological investigation suggests this did not take place until much later in the Age of Mammals (*Eocene*).[90,234] Again, Europe and Africa were to have parted near the middle of the Age of Reptiles, but biological data suggests that it was the end of that age.[90] The older, more primitive organisms are found in the higher latitudes, when evolutionary theory says that it should be the tropics.[90,160]

The major problem is that natural selection or "survival of the fittest" is not the all-powerful force. Nor is adaptation, as determined by habitat, geographic isolation or reproductive capability (*extrinsic or intrinsic barriers*), the force. Rather, it is geographic proximity to ionizing radiation as determined by Field position, and that a given species is "fit to survive."

Many examples of this can be found throughout the fossil record. Though Africa and North America were definitely separated in the Age of Mammals (Early Tertiary) diversity is still quite similar within the two continents.[234] Plants evolve shortly after major extinction events.[38,40] Distribution of flowering plants, and their earlier distribution and extinction, do not conform to plate movements.[448,685,702,741] The fossil record does not confirm that the continents separated, as would be expected from the tenets of natural selection (geographic isolation).

Plant communities that were geographically isolated should not evolve along the same lines, but the fossil record shows that they have. However, when considering isolated populations being affected by similar levels of ionizing radiation, producing similar genetic mutation, no conflicting data exists. In fact, the distribution of plants today in Australia is very similar to those of Japan, while each has been separated for an exceptionally long time and may have never been connected.[685] Meanwhile, their latitudes are very similar, but in different hemispheres, and they are near Fields supplying the ionizing radiation, which produces similar mutations.

Plants in India and Malaysia are identical, yet they have been separated for a long time or may have never been geographically connected. Eastern Asian and Western North American plants require a landbridge like today's Bering Straits long before plate tectonic data indicates that they were connected, and in fact, were several thousand kilometers apart.[685] All of the enigmas surrounding flowering plants (*Angiosperms*) can easily be solved with cycles of ionizing radiation.[38,40,258,645] Other evolutionary mysteries such as the coevolution of pollinating insects with flowering plants, fungi with plants, animals with plants and so forth can be resolved (some of these enigmas were discussed in section 23.4).

An idea presented more than a century ago, called Wallace's Line, describes an imaginary line that separates the plant and animal communities of the Orient and Australia. Meanwhile, it is a locality that basically embraces the confluence of the Japanese, and East and West Australian Fields.[235] The proposed land bridge for the transfer of plants and animals has not yet been shown to exist, and probably never existed.[33] The long-held idea that the midpoint, the Malay Archipelago and connecting islands, as a cradle of similar life-forms is "now seen to be improbable."[235]; p.2. Certain palms (*Cyrostachys, Licuala, Livistoia, and Rhopaloblaste*) exist on both sides of Wallace's Line, but are absent from the proposed connecting point, the island of Celebes.[183] Commenting on this, a botanist says: "There remains some genera the distribution of which seems at present quite inexplicable in terms of the gradual evolution of Malesia."[183]; p.56.

Many plants show distributions in both hemispheres that cannot be explained with the limited perspectives currently available in scientific hypotheses. This is particularly true with respect to geographic isolation.[785,786] Similarly, ionizing radiation is necessary for the rise of life from nonliving substances[57,645] (this does not preclude a non-chance phenomena, as will be discussed).

In the wider scope, the entire fossil record demonstrates that life is, so to speak, "is in the cards," it had to arise, it is inevitable. All the evolutionary changes occurred in order for life to survive and become more stable. This is why there is greater diversity, and hence more stability, as time progresses. One might refer to it as "purposeful" or "goal-directed" evolution in the maintenance of the living Earth.[451] The topic of molecular evolution has been discussed in the Tome Two, Chapters 22 and 23, and offers further support for this interpretation. The forces responsible for evolution include ionizing radiation, electrostatic fields and magnetic reversals, among others, which rearrange genetic material.

CHAPTER 29

Drifting Continents?
(Plate Tectonics:
The Roles of the Magnetic Properties
of Minerals, Electrostatic Forces and
Field Dynamics)

Plate tectonics, more commonly known as continental drift, are controlled by FEM. The descending limbs of the Fields regulate the plate boundaries. This is also why all cycles in Earth history include episodes of continental displacement, changes in sea level, and highly disturbed oceans and atmosphere.

In the early 20th Century a German meteorologist developed a theory known as continental drift.[777] This theory embodied the idea that beginning in the Age of Reptiles (Mesozoic), and continuing into the present, a huge primeval super continent, Pangea—a Greek word meaning "all land"—had broken (rifted), and its pieces have been separating ever since, forming the present-day's continents. Noting the geographic distribution of distinctive types of fossils, rocks, ancient climatic zones and ice sheets lent a great deal of support to the theory. Adding to this "proof" was the celebrated "jigsaw fit" of the continents bordering the Atlantic. Due to an inadequate mechanism for continental drift and the conventional "stabilist" interpretation at the time, it received a generally hostile reception, which unfortunately is the status quo for new ideas in most scientific disciplines.[431,771]

Through time the continental drift theory was scrutinized, leading to so much revision it is now more accurately named plate tectonics. The Earth's surface is understood to consist of plates made up solely of ocean floor, or both continent and ocean floor. The mechanism behind the movement of these plates involves seafloor spreading and subduction, which are believed to be the result of convection cells.

29.1 Convection Cell Hypothesis

A circulation pattern that is established within a mass of material because of temperature differences in the material is called a convection cell. For instance, air that is warmed tends to rise and spread out, cools and then sinks around the central region of rising warmer air. Theoretically it is said that convection cells develop within the Earth's molten boundary layers. Where the molten rock rises near the surface it produces ridges, usually in the middle of oceans. Where the convection moves downward from the Earth's surface it develops a subduction zone. A subduction zone is where the crust of the Earth plunges down into lower levels (*mantle*) and returns to a molten state.

As the mid-ocean ridges receive new material and form new ocean floor, in a process known as seafloor spreading, they create magnetic stripes on both sides of the ridge. These magnetic stripes demonstrate that the ocean floors of the world have all been created since the middle of the Age of Reptiles, in the Early Jurassic, and that the continents must have separated to accommodate the new ocean floor. At the same time, an equal amount of ocean floor was believed to have sunk into lower layers and melted in the subduction zones. The entire process of plate movement according to this theory is believed to be due to convection cells.[63,134,143,161,276, 510,555,627]

Meanwhile, this proposed mechanism has many flaws. For one, convection in the layer just below the crust (*mantle*) may be physically impossible, because it may not be fluid enough to permit it.[97,105,107,196,469,780] In addition, there is heat-flow (thermal) evidence that the continents have not moved over the underlying mantle, as is proposed in theory.[105,190]

Convection is never a regular process, and stable convection cells of the prescribed size have not been demonstrated. For one, the major descending currents are continuous, but the major ascending ones are not.[138] Furthermore, the zig-zag pattern of the rift-ridge system in mid-ocean is not what would be created by such a process.[98,411] "Cracks" that develop at right angles to the ridge (*transform faults*) are not possible with convection either.[196,411] The sediments of the ocean basins are horizontally regular over vast distances, which also discredits convection, because convection is not a regular process.[411]

A subdivision (*Mohorovic Discontinuity or Moho*) of the Earth's interior separating the crust from a lower layer (*mantle*) also presents some problems for convection. If the (*Moho*) subdivision is a phase transition of the same material then it is too shallow in oceanic regions. If the subdivision is a chemical change then convection is impossible.[107]

Convection cell models still show the convection cells aligned grossly parallel to the Earth's axis, which does not fit actual observations.[469,555] All of the convection cells should be oriented along longitudes, which does not explain how latitudinal features develop, such as the ridges connecting the Mid-Atlantic and Mid-Indian Ridges, or the Mid-Indian and East Pacific Ridges. Convection cells are the result of heat transfer, and with thin ocean (*abyssal*) plains new cells should develop, but have not.[469]

Plate motions (*and mass balance*), heat flow patterns, and irregularities in the Earth's shape (*large geoidal anomalies*) do not fit interpretations derived from convection either.[114] For example, the Pacific High and its associated convection should either extend or compress the

Pacific plate in a north-south direction, but neither has been observed.[114,617]

Most plates are in a state of compression, except Africa which is under tension. Meanwhile, present theories insist upon tension at plate edges, not mid-plate as the observations have indicated.[442,617,811,812] Furthermore, drill hole research near the San Andreas Fault, at Cajon Pass, has revealed an absence of large amounts of drag at plate-plate boundaries.[811] Tension and drag should exist at plate edges if convection cells were responsible.

Likewise, the theory of convection cell driven plate tectonics cannot explain mid-plate volcanic activity. Volcanic activity mostly occurs along the ridges, and where two or more plates converge (usually subduction zones), and convection cells could explain these features. However, the presence of volcanoes at mid-plate does not fit the convection cell hypothesis, as it would require a flow that is essentially perpendicular to, and has its roots at the depth of, the convection cell that is purported to drive the plate.[138]

Another contradiction to today's plate tectonic theory is what might be called the "jigsaw" Earth, where plates are made up of pieces that came from regions that appear to have never been in contact or seem to require the complete reversal of convection cell flow.[276,542] These and numerous other facts call for more complete plate tectonics theories. One scientist, whose work criticizes plate tectonic theory as it is presently theorized, states: "I conclude that convection cells, based on the arguments presented here, are products of the imagination and I urge that they be relegated to that realm."[469]

Many observations stand as clear contradictions or enigmas to the proposed model, such as those of the mid-oceanic ridges. It is not possible to generate the configuration of the Mid-Atlantic Ridge by convection cells.[105,469] The Mid-Atlantic Ridge still has the outline of the continents it is purported to have separated. It is extremely difficult to imagine how the ridge could retain this shape after a long period of stop-and-go movement, and changing convection cell flow (mostly Mid-Cretaceous to the present).[469] No one has even attempted to explain why the ridges have not been obstructed by thickening, nor why new ridges have not formed. It is against the laws of physics to consider that convection cells would retain their exact position in spite of the fact that seafloor spreading is often rapid, and is a stop and go process worldwide.[13, 411] Furthermore, the Earth's magnetic field, purported to be generated by the core which is also supposed to generate convection cells, has repeatedly changed position while the ridges retained their shape and location. In addition, evidence shows that the core is not in the theorized state.[21]

Regardless of the spreading rate, which has changed repeatedly, the thickness of new crust is constant over at least 56% of the Earth's surface.[503] The ocean surface becomes increasingly subdued away from the ridges, becoming the flat surfaces of abyssal plains.[469] Again, these observations indicate a very regular process that is not possible with convection. These and other disparities have lead a number of scientists to cast doubt on convection cell models of seafloor spreading and plate tectonics.[51,97,105,107,411,469,503,524,754,773]

Problems with seafloor spreading are quite devastating to the proposed convection-cell mechanism of plate motion. Spreading has varied through time, beginning in the late Cretaceous, then slowed, eventually stopped (*Miocene*) and then resumed more recently (*Quaternary*). In spite of this stop and go behavior, the spreading on each side of the ridge (axis) has been exactly half, so that throughout the entire formation of the Atlantic Ocean the ridge has always been down the middle. No known reason for this physical bias is available with present theory, yet it is necessary for the formation of the magnetic stripes observed.[51] In fact, two parallel ridges that have migrated away from each other seems necessary or the molten rock could not have been injected between them.[51,253] Furthermore, the magnetic stripes are of equal intensity, shape and width on either side of the ridge, and almost everywhere parallel to the ridge.[51] To suspect that convection cells, with all their irregularities, could create these observations sounds more like science fiction than fact.

Other observations defy the idea that spreading has occurred in some situations or not at the proposed rate. It is not unusual to find rocks formed long before the spreading occurred at the peak or crest of a ridge (*Jurassic, Cretaceous or Paleozoic fossiliferous rocks, even Cambrian are noted*).[411] The opposite was noted for an eight-kilometer (5-mile) region, which was traced without a break, that shows spreading has not occurred or is much less than had been assumed.[198,199] One ridge (*Reykjanes*), according to its age, should have spread twenty kilometers (12.5 miles), but has not.[198] The Mid-Atlantic Ridge began spreading at the end of the Cretaceous, yet there are much younger fossils (*Miocene foraminifera*) and sediments at its crest.[122] Iceland is immediately situated upon a ridge and with spreading it should have split in two, but has not. India was once attached to Africa and drifted until it smashed into Asia. This process is claimed to have built the Himalayas, some of the highest mountains on Earth, but a minor force such as seafloor spreading would not be capable of causing this massive uplift. Furthermore, according to fossil and sedimentary evidence, northern India was in its present position in the Late Cretaceous, but the oldest ridge in the Indian Ocean is much younger (*Eocene-Oligocene Boundary*).[105, 578] A number of other observations contradict present theory.

29.2 Subduction Zones

Subduction, the plunging down of a plate into the Earth, does not support present theory either. For one, the amounts of old crust consumed in subduction do not tally with the production of the crust at the ridges.[411,503] The Pacific Ocean floor has greatly increased, when it should have grossly shrunk, while the other oceans grew.[105,480,481,790] In fact, this is one of the main reasons some scientists claim that the Earth has "expanded."

Plate theory demands a subduction sink that has swallowed a tract at least the size of the African continent somewhere in the African plate, but this subduction sink does not exist. If the subduction zone on the west side of South America (*Peru-Chile Trench*) is to be considered the sink, it would have to consume 7,000 kilometers (4,340 miles) of crust. However, plate-tectonic geophysicists indicate the figure is 4,500 kilometers (2,790 miles). This relationship with the African Plate and Mid-Atlantic Ridge expansion defies explanation with the convection cell hypothesis.[105,480,481,790]

Furthermore, ocean basin studies of subduction zones (*island arch trench sediments in fracture zones*) show that they are not deformed as predicted, and therefore, subduction, per say, may not occur.[105,503] In addition, no accumulated scrapings of ocean sediments can be found, which should exist due to a plate plunging into the Earth's lower layer (*mantle*).[105] Plates are now known to penetrate to much greater depths (1,000-1,200 kilometers; 620-750 miles), making the aforementioned problems with subduction worse, and also discrediting convection.[237,778] Most of these inconsistencies go generally without discussion when describing plate tectonics.

Subduction has been claimed to be the source of major earthquakes. However, subduction cannot explain why some of the most powerful earthquakes take place in mid-continent, such as occurred in Missouri in 1811 to 1812, and Illinois on 11 June 1987. The Missouri quake was so powerful that it was felt from the Great Lakes to Florida and rang church bells 1,130 kilometers (700 miles) away in Washington, D.C.[748] One theory is that convection currents start underneath a continent and try to pry (rift) it apart, but this requires that an area beneath mid-continent become hotter than the expanse around it. No known mechanism is capable of this, especially convection cells. Furthermore, observations of heat flow contradicts this explanation and models for thermal evolution have not been reconciled with the geologic record.[62] A number of continents show stress in mid-continent, while plate tectonic models as they currently exist indicate that stress should occur at plate edges where mountains form.[443,617,812]

A scientist studying recent developments in subduction comments on the problem: "A lot of people believe that we're in a crisis in the geosciences right now, because the Earth has gotten so complicated that we can't say anything about it anymore."[778] The reason for this is that the new model of the Earth is not being considered, but the enigmas, such as observations of plate tectonics and earthquakes, are resolved with FEM, as will be shown.

No one has suggested a model that explains plate motions that is acceptable to anyone else. Present theory has not provided a universal framework that accommodates all tectonic problems. When it comes to some local problems, such as comparing a continent with adjacent seafloor, "scientists might as well be mapping on two different planets," comments a geologist.[144] Today, even though plate tectonics theory is more than a half-century old: "plate tectonic motions still lack a dynamic theory."[411]; p.121.

29.3 Magnetic Properties of Minerals

Meanwhile, all materials in Nature are magnetic and many tectonics features are the result of the magnetic properties of minerals. Materials are attracted or repelled by magnetic fields, and in most cases the forces are extremely small. Another force exerted on minerals is electrostatic, particularly if the force changes with time (time-varying; i.e., *Faraday's Law applies*). Electrostatic forces can be purely repulsive, so that two bodies always repel regardless of their relative orientations, such as the two sides of a ridge. The ocean floor and ocean water, including its life forms (organic compounds), tend to meet the characteristics of certain classes of magnetic minerals (*diamagnetic and antiferromagnetic*). The crust and the Earth's interior tend to meet the characteristics of other classes of magnetic materials (*paramagnetic and ferromagnetic*). Both share a fifth class (*ferrimagnetic*), particularly with regard to the mantle (*garnet*).[81,357,406] It is this class of material that produces an axis like that observed along the ridge (*uniaxial anistropy, not perpendicular anistropy*), and it is this material (*basalt or gabbro; i.e., garnet*) that exudes at the ridges.[311] These class distinctions in the magnetic properties of minerals allow for the development of ridge systems and subduction zones in the oceans, while the crust experiences mountain building (faulting, etc.).

Electrostatic levitation is being employed in physics for frictionless transport of monorails and other devices, because electrostatic forces can overcome gravitational forces.[85] The Fields bow down on the planet, creating the plate boundaries through the influence of electrostatic forces, which exert a force allowing for the magnetic alignment of minerals. Computer modeling of plate tectonics demonstrates that the mechanism requires the addition of other minor forces not accounted for by present theory.[402] In the case of plate tectonics the levitation is vertical, allowing the ridges to spread, and plate-plate boundaries to exist without large amounts of drag, and also causes plates to be compressed (see Figure 3.1).

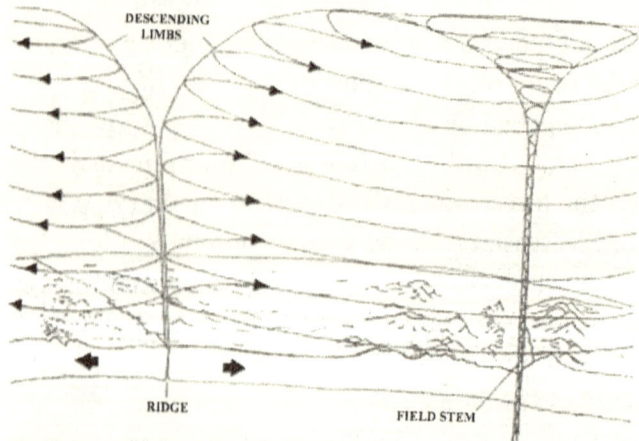

Figure 3.1. Field Interaction in Plate Tectonics. The narrow stem of the Field shown on the right is the point where there is the release of particles.

They accelerate along the contours of the Field where there is interaction with another Field along the ridge axis. These electrostatic forces bring about vertical levitation, causing the ridge to form and spread.

When the mid-ridge magnetic anomalies were first noted in the 1960s, it became apparent that their intensities decreased as they spread away from the ridge. Not only does a central, strong magnetic anomaly exist right over the axis of the spreading ridge, but there is also a stronger anomaly within it. Moreover, there is a very fine structure in the magnetic stripes that cannot be due to the proposed coarse process of convection cell driven seafloor spreading.[138] This can easily be explained by a field acting on the ridge, while conventional theory is at a loss for an explanation.

The magnetic properties of minerals under the influence of the Fields can explain a number of geophysical features on the Earth, including its main magnetic dipole field. The materials in the Earth's interior are of the class (*ferromagnetic*) that acquire a large magnetization in relatively weak magnetic fields. The molecular field in such material always forms parallel to the magnetization. If the magnetization direction rotates under the action of an applied field, the direction of the molecular field rotates with it. An applied field is necessary for maintaining the strength of the Earth's magnetic field (*ferromagnetic saturation*), and it does not retain its strength (*saturation*) when the applied field is removed.

In contrast, dynamos only reinforce existing fields, and therefore, do not explain the origin of the Earth's magnetic field. The Fields of FEM were first generated by the mechanical forces present in the spinning planetary nebula (i.e., Coriolis and centrifugal forces), and this reinforced mineral arrangements. Herein is the basis for the formation of the Earth's dipole magnetic field, the cause behind times of reduced strength, and polar wandering and reversals resulting from a change in direction under the influence of the Interplanetary Magnetic Field (*GMF-IMF coupling and electromagnetic induction*).

29.4 Global Field Dynamics

FEM's field system is a result of the mechanical forces present in the condensing planetary nebula and its interaction with the Sun's Interplanetary Magnetic Field. Thereby, a solar-terrestrial linkage should be detectable in various observations, which it is. Changes in the Earth's rotation or length of day (LOD), and earthquakes are correlated. Also the center of mass of the Solar System has a triggering effect on earthquakes. This has led the American Geophysical Union to state that a solar-terrestrial linkage exists.[203] A correlation between solar motion, geophysical phenomena and climate exists as well (see Volume Two for other discussions).

Plate motions have been observed to follow solar activity at 71 stations around the world. The plates move back and forth, while the 11-year cycle goes up and down.[142] This is exactly what could be expected from varying amounts of particle flow that accompany the solar cycle and a global system of fields (i.e., solar-wind plasma interacting with FEM Fields, which activate plate motion).

In combination, these and other facts indicate that there is a global system of fields that regulates plate motion, and is interrelated with the GMF, IMF and solar activity. As a result, gravitational effects are not the only influence, but electrostatic time-varying effects also play a role. That is, relativistic physics, not Newtonian physics play their part.

Gravitational forces are indistinguishable from the mechanical forces in a concept called the Einstein Equivalence Principle. Gravitational mass is identical with inertial mass and mass is equivalent to energy. The spiraling planetary nebula produced the main (dipole) field, and on further condensation the Coriolis force produced the mid-latitude Fields. The forces were present during the formation of the Earth and guided the alignment of minerals. It is a case of the weak and electromagnetic forces—the electroweak force—controlling gravitational forces. Evidence indicates that there is an interaction between gravitational and electromagnetic fields in accord with general relativity. And non-gravitational forces are evident in the Earth-Moon system. All of these observations result from electrostatic forces along Field lines.

In Tome Five of Volume Two, evidence will be shown that correlates earthquakes with solar activity, and lunar phases and cycles, which also confirms this electrostatic effect (see Volume Two, Chapter 16). A global network of earthquakes reveals a new model of the Earth that includes electrostatic forces and global Field dynamics. Illustrating the overall effects of this Field-dynamical electrostatic force are correlations between earthquake activity, and the polar (Chandler) wobble, lunar tidal forces, and solar activity. A field-dynamical mechanism produces forces (*bow waves, a plasma torus, potential gradients, and electrostatic repulsion*) that allow for solar and lunar triggering of earthquakes, not gravitational forces.

Contrary to expectations, a substantial loss (*degassing*) of helium is taking place through the continental crust and deep ocean.[569,767] The helium (^3He) released to the oceans at the ridges is about eight times higher than the atmosphere. This indicates that heat flux and helium release are the result of an unknown source from the core (*nonradiogenic degassing*). Scientists examining the helium and heat flux indicate one possibility: "Present models for the chemical and thermal evolution of the Earth are seriously in error."[569] The source of the helium and heat is the hydrogen fusion by-products that include helium and ionizing radiation producing heat, and this would be strongest along Field lines, such as along the ridges. With FEM one could predict these observations, while present theory considers them enigmatic.

As briefly discussed in Tome One, the Fields were shown to influence weather, as well as plate boundaries. Particle flow consisting of ionizing radiation (high-energy neutrons, protons, electrons, helium nuclei, etc.)

creates a vacuum. As a result, the upper, colder regions of the atmosphere flow into lower regions, producing winds and other weather phenomena, such as pressure changes, storms and precipitation.

Observations of earthquakes often display weather phenomena. Consider some examples. For a month to six weeks before earthquake activity, large recurring patterns of high pressure developed off the coast of California. The patterns of high pressure even outlined the San Andreas fault hundreds of kilometers off the coast.[529] Some types of quakes (*dip-slip*) have been known to occur along with pressure waves high in the atmosphere, and have been referred to as "ionoquakes."[679] The Fields are in the oceans and an increase of earthquakes along the seafloor near Easter Island (*East Pacific Rise*) was noted during the dramatic ocean current and weather changes associated with El Niño.[769] On 12 April 1978, Chatanika, Alaska was hit by an earthquake. Prior to the quake's onset, ionospheric particles were observed to flow vertically in contrast to the usual horizontal flow. Acoustic gravity-wave theory fails to explain the observations below 300 kilometers (185 miles), which require an unrecognized heat source.[390] Again, the evidence cannot be explained by present theory, but is fully in accord with FEM.

29.5 Evidence of Electrostatic Forces

As discussed in Tome One (Chapter 11), animal response to pending earthquakes is so common that it is used to predict them, and is said to be due to static electricity (i.e., electrostatic forces). Other phenomena include rain, winds, electrical discharges, ball-lightning, and strange crackling sounds.[138] All of these phenomena are explainable by particle flow, but not by present plate tectonic theory.

For many centuries, observations of earthquake lights and other similar phenomena have been reported. However, it is only recently that such controversial phenomena have been linked with tectonic processes. Theories have been without any real success when attempting to describe the origin of earthquake lights (EQL). As discussed, earthquake lights appear aurora-like, which is the result of air ionization due to particle flow along Field lines.

For instance, EQL sightings were recently reported between 1 November 1988 and 21 January 1989 in the Saguenay region, 200 kilometers (125 miles) north of Quebec City, Canada. Three types of EQL were correlated with 54 seismic shocks. These included: (1) sparkings without accompanying sound; (2) diffuse light, such as heat-lightning intensity at sunset or sunrise illuminating a large part of the sky; and (3) vertical and horizontal aurora-like stripes. A key observation is that the first type is more closely related to the atmosphere than to the Earth's surface. Fireballs often popped out of the ground repeatedly, and others were seen several hundred meters up in the sky. Some observers described dripping luminescent droplets under stationary fireballs. Orange, yellow, white and green were the colors most often identified with the various types. The greatest number of sightings were highly synchronous with the timing of two main shocks.[566]

None of the several theories that have been proposed to explain these luminous seismotectonic electromagnetic emissions is fully satisfactory.[507,566,802,813] One report seems to have touched upon an aspect of the luminosity, which involves the instability of water drops in an electric field.[271] However, the origin of the electric field is unknown in conventional thought, but indicate what could be expected of electrostatic forces along Field lines.

Mountain lights have been seen in the Andes, Alps, Mexico and Lapland even under cloudless skies and very low humidity. The effect was not due to lightning, but a potential gradient.[474] These mountain lights are sometimes visible far out at sea. Due to numerous sightings, the Andes have been described as a giant lightning rod. They have a constant glow in late spring to autumn with occasional outbursts, particularly during earthquakes, such as the great quake of August 1906. This aurora-like glow has long been noted on other mountains as well.[16]

Earthquake lights are most frequent when the Moon has passed its closest approach, and thereby, occur during a decrease in the lunar tide.[766] The piezoelectric effect is, as discussed in section 11.1, the luminous effect of quartz crystals under pressure. If the piezoelectric effect were producing these lights, then the opposite would be true (i.e., occurring with an increase of lunar tide). Likewise, the observation of EQL over the ocean is incompatible with this theory. Many lighted-displays could never be produced by the piezoelectric effect, or any other luminous effect due to rock fracture or pressure, and this is why these theories have remained controversial.

Anomalous electromagnetic changes preceded several quakes and an undersea volcanic eruption that shook the central coast of Japan in July 1989. On 5 July 1989, a magnitude 4.9 tremor struck off the coast of Ito. About six and again four hours before the earthquake a monitoring system detected electromagnetic bursts in the extremely and very low frequency ranges (*between about 1 and 9 kilohertz*). Sporadic bursts also took place four days later when a 5.5 magnitude quake occurred, and again the day before an undersea volcanic eruption on 13 July 1989.[226] Likewise, hours before the San Francisco (i.e., Loma Prieta) earthquake on 17 October 1989, anomalous ultra-low frequency (*0.01 to 10.0 hertz*) magnetic signals were observed.[221]

Magnetic fluctuations and radio emissions at or near the quake zone are common. Changes in magnetic field characteristics during and after quakes can be local or even global.[164,325,326,364-367,389,509,514,518,696] Radio emissions have been observed, for instance, during the Chilean quake of May 1960, which was picked up by cosmic radio noise monitors across the United States.[254,772] Electromagnetic precursors to large earthquakes have been observed in Japan, Iran, Chile, Greece, and the former Soviet Union. Radio waves undergo a sudden drop one to six days prior to an earthquake. Electrical conductivity increases as the rocks become stressed and short pulses of radio signals (time-varying acceleration) are observed just prior to earthquakes.[726] Electric and magnetic field effects have been repeatedly observed near active faults and volcanoes in the United States.[365] Magnetic fluctuations, increases in electrical conductivity, and radio emissions are indicative of particle flow (electrostatic forces), and fields (time-varying acceleration) not recognized in present plate-tectonic theory, but are fully in accord with FEM.

Another indication of electrostatic forces are observations of isotope fluctuations. Coseismic changes in radon levels in groundwater took place with earthquakes of 6.0 magnitude and greater in Japan. A total of eleven coseismic changes were noted in the period from January 1984 to July 1988, displaying downward spike-like decreases. As could be predicted, present physical theory offers no explanation and the mechanism is unknown.[768] A gradual rise and then a sudden fall in radon concentrations have also been observed before some Hawaiian quakes by the University of Hawaii's Center for the Study of Active Volcanoes, in Oahu. Groundwater radon anomalies have also been associated with earthquakes in other studies.[345] Radon and helium anomalies were observed for earthquakes from the Northern Apennines, Italy[259], the Phlegraean Fields, Central Italy[455], and Central California.[609] About 32 hours before the Loma Prieta, 17 October 1989 quake, a soil-monitoring station on the San Andreas fault recorded a spike in helium levels at Stone Canyon, 60 kilometers (37 miles) southeast of the epicenter.[610] Likewise, hydrocarbon (methane) emissions have been noted during earthquakes.[138]

Reports of luminous seas during earthquakes are also indicative of particle flow. Energetic particles are capable of producing thermoluminescence, and thereby, luminous seas follow.[528,809] Furthermore, the geometric patterns in luminous seas requires an applied field. This same effect is also why luminescent droplets were seen, for example, dripping from stationary fireballs in the Saguenay, Canada sightings.[566] In accord with this understanding is the discovery that the instability of water droplets in an electric field can produce earthquake lights (EQL).[271]

29.6 FEM's Dynamics

The Fields are all interrelated, and therefore, observations should show a global system of earthquakes. Many geophysicists have noted a correlation between the polar or Chandler Wobble and earthquakes.[17,52,468,523] A study of 234 quakes for the period of 1901 to 1970 demonstrates that their occurrence closely resembles the curve of the Chandler Wobble.[548] It was noted that major earthquake belts experience quakes at times of changes in the Earth's rotation, the polar wobble, and the drift of the geomagnetic field. Their comment could be expected from an understanding of FEM: "The patterns which emerged suggest that all of these diverse phenomena are related."[589] Another study correlated polar motion (tides) and earthquake occurrence, which indicates a "common excitation source."[581] Even the speeding up and slowing down of the Earth's rotation has been shown to be interconnected with earthquake occurrence.[112,464] Polar tides and seismic energy are correlated in such a way that there is a relationship between polar motion and quakes.

Different earthquake belts have nearly common active periods, indicating that they are strongly coupled on a global scale.[116,508] The number of moderately large earthquakes decrease when the number of very large earthquakes increase, which is "suggestive of a causal relationship between these two groups of quakes."[372] Likewise, there is a remarkable similarity in the curves of the annual number of large quakes, and large intermediate and deep focus earthquakes.[1] Such observations could be predicted if a global system of fields existed that is triggered by varying amounts of particle flow involving an electrostatic mechanism.

Because the Field-dynamical Model is electromagnetic and the particle flow is electrostatic, it would produce local to global magnetic fluctuations, which have been observed.[225,366,509,510] Two scientists wonder whether this points to some "behind-the-scenes" process not yet recognized, and ask the inevitable: "Is the occurrence of earthquakes and the position of the rotational pole dynamically related in some unseen way? Just as plate tectonics itself is not 'explained,' it is unclear why earthquakes exhibit the behavior found."[112]

Recent observations show that seismic or shock waves produced by earthquakes travel faster when going in a north-south rather than east-west direction through the Earth. This has left geophysicists baffled by what they consider "outrageous" theories, such as: "Another possibility is that the core contains a super magnetic field which would slow down seismic waves moving parallel to the equator. But no such magnetic field has ever been detected. Also, it would have to be as strong as magnetic fields in some stars."[21]

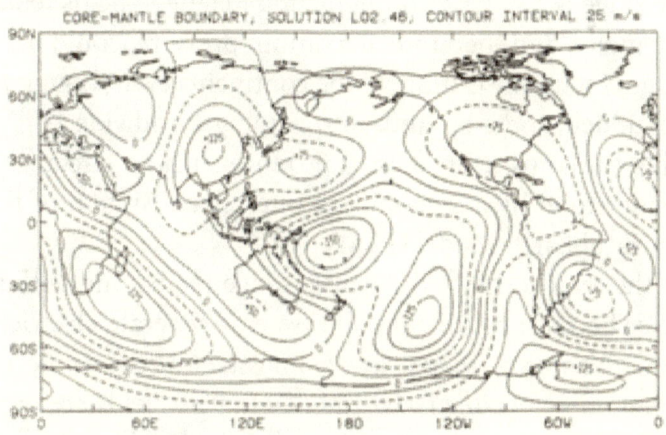

Figure 3.2. Seismic Velocity Anomalies at the Core-Mantle Boundary. Again, the Field locations are evident in this figure of earthquake velocity anomalies. This confirms the dynamic structural arrangement of the Fields is evident even to the depth of the core, and that they play a role in earthquake phenomena. [from Dziewonski, A.M. (1984) Mapping the Lower Mantle: Determination of Lateral Heterogeniety in P Velocity Up to Degree and Order 6. *Journal of Geophysical Research* 89B: 5929-5952; Copyright 1984, American Geophysical Union].

Certainly that is an adequate description of FEM, but conventional training has made this geophysicist resist his own conclusion. The core is purported to produce convection cells, while the evidence indicates that the core is not in the proposed state. Hence, convection cell theories are seriously flawed on the most fundamental level. See Figure 3.2 which shows earthquakes wave velocity anomalies at the core-mantle boundary.

Another indication that the Fields control plate dynamics is observations of the mid-ocean ridges. A concentration of earthquake epicenters occur along the axis of the ridges, which are generally shallow and of relatively small magnitude.[524] A large magnetic anomaly marks the axis of any ridge, such as the Mid-Atlantic Ridge.[304,305] The maximum pressure at the ridge is 45°.[411] Two interacting magnetic fields tend to produce right angles, and therefore, maximum pressure would likely occur at 45°. Ridges retain the outline of the continents even though there is a history of sudden, and stop-and-go spreading worldwide.[411,469,765] It is *not* possible for convection cells to produce such an effect, and it can only be understood by a dynamically interrelated system of fields; the Field-dynamical Earth Model (FEM).

In fact, the reason the Mid-Atlantic Ridge retains the outline of the continents is because continental life systems generate electrical stability to Field dynamics (*biogeoelectric-magnetohydrodynamic interrelationships maintained by the terrestrial biota*). The Fields are responsible for the magnetic stripes that form parallel to the ridge, and typically, equal amounts of seafloor form on both sides of the ridges.[51,462] This is also the reason why the ridge appears to be formed along coupled sliding (*shear*) directions in the upper crust.[295] Spreading in two different directions, such as occurs in the northeastern corner of the Pacific Ocean (eastwards and northwards), or under North America (east) and the Bering Sea (north), cannot be explained by present theories of seafloor spreading induced by convection cells. Tectonic stress patterns either flow towards or away from the Field areas (see Figure 3.3)

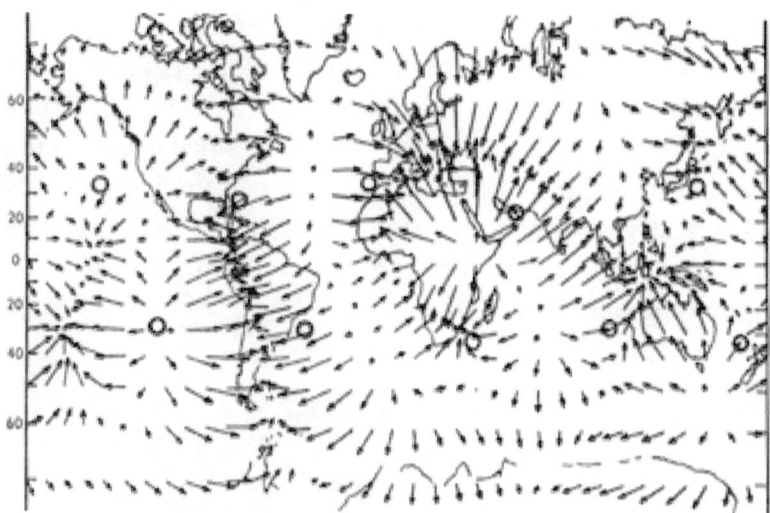

Figure 3.3. Tectonic Stress Patterns. These stress features and the Field positions (marked by circles) show definite relationships. Note how the lines of stress flow into or away from the Fields' bases. Other stress patterns are created at the plate boundaries where the Fields' descending limbs bow onto the Earth's surface. [from reference 135]

Likewise, abnormally heated regions of the ocean floor are confined to a very narrow zone along the crests of the ridges, and heat flow is already normal or below normal on the slopes of the ridges. That is, the ocean floor (*lithosphere*) cools extremely close to the ridge axis. This is hardly what could be expected from convection cells, which should produce a gradual temperature change (*i.e., a thermal gradient*).[51,105,411] The South Pole, which is nearly ringed by ridges, offers a clear example of the Field bowing down and creating a plate. Even if convection cells are necessary for plate tectonics then an applied field is required to stabilize and maintain their positioning and dynamics (see Figure 3.4).

Many of the anomalies encountered exhibit the effects of the Fields. According to the convection cell model, seafloor spreading should be at a maximum near the equator and a minimum at the poles. For one, the

Mid-Atlantic Ridge displays the exact opposite.[469] In fact, it is near the equator (3° North) and along the 30° to 40° North Latitudes (30° and 42°) that seafloor spreading has ceased for a long time (*since the Miocene*).[107,208-210,640] The powerful earthquakes that occurred in mid-continent in Missouri and Illinois were between the 35° to 40° North Latitudes, and were approximately midway between the stem of the North Atlantic Field and its descending limb (San Andreas and other faults along the west coast).[748]

Figure 3.4. Ferromagnetic Forces Shaping the Poles. Here we see the Arctic overlaid by Antarctica (hatched area). They appear as close mirror images of each other. A natural explanation for this phenomenon involves the effects of the main dipole field on the magnetic properties of minerals beginning from the protoplanetary stage onward. That is, the positive and negative charges of the opposite poles constrain minerals arrangements differently. This is also why one pole is dominated by land and the other by water. As we will see in Volume Two, Mars also has mirror image polar arrangements in a pinwheel shape, and all the planets have similar ferromagnetic-induced arrangements, including polar flatting. Likewise, the Fields on the Earth and other planets are responsible

for the plates in plate tectonics. This was also noted in the mirror image of simultaneous aurora at each pole shown in Plate 1.1.

Island clusters, branch ridges and continental masses that are without explanation in the framework of conventional plate tectonic theory become understandable with FEM. Seychelles is near the South African Field, the Canaries near the Mediterranean Field, the Chile Rise extends from the 30° to 40° South Latitudes, and the Faeroe Islands and Iceland are between the North Polar Field and the Mediterranean Field, among other examples. Observations (*spherical harmonic analysis*) of major tectonic features display tension in relation to the Fields.[135] Gravity and magnetic anomalies occur at the ridges and faults, while they are also precursors to individual earthquakes and volcanic eruptions, which is expected of FEM.

According to recent computer modeling, the plates work best if they somehow provide their own motive force.[402] The model does not work very well without the addition of some minor forces, such as electrostatic levitation induced by particle flow along Field lines. For instance, drill hole research near the San Andreas Fault at Cajon Pass reveals the unexpected absence of large amounts of drag at plate-plate boundaries.[811] The descending limbs of the Fields make contact, which leads to a ridge or a fault without drag due to electrostatic levitation acting on the magnetic properties of minerals at plate edges.

The shapes and positions of the ridges and trenches can only be created by the Fields. For example, the Mid-Atlantic Ridge has retained an "S" shape throughout its long history. The Indian Ocean is an inverted "Y" shape, and the Pacific Ocean an "X" or "Z" shape, or in other words, anything but the longitudinal type expected from convection. The Mid-Atlantic Ridge has retained its stationary position, while the Mid-Indian Ridge moved eastward from Africa, but both remained joined through their long histories. Likewise, Africa is perfectly stationed between both of these ridges and their connection to the south of

Africa. Africa was propelled and rotated through its tectonic history so that it is in a "three-fold" median position in spite of its varying tensions. Furthermore, all of the mid-ocean ridges have different spreading rates, but still retain their junctions.[469] Certainly a mutually interacting field system can easily produce a ridge system that retains its shape and connections through time and different rates of spreading; in contrast, its ludicrous to think that convection cells could be responsible (see Figure 3.5).

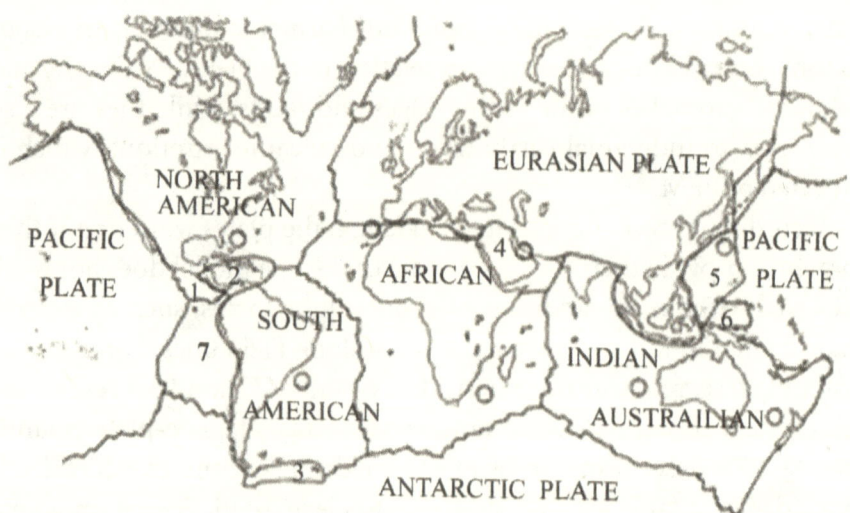

Figure 3.5. Tectonic Plates. These are the various plates of the Earth. Note how each Field, marked by a circle, shows a dynamic relationship to each plate. The small plates are near the stem or base of the Fields. The small plates are as follows: (1) Cocos Plate, (2) Caribbean Plate, (3) Scotia Plate (4) Arabian Plate, (5) Philippine Plate, and (6) Caroline Plate. The Nasca Plate (7) may be a part of the South American Plate, which it dives under in a subduction zone.

Likewise, the tension across continents (*compressional axis*) is nearly mutually perpendicular (*orthogonal*) to the coastline, which often mirrors the ridges and trenches.[442,812] As could be predicted by the new model: "The nature of the regional compression remains obscure. The compression is a consequence of forces acting on the plate margins, and on the base of the lithosphere."[442] The Fields bow down, creating the plate margins, and through electrostatic levitation compress the plate edges and indirectly the lithosphere, the rock-surface of the Earth.

Magnetic anomalies actually dive deep beneath continents in trenches or subduction zones.[503] This is impossible to explain with convection and seafloor spreading models, but predictable with an electromagnetic field and particle flow (*magnetohydrodynamic and mechanical effects beginning with the planetary nebula*). Earthquakes have not been known to occur beneath trenches or subduction zones at depths greater than 680 kilometers (425 miles). Recently, however, they have been reaching depths of as much as 1,200 kilometers (750 miles), which can only be explained by a progressively weakening Field bowing down on the planet to greater depths and increased particle flow (i.e., due to the present high level of solar activity). This is why earthquake waves have been observed to bend as if there were a lens bending a light wave (as indicated by stress and gravity).[237,778]

In fact, the deep focus earthquake is an anomaly with regard to present models of the Earth. Below 60 kilometers the rocks should be so hot that they become elastic—instead of breaking under stress—they just deform. Therefore, conditions for earthquakes should not exist at depths greater than this. Notwithstanding, since 1964 there have been more than 60,000 earthquakes recorded below 70 kilometers (43.5 miles), and a number to depths of 700 kilometers (435 miles) or more. This is yet another indication that the region between the crust and the core has much more structure than we believe, and a new model of the Earth is required.

Likewise, with the presently accepted, layered Earth model, there should be no mixing between the upper and lower mantle. Subducted plate slabs below this 200-kilometer (124-mile) depth severely challenges this layered-mantle hypothesis. Yet, there are a number of examples of slabs penetrating deeper than this. North of New Zealand they reach over 670 kilometers (416 miles), off Kamchatka Peninsula (northeast Siberia) the depth is at least 1,000 kilometers (621 miles), and so on.[138]

Trenches have no sediment fill as they should if it were like the "bulldozer scrapings" of one plate riding over another.[51,105] A gap (discontinuity) occurs below the zone beneath the region of normal quakes and the cluster of deeper quakes (*Benioff zone*).[105] Satellite data indicates grave contradictions between seafloor spreading and the distribution of land masses on Earth (*mantle*).[661] Seafloor spreading and convection cell theories cannot explain any of these observations, while an understanding of FEM would lead us to predict them.

When examining the ocean floor and the continents for tectonic and topographical features, Field relationships become quite apparent. The Fields are typically situated in basins surrounded by rises, trenches, banks, seamounts, island groups, fracture zones, plateaus and, though rare, rifts and ridges. These features have been created by tension at the base of the Fields' stems.

The rare instances of ridges close to the base, creating mini-plates, such as the Philippine Plate, Cocos Plate and Arabian Plate, are the result of weaker Fields that have two or more descending limbs (due to the helical Field structure unraveling). The Fields producing these mini-plates are located near areas where life has experienced crises in the past and present. The life system has not contributed sufficient electrical energy to maintain the dynamics of the Field (*loss of biogeoelectric, bioelectrostatic, and semiconducting, conducting and superconducting biological components*). These mini-plates are also in or border a region with high-velocity earthquake or seismic waves.

This scenario is in direct contrast to the observations made of the Pacific Plate. The Pacific Plate is the largest plate with the smoothest tectonic features, topography, and gravity field. The Pacific Ocean is the least disturbed life system on Earth, and so the Fields there are stronger, allowing all of these parameters to develop more physically balanced. The Circum-Pacific "Ring of Fire," the zone bordering the Pacific where there is a high frequency of earthquakes, and trenches and subduction zones, is the result of the strong Fields in the Pacific interacting with the weaker Fields close to continents where life crises have occurred. The interaction between weakened Fields, where their descending limbs meet, creates ridges, mountains (faulting, thrusting, etc.), trenches and fracture zones (the Antarctic Plate is a good example).[167,206]

29.7 Composition-Dependent Forces

The distribution of heavy metals, and gas, oil, and coal (i.e., carbon and hydrocarbon) deposits display distinct relationships to the Fields in their present and past (*paleogeographic*) locations. For the most part, gas and oil (hydrocarbons) are located near the base of the Field stem and the descending limb, where there is a sedimentary formation for its accumulation. For example, the world's richest deposit of oil is in the region of the Persian Gulf Field, and North America has richer deposits in the Southeast, near the Field base, and along the West Coast, where the descending limb bows down. However, the Americas display a more widespread distribution because of the opening of the Atlantic Ocean so that the Fields have been on both sides of the continent through time. In some locations, such as Southern Asia's Tarim and Shensi Basins, oil deposits can be found around the 40° North Latitudes, but have migrated away from the proximity of the Field because of continental breaking and buckling (*Altyn Tagh Fault and Kansu Thrust*).

Similar observations can be made of heavy metals and radioactive elements. Conducting and other heavy metals are near the Field stem, or between them where the descending limbs meet, particularly trenches and ridges. The distribution of uranium and its decay product, lead, is similar. However, in the case of metals and uranium, the distribution is not as strict as with hydrocarbons. The Arctic region encompassing the North Polar Field is a particularly good example being surrounded by heavy conducting metals, lead, uranium, and gas, oil and coal.[167,673] These relationships are the result of ionization of the atmosphere creating carbon and hydrocarbon deposits, and the photoelectric effect on elements (metals and radioactive elements) from particle flow (hydrogen fusion by-products and relativistic particles). Ionization along Field lines has been present from the formation of the planetary nebula onwards, leading to the observed geographically restricted hydrocarbon, and metal and radioactive element deposits.

This relationship is also apparent in the fact that plate tectonic episodes and hydrocarbon deposits are intimately related. Fluid hydrocarbons, such as oil, require organic-rich beds within a sedimentary base. Mass extinctions, producing organic-rich beds, hydrocarbon deposits and plate tectonic episodes all occur together in cycles or episodes. Major hydrocarbon accumulations are "genetically" related to continental break-up, especially during the Age of Reptiles, and the Age of Mammals (Mesozoic and Cenozoic).[172-174]

29.8 The Need for New Perspectives

Major inconsistencies exist between theory and observation when examining Mid-Ocean Ridge formation, sedimentary layers, and subduction; yet, they can be explained by the magnetic properties of minerals and FEM. Along the northern latitudes of the Mid-Atlantic Ridge there has not been any spreading for about one third of the time in which the Atlantic formed (*from Lower Miocene*). As could be suspected from the effects of the Fields, the absence of spreading is observed along the equator, and the 30° to 40° latitudes.[51,98,107,122,208,579,640,754] Iceland is upon an extension of the Mid-Atlantic Ridge called the Reykjanes Ridge; spreading that occurred elsewhere on the ridge has not rended Iceland in two.[469,513] Mount Cobb on the crest of the South Pacific Rise has not spread for quite some time (*since the Oligocene*).[306] Ancient rocks (*Jurassic, Cretaceous and even Paleozoic, including Cambrian*) are widespread near the crests of ridges.[411] According to the proposed mechanism of subduction, sediments should have been pushed down and melted, or accumulated in trenches to great depths, but many observations show that this is not the case.[51,105,208,411] Meanwhile, observations like these can be readily understood as electrostatic influences on various materials due to their magnetic properties.

Plate motions can be incredibly complex and unexplainable in terms of present theory, as one scientist comments:

> "Within the framework of drift's super-unreality, our planet might be called the planet of miraculous pinpoint encounters—the planet of pinpoint docking techniques, where continuous tectonic features are rent asunder, displaced out of context one or more times, only to find themselves, in the end, right back where they started from! Is it conceivable that so many diverse spatial and temporal elements, starting independently at different stages, and traveling different routes at different speeds (adjusted by drifters to rationalize specific situations) could pinpoint the geologic moment of their arrival and dock into a coherent, unitary system around the entire Earth? Is such a phenomenon really *possible*? This indeed is super-irreality!"[469]

One observation that encourages disbelief in present tectonic theory is what has been called the Gibraltar "reference" or "pivotal" point. The drifts of earlier groupings of continental mass (*Gondwanaland*) were westward for the Americas, but it was radially for the remainder of the continents from this point. During the Cretaceous, Spain rotated around this pivotal point forming the Alps, Apennines, Pyrenees and other mountain ranges. For instance, the Alps were generated by an eastward movement of Africa relative to Europe.[331,469]

South America was to have rotated about the Gibraltar pole of rotation, too. The migrating Mid-Atlantic Ridge is centered on Gibraltar. Africa is said to have moved eastward from the Mid-Atlantic Ridge and westward from the Mid-Indian Ridge, and rotated counterclockwise with respect to Eurasia, around a pole of rotation centered near Gibraltar.

In addition, in relation to an ancient (*Tethyan*) sea between Africa and Eurasia, it is said to have moved toward the east, then toward the west, and finally toward the north. Yet, with so many movements and counter-movements, Africa's geology perfectly matches its geologic counterparts in Eurasia! Likewise, Spain was to have rotated only to perfectly match-up with where it started.[332] Its connection with the new model of the Earth is clearly evident, because the Mediterranean Field, at the mouth of the Mediterranean, is centered near Gibraltar.

Africa itself reveals the inconsistencies of present plate tectonic models. It is purported to be rotating counterclockwise with respect to Eurasia. Meanwhile, Africa is also said to be rotating eastward from the Mid-Atlantic Ridge. A discrepancy, which has never been reconciled, is that Africa is said to have moved eastward away from the Mid-Atlantic Ridge since early in the Age of Reptiles (*Triassic*), and westward away from the Mid-Indian Ridge since early in the Age of Mammals (*Eocene*).[332,411] How could Africa be displaced in so many different directions at once without subduction zones swallowing up the newly formed seafloor? No subduction zones exist around Africa to do the job! Furthermore, certain geologic structures (*Afar Triangle*) cannot account for the release of this tension. To top it off, it is said that the African Plate has been at rest since the Middle of the Age of Mammals (*Early Miocene*), while both spreading mid-ocean ridges retained their shape the entire time! Isn't that entire scenario, with respect to convection cell theory, quite ridiculous?

The origin of the Himalayas is another tectonic enigma with respect to present plate tectonic theory. It is purported that the Himalayas are the result of a collision between India and Asia. The slow process of convection-cell driven, seafloor spreading was supposed to have pushed a huge part of the India subcontinent beneath the Himalayan region, crushed it, uplifted it, then pushed it past present-day Tibet, and finally, lifted Tibet thousands of meters into the air. Such a feat requires the sinking of India approximately 30 kilometers (19 miles) into the upper layers (*lithosphere and asthenosphere*) of the Earth and worse, not lose its rigidity, which it should had it reached that depth.

Contradicting this entire scheme is the geologic evidence that northern India has been in its place at least since the Late Cretaceous, while the Indian Ocean had begun at a much later time (*Eocene/Oligocene Boundary*).[331] Furthermore, the Indian Ocean's growth does not correlate with the rate of the Himalayas' uplift.[105] Theory indicates that plates move towards voids, not barriers, but India is said to have done the opposite, producing the gigantic Himalayas on the meager power of a single spreading ridge! A geophysicist rightfully asserts: "This strikes me as a sort of monstrous geophysical absurdity."[469]

The Americas also have performed some astounding feats in tectonics according to theory. North America began its drift first, followed some time later by South America. Yet, in spite of their separate histories, the ridge is exactly midway down the Atlantic and they dock so that their geologic histories and formations are re-composed into one total, unified profile.[503]

Another astounding maneuver is North America's drift westward toward the Pacific Ocean where it ends up making a perfect "pin-point" encounter with Siberia, producing a unified geologic unit.[331] Here we have the faster spreading, enormous Mid-Atlantic Ridge pushing the huge North American continent so that it smashes against Asia with only a pin-point encounter, while the relatively small Indian subcontinent, and meager Mid-Indian Ridge supposedly produced the Himalayas in much less time with much less force!

Other proposed events seem just as miraculous. The rotation of Africa was to have pushed the Arabian section precisely into a pinpoint "excavated slot" of the Iranian coast across the Strait of Oman (*Hormuz*) forming the Persian Gulf.[331] Clearly, the Persian Gulf is a single tectonic unit, because of the Persian Gulf Field, and so, it has not been affected by "drift."

The positions retained by the mid-ocean ridges through time is another enigma for present theory. India traveled northeastward, southeast Asia remained stationary, Australia drifted east then north, Africa rotated into drifting southwest Asia, and the Indian Ocean itself was pushed eastward. Yet, the Mid-Indian Ridge system retained a median position. Likewise, the Mid-Atlantic Ridge retained its median position, centered at Gibraltar, in spite of the different drifts of Eurasia, Africa, and North and South America. A geophysicist discusses the phenomenon: "For a traveling ridge to 'medianize' itself between traveling, nontraveling, countertraveling, supertraveling, and dispersing elements is sheer hallucination."[469]

However, this comment refers to the convection cell model, not the Fields retaining the position of the ridge regardless of the movement of the continents. This is also why ocean floor spreading at different rates produces the same thickness of seafloor everywhere, and volcanic islands still maintain their roots after plates have migrated and the poles have drifted or reversed.[13,51] A host of other problems exist with plate tectonics as proposed that can only find a solution with FEM.

The Earth's shape (*geoid*) has been referred to as having a "memory."[13] Using satellite observations some peculiarities (*geoid anomalies*) correlate with tectonic features, such as island chains, oceanic ridges, continents, hotspots of volcanism (seamounts and subduction zones), and features approximately perpendicular to the ridges (*transform faults*).[13,150,151,302,774] The correlation between subduction (*slabs*) and hotspots with the Earth's shape (*geoid*) suggests that the locations of subduction zones and hotspots are related to the Earth's spin axis or poles.[13,151, 380,557,558] The drift of the Earth's rotational pole has been shown to correlate with changes in hotspot location.[114,380] Hotspots are not simply cracks where volcanic material surfaces, but are the surface manifestations of processes related to the Earth's interior.[151] Even the bulge at the equator cannot be considered a holdover from a former faster rotation rate, because it shifts with the wandering pole.[257] The bend in the Hawaiian Island chain has also been shown to be the result of polar wander.[638] Convection cannot produce such accuracy with respect to structure relative to the main dipole magnetic field; meanwhile, a system of Fields interrelated with the geomagnetic field can (i.e., FEM).

Continental displacement, volcanic (*magmatic*) activity, and polar wander and reversal occur rapidly together followed by long periods of stability. As commented on previously, all tectonic phenomena take place in episodes that occur along with mass extinctions and speciation, sea level fluctuations, climate transitions, hydrocarbon deposits, and so forth. Geophysicists attempt to explain these correlations as the result of changes in the Earth's shape and internal imbalances (*geoid mass anomalies, tumbling of the mantle, etc.*) without much success.[59,139] Meanwhile, it is the shifting and varied levels of activation of the Fields (*magnetohydrodynamics and particle density*) that produce these observations.

The Fields control plate boundaries and motion during continental breakup, leading to what has been referred to as the "jigsaw" or "patchwork" Earth.[718,779] Many very old (*Paleozoic and Precambrian*) continental borders exist along the Pacific (*Circum-Pacific periphery*).[100,117,131,158,297,480,481,542,686] These areas include western North America[172,287,368,660], Alaska[77,511,616], the Andes[131,717,789], and eastern Siberia.[288,485,486] More-recent fragments (*Mesozoic and Early Cenozoic*) are found in Japan[297,611], New Zealand[67], the central Andes[350], and Alaska.[324,571,710]

The geology of Japan resembles the eastern and northern Far East regions.[405] The Sea of Japan is relatively young, and the Japanese landmass was originally connected to the Far East during the Age of Reptiles

(*Mesozoic*).[410,505,735] A sharp bend occurred in the main island of Japan at the boundary.[380]

The Far East region was also separated from Siberia until the end of the Age of Reptiles (*Mesozoic*).[103,119-121,173,227,409,485,486,542,814] Almost all of China south of the 40° North Latitude (the Field latitude) and Korea are from other locations.[340,542] Again, the sectioning is consistent with the latitude restrictions of FEM and is not explainable by convection.

North America shares fossils from the Mesozoic with New Zealand, Caledonia, Antarctica and Chile. In northwestern North America, a tract stretching from Alaska to Oregon (approximately 2,000 km. or 1,240 mi.) has enormous quantities of basalt of unknown origin.[368,542] The Caribbean region was once part of the Pacific Plate, and rotated clockwise and counterclockwise in the Cretaceous, finally ending up as the Caribbean Plate.[260]

The early evolution of Eastern Australia and New Zealand were not the result of collision, but subduction and extension.[443,636,674,701] Florida and coastal New England were part of South America, parts of New Foundland were once part of Georgia, Nova Scotia was connected to part of Africa, and Yucatan was once part of the Mississippi Valley.[718] India has close ties with Antarctica, Madagascar, Africa, China, Tibet, Kazkhistan, Afghanistan, Iran and Australia, which is absurd when considering a convection cell model of plate tectonics.[105] In attempting to explain these observations, two different supercontinents (*Panagea A and B*) have been proposed, with continents colliding and separating a number of times, and a lost continent, Pacifica, whose fragments are part of the continents bordering the Pacific.[516] These observations are without an explanation if a convection cell model is applied.

Other observations can only be explained by FEM. Generally, most continental plates are in a state of compression, except Africa which is under tension.[442,617,812] North America, Europe and Siberia have all been converging on the Arctic, which should have compacted the Arctic plate by about 5,000 kilometers (3,100 miles), but instead the plate has extended.[105] The plates would be expected to display observations like these if they were controlled by Fields and particle flow producing vertical, electrostatic levitation.

FEM offers a new understanding and resolves present model enigmas, and so-called, anomalies. Here we have the means by which global seafloor spreading can occur in spurts, allowing for rapid, global continental breakup and dispersal. Ridges maintain their positions in spite of these continental and plate movements. Again we are left with the question of how rapidly the process takes place and exactly when the events occurred. This will allow for the deduction of the relative time span for some events, and thereby, begin to resolve the problems associated with radiometric dating. FEM also accounts for the observed cycles and their global synchronism in a wide variety of geologic and fossil records.

CHAPTER 30

Time and Time Again
(The Problems of Radiometric Dating and Other Assumptions)

In 1905 to 1907, an American chemist found that in the decay of uranium (U) to lead (Pb), the ratio Pb/U was consistently greater for older rocks. Most rocks and minerals contain radioactive atoms of certain chemical elements that exist in Nature. The rate of decay for each radioactive isotope is characterized by a particular half-life. A half-life is the time required for half of the radioactive "parent" to decay into its more stable "daughter."

For example, assume a certain radioactive substance has a half-life of a million years. After a million years, half of the parent element will remain, another million years a quarter of the original quantity will remain, after yet another million years an eighth will persist, and so on, virtually to infinity. The age of a given rock formation is calculated by estimating the amounts of parent and daughter elements that were originally in the rock, measuring present ratios, and deciphering the number of half-lifes that appear to have passed since its formation. Dating rocks by utilizing radioactive isotopes is called radiometric dating and involves the decay of isotopes of uranium to lead, rubidium to strontium, potassium to argon, carbon to nitrogen, and others.

30.1 The Dating Dogma

A major assumption is that this entire process involves a *closed system*. This closed-system concept assumes that the radioactive decay is continually produced by nuclear reactions caused by the interaction of a *constant* stream of ionizing radiation (usually referred to as cosmic radiation).[194,286] The basic underlying questions, which are simply ignored: Is it proper reasoning to assume that a constant level of radiation is maintained throughout time? Likewise, does a closed system apply to the Earth and its biosphere?

Herein lies one of the major problems in geology, because the evidence indicates a necessity for an open system and fluctuations in background radiation. Evidence is plentiful for supporting the idea of changes in the level of radiation. Further support for these fluctuations was presented at the molecular and thermodynamic levels, in Tome Two.

Furthermore, the fossil record shows that life's evolution has undergone increasing order, complexity and build up through time. In contrast, the Second Law of Thermodynamics, entropy (decay, etc.) indicates that a closed system should do the exact opposite; that is, increase disorder, randomness and breakdown. For instance, a house will eventually fall to dust if not maintained. Evolutionary steps work in the opposite direction with increasing order, decreasing randomness and build up as observable characteristics. Most scientists will simply state that this is because evolution is an open system and the evidence does confirm that viewpoint (i.e., increasing complexity and diversity in evolution, fluctuations in iridium levels and stable isotopes, periodicity of the geological record, etc.).

However, those same scientists will often turn to the geologic time scale and say that the events took place so many millions of years ago. Yet, that dating system is based on the assumption that it is a closed system. Herein lies one of the great contradictions or paradoxes in geology: How can evolution be an open system, with confirming evidence, and then be dated by a closed system?

The evidence for periods of increased radiation is abundant. Various scientists have expounded on the necessity for such a scenario in order to explain what appears in the fossil record. Most scientists have indicated the need for an open system. Increased radiation in an open system translates into the fact that radioactive decay is not constant. Geological events also occur, generally, in a much shorter time than suggested by the proposed constant. The closed system that has been used has exaggerated the dates of events. This is especially true if we consider that rock formations were being laid down as increased radiation was present, producing daughter elements at the onset. The idea that geologic events take place rapidly and on a much shorter time scale is something that geologists have had a history of resisting. After all, there has been more than a century of theories that *assume* millions and billions of years are responsible for all of the evidence.

It seems quite evident, as supported by the numerous observations discussed and those which follow, that ionizing radiation has been cyclically or periodically present in the environment. This brings about the necessity of questioning the validity of geology's unquestioned assumption that a closed dating system, radiometric dating, is fairly accurate. The new model of the Earth can overcome the physical constraints and paradoxes deduced from observations made with the intent to prove predetermined theories; this is especially true of time (relativity).

Our dating systems give exaggerated dates to the geologic time scale. Geochronology is based on the assumption that everything through time is exactly as it appears this very minute: a relatively closed system. Cosmic radiation levels impinging on the Earth are assumed to be constant, and striking solid minerals cause the release of subatomic particles at a constant rate. In addition, weak atomic forces will eventually give way, allowing the release of subatomic particles. Again there is an unproven assumption that we know all about physical laws, which scientists admittedly state we do not.

The dating system works on the hypothesis of this limited perspective, because the release of subatomic particles changes one mineral into another, and it is the ratio between the original mineral and the newly formed one that yields the date of an event. For example, rubidium, if bombarded by a constant level of cosmic radiation, will release particles changing it into strontium, and it is the ratio of rubidium to strontium that allows the calculation of the apparent age. However, if more radiation than usual occurs at some point then the calculated age will be exaggerated. Our observation of iridium layers, selective and stepwise extinctions followed by sudden blossomings of new species, fluctuations in stable isotopes, and other observations indicate what could be expected by the cyclic or periodic presence of increased radiation.

Other factors that can alter decay rates are clearly manifested in the fossil and sedimentary records as well. These include, but are not limited to, changes in magnetic intensity, alterations in atmospheric temperature and pressure, increased levels of ionizing radiation, the presence of microwaves, and so forth. More evidence in this regard has been presented in Tome Two. These facts are not considered for bringing about an accurate dating system. The question that remains is beyond the scope of this work: When did these events take place? The answer will indicate that these events did not take place anywhere as far back as we often assume.

The entire problem had originated more than a century ago along with the theories proposed by Darwin and Lamarck. About three decades later the idea of radiometric dating surfaced which had to conform to these theories, and therefore was *assumed* to be a closed system. Since that time, geologists had been insisting upon proving that slow gradual changes are responsible for the geologic record. The words of one geologist reflect this century-old problem: "We are told that geological processes only take place as quickly as one's fingernails grow and that therefore it is pointless to look for empirical evidence for events that might have taken place more rapidly."[493]; p.43. In fact, some theories are judged as inadequate simply because they are not conventional and suggest a shorter time scale.[714] Such theories may be completely wrong, concluding them as erroneous in this manner is dogmatic, and involves circular reasoning.

If radiometric dating is inaccurate then observations should show out-of-phase relationships the more one extends back in time. This would be especially true if there were periods that experienced greater radiation, such as at the end of the Cretaceous (i.e., K/T Boundary). This relationship is evident in computer modeling of plate tectonics. Everything works well until the data approaches the Cretaceous when resistive and driving forces could not be brought anywhere near a balance point.[402] Likewise, polar reversals become more and more frequent as the present is approached. Conversely, the further back in time one goes the greater the apparent time span there is between reversals.[328] Similarly, mass extinctions and polar reversals move out of phase backward in time.[598] It is a well known fact that when examining the periodic nature of cycles or episodes in extinctions, polar reversals, tectonics, cratering and so forth, they tend to become less in phase further back in time.[11,104,159,216,233,280,298,349,392,413,450,456,467,484,515,594,595,604,672,750-752,762,764]

Moreover, the stratigraphic record is very incomplete. A few gaps, here and there, could be explained away a due to erosion or the cessation of deposited materials. However, the gaps are often so large in terms of time-span, volume, and geographic area, that it alone brings into question the accuracy of radiometric dating.[138]

The variety of phenomena accompanying each period, if occurring in the presence of intensified radiation would skew the dates. Other factors involve location and mineral composition, so that the cycles appear to be episodic or clustered rather than periodic. Such is the case with cratering in general[403], the length of geological cycles[594,595,602,603], and the objections raised against cycles.[280,327,399,414,415,758] Meanwhile, cycles appear very likely and have not been disproved. Individual events would display evidence that certain features are not in phase, such as the iridium layer and polar reversal at the end of the Cretaceous.[19,393] As predicted by the theory of relativity, there is a warping of space-time with the Unified Field, which is even apparent in the geological record.

30.2 <u>Revolutions</u> <u>in</u> <u>the</u> <u>Earth</u> <u>Sciences</u>

Revolutions away from dogma in the Earth sciences have continually taken place, and we are due for another. About 300 BC and again in the 15th Century, concepts developed away from a flat Earth to a spherical one. In the mid-16th Century a concept with the Earth at the center of the Solar System was replaced with one that had the Sun at the center. Fossils as the remains of former life was only recognized around 1800. The reality of an ice age was resisted until about 1830. In 1966 the revolution involved the acceptance of mobile continents. More recently the revolution shifted to plate tectonics.

In retrospect, it can be plainly seen that many basic understandings have only occurred very recently, which encourages the idea that our enlightenment still has some way to go. Frequently, geological processes have been observed and described, but the physics and chemistry involved have not been understood. As a result, dogma surfaced in the compromise and was subsequently maintained, but was later proven wrong by empirical evidence.

Problems with established professionals accepting new ideas has a long history. One of the earliest examples of such reproach was when Copernicus theoretically attempted to put the Sun in its rightful place. A prominent astronomer at the time, Brahe, discredited him, causing Copernicus to be scorned, cast from society as a heretic, and eventually jailed. More recently, Wegner's theory of continental drift received a generally hostile reaction in spite of its observational support. Jeffereys, a respected geologist at the time, even "proved" the whole idea was impossible. Yet, it withstood the test of time in a revised state now labelled plate tectonics.

Often professionals will use conventional theory as the groundwork to discredit unorthodox theory. The history of science shows that this approach has often hindered the progress of science and caused the loss of important observational facts. As stated by Francis Bacon, those who observe the world with preconceived laws in mind will only recover those laws and nothing fundamentally new will be uncovered. The facts are what must determine reality, not preconceived notions.

Facts that do not support conventional theory are always surrounded by controversy and the inability of professionals to examine them thoroughly enough to come to a sound conclusion about what they mean. It is all to easy to accept those facts that fit preconceived theory, and ignore the ones that do not. After more than a decade of training in a university that instills orthodox perceptions into each of its doctoral candidates, the scientific community has rigorously perpetuated limited horizons for progress in developing new concepts or paradigms. After all, if a fact or new theory does not support a lifetime of learning that one has identified with and defended, then that person is not going to give in without a fight. Often this approach wins for a time, and it is those in the profession the longest that offer the most resistance. Hence, not wholly accepting new understandings, complete with defensive antagonism, is unfortunately the groundwork of scientific revolutions.[431] Hopefully, open minds and an understanding of the history of problems with accepting new scientific theories will be realized, preventing the unscrutinized out-right rejection of a new paradigm simply because it is not the established theory.

30.3 In Retrospect: A New Mind-Set is Needed

This open-mindedness has never been more important. One of the clear lessons revealed by the fossil record is that life crises preceded mass extinctions and catastrophes. As will be uncovered further in Volume Two is the fact that life stabilizes the physical world in many ways. Today a life crisis exists that is comparable with that of the end of the Cretaceous.[191,682,743] One scientist comments on this in an essay titled "The Rise and Fall of *Homo sapiens sapiens*":

> "Natural selection has not equipped us with a long-term sense of self-preservation. Our population cannot continue to expand at its present rate for much longer, and the examples of many species suggest that expansion can end in catastrophic collapse.
>
> Survival beyond the next century in a tolerable state seems most unlikely unless all religions and economies begin to take account of the facts of biology. The possibility of human extinction has certainly been suggested of late because of our destruction of our own planet. In particular, it has been suggested that we are sowing the seeds of our own destruction by destroying so many other species; that we need a planet that is in ecological 'balance.'"[743]

It is the winter of inadequate theories, a time when, like annual plants, they must die to their root. From those roots—the facts—must spring the truth. It is the dawn of a new season where misconception must decay, leading to the soil of new intellectual growth. One paradigm that puts into perspective all facts is essential, not only those that fit predetermined theories. We must let the facts circumscribe what is the truth. It might be said that we have gone through the fall and winter of human thought in these matters. Ahead is the spring of rebirth, and the endless summer of abundant life throughout the cosmos!

If the Field-dynamical Model is a reality, then it should be observable in other celestial objects. In Volume Two, we will take a look at the other planets in our solar system, comets, galaxies, galactic clusters and super-clusters, among other objects. In spite of the very real differences in appearance, they all show phenomena that can be explained by this model, and in some cases, only with this model. Furthermore, the observations suggest that life is not only pervasive, but that life is the primary objective of the Universe. Let's take a journey beyond the

Earth, and see some of the details of the macrocosm by perusing Volume Two.

References

Tome One

1. Abbe, C. (1915) Mistpoeffer, Uminari, Atmospheric Noises. Mon Weather Rev 43:314.
2. Adeishvili, T.G., et al. (1985) Glow of the Night Ionosphere Due to Energetic Ion Beams. Adv Space Res 5(4):251-256.
3. Alerstam, T., Hogstedt, G. (1983) Role of the Geomagnetic Field in the Development of Bird's Compass Sense. Nature 306:463-465.
4. Alexander, J.K., Williams, D.J. (1979) Origin of Plasmas in the Earth's Neighborhood. Final Rept Sci Definition Working Group. Goddard Space Flight Center (April) NASA.
5. Anderson, C.J. (1973) Animals, Earthquakes, and Eruptions. Bull Field Mus Nat Hist 44(5):9-11.
6. Anderson, D.L. (1984) Earth as a Planet: Paradigms and Paradoxes. Science 223:347-354.
7. Anonymous (1860) Mysterious Music on the Gulf Shore. Sci Amer 3:51.
8. Anonymous (1870) Illumination of the Sea. Nature 2:165.
9. Anonymous (1870) Remarkable Illumination of the Sea. Sci Amer 23:102.
10. Anonymous (1875) Finny Musicians. Sci Amer 32:102.
11. Anonymous (1880) A Milky Sea. English Mechanic 31:153.
12. Anonymous (1880) The Milky Sea. Pop Sci Mon 17:573.
13. Anonymous (1883) Musical Fishes. Pop Sci Mon 23:571.
14. Anonymous (1896) Seismic Wave in Japan. Sci Amer 75:186.
15. Anonymous (1897) Mysteries of the Persian Gulf. Sci Amer 76:331.
16. Anonymous (1898) Anomalous and Sporadic Auroras. Mon Weather Rev 26:260.
17. Anonymous (1903) Nature 67:423.
18. Anonymous (1911) Nature 86:90.
19. Anonymous (1912) Curious Light Phenomenon of the Indian Seas. Sci Amer 106:51.
20. Anonymous (1913) Earthquake Sounds in Haiti. Geographical J 41:389.
21. Anonymous (1917) How High is the Aurora? Sci Amer 117:378.
22. Anonymous (1923) Remarkable Atmospheric Phenomenon Observed at Sea. Geographic J 61:66.
23. Anonymous (1924) Science 59:sup xiv.
24. Anonymous (1930) Phosphorescent Sea. Nature 125:513.
25. Anonymous (1931) Low Aurora and Its Effects on a Radio Receiver. Nature 127:108.
26. Anonymous (1932) Earthquakes, Fisheries and Flower Fall. Nature 130:28.
27. Anonymous (1934) Celestial Phenomenon Seen at Liverpool. Nature 134:709.

28. Anonymous (1934) Earth-Sounds in the East Indies. Nature 134:769.

29. Anonymous (1935) Sounds Made by Fishes in the East Indies. Nature 135:426.

30. Anonymous (1961) Do Earthquakes Cause Glowing 'Wheels' on the Sea? New Sci 10:528.

31. Anonymous (1964) What Happened to the Earth's Helium? New Sci 24:631-632.

32. Anonymous (1963) Chilean Earthquake Disturbed Earth's Magnetism. New Sci 18:738.

33. Anonymous (1967) Luminous Wheels Puzzle Seamen. New Sci 33:447.

34. Anonymous (1967) Predicting Earthquakes by Changes in Rock Magnetism. New Sci 36:176.

35. Anonymous (1969) Low-Salinity Area in Sea. Sci News 96:15.

36. Anonymous (1973) Volcano Magnetism. Nature 243:190.

37. Anonymous (1978) Mystery Booms Haunt American Coast. New Sci 77:341.

38. Anonymous (1978) East Coast Booms: Pick a Theory. Sci News 113:181.

39. Anonymous (1979) Tass News Agency, 3 December.

40. Anonymous (1984) There's Life Among the Seeps, Too. Sci News 126:374-375.

41. Anonymous (1985) Comets and Geological Rhythms of the Earth. Sci News 127:24.

42. Arshinkov, I., et al. (1985) Intense Current Structures During Low Geomagnetic Activity and Their Relation to Small-Scale Magnetic Perturbations Seen by the Intercosmos Bulgaria—1300. Adv Space Res 5(4):127-130.

43. Baas-Becking, L.M.G. (1925) Studies on the Sulfur Bacteria. Ann Botany 39:613-650.

44. Babcock, H.W. (1961) Topology of the Sun's Magnetic Field and the 22-Year Cycle. Astrophys J 133:572-587.

45. Babcock, H.W. (1963) Sun's Magnetic Field. Ann Rev Astron Astrophys 1:41.

46. Babcock, H.W. (1967) Zeeman Effect in Astrophysics. Physica 33:102-121.

47. Baker, R. (ed) (1980) Mystery of Migration. NY, MacDonald-Futura/Viking.

48. Baker, R.R., Mather, J.G. (1982) Magnetic Compass Sense in the Large Yellow Underwing Moth, Noctua Pronuba L. Anim Behav 30:543-548.

49. Baldwin, I.T., Schultz, J.C. (1983) Rapid Changes in Tree Leaf Chemistry Induced by Damage—Evidence for Communication Between Plants. Science 221:277-279..

50. Banerjee, S.K. (1984) Polar Flip-Flops. Sciences 24(6):24-30.

51. Baranskii, L.N. , Naumenkov, N.L. (1960) Observations on Earth Currents at Mirnyy and Oasis Station in 1957. No 14, pp.120-123 in Soviet Antarctic Expedition, vol. II, 1964. NY, Elsevier.

52. Barber, N., Longuet-Higgins, M.S. (1948) Water Movements and Earth Currents: Electrical and Magnetic Effects. Nature 161:192-193.

53. Barnes, T.G. (1977) Recent Origin and Decay of the Earth's Magnetic Field. SIS Rev 2:42-6

54. Barrett, J.W. (1898) A White Sea. Nature 58:520.

55. Bates, D.R. (1973) Auroral Audibility. Nature 244:217.

56. Bates, D.R. (1974) Auroral Sound. Polar Record 17:103.

57. Beals, C.S. (1933) Audibility of the Aurora and Its Appearance at Low Atmospheric Levels. Roy Astron Soc Canada J 27:184.

58. Beals, C.S. (1933) Low Auroras and Terrestrial Discharge. Nature 132:245.

59. Beason, R.C. Nichols, J.E. (1984) Magnetic Orientation and Magnetically Sensitive Material in a Transequatorial Migratory Bird. Nature 309:151-153.

60. Bednorz, J.G., et al. (1987) Superconductivity in Alkaline Earth—Substituted L2CuO4-y.. Science 236:73-5.

61. Bellaev, V.S., et al. (1975) Study of the Relation Between the Conductivity-Fluctuation Characteristics of Water and Features of Vertical Temperature Profiles in the Ocean. Atmos Ocean Phys 11(10):680-683.

62. Bellingham, J.G., et al. (1986) Detection of Magnetic Fields Generated by Electrochemical Corrosion. J Electrochem Soc 133(8):1753-4.

63. Bertin, L. (1957) Eels; A Biological Study. Trans B. Roquerbe. NY, Philosophical Library.

64. Bird, C. (1975) Planetary Grid. New Age J 1(5):36-46.

65. Blakemore, R. (1975) Magnetotactic Bacteria. Science 90:377-379.

66. Blanchard, D. (1982) Sea Water: Electrification of the Atmosphere, vol 12. p 151 in McGraw-Hill Encyclopedia of Science and Technology, 5th ed. NY, McGraw-Hill.

67. Blanchard, D.C. (1963) Electrification of the Atmosphere by Particles from Bubbles in the Sea. Prog Oceanogr 1:71-202.

68. Bobrovnikoff, N.T. (1937) An Unusual Auroral Display. Pop Astron 45:299.

69. Bolt, B. (1982) Inside the Earth: Evidence from Earthquakes. SF, Ca, WH Freeman.

70. Bonnycastle, R.H. (1837) Auroral Appearance. Amer J Sci 1(32):393.

71. Bosqued, J.M. (1985) Ion Precipitation into the Ionosphere During Geomagnetic Storms. Adv Space Res 5(4):179-192.

72. Bosqued, J.M., et al. (1985) Evidence for Ion Energy Dispersion in the Polar Cusp Related to a Northward-Directed IMF. Adv Space Res 5(4):149-154.

73. Bott, M.H.P. (1982) Interior of the Earth: Its Structure, Constitution and Evolution. London, E Arnold.

74. Bowin, C., et al. (1982) Free-Air Gravity Anomaly Atlas of the World. Mass, Woods Hole Oceanographic Instit.

75. Brown, F.C. (1941) One Unusual Observation in the Auroral Display of September 18. Science 94:562.

76. Brown, H.R., et al. (1979) Evidence that Geomagnetic Variations Can be Detected by Lorenian Ampullae. Nature 277:648-649.

77. Brown, Jr., F.A. (1965) A Unified Theory for Biological Rhythms. p 231 in (ed) J. Aschoff, Circadian Clocks. Amsterdam, North-Holland.

78. Bruning, K., et al. (1985) Why Does the Perpendicular Electric Field Increase at the Edge of Auroral Arcs? Adv Space Res 5(4):79-82.

79. Buchanan, F. (1821) Account of an Extraordinary Appearance of the Sea. Edinburgh Philos J 5:303.

80. Burch, J.L., Reiff, P.H. (1985) Field-Aligned Currents and Ion Convection at High Altitudes. Adv Space Res 5(4):23-40.

81. Burlaga, L.F. (1982) Magnetic Fields in the Interplanetary Medium. Adv Space Res 2(1):51-54.

82. Burton, E.T., Boardman, E.M. (1934) Audio-Frequency Atmospherics. Eos 15:155.

83. Cady, J.W. (1975) Magnetic and Gravity Anomalies in the Great Valley and Western Sierra Nevada Metamorphic Belt, California. Boulder, Colo, Geol Soc Amer.

84. Callis, L.C., Natarajan, M. (1986) Antarctic Ozone Minimum: Relationship to Odd Nitrogen, Odd Chlorine, the Final Warming and the 11-year Solar Cycle. J Geophys Res 91:771-796.

85. Carnegie, A.A. (1906) Remarkable Display of Phosphorescence. Roy Meteorol Soc Q J 32:280.

86.   Cevolani, G., et al. (1987) Luni-Solar Periodic Components in Precipitation Data. Geophys Res Lett 14:45-58.

87.   Chalmers, J.A. (1965) Generation of Electric Charges Outside Thunderclouds. Contrib E. Pierce. p 174 in (ed) S. Coroniti, Problems of Atmospheric and Space Electricity—Procc 3rd Intl Conf Atmos Space Electr, Montreux, Switz, 5-10 May 1963. NY, Elsevier.

88.   Chapman, S. (1931) Audibility and Lowermost Altitude of the Aurora Polaris. Nature 127:341.

89.   Chapman, S. (1932) Low Altitude Aurorae. Nature 130:764.

90.   Chapman, S., Bartels, J. (1962) Geomagnetism, 2 vols. Oxford, Clarendon.

91.   Chappel, C.R. (1982) Low Energy Particles in the Magnetosphere. Adv Space Res 2(1):33-38.

92.   Chen, F.F. (1984) Introduction to Plasma Physics and Controlled Fusion, vols 1 & 2. NY, Plenum.

93.   Chidsey, C. (1890) Mysterious Music of Pascagoula. Pop Sci Mon 36:791.

94.   Chinese Seismological Delegation (1976) Brief Summary of the Work of Premonitory Observation, Prediction and Precautionary Measures Before the Haicheng Earthquake, Liaoning Province, of Magnitude 7.3. Sci Geol, Sinica (2):120-123.

95.   Chiu, Y.T., et al. (1983) Auroral Plasmas in the Evening Sector: Satellite Observations and Theoretical Interpretations. Space Sci Rev 35:211.

96.   Chizhevskii, A.L. (1964) One Form of Specifically Bioactive or Z Emission of the Sun. p.342 in The Earth in the Universe. Mysl, Moscow.

97.   Chree, C. (1918) Auroral Observations in the Antarctic. Nature 101:114.

98.   Cladis, J.B., Francis, W.E. (1985) Transport of Ions in Presence of Induced Electric Field and Electrostatic Turbulence: Source of Ions Injected Into Ring Current. Adv Space Res 5(4):415-420.

99.   Clark, Jr., S.P. (1966) Handbook of Physical Constants. Geol Soc Amer Mem, no 97.

100. Clarke, W.B., et al. (1969) Evidence for Primordial He-3 in the Earth. Eos 50:222.

101. Cleland, J.B. (1909) Barisal Guns in Australia. Nature 81:127.

102. Cole, K.D. (1961) Airglow and the South Atlantic Anomaly. J Geophys Res 66:3064.

103. Cooke, W.E. (1908) Barisal Guns in Western Australia. Nature 78:390.

104. Cooper, W.S. (1896) Barisal Guns. Sci Amer 75:123.

105. Cope, E.J. (1896) Aurora of March 4, 1896. Brit Astron Assoc J 6:295.

106. Cope, F.W. (1973) Biological Sensitivity to Weak Magnetic Fields Due to Biological Superconductive Josephson Junctions. Physio Chem Phys 5(3):173-176.

107. Cope, F.W. (1974) Enhancement by High Electric Fields of Superconduction in Organic and Biological Solids at Room Temperature and a Role in Nerve Conduction. Physiol Chem Phys 6(5):405-410.

108. Corlin, A. (1931) Low Altitude Aurora of Nov. 16, 1929. Nature 127:928.

109. Corlin, A. (1931) Observations of a Low Altitude Aurora and Simultaneous Phenomena. Nature 127:553.

110. Corliss, W.R. (comp) (1983) Earthquakes, Tides, Unidentified Sounds and Related Phenomena. A Catalog of Geophysical Anomalies. [and] (1990) Neglected Geological Anomalies. Glen Arm, MD, Sourcebook Project.

111. Corliss, W.R. (comp) (1982) Lightning, Auroras, Nocturnal Lights and Related Luminous Phenomena. A Catalog of Geophysical Anomalies.[and] (1991) Inner Earth: A Search for Anomalies. Glen Arm, MD, Sourcebook Project.

112. Cotton, F.A., Wilkinson, G. (1980) Advanced Inorganic Chemistry: A Comprehensive Text (4th ed). NY, J. Wiley.

113. Creager, K.C., Jordan, T.H. (1986) Spherical Structure of the Core-Mantle Boundary from PKP Travel Times. Geophys Res Lett 13:1497-1500.

114. Crutcher, H.L., Davis, O.M. (1969) Navy Marine Climatic Atlas of the World, vol 8, Navair 50-IC-54. US Naval Weather Service Command.

115. Darwin, G.H. (1895) 'Barisal Guns' and 'Mistpouffers'. Nature 52:650.

116. Davies, F.T., Currie, B.W. (1933) Audibility of the Aurora and Low Auroras. Nature 132:855.

117. Davis, P.M., et al. (1973) Kilauea Volcano, Hawaii: A Search for the Volcano-Magnetic Effect. Science 180:73.

118. Dawson, E., Newitt, L.R. (1982) Magnetic Poles of the Earth. J Geomag Geoelect 34(4):225-240.

119. Dawson, E., Newitt, L.R. (1984) Magnetic Declination in Canada from 1750 to 1980. Canad Surv 38(1):35-40.

120. Dennehy, C. (1870) Strange Noises Heard at Sea Off Grey Town. Nature 2:25.

121. Dessler, A. (1959) Effect of Magnetic Anomaly on Particle Radiation Trapped in the Geomagnetic Field. J Geophys Res 64:713-715.

122. Dessler, A.J., et al. (eds) (1986) [Entire Issue on Antarctic Ozone Depletion] Geophys Res Lett 13(12).

123. Dewey, E.R. (1985) What Forces Could Cause Cycles? Cycles 35(3):68-69.

124. Dickerson, R.E. (1978) Chemical Evolution and the Origin of Life. Sci Amer 239:70-86.

125. Dolezalek, H. (1972) Discussion of the Fundamental Problem of Atmospheric Electricity. Geophysics 100:8-42.

126. Dubrov, A.P. (1969) Effect of Geophysical Factors on Membrane Permeability and Diurnal Rhythm of Excretion of Organic Substances by Plant Roots. Dokl Akad Nauk SSSR 187(6):1429.

127. Dubrov, A.P. (1978) Geomagnetic Field and Life: Geomagnetobiology. NY, Plenum.

128. Duppa-Crotch, W. (1891) A Rare Phenomena. Nature 44:614.

129. Dziewonski, A.M., Woodhouse, J.H. (1987) Global Images of the Earth's Interior. Science 236:37-48.

130. Eastman, T.E., Frank, L.A. (1982) Hot Plasmas in the Magnetosphere. Adv Space Res 2(1):39-42.

131. Edgell, J.A. (1926) Illumination of the Sea. Marine Obs 3:132.

132. Eucken, A. (1944) [Physio-Chemical Observations About the Early Formation of the Earth in Its Developmental History] Physikalisch-Chemische Betrachtungen uber die Fruheste Entwicklungsgeschichte der Erde. H1:pp.1-25 in Akad Wiss Gottingen, Mat-Phys Kl, Nachr Jg.

133. Eve, A.S. (1936) Northern Lights. Smithsonian Inst Ann Rept, pp.145 & 149.

134. Evans, F.J. (1870) Strange Noises Heard at Sea Off Grey Town. Nature 2:46.

135. Evens, E.H. (1935) Magnetic Disturbance. Marine Obs 12:144.

136. Fainberg, E.B. (1980) Electromagnetic Induction in the World Ocean. Geophys Surv 4:157-171.

137. Falthammar, C.-G.(1982) Electric Fields in the Magnetosphere. Adv Space Res 2(1):19-24.

138. Farman, J.C., et al. (1985) Large Losses of Total Ozone in Antarctica Reveal CLOx/NOx Interaction. Nature 315:207-210.

139. Fischer, G. (1979) Electromagnetic Induction Effects at an Ocean Coast. Proc IEEE 67:1050-1060.

140. Fischer, H.J. (1977) [Atmospheric Electric Field's Relationship with Air Pollution and Weather Conditions] Das Luftelektrische Feld in Abhangigkeit von Luftverunreinigug und Wetterlage. Prometheus 7(2):4-12.

141. Flemming, J.A. (1949) Physics of the Earth, Vol 8: Terrestrial Magnetism and Electricity. NY, Dover.

142. Fonarev, G.A. (1982) Electromagnetic Research in the Ocean. Geophys Surv 4:501-508.

143. Frank, L.A., et al. (1985) Images of the Earth's Aurora and Geocorona from the Dynamics Explorer Mission. Adv Space Res 5(4):53-68.

144. Fraser, D.C. (1965) Magnetic Fields of Ocean Waves. Nature 206:605-606.

145. Fraser-Smith, A.C. (1978) ULF Tree Potentials and Geomagnetic Pulsations. Nature 271:641-642.

146. Gaigerov, S.S., et al. (1974) [Results of Upper Atmosphere Research in Antarctic] Nekotorye Rezul'taty Issledovaniia Vysokikh Sloev Atmosfery v Antarktike. Antartika: Doklady Komissii (3):137-146.

147. Ganguli, G., Palmadesso, P. (1985) Role of Nonlocalities in Magnetospheric-Ionospheric Coupling Processes. Adv Space Res 5(4):19-22.

148. Garber, C.M. (1933) On the Audibility of the Aurora Borealis. Science 78:213.

149. Gassmann, G.J. Pike, C.P. (1966) On the Observation of Ionospheric Effects Due to Dumping of Trapped Particles. pp.378-385 in (ed) B. McCormac, Radiation Trapped in the Earth's Magnetic Field. Boston, Reidel.

150. Gates, H. (1931) Audibility and Lowermost Altitude of the Aurora Polaris. Nature 127:486.

151. Gates, R.R. (1931) Audibility and Lowermost Altitude of the Aurora Borealis. Nature 127:486.

152. Geballe, T.H. (1971) New Superconductors. Sci Amer225(5):22.

153. Gendrin, R., Domingo, V. (1982) Consequences of Solar-Related Plasma Processes on the Earth's Environment and Man's Devices. Adv Space Res 2(1):71-78.

154. Geological Bureau (1975) Earthquake Questions and Answers, 132. Peking, China, Geology Press.

155. Giardini, D., Woodhouse, J.H. (1986) Horizontal Shear-Flow in the Mantle Beneath the Tonga Arc. Nature 319 (6054):551-555.

156. Ginzburg, V.L., et al. (1962) Investigation of Charged Particle Intensity During the Flights of the Second and Third Space-Ships. Planet Space Sci 9:845.

157. Gish, O.H. (1923) General Description of the Earth-Current Measuring System at the Watheroo Magnetic Observatory. Terres Magnetism Atmos Electr 28:89-108.

158. Gish, O.H. (1936) Electrical Messages from the Earth: Their Reception and Interpretation. J Wash Acad Sci 26:267-289.

159. Gish, O.H. (1977) Terrestrial Electricity, vol 13, pp.524-528 in McGraw-Hill Encyclopedia of Science and Technology, 4th ed. NY, McGraw-Hill.

160. Gish, O.H., Rooney, W.J. (1925) Measurements of Resistivity of Large Masses of Undisturbed Earth. Terres Magnetism Atmo Elec 30:161-188.

161. Gledhill, J.A. (1976) Aeronomic Effects of the South Atlantic Anomaly. Rev Geophys Space Phys 14:173-187.

162. Gloersen, P., et al. (1975) Microwave Maps of the Polar Ice of the Earth. pp.407-414 in 24th Alaska Science Conf, Univ Alaska, 15-17 Aug 1973, Climate of the Arctic. Fairbanks, Univ Alaska.

163. Gnevyshev, M.N., Ol', A.I. (eds) (1977) Effects of Solar Activity on the Earth's Atmosphere and Biosphere (Acad Sci USSR-Astronomical Council). Nauka, Moscow. Jerusalem, Israel Prog Sci Trans.

164. Goertz, C.K., Bruning, K. (1984) Field-Aligned Currents Observed in the Vicinity of a Moving Auroral Arc. IMS-Symposium, ESA-SP.

165. Gogoshev, M.M., et al. (1985) Observations in the South Atlantic Geomagnetic Anomaly with Intercosmos-Bulgaria-1300 During a Geomagnetic Storm. Adv Space Res 5(4):213-6.

166. Gokhberg, M.G., et al. (1982) Experimental Measurement of Electromagnetic Emissions Possibly Related to Earthquakes in Japan. J Geophys Res 87b:7824.

167. Gold, T., Soter, S. (1979) Brontides: Natural Explosive Noises. Science 204:371.

168. Goldfein, S. (1974) Some Evidence for High-Temperature Superconductivity in Cholates. Physiol Chem Phys 6:261-268.

169. Goncharov, N., Morozov, V., Makarov, V. (197?) Is the Earth a Large Crystal? Khimiya i Zhizn' [in Russian].

170. Gonzales, C.A., et al. (1983) On the Latitudinal Variations of the Ionospheric Electric Field during Magnetospheric Disturbances. J Geophys Res 88(11):9135-9144.

171. Gould, J.L., et al. (1978) Bees Have Magnetic Remanence. Science 201:1026-1028.

172. Greenspan, J.A., Stone, C.A. (1964) Longitudinal Night Airglow Intensity in the Region of the South Atlantic Magnetic Anomaly. J Geophys Res 69:465.

173. Greiner, W., et al. (1985) Quantum Electrodynamics of Strong Fields. NY, Springer-Verlag.

174. Gringauz, K.I. (1985) Structure and Properties of the Earth's Plasmasphere. Adv Space Res 5(4):391-400.

175. Gringel, W., et al. (1986) Electrical Structure from 0 to 30 Kilometers. pp.166-182 in The Earth's Electrical Environment. NRC Geophys Study Comm, Wash, DC, Natl Acad.

176. Grudinski, U. (1975) [Do Animals Announce the Coming of Earth Tremors?] Verrat die Tierwelt den Kommenden Erdstoss? Frankfurter Allgemeine Zeitung, 5 March.

177. Hadden, D.E. (1902) Auroral Phenomena at Alta Iowa. Pop Astron 10:249.

178. Hadden, D.E. (1902) Auroral Phenomena or Zodical Light? Pop Astron 10:388.

179. Hagiwara, Y. (1977) Gravity Changes Associated with Seismic Activities. pp.137-146 in (eds) C. Kisslinger, Z. Suzuki, Earthquake Precursors. Japan, Sci Soc Press.

180. Hale, L.C., Croskey, C.L. (1979) An Auroral Effect on the Fair Weather Electric Field. Nature 278:239.

181. Halpern, E.H. (1976) Search for Organic Superconductors. US NTIS, AD-A026236, Wash, DC, USGPO.

182. Halpern, E.H., Wolf, A.A. (1972) Speculations of Superconductivity in Biological and Organic Systems. Adv Cryog Eng 17:109-15.

183. Harder, E.C. (1919) Iron-Depositing Bacteria and Their Geologic Relations. US Geol Surv Prof Pap, 113.

184. Hargreaves, J.K. (1979) Upper Atmosphere and Solar-Terrestrial Relations. London, Van Nostrand-Reinhold.

185. Harries, H. (1896) Barisal Guns and Similar Sounds. Nature 53:295.

186. Hatai, S., et al. (1932) Earth Currents in Relation to the Responses of Catfish. Proc Imp Acad Japan 8:478-481.

187. Heelis, R.A., Reiff, P.H. (1985) Observations of Magnetospheric Convection from Low Altitudes. Adv Space Res 5(4):349-362.

188. Henderson, J.P. (1970) Effect of Earth Tremors. New Sci 46:300.

189. Henriksen, K., et al. (1977) Lunar Influence on the Occurrence of Aurora. J Geophys Res 82:2842.

190. Herman, J.R. (1985) Sun, Weather and Climate. Earth Sciences Series. NY, Dover.

191. Hewson-Browne, R.C., Kendall, P.C. (1981) Electromagnetic Induction in the Earth in Electrical Contact with the Oceans. Geophys J R Astron Soc 45:527-542.

192. Hill, T.W., Wolf, R.A. (1977) Solar-Wind Interaction. pp. 25-41 in The Upper Atmosphere and Magnetosphere, NRC Geophys Study Comm. Wash, DC, Natl Acad Sci.

193. Hirschberg, J., et al. (1967) Long Period Geomagnetic Fluctuations After the March 1964 Alaska Earthquake. Eos 48:80.

194. Hirschberg, J., et al. (1967) Long Period Geomagnetic Fluctuations After the March 1964 Alaska Earthquake. Earth Planet Sci Lett 3:426.

195. Hobbs, J.E. (1980) Applied Climatology: A Study of Atmospheric Resources. Boulder, Colo, Westview.

196. Hollister, C.D., et al. (1984) Dynamic Abyss. Sci Amer 250(3):42-53.

197. Holzworth, R.H., Mozer, F.S. (1979) Direct Evidence of Solar Flare Modification of Stratospheric Electric Fields. J Geophys Res 84:363-367.

198. Honkura, Y. (1981) Electric and Magnetic Approach to Earthquake Prediction. pp.301-383 in (ed) T. Rikitake, Current Research in Earthquake Prediction. Tokyo, Japan, Center Acad, Boston, Reidel.

199. Hoppel, W.A., et al. (1986) Atmospheric Electricity in the Planetary Boundary Layer. pp.149-165 in Earth's Electrical Environment. NRC Geophys Study Comm. Wash, DC, Natl Acad.

200. Horita, R.E., et al. (1985) Counterstreaming Hydrogen and Oxygen Ions Observed in the Magnetosphere on ISEE-1. Adv Space Res 5(4):421-424.

201. Hoseason, W.S. (1902) Remarkable Phosphorescent Phenomenon Observed in the Persian Gulf, April 4 and 9, 1901. Roy Meteorol Soc Q J 28:29.

202. Howard, R. (1967) Magnetic Field of the Sun (Observational). Ann Rev Astron Astrophys 5:1.

203. Howell, B.F. (1959) Introduction to Geophysics. NY, McGraw-Hill.

204. Hughes, W.J., et al. (1985) Multisatellite Investigations of Substorm Onsets. Adv Space Res 5(4):159-162.

205. Hutton, V.R.S. (1976) Electrical Conductivity of the Earth and Planets. Rep Prog Phys 39:487-572.

206. Imhof, L.M., et al. (1985) Localized Electron Precipitation Events at High Latitudes Studied with X-Ray Imagery from a Satellite. Adv Space Res 5(4):69-72.

207. Israel, H. (1973) Atmospheric Electricity, 2 vols. Trans from German. Jerusalem, Israel Prog Sci Trans.

208. Jacobs, J.A. (1977) Geomagnetic Micropulsations. Berlin, Springer-Verlag.

209. Jannasch, H.W., et al. (1971) Microbial Degradation of Organic Matter in the Deep Sea. Science 171:672-675.

210. Jephcoat, A., Olson, P. (1987) Is the Inner Core of the Earth Pure Iron? Nature 325:332-35

211. Jezek, K.C., et al. (1978) Dielectric Permitivity of Glacier Ice Measured in Situ by Radar Wide-Angle Reflection. J Glaciol 21(85):315-329.

212. Johnson, J.H. (1927) On the Altitude of the Aurora. Astron Soc Pacific Pubs 39:347.

213. Johnston, M. (1978) Tectonomagnetic Effects. Earthq Info Bull 10:82.

214. Johnston, M.J.S., et al. (1976) Tectonomagnetic Experiments and Observations in the Western U.S.A. J Geomag Geoelect 28:85-97.

215. Jones, D. (1982) Plasma Waves in the Earth's Magnetosphere. Adv Space Res 2(1):25-32.

216. Jordan, F.C. (1916) Peculiar Aurora. Pop Astron 24:401.

217. Josephson, B.D. (1969) Weakly Coupled Superconductors. pp. 423-447 in (ed) R. Parks, Superconductivity, vol 1. NY, Marcel Dekker.

218. Kabachenko, V. (1973) Discovering Invisible Links. [in Russian] Tekhnika i Molodezhi (Sept).

219. Kalmijn, A.J. (1966) Electro-Perception in Sharks and Rays. Nature 212:1232-3.

220. Kalmijn, A.J. (1978) Electric and Magnetic Sensory World of Sharks, Skates and Rays. In (eds) H.S. Hodgson, R.F. Matthewson. Wash, DC, USGPO.

221. Kapitza, P.L. (1979) Plasma and Controlled Thermonuclear Reaction. Science 5:959-964.

222. Keller, G.V. (1982) Electrical Methods of Geophysical Prospecting. NY, Pergamon.

223. Keller, G.V., et al. (1972) Magnetic Noise Preceding the August 1971 Summit Eruption of Kilauea Volcano. Science 175:1457.

224. Kelley, F.C. (1934) Audibility of Auroras and Low Auroras. Nature 133:218.

225. Kelley, M.C., et al. (1979) An Explanation for Anomalous Equatorial Ionospheric Electric Fields Associated with a Northward Turning of the Interplanetary Magnetic Field. Geophys Res Lett 6:301-304.

226. Kendall, P.C., Quinney, D.A. (1983) Induction in the Oceans. Geophys J R Astron Soc 74:239-255.

227. Kerr, R.A. (1977) Oceanography: A Closer Look at Gulf Stream Rings. Science 198:387-89

228. Kerr, R.A. (1979) East Coast Mystery Booms: Mystery Gone but Booms Linger On. Science 203:256.

229. Khvedelidze, M.A., et al. (1973) Bionic Aspects of Magnetoelectric Effects. p 196 in Problems of Bionics. Nauka, Moscow.

230. Kingsley, C. (1870) Strange Noises Heard At Sea Off Grey Town. Nature 2:46.

231. Klocker, N., et al. (1985) Ground Observations of Kinetic Alfven Waves. Adv Space Res 5(4):237-242.

232. Knudsen, W.C., Sharp, G.W. (1968) F2 Region Electron Concentration Enhancements from Inner Radiation Belt Particles. J Geophys Res 73:6275-6283.

233. Kotsch, W.J., Henderson, R. (1984) Heavy Weather Guide. Annapolis, MD, Naval Instit.

234. Kuhn, W., Rittman, A. (1941) [About the State of the Earth's Interior and Its Early Formation Away from its Primitive State] Uber den Zustand des Erdinnern und Seine Entstehung aus Einem Homogenen Urzustand. Geol Rundschau 32(3):215-256.

235. Kunzi, K.F., et al. (1976) Snow and Ice Surfaces Measured by the Nimbus 5 Microwave. J Geophys Res 81:4965-4980.

236. Kuo, J.T. (ed) (1983) International Symposium on Earth Tides. Proc 9th Intl Symp Earth Tides. NY, Stuggart.

237. Kutina, Ya. (1974) Planetary Network of Faults and Its Significance in the Prognosis of Ore Deposits. Intl Volcanological Symp, Sept 1973, Bucharest, Rumania.

238. Kvenvolden, K.A. (1974) Natural Evidence for Chemical and Early Biological Evolution. Origins of Life 5:71-86.

239. Lanzerotti, L.J., Gregori, G.P. (1986) Telluric Currents: The Natural Environment and Interactions with Man-Made Systems. pp.232-257 in Earth's Electrical Environment. NRC Geophys Study Comm. Wash, DC, Natl Acad.

240. Larkin, R.P., Sutherland, P.J. (1977) Migrating Birds Respond to Project Sea Farer's Electromagnetic Field. Science 195:777-779.

241. Larsen, J.C., Sanford, T.B. (1985) Florida Current Volume Transports from Voltage Measurements. Science 227:302-304.

242. La Touche, T.D. (1890) On the Sounds Known as the "Barisal Guns", Occurring in the Gangetic Delta. Rept Brit Assoc, p.800.

243. Lavander, F.C. (1891) Rare Phenomenon. Nature 44:519.

244. Lavely, E.M., et al. (1986) Scales of Heterogeneity Near the Core-Mantle-Boundary. Geophys Res Lett 13:1505-1508.

245. Lazutin, L.L., et al. (1985) SAMBO-GEOS: On Three-Dimensional Substorm Dynamics — A Case Study for 4 March 1979. Adv Space Res 5(4):171-174.

246. Leggett, W.C. (1977) Ecology of Fish Migration. Ann Rev Ecol Syst 8.

247. Lemaire, J. (1982) Brief Panorama. Adv Space Res 2(1):3-10.

248. Lemaire, J., Rycroft, M.J. (eds) (1982) Solar System Plasmas and Fields. Comm Space Research. Adv Space Res 2(1).

249. Lennartsson, W., Sharp, R.D. (1985) Relative Contributions of Terrestrial and Solar Wind Ions in the Plasma Sheet. Adv Space Res 5(4):411-414.

250. Liboff, A.R., et al. (1984) Time-Varying Magnetic Effects: Effect on DNA Synthesis. Science 223:818-820.

251. Ling-Huang, S. (1978) Can Animals Help to Predict Earthquakes? Earthq Info Bull, Nov-Dec.

252. Lissman, H.W. (1951) Continuous Electric Signals from the Tail of a Fish, *Gymnarchus Niloticus Cuv.* Nature 167:201.

253. Lissman, H.W. (1958) On the Function and Evolution of Electric Organs in Fish. J Exp Biol 35:156-191.

254. Lissman, H.W., Machin, K.E. (1958) Mechanism of Object Location in *Gymnarchus Niloticus* and Similar Fish. J Exp Biol 35:451.

255. Little, W.A. (1964) Possibility of Synthesizing an Organic Superconductor. Phys Rev 134(6A):1416.

256. Longuet-Higgens, M.S., et al. (1954) Electrical Field Induced by Ocean Currents and Waves, etc. Woods Hole Oceanogr Instit Contrib, no 690.

257. Lord, L.W. (1896) Barisal Guns and Mist Pouffers. Sci Amer 75:22.

258. Lovelock, J.E. (1979) Gaia—A New Look at Life on Earth. NY, Oxford Univ..

259. Lowenstam, H.A. (1967) Lepidocrocite, an Apatite Mineral and Magnetite in Teeth of Chitons (*Polyplacophora*). Science 156:1373-1375.

260. Lyons, L.R., Williams, D.J. (1984) Quantitative Aspects of Magnetospheric Physics. Boston, Riedel.

261. Maekawa, K., Maeda, H. (1978) Electric Fields in the Ionosphere Produced by Polar Field-Aligned Currents. Nature 273:649-650.

262. Maeno, N. (1975) Electrical Properties of Antarctic Ice. Ms Submitted Intl Symp Isotopes & Impurities in Snow & Ice, Grenoble, France, Aug 28-30.

263. Markert, M. (1976) Earthquake. New Sci 70:488.

264. Markham, T.P., Anctil, R.E. (1966) Airborne Night Airglow Measurements in the South Atlantic Magnetic Anomaly. J Geophys Res 71:997.

265. Marklund, G. (1984) Auroral Arc Classification Scheme Based on the Observed Arc-Associated Electric Field Pattern. Planet Space Sci 32:193-211.

266. Marton, J.P. (1973) Conjectures on Superconductivity and Cancer. Physiol Chem Phys 5(3):259-270.

267. McCormac, B.M. (1983) Weather and Climate Responses to Solar Variations. Boulder, Colo Assoc Univ.

268. McCormac, B.M., Seliga, T.H. (eds) (1979) Solar-Terrestrial Influences on Weather and Climate. Boston, Reidel.

269. McDowell, S.E., Rossby, H.T. (1978) Mediterranean Water: An Intense Mesoscale Eddy Off the Bahamas. Science 202:1085.

270. Meelis, J. (1977) On an "Unexpected Anomaly" in the Maximum of the 11-Year Sunspot Cycle. J Interdisc Cycle Res 8:205-6.

271. Meigs, W.M. (1890) Mysterious Music of Pascagoula. Pop Sci Mon 37:410.

272. Meloni, A.L., et al. (1983) Induction of Currents in Long Submarine Cables by Natural Phenomena. Rev Geophys Space Phys 21:795-803.

273. Miller, B.I. (1967) Characteristics of Hurricanes. Science 157:1389-1399.

274. Milyaev, N.A. (1961) Simultaneous Magnetic Variations in the Antarctic and the Arctic. pp.366-369 in Soviet Antarctic Expedition, vol III, 1965. NY, Elsevier.

275. Monro, E.M. (1938) Aurora: Its Audibility. Roy Astron Soc Canada J 32:435.

276. Moore, B.R. (1980) Is the Homing Pigeon's Map Geomagnetic? Nature 285:69.

277. Moore, F.R. (1977) Geomagnetic Disturbance and the Orientation of Nocturnally Migrating Birds. Science 196:682-4.

278. Moore, G.W. (1964) Magnetic Disturbances Preceding the 1964 Alaska Earthquake. Nature 203:508.

279. Moriarty, C. (1978) Eels: A Natural and Unnatural History. NY, Universe Books.

280. Moss, E.L. (1879) Report of an Unusual Natural Phenomenon Observed at Sea. Nature 20:428.

281. Murray, R.W. (1957) Evidence for a Mechanoreceptive Function of the Ampullae of Lorenzini. Nature 179:106-7.

282. Murray, R.W. (1960) Electrical Sensitivity of the Ampullae of Lorenzini. Nature 187:957.

283. Nalivkin, D.V. (1982) [Hurricanes, Storms and Tornadoes: Geographic Characteristics and Geological Activity] Uragany, Buri i Smerchi: Geograficheskie Osobennosti i Geologicheskaya Deyatel'nost'. Acad Sci USSR—Dept Earth Sci. Leningrad, Nauka Pubs,, c1969. Trans from Russian. New Delhi, Amerind.

284. National Aeronautics and Space Administration (1972) Animal Orientation and Navigation, NASA Rept SP-262, Wash, DC, USGPO.

285. Newitt, L.R., Dawson, E. (1984) Secular Variation in North America During Historical Times. Geophys J Roy Astron Soc 78(1):277-289.

286. Nishida, A. (1978) Geomagnetic Diagnosis of the Magnetosphere. NY, Springer-Verlag.

287. Nishida, A. (ed) (1982) Magnetospheric Plasma Physics. Center for Academic Pubs, Tokyo. Boston, Reidel.

288. Nordenskjold, A.E. (1884) Nordenskjold's Greenland Expedition of 1883. Sci Amer Supp 17:6740.

289. Oae, S. (ed) (1977) Organic Chemistry of Sulfur. NY, Plenum.

290. Olcott, H.S. (1895) Barisal Gun. Nature 53:130.

291. O'Brien, T. (1896) 'Barisal' Guns in Gippsland, Australia. Sci Amer 75:143.

292. Oliver, S.P. (1871) Noises at Sea Off Grey Town. Nature 4:26.

293. Olsen, W.P. (1982) Geomagnetic Field and Its Extension into Space. Adv Space Res 2(1):13-18.

294. Owen, H. (1983) Atlas of Continental Displacement, 200 Million Years to the Present. NY, Cambridge Univ.

295. Owen, H. (1984) The Earth is Expanding and We Don't Know Why. New Sci 104(14317):27-29.

296. Page, D.E. (ed) (1974) Correlated Interplanetary and Magnetospheric Observations. Boston, Reidel.

297. Palmer, A.R. (1920) White Water. Nature 104:563.

298. Paren, J.G. (1973) Electrical Behavior of Polar Glaciers. Symp Phys Chem Ice, Ottawa, Canada, Aug 14-18. pp.262-267 in (eds) E. Whalley, et al., Physics and Chemistry of Ice. Ottawa, Roy Soc Canada.

299. Parker, E.N. (1975) The Sun. Sci Amer 233(3):42-50.

300. Parker, E.N. (1983) Magnetic Fields in the Cosmos. Sci Amer 249(2):44-65.

301. Parkes, C.E. (1935) Abnormal Magnetic Variation. Marine Obs 12:96-97.

302. Parkhomenko, E.I. (1967) Electrical Properties of Rocks. NY, Plenum.

303. Parkhomenko, E.I. (1971) Electrification Phenomena in Rocks. NY, Plenum.

304. Parkinson, W.D. (1982) Introduction to Geomagnetism. Amsterdam, Elsevier.

305. Paulikas, G.A., Blake, J.B. (1982) High Energy Particles in the Magnetosphere. Adv Space Res 2(1):43-46.

306. Peckover, R.S. (1973) Oceanic Electric Currents Induced by Fluid Convection. Phys Earth Planet Inter 7:137-142.

307. Persinger, M.A., Lafreniere, G.F. (1977) Space-Time Transients and Unusual Events. Chicago, Nelson-Hall.

308. Pidegon, D. (1898) A White or Milky Sea. Nature 58:520.

309. Pliny the Elder (1st cent.) [Natural History] Historia Naturalis 2:25-37,81-86,8:183,18:87-8

310. Podshibyakin, A.K., Smirnov, R.V. (1967) Group Features of Bioelectric Anticipation of Geomagnetic Disturbances by Man. p.18 in Abstr Papers Commun 23rd All-Union Sci Session, Sci-Tech Soc Radio-Eng Electrocomm. Moscow.

311. Pokrovskaya, T.V. (1977) Solar Activity and Climate. pp.1-24 in (eds) M. Gneyshev, A. Ol', Effects of Solar Activity on the Earth's Atmosphere and Biosphere. Nauka, Moscow. Jerusalem, Israel Prog Sci Trans.

312. Prakash, S., Pal, S. (1985) Electric Fields and the Electron Density Irregularities in the Equatorial Electrojet. Adv Space Res 5(4):205-208.

313. Presti, D., Pettigrew, J.D. (1980) Ferro-Magnetic Coupling to Muscle Receptors as a Basis for Geomagnetic Field Sensitivity in Animals. Nature 285:99-101.

314. Pringle, J.E. (1879) Report of an Unusual Phenomenon Observed at Sea. Nature 20:291.

315. Procunier, R.W. (1971) Observations of Acoustic Aurora in the 1-16 Hz Range. Geophys J 26:183.

316. Racey, R.R. (1938) Sound from the Aurora. Roy Astron Soc Canada J 32:396.

317. Rampino, M.R., Stothers, R.B. (1984) Geological Rhythms and Cometary Impacts. Science 226:1427-1431.

318. Reedy, F. (1983) Celestial Winds, Polar Lights. Astronomy 11(18):9-15.

319. Rees, D. (1985) Response of the High-Latitude Thermosphere to Geomagnetic Activity. Adv Space Res 5(4):267-282.

320. Reid, G.C. (1986) Electrical Structure of the Middle Atmosphere. pp.183-194 in The Earth's Electrical Environment. NRC Geohys Study Comm. Wash, DC, Natl Acad.

321. Reynolds, J.M., Paren, J.G. (1980) Recrystallisation and Electrical Behavior of Glacier Ice. Nature 283:63-64.

322. Richmond, A.D. (1986) Upper-Atmosphere Electric-Field Sources. pp.195-205 in The Earth's Electrical Environment. NRC Geohys Study Comm. Wash, DC, Natl Acad.

323. Riley, D. (1974) World Weather and Climate. NY, Cambridge.

324. Rix, H. (1891) A Rare Phenomenon. Nature 44:541.

325. Roble, R.G., Tzur, I. (1986) Global Atmospheric-Electrical Circuit. pp.206-231 in Earth's Electrical Environment. NRC Geohys Study Comm. Wash, DC, Natl Acad.

326. Rokityansky, I.I. (1982) Geoelectromagnetic Investigation of Earth's Crust and Mantle. Trans N. Chobotova. NY, Spring-Verlag.

327. Rooney, W.J. (1949) Earth-Current Results at Tucson Magnetic Observatory, 1932-1942. Carnegie Instit Wash Pub, no 175.

328. Rosenbauer, H. (1982) Solar Wind Plasma. Adv Space Res 2(1):47-50.

329. Roses-Innes, A.C., Rhoderick, E.H. (1969) Introduction to Superconductivity. Oxford, Pergamon.

330. Russell, C.T., et al. (1985) On the Source Region of Flux Transfer Events. Adv Space Res 5(4):363-368.

331. Russell, R. (1898) Aurora of September 9. Nature 58:496.

332. Sanders, R. (1961) Effect of Terrestrial Electromagnetic Storms on Wireless Communication. IRE Trans Commun Syst (Dec):367-377.

333. Sastri, J.H. (1985) IMF Polarity Effects on the Equatorial Ionospheric F-Region. Adv Space Res 5(4):199-204.

334. Saunders, M.A., et al. (1984) Flux Transfer Events: Scale Size and Interior Structure. Geophys Res Lett 11:131-134.

335. Sauvaud, J.A., et al. (1985) Positive Ion Distributions in the Morning Auroral Zone: Local Acceleration and Drift Effects. Adv Space Res 5(4):73-78.

336. Scherer, J. (1912) Notes on Remarkable Earthquake Sounds in Haiti. Seismol Soc Amer Bull 2:230.

337. Schubert-Soldern, R. (1962) Mechanism and Vitalism; Philosophical Aspects of Biology. Ind, Univ Notre Dame.

338. Schulz, M., Lanzerotti, L.J. (1973) Particle Diffusion in the Radiation Belts. Berlin, Springer-Verlag.

339. Schunk, R.W. (1986) Solar-Terrestrial Physics: A Space Age Birth. Logan, Utah State U.

340. Scott, G.B. (1896) Barisal Guns. Nature 53:197.

341. Sehra, P.S. (1974) Upper Mesospheric Wind Structure in Antarctia. Nature 242:683-686.

342. Senning, A. (1976-79) Topics in Sulfur Chemistry, vols. 1-4. Stuttgart, George Thieme.

343. Shapley, D. (1978) East Coast Mystery Booms: A Scientific Suspense Tale. Science 199:416.

344. Shelly, E.G. (1985) Circulation of Energetic Ions of Terrestrial Origin in the Magnetosphere. Adv Space Res 5(4):401-410.

345. Shelley, E.G., et al. (1976) Satellite Observations of an Ionospheric Acceleration Mechanism. Geophys Res Lett 3:654-656.

346. Silverman, S.M., Tuan, T. (1973) Auroral Audibility. Adv Geophys 16:155.

347. Simpson, G.C. (1918) Auroral Observations in the Antarctic. Nature 102:24.

348. Simpson, G.C. (1931) Low Altitude Aurora. Nature 127:663.

349. Singh, R.N., Prasad, R. (1985) Effect of Spectral Distribution of Energetic Electrons on Wave-Particle Interaction and Precipitated Energy Input. Adv Space Res 5(4):247-250.

350. Small, L.M., et al. (1985) Modeling Studies of Ionospheric Convection in Northern and Southern Polar Regions. Adv Space Res 5(4):41-46.

351. Smith, B.E., Johnston, M. (1976) Tectonomagnetic Effect Observed Before a Magnitude 5.2 Earthquake Near Hollister, California. J Geophys Res 81:3556.

352. Smith, P.J. (1976) Central Anomalies: Why So Strong? Nature 260:486.

353. Sonnemann, G., et al. (1985) Do There Exist Effects in the Thermospheric Plasma, Arising from Dynamic Variations in the Middle Atmosphere? Adv Space Res 5(4):299-304.

354. Southwood, D.J.(1985) Theoretical Aspects of Ionosphere-Magnetosphere-Solar Wind Coupling. Adv Space Res 5(4):7-14.

355. Sprowl, D.R., Banerjee, S.K. (1985) High-Resolution Paleomagnetic Record of Geomagnetic Field Fluctuations from the Varved Sediments of Elk Lake, Minnesota. Geology (Boulder) 13(8):531-533.

356. Sprowl, D.R., Banerjee, S.K. (1985) Limits of Confidence on Sediment Paleomagnetic Data; A Cautionary Note. Eos 66:260.

357. Stacey, W.M. (1984) Fusion. An Introduction to the Physics and Technology of Magnetic Confinement. NY, J Wiley.

358. Stasiewicz, K. (1985) On the Formation of Auroral Arcs. Adv Space Res 5(4):83-86.

359. Stephenson, A. (1975) Auroral Sound. Polar Record 17:413.

360. Stern, D.P. (1977) Large-Scale Electric Fields in the Earth's Magnetosphere. Rev Geophys Space Phys 15:156-194.

361. Stierman, D.J., et al. (1981) Natural Explosive Noises. Science 212:1296.

362. Stormer, C. (1938) Photographic Measurements of the Great Aurora of January 25-26, 1938. Nature 141:955.

363. Straus, J.M., et al. (1985) Response of the High-Latitude Thermosphere to Geomagnetic Substorms. Adv Space Res 5(4):289-292.

364. Sulman, F.G. (1980) Effect of Air Ionization, Electric Fields, Atmospherics and Other Electric Phenomena on Man and Animal. Springfield, Ill, C Thomas.

365. Suyehiro, Y. (1971) Unusual Behavior of Fish Before Earthquakes—Second Report. Scii Rep Keikyu Aburatsubo Marine Park Aquarium 4:13-14.

366. Sverdrup, H.U. (1931) Audibility of the Aurora Polaris. Nature 128:457.

367. Talman, C.F. (1913) Brontidi, Mistpoeffers, or Barisal Guns. Sci Amer Supp 75:47.

368. Tarcsai, Gy. (1985) Ionosphere-Plasmasphere Electron Fluxes at Middle Latitudes Obtained from Whistlers. Adv Space Res 5(4):155-158.

369. Templeton, E.C. (1915) Subterranean Sounds Heard in the West Indies. Seismo Soc Amer Bull 5:171.

370. Terada, T. (1915) Oceanic Noises: Uminari. Mon Weather Rev 43:315.

371. Terada, T. (1932) Bull Earthq Res Instit 10:29-35.

372. Terada, T. (1932) On Luminous Phenomena Accompanying Earthquakes. Earthq Res Instit Bull 9:253.

373. Terada, T. (1932) On Some Probable Influence of Earthquakes Upon Fisheries. Bull Earthq Res Instit, Tokyo Univ 10:393-401.

374. Terada, T. (1932) Proc Imp Acad, Tokyo 8:38-86.

375. Tesch, F.-W. (1977) Eel: Biology and Management of Anguillid Eels. London, Chapman & Hall [and] NY, J Wiley.

376. Thomson, K.S. (1983) Marginalia: The Sense of Discovery and Vice-Versa. Amer Sci 71:522-524.

377. Tomlinson, C. (1896) Barisal Guns and Similar Sounds. Nature 53:295.

378. Toptygin, I.N. (1985) Cosmic Rays in Interplanetary Magnetic Fields. Trans D. Yakovlev from Russian. Boston, Reidel.

379. Torr, D.G., et al. (1975) Particle Precipitation in the South Atlantic Geomagnetic Anomaly. Planet Spac Sci 23:15.

380. Treilhou, J. P., et al. (1985) SAMBO-GEOS: Electric Field Measurements in the Disturbed Ionosphere and Magnetosphere. Adv Space Res 5(4):163-170.

381. Tributsch, H. (1978) Do Aerosal Anomalies Precede Earthquakes? Nature 276:606.

382. Tributsch, H. (1978) [When the Snakes Awake: Mysterious Earthquake Signs in Germany] Wenn die Schlangen Erwachen—Mysteriose Erdbebenvorzeichen Deutsche. Stuttgart,, Verlags Anstalt.

383. Tributsch, H. (1982) A Seismic Sense. Sciences 22(9):24-28.

384. Tributsch, H. (1982) How Life Learned to Live: Adaptation in Nature. Trans M. Varon. Cambridge, Mass, MIT Press.

385. Tributsch, H. (1982) When the Snakes Awake: Animals and Earthquake Prediction. Cambridge, Mass, MIT Press.

386. Tsigel'nitskii, I.I., et al. (1979) [Structure of Jet Streams and Discontinuous Tropause Zones] Struktura Struinykh Techenii i zon Razryva Tropopauzy. Leningrad. Articheski i Antarkticheskii Nauchnoissledovatel'skii Institut. Trudy 360:58-65.

387. Turner, H.H., et al. (1896) Note on a Curious Light (The Zodiacal Light?) as Seen at Oxford, 1896, March 4. Roy Astron Soc Mon Not 56:332.

388. US Geological Survey (1976) Conference I, Abnormal Animal Behavior Prior to Earthquakes. Wash, US Geol Survey.

389. Van den Broeck, E. (1895) Curious Aerial or Subterranean Sounds. Nature 53:30.

390. Vogt, P.R., Borenza, J.M. (1985) Global Fluctuations in Plate Speeds and Correlated Hotspot and Island Arc Volcanic Episodicity in the Last 10-20 Ma. Eos 66:369.

391. Volland, H. (1984) Atmospheric Electrodynamics. Physics and Chemistry in Space, Vol.11. NY, Springer-Verlag.

392. Vonnegut, B. (1960) Electrical Theory of Tornadoes. J Geophys Res 65:203-212.

393. Vonnegut, B., Weyer, J. (1966) Luminous Phenomena in Nocturnal Tornadoes. Science 153:1213-1220.

394. Voss, H.D., Smith, L.G. (1980) Global Zones of Energetic Particle Precipitation. J Atmos Terr Phys 42:227-239.

395. Voss, H.D., et al. (1985) Energetic Particles in the Night-Time Middle- and Low-Latitude Ionosphere. Adv Space Res 5(4):175-178.

396. Wadsworth, J. (1950) Barisal Guns. Weather 5:293.

397. Wagner, W.S., Visvanathan, T.R. (1978) Earthquake Lights—A Potential Aid inn Earthquake Forecasting. Eos 59:329.

398. Walcott, C., Green, R.P. (1974) Orientation of Homing Pigeons Altered by a Change in the Direction of an Applied Magnetic Field. Science 184:180-182.

399. Walcott, C., et al. (1979) Pigeons Have Magnets. Science 205:1027-1028.

400. Walker, M.M., et al. (1984) Candidate Magnetic Sense Organ in the Yellowfin Tuna, *Thunnus Albacares*. Science 224:751-753.

401. Walker, R.A. (1963) Some Low-Frequency, High Amplitude Underwater Noise Pulses of Wide Geographic Distribution Apparently of Biological Origin. Eos 44:59.

402. Warwick, J., et al. (1982) Radio Emission Associated with Rock Fracture: Possible Application to the Great Chilean Earthquake of May 22, 1960. J Geophys Res 87B:2851.

403. Weisburd, S. (1985) Earth's 'Pulses' Tied to Plate Rates. Sci News 127:324.

404. Weisburd, S. (1985) Modeling Magnetism: The Earth as a Dynamo. Sci News 128:220.

405. Weisburd, S. (1985) The Earth's Magnetic Hiccup—Something Strange Happened to thee Geomagnetic Field in 1969: It Jerked. Sci News 128:218-220.

406. Weisburd, S. (1986) One Ozone Hole Returns, Another is Found. Sci News 130:215.

407. Weisburd, S. (1986) Plunging Plates Cause a Stir. Sci News 130:106-109.

408. Weisburd, S. (1986) Seismic Journey to the Center of the Earth. Sci News 130:10-11.

409. Werner, A. (1974) Russian Scientists Use Animals to Predict Natural Disasters. pp.40-43 in (ed) P. James, California Superquake. Hicksville, NY, Exposition.

410. Wertheim, G.K. (1954) Studies of Electrical Potential Between Key West, Florida, and Havana, Cuba. Eos 35:872.

411. Whitfield, M., Jagner, D. (1981) Marine Electrochemistry. NY, J Wiley.

412. Wilhelm, K. (1981) On the Relation Between Magnetic Field-Aligned Electrostatic Electron Acceleration and the Resulting Energy Flux. J Geophys 49:69-73.

413. Williams, Q., et al. (1987) Melting Curve of Iron to 250 Gigapascals: A Constraint on the Temperature at Earth's Center. Science 236:181-182.

414. Wiltschko, W., et al. (1983) Growing up in an Altered Magnetic Field Affects the Initial Orientation of Young Homing Pigeons. Behav Ecol Sociobiol 12:135-42.

415. Winckler, J.R., et al. (1959) Auroral X-Rays, Cosmic Rays, and Related Phenomena During the Storm of February 10-11, 1958. J Geophys Res 64:597-610.

416. Winningham, J.D., Heikkila, W.J. (1974) Polar Cap Auroral Electron Fluxes Observed with ISIS 1. J Geophys Res 79:949.

417. Woodhouse, J.H., et al. (1986) Evidence for Inner Core Anistropy from Free Oscillations. Geophys Res Lett 13:1549-1552.

418. Woods, D.V., Lilley, F.E.M. (1980) Anomalous Geomagnetic Variations and the Concentration of Telluric Currents in South-West Queensland, Australia. Geophys J R Astron Soc 62:675-689.

419. Wylie, T.A. (1882) Auroral Phenomenon. Sci Amer 47:117.

420. Yanagihara, K., Yokouchi, T. (1965) Local Anomaly of Earth-Resistivity [in Japanese]. Mem Kakioka Magn Obs 12:105-113.

421. Yelkin, H.J. (1976) In Izvestia, December.

422. Yungul, S.H. (1982) Geoelectricity, vol 6. pp.171-175 in McGraw-Hill Encyclopedia of Science and Technology, 5th ed. NY, McGraw-Hill.

423. Yungul, S.H. (1977) Telluric Methods in the Study of Sedimentary Structures: A Survey. Geoexplor 15:207-38.

424. Zoeger, J., et al. (1981) Magnetic Material in the Head of the Common Pacific Dolphin. Science 213:892-894.

425. Zwally, H.J. (1979) NASA Mapping Projects (Summary). Geol Data, GD-5 (May):109-111.

426. Zwally, H.J. (1981) Antarctic Sea Ice Cover from Satellite Passive Microwave. Glaciol Data, GD-11 (Oct):79-85.

Tome Two

1. Abdel-Ghaffer, A.S., et al. (1977) Effects of Organic Matter and the Activity of Some Soil Enzymes. pp.319-324 in Soil Organic Matter Studies. Vol. II. Vienna, Intl Atomic Energy Agency.

2. Abelson, P.H. (1989) Effects of Electric and Magnetic Fields. Science 245:241.

3. Abo-Hegazi, A.M.T., et al. (1973) Breeding for Improved Protein in Pulses Using Radiation-Induced Mutations. pp.265-268 in FAO/IAEA Division of Atomic Energy in Food and Agriculture & Gesellschaft fur Strahlen- und Umweltforschung, 1972. Vienna, Intl Atomic Energy Agency.

4. Abo-Hegazi, A.M.T. (1979) High-Protein Lines in Field Beans Vicia Faba from a Breeding Programme Using Gamma Rays 1. Seed Yield and Heritability of Seed Protein Increase. pp.33-36 in FAO/IAEA/Gesellschaft fur Strahlen- und Umweltforschung Seed Protein Improvement in Cereals and Grain Legumes. Vol.2, 1978. Vienna, Intl Atomic Energy Agency.

5. Abrahamson, E. M. (1977) Body, Mind and Sugar. NY, Avon.

6. Abros'kin, V. (1971) Geomagnetic Field (GMF) and Sex Determination of the Human Embryo. 18th Conf Physiol South RSFSR. Vol.2. Voronezh.

7. Achkasova, Yu, N. (1978) Sector Structure of the Interplanetary Magnetic Field and Reproduction of Bacteria in Laboratory Tests. Trans Acad Sci USSR (1):99-102.

8. Achkasova, Yu, N., et al. (1975) Sectorial Boundaries of Interplanetary Magnetic Field and Vital Activity of Bacteria. pp.71-72 in Proc 3rd All-Union Symp Effect of Magnetic Fields on Biological Effects. Kaliningrad State Univ.

9. Adey, W.R. (1977) Experiment and Theory of Long-Range Interaction of Electromagnetic Fields at Brain Cell Surfaces. Neurosci Res Prog Bull 15:1-141.

10. Adey, W.R. (1979) Long-Range Electromagnetic Field Interactions at Brain Cell Surfaces. pp.57-80 in (ed) T.S. Fenforde, Magnetic Field Effect on Biological Systems. NY, Plenum.

11. Adey, W.R. (1981) Tissue Interaction With Non-ionizing Electromagnetic Fields. Phys Rev 61:435-500.

12. Adey, W.R. (1983) Evidence for Cooperative Mechanisms in the Susceptibility of Cerebral Tissue to Environmental and Intrinsic Electric Fields. pp.108-116 in (ed) J.M. Osepchuk, Biological Effects of Electromagnetic Radiation. NY, IEEE Press.

13. Adey, W.R. (1983) Biological Effects of Low Energy Electromagnetic Fields. pp.359-381 in (eds) M. Grandolfo, et al., Biological Effects and Dosimetry of Non-ionizing Radiation. NY, Plenum.

14. Adey, W.R. (1988) Cellular Microenvironment and Signaling Through Cell Membranes. pp.81-106 in (eds) M.E. O'Connor, R.H. Lovely, Electromagnetic Fields and Neurobehavioral Function. NY, Alan R. Liss.

15. Air Pollution Control Association (1970) Recognition of Air Pollution Injury to Vegetation: A Pictorial Atlas. Pitts, Penn, Air Pollution Control Assoc.

16. Alabovskii, Yu, I., Babenko, A.N. (1977) Mortality from Vascular Diseases of the Brain in Years With Different Levels of Magnetic Activity. pp.213-214 in (eds) M.N. Gnevyshev, A.I. Ol', Effects of Solar Activity on the Earth's Atmosphere and Biosphere. Jerusalem, Israel Prog Sci Trans.

17. Alavania, M.C., et al. (1987) Occupational Cancer Risk Associated With the Storage and Bulk Handling of Agricultural Foodstuff. J Toxicol Environ Health 22:247-54.

18. Albritton, C.C. (1989) Catastrophic Episodes in Earth History, NY, Chapman & Hall.

19. Allam, B.F., et al. (1981) Relative Vitamin D Deficiency in Paget's Disease. Lancet I(8216):384-385.

20. Allaway, W.H. (1975) Effect of Soils and Fertilizers on Human and Animal Nutrition. Agriculture Info Bull No 378, Wash, DC, USGPO.

21. Allen, J.J. (1980) Cellular Membranes. Experientia 148:1268-1271.

22. Allen, M.J. (1982) Bioelectric Behavior of Plant Leaf Membranes—The Effects of Bioreactive Substances on a Semiconductive Property. J Bioelect 1:161-172.

23. Allison, F.E. (1973) Soil Organic Matter and Its Role in Crop Production. NY, Elsevier.

24. Almendras, A.S., Bottomlay, P.J. (1987) Influence of Lime and Phosphate on Nodulation of Soil-Grown Trifolium Subterraneum L. by *Indigenous Rhizobium Trifolii*. Appl Environ Microbiol 53:2090-7.

25. Alvery, O.M., et al. (1982) Healing of Partial Thickness Wounds is Stimulated by External Electric Current. J invest Derm 78:353.

26. Amirov, R.M., et al. (1987) Effect of Pesticides on Bacterial Membranes [English Abstract]. Prikladnaia Biokhimiia i Mikrobiologiia 23:398-8.

27. Anand, M., et al. (1986) Endosulfan and Cholinergic (Muscarinic) Transmission: Effect on Electroencephalograms and [3H]Quinuclidinyl Benzilate in Pigeon Brain. Environ Res 40:421-6.

28. Andersen, I., Vad, E. (1965) Influence of Electric Fields on Bacterial Growth. Intl J Biometeor 9:211-218.

29. Anderson, P.W. (1972) More is Different: Broken Symmetry and the Hierarchial Structure of Science. Science 177:393-396.

30. Andronova, T.I. (1971) Effect of Heliogeophysical Factors on Cardiovascular System of Healthy People in Northern Conditions. pp.15-20 in Adaptation and Health in Man in the Far North, Krasnoyarsk.

31. Anisimova, L.A., et al. (1987) Immunity Status in Persons Exposed to Pesticides. Gigiena Truda i Professionalnye Zabolevaniia (6):15-18.

32. Anonymous (1961) The Humming Earth. New Sci 9:763.

33. Anonymous (1964) Those Tired Children. Time (6 Nov).

34. Anonymous (1975) Pesticide Perspective. Nature 258:282.

35. Anonymous (1976) Folic Acid and the Nervous System. Lancet II(7990):836.

36. Anonymous (1976) Tryptophan and Depression. Brit Med J 1(6004):242-243.

37. Anonymous (1983) Food is Being Garnished With Cadmium—Warning on Fertilizers. Chem Indust 18:686.

38. Anonymous (1984) Acid Rain Annual Report. Sci News 125:392.

39. Anonymous (1984) Childhood Poisonings Decrease But Remain Serious Threat in U.S. Amer Family Physician 29:394.

40. Anonymous (1984) Detecting Toxic Microdose Exposures. Sci News 126:329.

41. Anonymous (1984) Elderberry Poisoning. Amer Family Physician 29:346.

42. Anonymous (1984) Ethylene Dibromide is a Real Cancer Causer—But How? Chem & Eng News 62:15.

43. Anonymous (1984) New Health Hazard: Treated Wood. Registered Nurse 47:83-4.

44. Anonymous (1984) Pesticides May Penetrate Protective Clothes. New Sci 102:4.

45. Anonymous (1984) Third World Poisonings on the Rise. Sci News 126:25.

46. Anonymous (1984) Two Pesticides Workers Killed by EDB. Sci News 126:296.

47. Anonymous (1985) Acute Convulsions Associated With Endrin Poisoning— Pakistan. JAMA 253:334-5.

48. Anonymous (1985) From Air Pollution to Fertilizer? J Environ Health 48:41.

49. Anonymous (1985) Gene-Swapping Breaks Barriers in Evolutionary Theory. New Sci 105 (1444):19.

50. Anonymous (1985) Groups Petition EPA for Dioxin Controls. Intl Wildlife 15:27.

51. Anonymous (1985) Grow Your Own Pesticides. New Sci 108:10.

52. Anonymous (1985) Insecticides Growing in Trees. Sci News 128:168.

53. Anonymous (1985) Quantum Leaps. Sci Amer 253 (3):69.

54. Anonymous (1985) U.S. to Close Poisoned Refuge. Audubon 87:108.

55. Anonymous (1986) High Cost of Pesticide Subsidies. Sci News 129:24.

56.   Anonymous (1986) Lacewings vs. Aphids. New Sci 112:28.

57.   Anonymous (1986) Melon Deaths. New Sci 110:21.

58.   Anonymous (1986) Occupational Exposures to Chlorophenols. IARC Monogr Eval Carcinog Risk Chem Hum 41:319-56.

59.   Anonymous (1986) Pesticide Use in the Third World. J Environ Health 48:228.

60.   Anonymous (1986) Pesticides and Breast Milk. J Nutrition Education 18:116.

61.   Anonymous (1986) Pesticides in Estuary. New Sci 112:24.

62.   Anonymous (1986) Some Halogenated Hydrocarbons and Pesticide Exposures. IARC Working Gropu Lyon, 4-11 February 1986. IARC Monogr Eval Carcinog Risk Chem Hum 41:1-407.

63.   Anonymous (1986) Toxicity of TBT. Environment 28:18-19.

64.   Anonymous (1986) Unseen Toll of Pesticide Poisonings. New Sci 111:18.

65.   Anonymous (1987) Chemical-Free Farming: An Idea That's Catching On. Natl Wildlife 26:29

66.   Anonymous (1987) EEC To Clash With Farmers Over Nitrates. New Sci 114:24.

67.   Anonymous (1987) EPA Failed to Protect Rare Species from Pesticides. Natl Wildlife 25:29

68.   Anonymous (1987) Fatalities Resulting from Sulfuryl Fluoride Exposure After Home Fumigation—Virginia. JAMA 258:2041ff.

69.   Anonymous (1987) Federation Calls for Pesticide Ban. Intl Wildlife 17:26.

70.   Anonymous (1987) Fertilizer Seen as Key to Poor Gulf Water Quality. Environment 219:17-18.

71.   Anonymous (1987) Garden Chemicals May Cause Leukemia. New Sci 115:29.

72.   Anonymous (1987) Inert Pesticide Ingredients, Too. Sci News 131:361.

73.   Anonymous (1987) Magnetic Anomaly Upsets Migrating Birds. New Sci 116 (1585):32.

74.   Anonymous (1987) More Research Into Safety Urged. BMA News Rev (April):11.

75.   Anonymous (1987) Pesticide Controls Too Lax, Say Environmentalists. New Sci 113:15.

76.   Anonymous (1987) Pesticidial Rains. Sci News 131:360.

77.   Anonymous (1987) NAS Reports on Pathogens in Poultry and Pesticides in Food. Sci News 131:361.

78.   Anonymous (1987) Tropical Forests: A Plan for Action. Ecologist 17:129-133.

79.   Anonymous (1987) Wrangle Blocks Nitrate Solution. New Sci 114:17.

80.   Anonymous (1988) Magnetic Fields Maybe Hazardous. Electrical Const Maint 87(4):16 & 21

81.   Anonymous (1988) Toll Free Number Has Pesticide Information. J Environ Health 50:215.

82.   Antoniou, L., Shalhoub, R.J. (1980) Zinc and Sexual Dysfunction. Lancet II(8202):1034.

83.   Arango, M.A., Persinger, M.A. (1988) Geophysical Variables and Behaviour: LII. Decreased Geomagnetic Activity and Spontaneous Telepathic Experiences from the Sidgwick Collection. Perceptual & Motor Skills 67:907ff

84.   Arden-Clarke, C., Hodges, D. (1987) Soil Erosion: The Answer Lies in Organic Farming. New Sci 113:42-3.

85.   Armstrong, D., et al. (eds)(1984) Free Radicals in Molecular Biology. Aging and Disease. NY, Raven.

86.   Armstrong, W.C. (1987) Lightning Triggered from the Earth's Magnetosphere as the Source of Synchronized Whistlers. Nature 327:405-407.

87. Asami, T. (1984) Pollution of Soils by Cadmium. pp.95-112 in (ed) J.O. Nriagu, Changing Metal Cycles and Human Health. NY, Springer-Verlag.

88. Assael, M., et al. (1974) Influence of Artificial Air Ionization on the Human Electroencephalogram. Intl J Biometerology 18(4):306-312.

89. Athenstaedt, H. (1972) Pyroelectric Polarization in Cells, Tissues and Organs of Plants. Pflanzenphysiol 68:82-91.

90. Athenstaedt, H. (1974) Pyroelectric and Piezoelectric Properties of Vertebrates. Ann N Y Acad Sci 238:68-94.

91. Auerbach, C. (1976) Mutation Research: Problems, Results and Perspectives. London, Chapman & Hall.

92. Austin, R.B., et al. (1986) Molecular Biology and Crop Improvement. NY, Cambridge Univ Press.

93. Avnimelech, Y. (1986) Organic Residues in Modern Agriculture. pp.1-10 in (eds) Y. Chen, Y. Avnimelech, Role of Organic Matter in Modern Agriculture. Boston, Martinus Nijhoff.

94. Awasthi, M.S., Punita Duhale, M.S. (1984) Metabolic Changes Induced by Organophosphates in the Piscine Organs. Environ Res 35:320-5.

95. Axelson, O. (1987) Pesticides and Cancer Risks in Agriculture. Med Oncol Tumor Pharmacother 4(3-4):207-17.

96. Bach, W., Pankrath, J. (eds) (1979) Man's Impact on Climate. Proc Intl Conf, Berlin, 14-16 June, 1978. NY, Elsevier.

97. Bajaj, Y.P.S. (ed) (1986) Biotechnology in Agriculture and Forestry. Trees I. NY, Springer Verlag.

98. Baker, B., et al. (1974) Electrical Stimulation of Articular Cartilage Regeneration. Ann N Y Acad Sci 238:491-499.

99. Baker, H. et al. (1979) Vitamin Profiles in Elderly Persons Living at Home or in Nursing Homes Versus Profile in Healthly Young Subjects. J Amer Geriatrics Soc 28:444-450.

100. Baker, R.R. (1988) Human Magnetoreception for Navigation. pp.63-80 in (eds) M.E. O'Connor, R.H. Lovely, Electromagnetic Fields and Neurobehavioral Function. NY, Alan R. Liss.

101. Baker, R.R. (1988) Integrated Use of Moon and Magnetic Compasses by the Heart-and-Dart Moth, *Agrotis Exclamationis*. Anim Behav 35:94-101.

102. Baker-Blocker, A. (1981) Winter Weather and Mortality from Heart Attack in Houston and Minneapolis—St. Paul. pp.90-93 in Conf Agric & Forest Meterolo (15th) and Biometeorol (5th). Boston, Mass, Amer Meteor Soc.

103. Balcomb, R.D., et al. (1984) Effects on Wildlife of At-Planting Corn Applications of Granular Carbofuran. J Wildlife Mgnt 48:1353-9.

104. Balfour, E.B. (1976) The Living Soil and the Haughley Experiment. NY, Universe Books.

105. Balk, I.F., Koeman, J.H. (1984) Future Hazards from Pesticide Use. Intl Union Conserv Nature Natural Resources (IUCN). Comm Ecology, Paper No 4

106. Baltimore, D. (1981) Gene Conversion: Some Implications for Immunoglobulin Genes. Cell 24:592-4.

107. Baranova, R.P. (1970) A Characterization of the Biological Activity of Some Soils of Forest—Steppe Transuralia. Trudy Sverdlovskogo 19,38.

108. Barcelona, M.J., Naymik, T.B. (1984) Dynamics of Fertilizer Contaminant Plume in Groundwater. Environ Sci Tech 18:267-81.

109. Barnhart, E. (1982) Food Forest: An Agricultural Strategy for the Northeast U.S.A. pp.81-84 in (ed) S. Hill, Basic Technics in Ecological Farming. Basel, Switz, Birkhauser Verlag.

110. Barton, K. (1984) Pesticides and Famine. Environment 26:24.

111. Bassett, C.A.L., et al. (1974) Acceleration of Fracture Repair by Electromagnetic Fields. A Surgically Noninvasive Method. Ann N Y Acad Sci 238:242-262.

112. Bauer, W. (1983) Neuroelectric Medicine. J Bioelect 2(2&3):159-180.

113. Baum, J.W., et al. (1979) Tests in the Plant *Tradescantia* for Mutagenic Effects of Strong Magnetic Fields. pp.22-24 in (ed) T.S. Tenforde, Magnetic Field Effect on Biological Systems. NY, Plenum.

114. Baum, R.M. (1984) Hazardous Waste Finds Use as Low-Cost Fertilizer. Chem & Eng News 82:21-2.

115. Bawin, S.M., Adey, W.R. (1976) Sensitivity of Calcium Binding in Cerebral Tissue to Weak Environment Electric Fields Oscillating at Low Frequency. Proc Natl Acad Sci (USA) 73:1999-2003.

116. Beal, V.A. (1986) Nutrition in the Life Span. NY, Wiley.

117. Bean, R.C., et al. (1960) Changes in the Electrical Characteristics of Avocados During Ripening. Yearb Calif Avocado Soc 44:75-78.

118. Bear, F.E. (1986) Earth: The Stuff of Life. Norman, OK, Univ Oklahoma.

119. Beaton, G.H., Bengoa, J.M. (eds)(1976) Nutrition in Preventive Medicine, Part 2. Epidemology, pp.161-212. Geneva, WHO.

120. Beck-Friis, J., et al. (1985) Rebound Increase of Nocturnal Serum Melatonin Levels Following Evening Suppression by Bright Light Exposure in Healthy Men: Relation to Cortisol Levels and Morning Exposure. Ann N Y Acad Sci 453:371-375.

121. Becker, G. (1963) [Magnetic Field Orientation of Diptera] Magnetfeld-Orientierung von Dipteren Naturwissenschaften 50(21):664.

122. Becker, G., Speck, U. (1964) [Understanding the Magnetic Field-Orientaion of *Diptera*] Untersuchungen uber die Magnetfeld-Orientierung von *Dipteren*. Z Vgl Physiol 49(3):301.

123. Becker, R.O. (1963) Biological Effects of Magnetic Fields. Med Electron Biol Eng 1:293.

124. Becker, R.O. (1969) Effect of Magnetic Fields Upon the Central Nervous System. p 207 in (ed) M. Barnothy, Biological Effects of Magnetic Fields. Vol.2. NY, Plenum.

125. Becker, R.O. (1972) Stimulation of Partial Limb Regeneration in Rats. Nature 235:109-111

126. Becker, R.O. (1982) Review Article: Electrical Control Systems and Regenerative Growth. J Bioelect 1:239-264.

127. Becker, R.O. (1984) Electromagnetic Controls Over Biological Processes. J Bioelect 3:105-118.

128. Becker, R.O. (1985) Theory of the Interaction Between DC and ELF Electromagnetic Fields and Living Organisms. J Bioelect 4:133-140.

129. Becker, R.O. (1988) Electromagnetism and Life. pp.1-15 in (ed) A.A. Marino, Modern Bioelectricity. NY, Marcel Dekker.

130. Beischer, D.E. (1965) Biomagnetics. Ann N Y Acad Sci 134(1):454.

131. Beischer, D.E., Miller, F.F. (1962) Exposure of Man to Low-Intensity Magnetic Fields. NSAM-823, NASA Order R-39, Pensacola, Fl.

132. Belford, R. (1989) Personal Communication. Dept Chemistry, Tempe, Arizona, Arizona State Univ.

133. Bellamy, D. et al. (1981) Effect of Long-Term Exposure of Laboratory Rats to Ionised Air. pp.180-181 in Conf Agric & Forest Meterol (15th) and Biometerol (5th). Boston, Mass, Amer Meterol Soc.

134. Bellingham, J.G., et al. (1986) Detection of Magnetic Fields Generated by Electrochemical Corrosion. J Electrochem Soc 133(8):1753-4.

135. Benko, F. (1985) Geological and Cosmogonic Cycles, as Reflected by the New Law of Universal Cyclicity. Budapest, Akademiai Kiado.

136. Bennie, E.H. (1975) Lithium in Depression. Lancet I (7900):216.

137. Benton, M.J. (1988) Mass Extinctions in the Fossil Record of Reptiles: Paraphyly, Patchiness and Periodicity. pp.269-294 in (ed) G.P. Larwood, Extinction and Survival in the Fossil Record. Oxford, Clarendon Press.

138. Berg, N. (1981) Comments: From the SCS Chief Soil and Water Conservation News, June.

139. Berger, M.J., Goldstein, D.P. (1980) Impaired Reproductive Performance in DES-Exposed Women. Obstet Gynecol 55:25.

140. Berger, P.A., et al. (1978) Cholinomimetics in Mania, Schizophrenia and Memory Disorders. in (eds) A. Barbeau et al., Nutrition and the Brain. Vol.5. NY, Raven.

141. Bernhard, M., Andreae, M.O. (1984) Transport of Trace Metals in Marine Food Chains. pp.143-168 in (ed) J.O. Nriagu, Changing Metal Cycles and Human Health. NY, Springer-Verlag.

142. Bertram, H.P., et al. (1986) Hexachlorobenzene Content in Human Whole Blood and Adipose Tissue: Experiences in Environmental Specimen Banking. IARC Sci Pub (77):173-82

143. Berry, M. (1987) Quantum Physics on the Edge of Chaos. New Sci 116 (19 Nov):44-47.

144. Besdine, R.W., et al. (1980) Senility Reconsidered. JAMA 244(3):259-263.

145. Bhaskaram, C., Sadasivan, G. (1975) Effects of Feeding Irradiated Wheat to Malnourished Children. Amer J Clin Nutr 28:130-135.

146. Bidwell, R.G.S., Bebee, G.P. (1974) Carbon Monoxide Fixation by Plants. Can J Bot 52:1841-1847.

147. Biederbeck, V.O. (1978) Soil Organic Sulfur and Fertility. pp.273-310 in (eds) M. Schnitzer, S.U. Khan, Soil Organic Matter. NY, Elsevier Scientific.

148. Biebl, H., Pfennig, N. (1977) Growth of Sulphate-Reducing Bacteria With Sulphur as Electron Acceptor. Archives Microbiol 112:115-117.

149. Birzele, K. (1966) [Solar Activity and Biorhythms of Humans] Sonnenaktivitat und Biorhythmus des Menschen. F Deuticke, Wien.

150. Black, D.I. (1967) Cosmic Ray Events and Faunal Extinctions at Geomagnetic Field Reversals. Earth Planet Sci Lett 3(3):225-236.

151. Black, J.D., et al. (1971) Electrical Stimulation and Its Effect on Growth and Ion Accumulation in Tomato Plants. Can J Botany 49:1809-1815.

152. Blackman, C.F. (1988) Stimulation of Brain Tissue in Vitro by Extremely Low Frequency, Low Intensity, Sinusoidal Electromagnetic Fields. pp.107-118 in (eds) M.E. O'Connor, R.H. Lovely, Electromagnetic Fields and Neurobehavioral Function. NY, Alan Liss.

153. Blackman, S., Catalina, D. (1973) Moon and the Emergency Room. Percept Motor Skills 37:624-626.

154. Blakemore, R.P. (1979) Magnetic Effects on Lower Organisms. pp.13-15 in (ed) T.S. Tenforde, Magnetic Field Effects on Biological Systems. NY, Plenum.

155. Blakemore, R.P. et al. (1988) Bacterial Biomagnetism and Geomagnetic Filed Detection by Organisms. pp.19-34 in (ed) A.A. Marino, Modern Bioelectricity. NY, Marcel Dekker.

156. Blanchet, M. (1976) Impact of Diet on Disease. Dimensions Health Sci 53(10):27-29.

157. Blumenthal, M.D. (1980) Depressive Illness in Old Age: Getting Behind the Mask. Geriatrics 35(4):34-43.

158. Blus, L.J., et al. (1984) Effects of Heptachlor- and Lindane-Treated Seed on Canada Geese. J Wildlife Mgnt 48:1097-1111.

159. Boeringa, R. (ed) (1980) Alternative Methods of Agriculture. NY, Elsevier.

160. Boersma, A. (1984) Campanian Through Paleocene Paleotemperature and Carbon Isotope Sequence and the Cretaceous-Tertiary Boundary in the Atlantic Ocean. pp.247-277. in (eds) W.A. Berggren, J.A. Van Couvering, Catastrophes and Earth History. The New Uniformiatarianism. NJ, Princeton Univ Press.

161. Bogacka, T. (1987) Pesticides in the Water of the Reda and Bolszewka Rivers [English Abstract]. Roczniki Panstwowego Zakladu Higieny 38(1):82-9.

162. Bogden, J.D., et al. (1980) Relatively High Selenium Concentrations in Cigarette Tobaccos from Low Lung Cancer Incidence Countries. Fed Proc 39(3):556.

163. Bohm, D. (1993) Causality and Chance in Modern Physics. Philadelphia, PA, Univ Penn Press.

164. Bohm, D. (1995) Wholeness and the Implicate Order. NY, NY, Routledge.

165. Bohm, D. (1988) A Realist View of Quantum Theory. pp.3-18 in (eds) A. Van Der Merw, et al. Microphysical Reality and Quantum Formalism. Boston, Kluwer Acad.

166. Bohr, N. (1958) Atomic Physics and Human Knowledge. NY, J Wiley.

167. Bolton, Jr., H., et al. (1985) Soil Microbial Biomass and Selected Soil Enzyme Activities: Effect of Fertilization and Cropping Practices. Soil Biol Biochem 17:297-302.

168. Bondi, A., Alumot, E. (1987) Anti-Nutritive Factors in Animal Feedstuffs and Their Effects on Livestock. Prog Food Nutr Sci 11(2):115-51.

169. Bone, S. (1985) Dielectric Properties of Biomacromolecules: Some Aspects of Relevance to Biological Systems. J Bioelect 4:389-418.

170. Bonnet, P. (1982) Diagnosis of Biochemical Disorders. pp.112-127 in (ed) L.J. Hippchen, Holistic Approaches to Offender Rehabilitation. Springfld,Ill, C Thomas.

171. Borojevic, K. (1980) Comparative Mutagenesis in Plants. Radiation vs. Chemicals. pp.23-31 in (ed) M. Alacevic, Progress in Environmental Mutagenesis. NY, Elsevier/North Holland Biomedical.

172. Botez, M.I., et al. (1978) Polyneuropathy and Folate Deficiency. Arch Neurol 35:581-584.

173. Boulter, M.C. et al. (1988) Patterns of Plant Extinction from Some Palaeobotanical Evidence. pp.1-36 in (ed) G.P. Larwood, Extinction and Survival in the Fossil Record. Systematics Assoc Special. Vol.34, Oxford, Clarendon Press.

174. Bowen, H.J.M. (1975) Trace Elements in Biochemistry. NY, Academic Press.

175. Bowrey, J.J. (1887) Fall of Peculiar Hailstones in Kingston, Jamaica. Nature 36:153.

176. Bram, S. et al. (1980) Vitamin C Preferential Toxicity for Malignant Melonoma Cells. Nature 284:629-631.

177. Brandes, R., Kearns, D.R. (1986) Magnetic Ordering of DNA Liquid Crystals. Biochemistry 25:5890-5895.

178. Braud, S. (1979) Human Harm to Human DNA. Co Evolutionary Q 21(Spr):4-21.

179. Braude, S. (1986) Limits of Influence: Psychokinesis and the Philosophy of Science. NY, Routledge & Kegan Paul.

180. Brazier, M.D. (1977) Electrical Activity of the Nervous System. Baltimore, Williams & Wilkins.

181. Brezowsky, H., Ranscht-Froemsdorff, W.R. (1966) Myocardial Infarcts and Atmospherics. Z. Angew. Bader.-U. Klimaheik 13:679-686.

182. Brighton, C.T., et al. (1979) Electrical Properties of Bone and Cartilage. NY, Grune & Stratton.

183. Briggs, J.C. (1984) Centres of Origin in Biogeography. Biogeographical Monographs, No 1 Leeds, Univ Leeds.

184. Briggs, J.C. (1987) Biogeography and Plate Tectonics. NY, Elsevier.

185. British Med Assoc, Board of Science and Education (1987) Irradiation of Foodstuffs. London, BMA.

186. Britten, R.J. (1986) Rates of DNA Sequence Evolution Differ Between Taxonomic Groups. Science 231:1393-8.

187. Brittin, W.E., et al. (1972) Air and Water Pollution. Boulder, Co, Associated Univ Press.

188. Broadbent, F.E. (1986) Effects of Organic Matter on Nitrogen and Phosphorus Supply on Plants. pp.13-27 in (eds) Y. Chen, Y. Avnimelech, Role of Organic Matter in Modern Agriculture. Boston, Martinus Nijhoff.

189. Broadhurst, M. (1970) Complex Dielectric Constant and Dissipation Factor of Foliage. Natl Bur Stand Rep 9592. Springfield, Va, Natl Tech Info Service.

190. Brodribb, A.J.M., Humphreys, D.M. (1976) Diverticular Disease: Three Studies. Part I—Relation to Other Disorders and Fibre Intake. Part II—Treatment With Bran, Part III—Metabolic Effect of Bran in Patients With Diverticular Disease. Brit Med J 1(6007):424-29

191. Brown, D. (1987) Effects of Colorants in the Aquatic Environment. Ecotoxicol Environ Safety 13(2):139-47.

192. Brown, F. (1971) Some Orientational Influences of Nonvisual Terrestrial Electromagnetic Fields. Ann N Y Acad Sci 188:224-241.

193. Brown, L.R. (1984) Global Loss of Topsoil. J Soil Water Conserv 39(3):162-165.

194. Brown, Jr, F.A. (1965) A Unified Theory for Biological Rhythms. p 231 in (ed) J. Aschoff, Circadian Clocks. Amsterdam, J. Aschoff.

195. Brown, R.L., et al. (1987) Optimization of Sweep Codistillation Apparatus for Determination of Coumaphos and Other Organophosphorous Pesticide Residues in Animal Fat. J Assoc Off Anal Chem 70:442-5.

196. Brown, R., Thacker, J. (1984) Nature of Mutants Induced by Ionizing Radiation in Cultured Hamster Cells. Mutat Res 129:269-281.

197. Bryant, R.G., Jarvis, M. (1984) Nuclear Magnetic Relaxation Dispersion in Protein Solutions. A Test of Proton-Exchange Coupling. J Phys Chem 88:1323-1324.

198. Bull, G.M. (1973) Meterological Correlates With Myocardial and Cerebral Infarction and Respiratory Disease. Brit J Preventive Soc Med 27:108-113.

199. Bullock, T.H. (1977) Electromagnetic Sensing in Fish. Neurosci Res Bull 15:17-22.

200. Bulusu, S.C.I. (1987) Effect of Subacute Administration of Three Organophosphorus Pesticides on the Hepatic Phosphates Under Various Nutritional Conditions. Environ Res 44:128-35.

201. Bunk, M.J., et al. (1980) Relationship of Selenium Dependent Glutathione Peroxidase Activity to Nutritional Pancreatic Atrophy in the Selenium Deficient Chick. Fed Proc 39(3):556.

202. Burack, G.D., et al. (1984) Effects of Prenatal Exposure to a 60-Hz-High-Intentsity Electric Field on Postnatal Development and Sexual Differentiation. J Bioelect 3:451-467.

203. Burch, G.E., Giles, T.D. (1977) Influence of Weather and Climate on Cardiovascular Disease. pp.52-60 in (ed) S.W. Tromp, Progress Biometerology. Amsterdam, Swets & Zeitlinger.

204. Burkitt, D.P. (1994) Some Diseases Characteristic of Modern Western Civilization. pp.1-28 In N.J. Temple, D.P. Burkitt (eds) Western Diseases: Their Dietary Prevention and Reversibility. Totowa, NJ, Humana Press.

205. Burkitt, D.P. (1978) Mechanical Effects of Fibre With Reference to Appendicitis, Hiatus Hernia, Hemorrhoids and Varicose Veins. J Plant Foods 3(1/2):35-44.

206. Burkitt, D.P., Trowell, H.C. (eds)(1975) Refined Carbohydrate Foods and Disease: Some Implications of Dietary Fiber. NY, Academic.

207. Burkitt, D.P., et al. (1972) Effect of Dietary Fibre on Stools and Transit-Times, and Its Role in Causation of Disease. Lancet II(7792):1408-1422.

208. Burns, F.J. et al. (eds)(1986) Radiation Carcinogenesis and DNA Alterations. NATO Advanced Study Instit Rad Carcinogensis and DNA Alterations, 1984. NY, Plenum.

209. Burr, H.S. (1945) Diurnal Potentials in Maple Tree. Yale J Biol Med 17:727-735.

210. Burr, H.S. (1947) Tree Potentials. Yale J Biol Med 19:311-8.

211. Burr, H.S. (1972) Blueprint for Immortality: The Electric Patterns of Life. London, Neville Spearman.

212. Busby, D.E. (1968) Space Biomagnetics. Space Life 1(1):23.

213. Buzas, M.A., Culver, S.J. (1984) Species Duration and Evolution: Benthic Foraminifera on Atlantic Continental Margin of North America. Science 225:829-830.

214. Cabral, Jr., S.P. (1986) Carcinogenic Activity of Hexachlorobenzene in Mice and Hamsters. IARC Sci Pub (77):411-6.

215. Calabrese, E.J. (1980-81) Nutrition and Environmental Health: The Influence of Nutritional Status on Pollutant Toxicity and Carcinogenicity. Vol.1: The Vitamins; Vol.2: Minerals and Macronutrients. NY, Wiley & Sons.

216. Calcinai, M., Sequi, P. (1977) Contribution of Organic Matter to Cation-Exchange Capacity of Soils. pp.63-68 in Soil Organic Matter Studies. Vol.I. Intl Atomic Energy Agency, Vienna.

217. Callahan, P.S. (1984) Ancient Mysteries, Modern Visions—The Magnetic Life of Agriculture. Ca, Acres.

218. Campbell, C.A. (1978) Soil, Organic Carbon, Nitrogen and Fertility. pp.173-271 in (eds) M. Schnitzer, S.Y. Khan, Soil Organic Matter. NY, Elsevier Sci.

219. Campbell, D.E., Beets, J. (1978) Lunacy and the Moon. Psych Bull 85:1123-1129.

220. Cannon, H.L., Hopps, H.C. (1972) Geochemical Environment in Relation to Health and Disease. AAAS Symp, Spec Paper 140, Boulder, Colo, Geol Soc Amer.

221. Caola, R.J. et al. (1983) Measurements of Electrical and Magnetic Fields in and Around Homes Near a 500Kv Transmission Line. IEEE Trans Power Apparatus Sys PAS-102:3338-3347.

222. Capel-Boute, C. (1955) [The Effects of Atmospheric Electric Fields and Lightning and the Cosmic Origins of Water] Wirkungen Elektrischer Felder und Strahlungen Atmospharischen und Kosmischen Ursprungs auf das Wasser. Arch Meteor Geophys Bioklim B7:146-155.

223. Capel-Boute, C. (1962) [Chemical Tests and Clinical Tests in the Study of Factors in the Geophysical Environment] Tests Chimiques et Tests Clinques Dans L'etude des Facteurs Geophysiques de L'ambiance. pp.239-261 in IX Convegno della Salute, Ereditarieta-Ambiente-Alimen Tazione, Ferrara.

224. Cappaert, I.M.J., et al. (1977) Degradation of Bark and Its Value As a Soil Conditioner. pp.123-129 in Soil Organic Matter Studies. Vol.I. Intl Atomic Energy Agency, Vienna.

225. Carelli, P., et al. (1983) Magnetoencephalography. pp.471-482 in (eds) S.J. Williamson, et al., Biomagnetism. NY, Plenum.

226. Carey, A.E., Kutz, F.W. (1983) Trends in Ambient Concentrations of Agrochemicals in Humans and the Environment of the United States. Environ Monitoring Assessment 5:155-163.

227. Carney, M.W.P., et al. (1979) Thiamine and Pyridoxine Lack in Newly-Admitted Psychiatric Patients. Brit J Psychi 135:249-254.

228. Carrasco, J.M., et al. (1987) Pesticide Residues in Lake Albufrra, Valencia Spain. J Assoc Off Anal Chem 70:752-3.

229. Carroll, K.K. (1977) Dietary Factors in Hormone-Dependent Cancers. pp.25-40 in (ed) M. Minick, Nutrition and Cancer. NY, J Wiley.

230. Carroll, M.E., et al. (1979) Food Deprivation Increases Oral and Intravenous Drug Intake in Rats. Science 205:319-321.

231. Carson, H.L. (1970) Chromosome Tracers of the Origin of Species. Science 168:1414-1418

232. Carter, C.H. (1970) Handbook of Mental Retardation Syndromes. Springfield, Ill, C.C. Thomas.

233. Carter, S.D., et al. (1987) Fungicide Methyl Z-Benzimidazole Carbamate Causes Infertility in Male Sprague-Dawley Rats. Biol Reprod 37(3):709-17.

234. Carter, V.G., Dale, T. (1974) Topsoil and Civilization. Norman, Univ Oklahoma Press.

235. Caster, W.O. (1976) Role of Nutrition in Human Aging. p 41 in (eds) M. Rockstein, M.L. Sussman, Nutrition, Longevity and Aging. NY, Academic.

236. Caufield, C. (1984) Pesticides: Exporting Death. New Sci 103(1417):15-17.

237. Chadwick K.H., Leenhouts, H.P. (1981) Molecular Theory of Radiation Biology. NY, Springer-Verlag.

238. Chahoussou, F. (1986) How Pesticides Increase Pests. Ecologist 16(1):29-35.

239. Chalmers, J.A. (1967) Atmospheric Electricity. NY, Pergamon.

240. Chaloner, W.G., Meyen, S.V. (1973) Carboniferous and Permian Floras of the Northern Continents. pp.169-186 in (ed) A. Hallam, Atlas of Palaeobiogeography. NY, Elsevier Sci.

241. Chambers, Y. (1982) Alterations in Liver Structure and Function Resulting from Chronic Insecticide Exposure. In (ed) Y. Chambers, Effects of Chronic Exposure to Pesticides on Animal Systems. NY, Raven.

242. Chang, A.C., et al. (1984) Accumulation of Heavy Metals in Sewage Sludge-Treated Soils. J Environ Qual 13(3):87-90.

243. Chang, S., et al. (1985) Volume and Shape Changes of Human Erythrocytes Induced by Electric Fields. J Bioelect 4:301-315.

244. Chapman, L.J., et al. (1987) Parkinson's Disease and Industrial Chemicals. Lancet 1 (8528):332-333.

245. Charig, A.J. (1973) Jurassic and Cretaceous Dinosaurs. pp.339-352 in (ed) A. Hallam, Atlas of Palaeobiogeography. NY, Elsevier.

246. Charlesworth, B. (1988) Driving Genes and Chromosomes. Nature 332:394-395.

247. Charry, J.M. (1984) Biological Effects of Small Air Ions: A Review of Findings and Methods. Environ Res 34:351-389.

248. Charry, J.M., Hawkinshire, F.B. (1981) Effects of Atmospheric Electricity on Some Substrates of Disordered Social Behavior. J Personal Soc Psych 41(1):185-197.

249. Checkoway, H., et al. (1987) Medical Life Style and Occupational Risk Factors for Prostate Cancer. Prostate 10(1):79-88.

250. Chen, I.I.H., Saha, S. (1984) Analysis of an Intensive Magnetic Field on Blood Flow. J Bioelect 3:293-298.

251. Chen, I.I.H., Saha, S. (1985) Analysis of an Intensive Magnetic Field on Blood Flow: Part 2. J Bioelect 4:55-61.

252. Chen, J.R, Anderson, J.M. (1979) Legionnaires' Disease: Concentrations of Selenium and Other Elements. Science 206:1426-1427.

253. Chen, X, et al. (1980) Studies on the Relations of Selenium and Keshan Disease. Biol Trace Element Res 2(2):91-107.

254. Chen, Y., Stevenson, F.J. (1986) Soil Organic Matter Interactions with Trace Elements. pp.73-116 in (eds) Y. Chen, Y. Avnimelech, Role of Organic Matter in Modern Agriculture. Boston, Martinus Nijhoff.

255. Cheng, B.T. (1977) Soil Organic Matter as a Plant Nutrient. pp.31-38 in Soil Organic Matter Studies. Vol.I. Vienna, Intl Atomic Energy Agency.

256. Chernyshev, V.B. (1972) Solar Activity, Disturbances of the Geomagnetic Field and Insect Behavior. p.87 in The Sun, Electricity and Life. Izd MGU, Moscow.

257. Chernyshev, V.B. (1977) Disturbance Level of the Geomagnetic Field and the Motor Activity of Insects. pp.247-258 in (eds) M.N. Gnevyshev, A.I. Ol', Effects of Solar Activity on the Earth's Atmosphere and Biosphere. Jerusalem, Israel Prog Sci Trans.

258. Cherskov, M. (1985) Grasshopper Plague Raises Concerns Over Pesticides. Amer Med News 28(3):7.

259. Chisci, G., Morgan, R.P.C. (eds) (1986) Soil Erosion in the European Community. Impact of Changing Agriculture. Proc Seminar Land Degradation Due to Hydrological Phenomena in Hilly Areas: Impact of Change of Land Use and Managment Cesana, 9-11 Oct 1985. Boston, A.A. Balkema.

260. Chiu, S.M., Oleinick, N.L. (1982) Sensitivity of Active and Inactive Chromatin to Ionizing Radiation-Induced DNA Strand Breakage. Intl J Radiat Biol 41:71-77.

261. Chizhevskii, A.L. (1930) Epidemic Catastrophes and Periodic Activity of the Sun. Izd. VOVG, Moscow.

262. Chopdar, A. (1979) Incidence of Ocular Changes Due to Vitamin A Deficiency in Western Orissa. Indian Pediatrics 16(9):787-790.

263. Chouinard, G., et al. (1977) Tryptophan-Nicotinamide Combination in Depression. Lancet I(8005):249.

264. Christoffel, T. (1985) Grassroots Environmentalism Under Legal Attack: Dandelions, Pesticides and a Neighbor's Right-To-Know. Amer J Pub Health 75:585-7.

265. Chu, Y., et al. (1986) Separation of Large DNA Molecules by Contour-Clamped Homogenous Electric Fields. Science 234:1582-1585.

266. Chu, Y., et al. (1984) Blood Selenium Concentration in Residents of Areas in China Having a High Incidence of Lung Cancer. Biol Trace Element Res 6(2):133-137.

267. Churchill, J.A., Henderson, W. (1974) Perinatal Factors Affecting Fetal Development— Twin Pregnancy. pp.69-76 in (ed) K.S. Moghissi. Birth Defects and Fetal Development. Springfield, Ill, Charles Thomas.

268. Chuvaev, P.P. (1969) Effect of an Extremely Weak Steady Magnetic Field on Seedling Root Tissues and on Some Microorganisms. p.252 in Materials of 2nd All-Union Conf Effect of Magnetic Field on Biological Objects, Moscow.

269. Clark, C.R. (1984) Effects of Noise on Health. pp.111-124. in (eds) D.M. Jones, A.J. Chapman. Noise and Society. NY, J Wiley.

270. Clark, D.L. et al. (1986) Conodont Survival and Low Iridium Across the Permian-Triassic Boundary in South China. Science 233:984-986.

271. Clemmens, W.A. (1986) Evolution of the Terrestrial Vertebrate Fauna During the Cretaceous-Tertiary Transition. pp.63-86 in (ed) D.K. Elliot. Dynamics of Extinction. NY, J Wiley.

272. Clinton, M. (1950) The Intoxications. In (ed) T. Harrison, Principles of Internal Medicine. NY, Blakiston.

273. Coakley, S.M. (1981) Relationship Between Weather Variables and Stripe Rust Epidemics on Winter Wheat. pp.3-4 in Conf on Agric & Forest Meterolo (15th) and Biometerolo (5th). Boston, Mass, Amer Meterolo Soc.

274. Coggon, D., et al. (1987) Mortality of Workers Exposed to Styrene in the Manufacture of Glass-Reinforced Plastics. Scand J Work Environ Health 13(2):94-9.

275. Cohen, D. (1975) Magnetic Fields of the Human Body. Phys Today (8):34-43.

276. Cohen, D. (1983) Introduction. pp.5-16 in (eds) S.J. Williamson, et al., Biomagnetism. NY, Plenum.

277. Cohen, D. (1983) Steady Fields of the Body. pp.327-340 in (eds) S.J. Williamson, et al., Biomagnetism. NY, Plenum.

278. Cohen, S., Spacapan, S. (1984) Social Psychology of Noise. pp.221-246 in (eds) D.M. Jones, A.J. Chapman. Noise and Society. NY, J Wiley.

279. Cole, C.V., et al. (1978) Trophic Interactions in Soils as They Affect Energy and Nutrient Dynamics. I-VI. Micro Ecology 4:381-387.

280. Collier, J. (1988) Dynamics of Biological Order. pp.227-242 in (eds) B.H. Weber. Entropy, Information and Evolution. Cambridge, Mass, MIT Press.

281. Combs, G.F., Peterson, F.J. (1980) Protection from Paraquat Toxicity by Dietary Selenium in the Chick. Fed Proc 39(3):556.

282. Comfort, A. (1979) Myth of Senility. Diagnosing Nonspecific Major Illness in the Elderly. Postgrad Med 65(3):130-142.

283. Commoner, B. (1977) Closing Circle: Nature, Man and Technology. NY, Alfred A. Knopf.

284. Connor, S. (1988) Protein Reveals Damage from Radiation. New Sci 117(1595):41.

285. Conti, P., et al. (1983) Reduced Mitogenic Stimulation of Human Lymphocytes by ELF Electromagnetic Fields. FEBS Lett 162:156-160.

286. Conti, P., et al. (1985) Effect of Electromagnetic Fields on Two Calcium Dependent Biological Systems. J Bioelect 4:227-236.

287. Conti, R., Nicolini, P. (1985) Possible Biological Effects of 50-Hz Electric Fields: A Progress Report. J Bioelect 4:177-193.

288. Cope, F.W. (1978) Man in a Gas of Tachyon Magnetoelectric Dipoles—A New Hypothesis. Part I. A Summary of Some Real But Unexplained Biocosmic Phenomena. Physiol Chem Phys 10:535-540.

289. Cope, F.W. (1978) Man in a Gas of Tachyon Magnetoelectric Dipoles—A New Hypothesis. Part II. Introduction to the Theory. Physiol Chem Phys 10:541-545.

290. Cope, F.W. (1978) Man in a Gas of Tachyon Magnetoelectric Dipoles—A New Hypothesis. Part III. Development of the Physics of the Theory, and Experimental Correlations: Cosmic Radiation Detection by Man, Auras of Magnetics and Man, Dowser's Grid Lines of Earth, Mediation of Interaction of Electromagnetic Waves with Man, Interactions with Solids, and Calculation of Magnetic Moment. Physiol Chem Phys 10:547-55

291. Cope, F.W. (1979) Man in a Gas of Tachyon Magnetoelectric Dipoles—A New Hypothesis. Part IV. Beta Rays May Add to Magnetoelectric Dipoles in Accord with Schwinger Condition to Yield Toxic Products Observed in Reich Oranur Experiment. A Possible Mechanism for Toxic Air Ion Effects. Physiol Chem Phys 11:87-91.

292. Cope, F.W. (1979) Superconducting Josephson Junctions—A Possible Mechanism for Detection of Weak Magnetic Fields and of Microwaves by Living Organisms. pp.87-88 in (ed) T.S. Tenforde, Magnetic Field Effects on Biological Systems. NY, Plenum.

293. Cope, F.W. (1982) Biological and Organic Superconduction at Physiological Temperatures. pp.99-124 in (ed) B. Lipinski, Electronic and Mechanoelectrical Transduction in Biological Materials. NY, Marcel Dekker.

294. Cope, M.J., Chaloner, W.G. (1985) Wildfire: An Interaction of Biological and Physical Processes. pp.257-277. In (ed) B.H. Tiffney. Geological Factors and the Evolution of Plants. New Haven, Conn Yale Univ.

295. Coquerelle, T., Hagen, U. (1978) Radiation Effects on the Biological Function of DNA—Introduction. pp.261-266 In (eds) J. Hutterman, et al., Effects of Ionizing Radiation on DNA. NY, Springer-Verlag.

296. Corbett, T. (1979) Cancer and Chemicals. Chicago, Nelson-Hall.

297. Correa, A.D., et al. (1986) Chemical Constituents in Amaranth Grains. Arch Latinoam Nutr 36:319-26.

298. Coster, W.O. (1976) Role of Nutrition in Human Aging. p 41 in (eds) M. Rockstein, M.L. Sussman, Nutrition, Longevity and Aging. NY, Academic.

299. Cott, M.D. (1964) Orthomolecular Approach to Learning Disabilities. pp.14-44 in A. Montagu, Life Before Birth. NY, New American Library.

300. Council Environ Quality & Dept State (1980) Global 2000 Report to the President of the U.S. Entering the 21st Century. Vol.2. NY, Pergammon.

301. Courtney, K., et al. (1985) Placental Transfer and Fetal Deposition of Hexachlorobenzene in the Hamster and Guinea Pig. Environ Res 37:238-48.

302. Cowgill, U.M. (1983) Distribution of Selenium and Cancer Mortality in the Continental United States. Biol Trace Element Res 5(4/5):345-361.

303. Cox, R. (1980) Comparative Mutagenesis in Cultured Mammalian Cells. pp.33-46 in (ed) M. Alacevic, Progress in Environmental Mutagenesis. NY, Elsevier/North Holland Biomedical.

304. Cragnolino, G., Tuovinen, O.H. (1984) Role of Sulphate-Reducing and Sulphur-Oxidizing Bacteria in the Localized Corrosion of Iron-Base Alloys—A Review. Intl Biodeterioration 20(1):9-26.

305. Crain, I.K. (1971) Possible Direct Causal Relation Between Geomagnetic Reversal and Biological Extinctions. Geol Soc Amer Bull 82:2603-2606.

306. Crain, I.K., Crain, P.L. (1970) New Stochastic Model for Geomagnetic Reversals. Nature 228:39-41.

307. Crammer, J. (1975) Lithium, Calcium, and Mental Illness. Lancet I(7900):215-216.

308. Cramp, W.A. (1978) Irradiation Effects on DNA Replication. pp.292-303. in (eds) J.Hutterman et al., Effects of Ionizing Radiation on DNA. NY, Springer-Verlag.

309. Cramp, W.A., et al. (1978) Biological Functions of DNA and Methods of Testing. pp.255-260 in (eds) J. Hütterman, et al., Effecting of Ionizing Radiation on DNA. NY, Springer-Verlag.

310. Crick, F. (1982) Life Itself: Its Origin and Nature. NY, Simon & Schuster.

311. Crider, A. (1979) Schizophrenia: A Biopsychological Perspective. Hillsdale, N.J., Lawerance Erlbaum.

312. Crowley, T.J., North, G.R. (1988) Abrupt Climate Change and Extinction Events in Earth History. Science 240:996-1002.

313. Cullen, J.M., Spadaro, J.A. (1983) Axonal Regeneration in the Spinal Cord: A Role for Applied Electricity (Review). J Bioelect 2(1):57-75.

314. Cullis, P.M. et al. (1987) Electron Transfer from Protein to DNA in Irradiated Chromatin. Nature 330:773-774.

315. Cummings, J., et al. (1980) Reversible Dementia. JAMA 243:2434-2439.

316. d'Ambrosio, G., et al. (1985) Chromosomal Aberrations Induced by ELF Electric Fields. J Bioelect 4:279-84.

317. d'Espagnat, B. (1979) Quantum Theory and Reality. Sci Amer 241(11):158-181.

318. Daft, J. (1987) Determining Multifumigants in Whole Grains and Legumes, Milled and Low-Fat Grain Products, Spices, Citrus Fruit and Beverages. J Assoc Off Anal Chem 70:734-9.

319. Dakshinamurti, K. (1977) B Vitamins and Nervous System Function. pp.245-318 in (eds) R.J. Wurtman, J.J. Wurtman, Nutrition and the Brain. Vol.1. NY, Raven Press.

320. Dainty, Jr., J. (1961) Ion Transport and Electrical Potential in Plant Cells. Ann Rev Plant Physiol 13:379-402.

321. Damodaran, K.V.R., et al. (1979) Vitamin B-Complex Deficiency and Visual Acuity. Brit J Nutr 41:27-30.

322. Darby, H.C. (1966) Clearing of Woodland in Europe in Man's Role. pp.183-186 in (ed) W.L. Thomas, Changing the Face of the Earth. Univ Chicago Press, c1956.

323. Dardanoni, L., et al. (1985) Millimeter-Wave Effects on Candida Albicans Cells. J Bioelect 4:171-176.

324. Darnell, J.E., Jr. (1978) Implications of RNA. RNA Splicing in Evolution of Eukaryotic Cells. Science 202:1257-1260.

325. Darwin, C. (1858) In a Letter to Asa Gray, 5 Sept 1857. Zoologist 16:6297-99.

326. Darwin, C. (1859) Origin of Species by Means of Natural Selection or the Preservation of Favoured Races in the Struggle for Life. 6th ed. London, John Murray, c1902.

327. Davies, D.M. (1985) Calcium Metabolism in Healthy Men Deprived of Sunlight. Ann N Y Acad Sci 453:21-27.

328. Davies, J.F. (1987) Changing Profile of Pesticide Poisoning. New Eng J Med 316:807-8..

329. Davies, P. (1995) Cosmic Blueprint. NY, Simon & Schuster.

330. Davis, J.B. (1963) Review of Scientific Information on the Effects of Ionized Air on Human Beings and Animals. Aerospace Med 34:35-42.

331. Dawkins, R. (1996) Selfish Gene. NY, Oxford Univ Press.

332. DeBoodt, M., et al. (1977) Mulching as Protection Against Erosion. pp.117-121 in Soil Organic Matter Studies. Vol.I. Intl Atomic Energy Agency, Vienna.

333. De Guzman, E.V. et al. (1982) Production of Mutants by Irradiation of In Vitro-Cultured Tissues of Coconut and Banana, and Their Mass Propagation by the Tissue Culture Technique. pp.113-138 in Induced Mutations in Vegetatively Propagated Plants II. FAO/IAEA Div Isotope & Radiation Applic Atomic Energy Food & Agric Develop, 1980. Vienna, Intl Atomic Energy Agency.

334. Delgado, J.M.R. (1985) Review Article: Biological Effects of Extremely Low Frequency Fields. J Bioelect 4:75-91.

335. Dempsey, J.L., Morley, A.A. (1986) Environ Mutagen 8:385.

336. Denbigh, K.G., Denbigh,J.S. (1985) Entropy in Relation to Incomplete Knowledge. NY, Cambridge Univ.

337. Denson, R. (1962) Nicotinamide in the Treatment of Schizophrenia. Dis Nerv Sys 23:167-72

338. Denton, M. (1986) Evolution: A Theory in Crisis. London, Burnett Books.

339. De Ploey, J. (1986) Soil Erosion and Possible Conservation Measures in Loess Loamy Areas. pp.157-164 in (eds) G. Chisci, R.P.C. Morgan, Soil Erosion in the European Community. Impact of Changing Agriculture. Boston, A.A. Balkema.

340. Deryapa, N.R., et al. (1975) Effect of Geomagnetic Field on State of Adrenal Cortex Function. pp.77-78 in Physicomathematical and Biological Problems of Effect of Electromagnetic Fields and Ionization of the Air. Vol.2. Nauka, Moscow.

341. De Vault, D. (1984) Quantum-Mechanical Tunneling in Biological Systems. NY, Cambridge Univ Press.

342. Dewey, E.R. (1985) What Forces Could Cause Cycles? Cycles 36(3):68-69.

343. Diamond, J.M. (1984) Island Population Biology: Possible Effects of Unrestricted Pesticide Use on Tropical Birds. Nature 310:452.

344. Diamond, M.C., et al. (1980) Environmental Influences on Serotonin and Nucleotides in Rat Cerebral Cortex. Science 210:652-654.

345. Dickerson, J.W.T., Basu, T.K.(1977) Specific Vitamins Deficiencies and Their Significance in Patients with Cancer and Receiving Chemotherapy. pp.95-104 in (ed) M. Winick, Nutrition and Cancer. NY, Wiley.

346. Diez, J.A. (1973) Genetic Variation and the Role of Adrenal Function in the Regulation of Serotonin and Catecholamine Synthesis in Developing an Adult Mouse Brain. Phd Dissertation, Univ Conn.

347. Dill, D.B., et al. (eds) (1964) Adaptation to the Environment, Handbook of Physiology, Section 4. Wash, DC, Amer Physiol Soc.

348. Diprose, M.F., et al. (1984) Effect of Externally Applied Electrostatic Fields, Microwave Radiation and Electric Currents on Plants and Other Organisms, with Special Reference to Weed Control. Bot Rev 50:171-223.

349. Direnfeld, L.K. (1983) Genesis of the EEG and Its Relation to Electromagnetic Radiation. J Bioelect 2:111-121.

350. Diver, C. (1936) Present State of the Theory of Natural Selection. Special Symposium. Proc Roy Soc Brit 121:43-73.

351. Dlouhy, J. (1977) Vaxtprodukters Kvalitet vid Konventionell och Biodynamisk Odling. Reports of the Agricultural College of Sweden. Series A, Nr 272. Upsala, Sweden.

352. Dodge, P.R., et al. (1975) Nutrition and the Developing Nervous System. St. Louis, C.V. Mosby Co.

353. Doll, R., et al. (1985) Nitrates, Nitrites and Gastric Cancer in Great Britain. Nature 313:620-625.

354. Donini, B. et al. (1984) Spectrum of Mutant Characters Utilized in Developing Improved Cultivators. pp.7-31 in Selection in Mutation Breeding. FAO/IAEA Div Isoyope & Radiation Applications of Atomic Energy for Food & Agric. Vienna, Intl Atomic Energy Agency.

355. Doolittle, R.F. (1985) Genealogy of Some Recently Evolved Vertebrate Proteins. Trends Biochem Sci 10:233-237.

356. Doolittle, W.F. (1987) Origin and Function of Intervening Sequences in DNA: A Review. Amer Nat 130:915-928.

357. Dover, G.A. (1987) DNA Turnover and the Molecular Clock. J Mol Evol 26:47-58.

358. Downes, C.S. (1988) DNA Repair: Views of Unity and Diversity. Nature 332:208-209.

359. Drorbaugh, J.E., Moore, D.M. (1974) Effect of Maternal Diabetes on Development of Central Nervous System Function in the Child. pp.106-115 in (ed) K.S. Moghissi, Birth Defects and Fetal Development. Springfield, Ill, Charles Thomas.

360. Dubrov, A.P. (1970) Contemporary Heliobiology. Nauka i Zhizn' 9, 97.

361. Dubrov, A.P. (1971) Global Changes in Biochemical and Physiochemical Processes Due to the Geomagnetic Field. p 302 in Questions of Theory and Practice of Magnetic Treatment of Water and Aqueous Systems. Tsvetmetin-Formatsiya, Moscow.

362. Dubrov, A.P. (1973) Heliobiological Factors and Dynamics of Excretion of Organic Substances by Plant Roots. p 67 in Effect of Some Cosmic and Geophysical Factors on Earth's Biosphere. Nauka, Moscow.

363. Dubrov, A.P. (1973) Some Aspects of Heliobiological Responsibility for Rhythmicity in Elements of the Biosphere. p.233 In Lectures in Memory of L.S. Berg XV-XIX. Nauka, Leningrad.

364. Dubrov, A.P. (1975) Effect of Geomagnetic Field on Genetic Homeostasis. pp.168-175 In Investigation of Species Productivity in Range. Mintis, Vilnius.

365. Dubrov, A.P. (1978) The Geomagnetic Field and Life. Geomagnetobiology. NY, Plenum.

366. Dubrow, R., et al. (1988) Farming and Malignant Lymphonia in Handcock County, Ohio. Brit J Ind Med 45(1):25-8.

367. Dudley, H.C. (1975) Is There An Ether? pp.65-77 in (ed) R.R. Jones, Unsettled Earth. Mich, Ann Arbor Sci.

368. Düll, T., Düll, B. (1953) [Correlations Between Geomagnetic Disturbances and the Increased Occurrence of Fatal Accidents] Zusammerhange Zwischen Storungen Erdmagnetismus und Haufungen von Todesfallen. Dtsch Med Wschr 61(3):95.

369. Düll, T., Düll, B. (1953) [A New Understanding About the Connection Between the Number of Daily Fatal Accidents and the Character of Magnetic Disturbance] Neue Untersuchungen uber die Beziehungwischen den Zahl der Taglichen Todesfalle und dem Magnetischen Storungscharakter. Bioklimat Beibl 2(1):24.

370. Dunn, P.J. (1974) New Dynamics of Preventive Medicine. Vol. 2. Symposia Specialists.

371. Durkheim, E. (1952) Suicide. London, Routledge and Kegan Paul.

372. Dusheck, J. (1985) Plant Toxins: A Double-Edged Sword. Sci News 128:101.

373. Eaks, I.L., Ludi, W.A. (1964) Effect of Harvesting and Packing House Procedures on Rind Staining of Central California Washington Navel Oranges. Proc Amer Soc Hort Sci 85:245.

374. Ebeling, W. (1981) How Fertilizers Affect the Nutrient Balance in Plant Crops. J Appl Nut 33(2):138-155.

375. Ebrahim, G.J. (1979) Problems of Undernutrition. pp.13-130 in (ed) R.J. Jarrett. Nutrition and Disease. Baltimore, Univ Park Press.

376. Eckholm, E.P., Brown, L.R. (1977) Spreading Deserts—The Hand of Man. Worldwatch Paper 13. Wash, Worldwatch Instit.

377. Edmiston, J. (1975) Effect of Exclusion of the Earth's Magnetic Field on the Germination and Growth of Seeds of White Mustard (*Sinapis Alba*). Biochem Physiol Pflanzen 167(1):97-100.

378. Edmiston, S., Maddy, K.T. (1987) Summary of Illnesses and Injuries Reported in California by Physicians in 1986 as Potentially Related to Pesticides. Vet Hum Toxicol 29:391-397.

379. Edwards, D. (1961) Influence of Electrical Field on Pupation and Oviposition. Nature 191:976.

380. Edwards, D. (1973) Devonian Floras. pp.105-116 in (ed) A. Hallam, Atlas of Palaeobiogeography. NY, Elsevier Sci.

381. Edwards, D.D. (1987) ELF: The Current Controversy. Sci News 131:107-109.

382. Edwards, D.D. (1987) ELF Under Suspicion in New Report. Sci News 132:39.

383. Edwards, D.D. (1988) Cells Haywire in Electromagnetic Field? Sci News 133:216.

384. Edwards, D.K. (1960) Effects of Artificially Produced Atmospheric Electrical Fields Upon the Activity of Some Adult Diptera. Canad J Zool 38:899-912.

385. Edwards, D.K. (1960) Effects of Experimentally Altered Unipolar Air Ion Density Upon the Amount of Activity of the Blowfly, *Calliphora Vicina* R.D. Canad J Zool 38:1079-91.

386. Edwards, D.K. (1961) Influence of Electrical Field on Pupation and Oviposition in *Nepytia Phantasmania*. Nature 191:976-993.

387. Effendy, I., Krause, W. (1987) Environmental Risk Factors in the History of Male Patients of an Infertility Clinic. Andrologia (19 June)Spec No:262-5.

388. Efron, E. (1984) The Apocalyptics: Cancer and the Big Lie. NY, Simon & Schuster.

389. Ehling, U.H. (1984) Methods to Estimate the Genetic Risk. pp.292-318 In (ed) G. Obe, Mutations in Man. NY, Springer-Verlag.

390. Ehling, U.H., Randolph, M.L. (1962) Skeletal Abnormality in the Fl Generation of Mice Exposed to Ionizing Radiation. Genetics 47:1543-1555.

391. Ehrlich, P.R. (1986) Extinction: What is Happening Now and What Needs to Be Done. pp.157-164 in (ed) D.K. Elliot, Dynamics of Extinction. NY, J Wiley.

392. Ehrlich, P.R. et al. (1983) Long-Term Biological Consequences of Nuclear War. Science 222:1293-1300.

393. Ehrmann, V.W., et al. (1976) Therapy With ELF-Magnetic Fields. Physikalische Medizin 5:161-170.

394. Eichmeier, J., Buger, P. (1969) [The Influence of Lightning's Electromagnetic Effects Upon the Precipitation Reaction of Bismuth Choloride According to the Piccardi Test] Uber den Einfluss Elektromagneitscher Strahlung auf die Bismuthchlorid-Fallungsreaktion nach Piccardi. Intl J Biometeor 13:239-256.

395. El-Bassam, N., Tietjen, C. (1977) Municipal Sludge as Organic Fertilizer With Special Reference to the Heavy Metals Constituents. pp.253-258 in Soil Organic Matter Studies. Vol.II. Vienna, Intl Atomic Energy Agency.

396. Ellis, B.F., Messina, A. R. (1965-67) Catalogue of Index Foraminifera. NY, Amer Museum Nat History.

397. Elwood, P.C., et al. (1980) Magnesium and Calcium in the Myocardium: Cause of Death and Area Differences. Lancet I (8197):720-722.

398. Elwood, W.K., Smith S.D. (1983) Electroosmosis in Compact Bone. J Bioelect 2(1):37-56.

399. Emilia, G., et al. (1985) Effect of Low-Frequency Low-Energy Pulsing Electromagnetic Fields on the Response to Lectin Stimulation of Human Normal and Chronic Lymphocytic Leukemia Lymphocytes. J Bioelect 4:145-161.

400. Engel, K.S.R. (1983) Pesticides and Plagues. Sci Digest (Jan):58-59.

401. Engeman, R.M., Pank, L.F. (1984) Potential Secondary Toxicity from Anticoagulant Pesticides Contaminating Human Food Sources. New Eng J Med 311:257-8.

402. Epstein, S.S. (1974) Chronic Biological Hazards Due to Chemical Polutants. pp.136-166. in (ed) K.S. Moghissi. Birth Defects and Fetal Development. Springfield, Ill, Charles Thomas.

403. Erickson, J., et al. (1979) Skeletal Concentrations of Lead in Ancient Peruvians. New Eng J Med 26:946-951.

404. Ewertsen, H.W. (1979) Psychological Effect of Noise. Acta Otolaryngol (Suppl) (Stockh) 360:88-89.

405. Fackelman, K.A. (1989) Avian Altruism. Sci News 135:364-365

406. Fajer, E.D. et al. (1989) Effects of Enriched Carbon Dioxide Atmospheres on Plant-Insect Herbivore Interactions. Science 243:1198-1200.

407. Falandysz, J., Falandysz, J. (1986) Organochlorine Pesticides and Polychlorinated Biphenyls in the Adipose Tissue of Abattoir and Game Animals from the Northern Region of Poland, 1984 [in Polish with English Abstract] Roczniki Panstwowego Zakladu Higieny 37(6):487-93.

408. FAO/IAEA Division of Isotope and Radiation Applications of Atomic Energy for Food and Agricultural Development (1986) Quantification, Nature and Bioavailability of Bound 14C-Pesticide Residues in Soils, Plants and Food. Vienna, Intl Atomic Energy Agency.

409. Farrell, D.E. (1983) Assessment of Iron in Human Tissue: The Magnetic Biopsy. pp.483-499 in (eds) S.J. Williamson, et al., Biomagnetism. NY, Plenum.

410. Farrell, D.E. (1983) Magnetic Observation of Conduction System Activity in Normal Subjects. pp.275-284 in (eds) S.J. Williamson, et al., Biomagnetism. NY, Plenum.

411. Faust, V. (1977) Biometerologie. Stuttgart, Hippokrates.

412. Fedorov, S.M., Getling, Z.M. (1987) Electron Microscopy Changes in the Skin After Administration of Organophosphate Pesticides [English Abstract]. Vestnik Dermatologii Venerologii (5):24-8.

413. Fenici, R.R. (1983) Experimental Electrophysiology. pp.211-226 in (eds) S.J. Williamson, et al., Biomagnetism. NY, Plenum.

414. Fensom, D.S. (1957) Bio-Electric Potentials of Plants and Their Functional Significance—I. An Electrokinetic Theory of Transport. Can J Bot 35:573-582.

415. Fensom, D.S. (1985) Electrical and Magnetic Stimuli, In Hormonal Regulation of Development III, pp.625-651 in (eds) R.P. Pharis, D.M. Reid, Endocrinology of Plant Physiology. Berlin, Springer Verlag.

416. Ferrando, M.D., et al. (1987) Acute Toxicity of Organochlorinated Pesticides to the European Eel Anguilla Anguilla: The Dependence on Exposure Time and Temperature. Bull Environ Contam Toxicol 39:365-369.

417. Feyerabend, P. (1965) Problems of Empiricism. pp.145-260 in (ed) R.G. Colodny, Beyond the Edge of Certainty; Essays in Contemporary Science and Philosophy. Englewood Cliffs, NJ, Prentice-Hall.

418. Fillippi, N., et al. (1986) Geomorphologic Features, Land Use and Early Results of the Pedologic Research in Cesena Pilot Area. pp.105-121 in (eds) G. Chisci, R.P.C. Morgan, Soil Erosion in European Community. Impact of Changing Agriculture. Boston, A.A. Balkema.

419. Fishbein, L. (1979) Potential Industrial Carcinogens and Mutagens. NY, Elsevier Sci.

420. Flavell, A. (1985) Introns Continue to Amaze. Nature 316:574-575.

421. Fletcher, J.L., Busnel, R.G. (eds) (1978) Effects of Noise on Wildlife. NY, Academic Press

422. Fleury, P., et al. (1987) Effects of Organic Fertilizer Application on Hay Meadow Quality in the French Northern Alps. pp.309-312 in (eds) H.G. Van Der Meer, et al., Animal Manure on Grasslands and Fodder Crops. Dordrecht, Netherlands, Martinus Nihoff.

423. Flispe, W.J., Bonner, F.T. (1985) Nitrogen-Isotope Ratios of Nitrate in Ground Water Under Fertilized Fields, Long Island, New York. Ground Water 23:59-67.

424. Flodin, U., et al. (1988) Chronic Lymphatic Leukaemia and Engine Exhausts, Fresh Wood and DDT: A Case-Referent Study. Brit J Ind Med 45(1):33-8.

425. Flohn, H. (1955) [Natural and Anthropogenic Climate Modification] Naturliche und Anthropogene Klimamodifik. Annalen der Meterologie NF 6:59-66.

426. Flynn, et al. (1981) Zinc Status of Pregnant Alcoholic Women: A Determinant of Fetal Outcome. Lancet I (8220):572.

427. Fontanesi, G., et al. (1984) Effect of Low Frequency Pulsing Electromagnetic Fields for the Treatment of Congenital and Acquired Pseudoarthroses. J Bioelect 3:155-175.

428. Ford, J. (1987) Quantum Chaos: A Beginning or an End? Nature 325:19-20.

429. Forstner, U. (1984) Metal Pollution of Terrestrial Waters. pp.71-94 in (ed) J.O. Nriagu, Changing Metal Cycles and Human Health. NY, Springer-Verlag.

430. Fredericks, C. (1976) Psycho-Nutrition. Phila, Pa, Grosset & Dunlap.

431. Freedman, B., Hutchinson, T.C. (1980) Pollutant Inputs from and Accumulation in Soils and Vegetation Near a Nickel-Copper Smelter at Sidbury, Canada. Can J Bot 58:108-132.

432. Freeman, J. (ed) (1975) Trace Element Geochemistry in Health and Disease. The Geological Society of America Special Paper 155. Boulder, Co.

433. Freese, A. (1978) The End of Senility. NY, NY, Arbor House.

434. Freier, S., Eidelman, A.I. (eds) (1980) Human Milk—Its Biological and Social Value. Amsterdam, Excerpta Medica.

435. Frey, A.H. (1983) Comment on "Microwaves and the Blood-Brain-Barrier". J Bioelect 2:83-88.

436. Frey, A.H. (1984) Possible Modification of the Blood-Vitreous Humor Barrier of the Eye with Electromagnetic Energy. J Bioelect 3:281-292.

437. Frey, A.H. (1988) Evolution and Results of Biological Research with Low-Intensity Non-ionizing Radiation. pp.785-838 in (ed) A.A. Marino. Modern Bioelectricity. NY, Marcel Dekker.

438. Frey, A.H., Eichert, E.S. (1985) Psychophysical Analysis of Microwave Sound Perception. J Bioelect 4:1-14.

439. Frey, A.H., et al. (1975) Neural Functioning and Behavior: Defining the Relationship. Ann N Y Acad Sci 247:433.

440. Frey, A.H., Wesler, L.S. (1982) Test of the Dopamine Hypothesis of Microwave Energy Effects. J Bioelect 1:305-312.

441. Frey, A.H., Wesler, L.S. (1983) Dopamine Receptors and Microwave Energy Exposure. J Bioelect 2:145-157.

442. Frey, A.H., Wesler, L.S. (1984) Morphine Effects Appear to be Potentiated By Microwave Energy Exposure. J Bioelect 3:373-383.

443. Friedenberg, Z.B., Brighton, C.T. (1974) Electrical Fracture Healing. Ann N Y Acad Sci 238:564-574.

444. Friedman, H., Becker, R.O. (1963) Geomagnetic Parameters and Psychic Hospital Admissions. Nature 200:626-628.

445. Fritz, J.C. (1972) Iron and Associated Trace Mineral Problems in Man and Animals. Geol Soc Amer Spec Paper 140:25-32.

446. Frost, D.V., Lakin, H.W. (1972) Selenium Accumulation in Soils: Discussion and Reply. Geol Soc Amer Spec Paper 140:55-56.

447. Fukada, E. (1974) Piezoelectric Properties of Organic Polymers. Ann N Y Acad Sci 238:7-25.

448. Fukada, E. (1982) Piezoelectricity of Biological Materials. pp.125-156 in (ed) B. Lipinski, Electronic Conduction and Mechanoelectrical Transduction in Biological Materials. NY, Marcel Dekker.

449. Fukuoka, M. (1985) Natural Way of Farming. Tokyo, Japan Pubs.

450. Fulton, J. P. et al. (1980) Electrical Wiring Configurations and Childhood Leukemia in Rhode Island. Amer J Epidemiol III:292-296.

451. Furst, A., Radding, S.B. (1979) Unusual Metal as Carcinogens. Biol Trace Element Res 1(2):169-181.

452. Gadalina, I.D., Ivanov, I.U.V. (1987) Age-Related Changes in the Hematopoietic Organs of Experimental Animals After Exposure to Pesticides [English Abstract]. Gigiena Sanitariia (6):74-5.

453. Gallicchio, V.S., et al. (1987) Inhibition of Human Bone Marrow: Derived Stem Cell Colony Formation (CFU-E BFU-E and CFU-GM) Following In Vitro Exposure to Organophosphates. Exp Hematol 15(11):1099-102.

454. Gamlin, L. (1988) Sweden's Factory Forests. New Sci 117(1597):41-45.

455. Garrels, R.M., et al. (1975) Chemical Cycles and the Environment: Assessing Human Influences. Los Altos, Calif, William Kaufman.

456. Gartner, S., McGuirk, J.P. (1979) Terminal Cretaceous Extinction, Scenario for a Catastrophe. Science 206:1272.

457. Gauger, J.R. (1985) Household Appliance Magnetic Field Survey. IEEE Trans Power Apparatus Sys PAS-104: 2436-2444.

458. Gavalos, R.J., et al. (1970) Effect of Low-Level Frequency Electric Fields on EEG and Behavior in *Macaca Menestrina*. Brain Res 18:491.

459. Geller, S.H., Shannon, H.W. (1973) Moon, Weather and Mental Hospital Contracts: Confirmation and Explanation of the Transylvania Effect. Toronto: Lakeshore Psychiatric Hospital. Unpub Manuscript.

460. Gellhorn, E., Loofbourrow, G. (1963) Emotions and Emotional Disorders: Anguro-Physiological Approach. NY, Harper & Row.

461. Gengerelli, J.A., Holter, N.J. (1941) Experiments on Stimulation of Nerves by Alternating Electrical Fields. Proc Soc Exp Biol (N.Y.) 46:523.

462. Gensler, W. (1974) Bioelectric Potentials and Their Relation to Growth in Higher Plants. Ann N Y Acad Sci 238:280-299.

463. Gensler, W. (1988) Apoplastic Electropotentials in Plants: Measurement and Use. pp.105-122 in (ed) A.A. Marino. Modern Bioelectricity. NY, Marcel Dekker.

464. Gilbert, E.F., et al. (1987) Genetic Aspects of Developmental Pathology. March of Dimes Birth Defects Foundation. Birth Defects: Original Article Series. Vol.23, No.1. NY, Alan Liss.

465. Gilbert, L.E., Raven, P.H. (1975) Coevolution of Animals and Plants: Symposium V, 1st Intl Cong Systematic & Evolutionary Biology, Boulder, Colo, Aug 1973. Austin, Univ Texas Press.

466. Gilbert, W. (1978) Why Genes in Pieces? Nature 271:501.

467. Gilbert, W. (1985) Genes-in-Pieces Revisited. Science 228:823-824.

468. Gillespie, J.H. (1984) Molecular Clock Maybe an Episodic Clock. Proc Natl Acad Sci 81:809-813. (Ck pgs)

469. Gillespie, J.H. (1986) Variability of Evolutionary Rates of DNA. Genetics 113:1077-1091.

470. Gipps, E., Kidson, C. (1981) Ionising Radiation Sensitivity in Multiple Sclerosis. Lancet I (8226):947.

471. Glass, B., Erickson, D. (1967) Geomagnetic Reversals and Pleistocene Chronology. Nature 216:437-442.

472. Glinsman, W.H., Mertz, W. (1966) Effect of Trivalent Chromium on Glucose Tolerance. Metab Clin Experi 15B:510-520.

473. Glotfelty, D.E., et al. (1987) Pesticides in Fog. Nature 325:802-805.

474. Gnevyshev, M.M., Ol', A.I. (eds) (1977) Effects of Solar Activity on the Earth's Atmosphere and Biosphere. Izd. Nauka, Moscow, Acad Sci, USSR, Astron Council. Jerusalem, Israel Prog Sci Trans.

475. Gnevyshev, M.N., et al. (1977) Sudden Death from Cardiovascular Diseases and Solar Activity. pp.201-210 in (eds) M.N. Gnevyshev, A.I. Ol', Effects of Solar Activity on the Earth's Atmosphere and Biosphere. Jerusalem, Israel Prog Sci Trans.

476. Godley, B.F., et al. (1985) Effects of Light on Retinal Dopamine in the Rat. Ann N Y Acad Sci 453:383-384.

477. Goggon, D. (1987) Are Pesticides Carcinogenic? Editorial. Brit Med J 294:725.

478. Goldberg, E.D. (1986) TBT an Environmental Dilemma. Environment 28:17-20+.

479. Golding, G.B. (1987) Nonrandom Patterns of Mutation are Reflected in Evolutionary Divergence and May Cause Some of the Unusual Patterns Observed in Sequences. in (ed) E. Loeschcke, Genetic Constraints on Adaptive Evolution. NY, Springer-Verlag.

480. Goldschmidt, R.B. (1940) Material Basis of Evolution. New Haven, Yale Univ Press.

481. Gonzalez, E.R. (1980) Vitamin E Relieves Most Cystic Breast Disease; May Alter Lipids, Hormones. JAMA 244:1077-1088.

482. Goodman, R., et al. (1985) Nucleic Acid and Protein Synthesis in Cultured Chinese Hamster Ovary (CHO) Cells Exposed to the Pulsed Electromagnetic Fields. J Bioelect 4:565-575.

483. Gopal, B. (1987) Water Hyacinth. Aquatic Plant Series, No 1. Amsterdam, Elsevier.

484. Gori, G.B. (1978) Diet and Nutrition in Cancer Causation. Nutr Cancer 1(1):5-7.

485. Gottlieb, B. (1979) Way They Ate Was A Crime. Prevention (May):64-68.

486. Gottlieb, B., Wiley, P. (1986) Pesticides Hit Home. Sierra 71:48+.

487. Gottlieb, L.D. (1984) Genetics and Morphological Evolution in Plants. Amer Naturalist 123:681-709.

488. Gottlieb, L.D., Ford, V.S. (1987) Genetic and Developmental Studies of the Sbsence of Ray Florets in *Layia Discoidea*. pp.2-17 in (eds) H.Thomas, D. Grierson, Developmental Mutants in Higher Plants. NY, Cambridge Univ Press.

489. Gottlieb, L.D., Jain, S.K. (1988) Plant Evolution Biology. NY, Chapman & Hall.

490. Gould, S.J. (1977) Ontogeny and Phylogeny. Cambridge, Mass, Belknap Press Harvard Univ.

491. Gould, S.J. (1977) Return of the Hopeful Monsters. Nat Hist 86(6):22-30.

492. Gould, S.J. (1978) This View of Life: An Early Start. Nat Hist 87:10-26.

493. Grainger, A. (1980) State of the World's Tropical Forests. Ecologist 10:6-54.

494. Grant, M.C., Lewis, W.M. (1980) Acid Precipitation in the Western United States. Science 207:176-177.

495. Grant, S. (1984) Beauty and the Beast: The Coevolution of Plants and Animals. NY, C Scribner's.

496. Grattarola, M., et al. (1985) Interactions Between Weak Fields and Biosystems: A Summary of Nine Years of Research. J Bioelect 4:211-225.

497. Graves, H.B., Long, P.D. (1979) Biological Effects of 60Hz Alternating Current Fields: A Cheshire Cat Phenomena? p 184 in Biological Effects of Extremely Low Frequency Electro-magnetic Fields, DOE-50. US Dept Energy, Wash, DC, USGPO.

498. Greaves, M.P., et al. (1981) Effects of Pesticides on Soil Microflora Using Dalapon as an Example. Arch Environ Contam Toxicol 10:437-449.

499. Gregori, S. (1978) Electron Transfer in [Gamma]-Irradiated DNA. pp.118-124 in J. Hutterman, et al., (eds) Effects of Ionizing Radiation on DNA. NY, Springer-Verlag.

500. Gregory, P.H. (1973) Microbiology of the Atmosphere. NY, NY, Wiley.

501. Grissom, R.E., et al. (1987) *In Vivo* and *In Vitro* Dermal Penetration of Lipophilic and Hydrophilic Pesticides in Mice. Bull Environ Contam Toxicol 38:917-924.

502. Grodsky, I.T. (1975) Possible Physical Substrates for the Interaction of Electromagnetic Fields With Biological Membranes. Ann NY Acad Sci 247:117-123.

503. Grosovsky, A.J., Little, J.B. (1985) Evidence for Linear Response for the Induction of Mutations in Human Cells by X-Ray Exposures Below 10 Rads. Proc Natl Acad Sci (USA) 82:2092-2095.

504. Growdon, J.H. (1979) Neurotransmitter Precursors in the Diet: Their Use in the Treatment of Brain Diseases. pp.117-182 in (ed) R.J. Wurtman, Nutrition and the Brain. Vol.3. NY, Raven.

505. Grünner, O. (1973) Effect of Geomagnetic Climatic Complex on Neuroactive Problems of Balnear Patients. Fysiat Reumat Vestn 51(1):21-29.

506. Grzimek, H.C. (ed)(1976) Grzimek's Encyclopedia of Evolution. NY, Van Nostrand Reinhold.

507. Gul'ko, A.G., et al. (1978) [Pesticide Residues Among Certain Population Groups of Moldavia] Moldavskii Nauchno-issledovatel'skii inst. Gigieny i Epidemiologii, Kishinev, Moldavian SSR. Gigiena i Sanitariya (4):36-41.

508. Gulyuk, N.G. (1965) Effect of Solar Activity, Variation of Terrestrial Magnetism, and Other Factors of Cosmic and Geophysical Origin on Labor Rhythm and Cyclicity of Menstruation in Women. p.245 in Topical Questions of Midwifery and Gynecology, Uzhgorod.

509. Gunn, W.J. (1987) Importance of the Measurement of Annoyance in Prediction of Effects of Aircraft Noise on the Health and Well Being of Noise Exposed Communities. pp.237-255 in H.S. Koelega (ed) Environmental Annoyance: Characterization, Measurement and Control. NY, Elsevier.

510. Gupta, Y.P. (1986) Pesticide Misuse in India. Ecologist 16:38-39.

511. Haake, J, et al. (1987) Effects of Organochlorine Pesticides as Inducers of Testosterone and Benzolavpyrene Hydroxylases. Gen Pharmacol 18(2):165-169.

512. Habicht, J.A. (1979) Paleoclimate, Paleomagnetism, and Continental Drift. Tulsa, Oklahoma, AAPG.

513. Habib, M.A., Bockris, J., O'M. (1982) Interpretation of Current-Potential Relationships Across Biological Membranes. J Bioelect 1:289-294.

514. Hadar, Y. (1986) Role of Organic Matter in the Introduction of Biofertilizers and Biocontrol Agents to Soils. pp.169-180 in (eds) Y. Chen, Y. Avnimelech, Role of Organic Matter in Modern Agriculture. Boston, Martinus Nijhoff.

515. Hafez, Y.S., et al. (1985) Effects of Gamma Irradiation on Proteins and Fatty Acids of Soybeans. J Food Sci 50:1271-4.

516. Hagen, U. (1978) Irradiation Effects on Transcription. pp.286-292 in (eds) J. Hutterman et al., Effects of Ionizing Radiation on DNA. NY, Springer-Verlag.

517. Haine, E. et al. (1964) Aphid Moulting Under Controlled Electrical Conditions. Intl J Biometer 7:265-276.

518. Hallam, A. (ed) (1973) Atlas of Palaeobiogeography. NY, Elsevier.

519. Hallenbeck, W.H., Cunningham-Burns, K.M. (1985) Pesticides and Human Health. NY, Springer-Verlag.

520. Halpern, M.H., Van Dyke, J.H. (1966) Very Low Magnetic Fields: Biological Effects and Their Implication for Space Exploration. Aerospace Med 37(3):281.

521. Halstead, L.B. (1988) Extinction and Survival in Jawless Vertebrates, the Agnatha. pp.257-258 in (ed) G.P. Larwood, Extinction and Survival in the Fossil Record. Systematics Assoc Special. Vol.34. Oxford, Clarendon Press.

522. Hamaker, J.D. (1982) Survival of Civilization. Lansing, MI, Hamaker-Weaver.

523. Hamer, J. (1969) Effects of Low-Level Low Frequency Fields on Human Time Judgement. Intl J Biometeorol 13(Suppl):129.

524. Hamza, V.M., Beck, A.E. (1972) Terrestrial Heat Flow, the Neutrino Problem, and a Possible Energy Source in the Core. Nature 240:343-344.

525. Haney, A.F. (1981) Causes of Infertility in Sixteen Women Exposed to Diethylstilberol in Utero. Fertil Steril 35:258.

526. Hansen, J.B. (1970) Relations Between Barometric Pressure and the Incidence of Peripheral Embolism. Intl J Biometerol 14:391-97.

527. Hansen, J.B. (1977) Influence of Weather and Climate on Duodenal and Gastric Ulcer. pp.95-99 in (ed) S.W. Tromp, Progress in Biometerology. Amsterdam, Swets & Zeitlinger.

528. Hansen, J.B., Pedersen, S.A. (1972) Relationship Between Barometric Pressure and the Incidence of Perforated Duodenal Ulcer. Intl J Biometerol 16:85-91.

529. Hansson, H-A. (1988) Effects on the Nervous System by Exposure to Electromagnetism Fields: Experimental and Clinical Studies. pp.119-134 in (eds) M.E. O'Connor, R.H. Lovely, Electromagnetic Fields and Neurobehavioral Function. NY, Alan Liss.

530. Hardell, L., et al. (1987) Exposure to Hair Dyes and Polychlorinated Dibenzo-P-Dioxins in AIDS Patients With Kaposi Sarcoma: An Epidemiological Investigation. Cancer Detect Prev 1(Suppl):567-70.

531. Harman, D. (1977) Hard and Soft Water and the Incidence of Sudden Death from Ischemic Heart Disease: Consideration of Copper, Magnesium, and Calcium. pp.1-7 in (ed) M.S. Seelig, Nutritional Imbalances in Infant and Adult Disease. NY, SP Books Div Spectrum Pubs.

532. Harman, W.W. (1981) Broader Implications of Recent Findings in Psychological and Psychic Research. pp.113-132 in (ed) R.G. Jahn, Role of Consciousness in the Physical World. AAAS. Boulder, Co, Westview.

533. Harrell, R.F. (1981) Can Nutritional Supplements Help Mentally Retarded Children? Proc Natl Acad Sci 78:574-578.

534. Harrison, C.G.A. (1968) Evolutionary Processes and Reversals of the Earth's Magnetic Field. Nature 217:46-47.

535. Harrison, G.A., et al. (1964) Human Biology. Oxford, Claredon Press.

536. Hart, F.X. (1982) Use of Time Domain Dielectric Spectroscopy to Characterize the Progress of Wound Repair. J Bioelect 1:313-328.

537. Hart, F.X. (1983) Time Domain Dielectric Spectroscopy of Plant Stems. pp.391-396 in Conf Electrical Insulation & Dielectric Phenomena. Buck Hills Falls, Pa, Oct 16-20, 1983. NY, IEEE Press.

538. Hart, F.X. (1985) Extremely Low Frequency Electrical Properties of Plant Stems. Bioelectromagnetics 6:243-256.

539. Hart, F.X. (1988) Mathematical Modeling of Electromagnetic Interactions with Biological Systems. pp.281-344 in (ed) A.A. Marino, Modern Bioelectricity. NY, Marcel Dekker.

540. Hart, F.X., Marino, A.A. (1982) ELF Dosage in Ellipsoidal Models of Man Due to High Voltage Transmission Lines. J Bioelect 1:129-154.

541. Hart, T.B. (1987) Parkinson's Disease and Pesticides. Lancet I(8523):38.

542. Harwell, M.A. (1984) Nuclear Winter: the Human and Environmental Consequences of Nuclear War. NY, Springer Verlag.

543. Hasted, J.B. (1985) Biomolecular Effects of Electromagnetic Radiation: Classical or Quantum Physics? J Bioelect 4:367-387.

544. Hawkins, D., Pauling, L. (eds) (1973) Orthomolecular Psychiatry—Treatment of Schizophrenia. SF, W.H. Freeman.

545. Hayes, P. (1987) U.S. Fake Hit-Miss Approach to Food Purity Rules. Milwaukee J 42B:12-13

546. Hays, J.D. (1971) Faunal Extinctions and Reversals of the Earth's Magnetic Field. Geol Soc Amer Bull 82(9):2433-2447.

547. Hays, J.D., Opdyke, N.D. (1967) Antarctic Radiolaria, Magnetic Field Reversals and Climatic Change. Science 158:1001-1011.

548. Heck, W.W., McLaughlin, S.B. (1986) Effects of Gaseous Air Pollutants on Terrestrial Vegetation. pp.325-336 in (eds) A.H. Legge, S.V. Krupa, Air Pollutants and Their Effects on the Terrestrial Ecosystems. NY, J Wiley.

549. Hedge, A., Eleftherakis, E. (1982) Air Ionization: An Evaluation of Its Physiological and Psychological Effects. Ann Occupational Hygiene 25:409-419.

550. Heidmann, W.A. (1986) Hexachlorobenzene Residues in Selected Species of Land and Sea Birds in Northern Germany. IARC Sci Pub (77):223-9.

551. Heisenberg, W. (1949) Physical Principles of the Quantum Theory. Trans C. Eckart, F. Hoyt. NY, Dover, c1930.

552. Heller, J., Teixeria-Pinto, A. (1959) New Physical Method of Chromosomal Aberrations. Nature 183:905-906.

553. Henderson, J., et al. (1986) Association Between Occupational Group and Sperm Concentration in Infertile Men. Clin Reprod Fertil 4:275-81.

554. Henis, Y. (1986) Soil Microorganisms, Soil Organic Matter and Soil Fertility. pp.159-168 in (eds) Y. Chen, Y. Avnimelech, Role of Organic Matter in Modern Agriculture. Boston, Martinus Nijhoff.

555. Henny, C.J., et al. (1985) Organophosphate Insecticide (Famphur) Topically Applied to Cattle Kills Magpies and Hawks. J Wildlife Mgnt 49:648-58.

556. Herroux, O. (1979) Circadian and Circannual Variations in Growth and Cold Resistance of Warm Acclimated Rats and Their Relationship With Geomagnetic Activity. Intl J Biometeor 23(1):51-62.

557. Herzberg, L., Herzberg, B. (1977) Mood Change and Magnesium: A Possible Interaction Between Magnesium and Lithium? J Nerv Mental Dis 165:423-426.

558. Higinbotham, N. (1973) Electropotentials of Plant Cells. Ann Rev Plant Physiol 24:25-46.

559. Hill, S.B. (1982) Steps to a Holistic Ecological Food System. pp.15-22 in (ed) S.B. Hill, Basic Technics in Ecological Farming. Basel, Switz, Birkhauser Verlag.

560. Hill, S.B. (ed) (1982) Basic Technics in Ecological Farming. Basel, Switz, Birhäuser Verlag.

561. Hill, T. (1958) Some Possible Biological Effects of an Electric Field Acting on Nucleic Acids or Proteins. J Amer Chem Soc 8:2142-2147.

562. Hippchen, L.J. (ed) (1978) Ecologic-Biochemical Approaches to Treatment of Delinquents and Criminals. NY, Van Nostrand Reinhold.

563. Ho, M.-H., et al. (1986) A New Paradigm for Evolution. New Sci 109(1497):41-43.

564. Hoffer, A. (1971) Vitamin B3 Dependent Child. Can Psychiatr Assoc J 16:499-504.

565. Hoffer, A. (1974) Hyperactivity, Allergy and Megavitamins. Can Med Assoc J 111:906-7.

566. Hoffer, A. (1982) Nutritional Therapy. pp.207-236 in (ed) L.J. Hippchen, Holistic Approaches to Offender Rehabilitation. Springfield, Ill, C Thomas.

567. Hoffer, A., et al. (1954) Schizophrenia: A New Approach. II. Results of a Year's Research. J Ment Sci 100:29-54.

568. Hogan, D.J., Lane, P. (1986) Dermetologic Disorders in Agriculture. State Art Rev Occup Med 1:285-300.

569. Holden, C. (1988) U.S.-Soviet Ecology Agreement. Science 242:1629.

570. Hollis, L.S. et al. (1985) Synthesis, Structure and Antitumor Properties of Platinum Complexes of Vitamin C. J Amer Chem Soc 107:274-276.

571. Hollwich, F. (1979) Influence of Ocular Light Perception on Metabolism in Man and Animal. NY, Springer-Verlag.

572. Holser, W.T. (1984) Gradual and Abrupt Shifts in Ocean Chemistry During Phanerozoic Time. pp.123-144 in (eds) H.D. Holland, A.F. Trendall, Patterns of Change in Earth Evolution. NY, Springer-Verlag.

573. Holzman, D. (1986) Catalytic RNA Provides Clues to Evolution. New Sci 109 (16 Jan):26.

574. Honorton, C. (1981) Psychophysical Interaction. pp.19-36 in (ed) R.G. Jahn, Role of Consciousness in the Physical World. AAAS. Boulder, Co, Westview.

575. Hook, J, Serbia, V. (1982) Effects of Pesticides on the Kidney. in (eds) J.E. Chambers, J.D. Yarbrough, Effects of Chronic Exposures to Pesticides on Animal Systems. NY, Raven.

576. Hoover, R., Fraumeni, Jr., J.F. (1975) Cancer Mortality in U.S. Counties with Chemical Industries. Environ Res 9:196-207.

577. Hopps, H.C. (1972) Ecology of Disease in Relation to Environmental Trace Elements— Particularly Iron. Geol Soc Amer Spec Paper 140:1-8.

578. Horvath, D.J. (1972) Availability of Manganese and Iron to Plants and Animals. Geol Soc Amer Spec Paper 140:33-44.

579. Hotz, G. (1978) Modification of Radiation Damage. pp.304-311 in (eds) J. Hutterman et al., Effects of Ionizing Radiation on DNA. NY, Springer-Verlag.

580. House, M.R. (1988) Extinction and Survival in the Cephalopoda. pp.139-154 in (ed) G.P. Larwood, Extinction and Survival in the Fossil Record. Systematics Assoc Special. Vol. 34. Oxford, Clarendon.

581. Howard, A. (1939) Speech at Crewe, 22 March 1939. Reprinted in New English Weekly 6 April.

582. Hoyle, F., Wickramsinghe, C. (1986) Case for Life as a Cosmic Phenomena. Nature 322:509-511.

583. Hoyle, F., Wickramsinghe, C. (1981) Evolution from Space. London, Dent & Sons.

584. Huai, C., et al. (1984) Effects of Microwave Expose at Various Power Densities on Mitochondrial Marker Enzymes in Mouse Brains. J Bioelect 3:361-366.

585. Huai, C., Min, W. (1984) Morphological Change in Mouse Sperm Following Microwave Exposure. J Bioelect 3:367-372.

586. Huai, C., et al. (1985) Review Article: Experimental Research in China on the Biological Effects of Microwaves. J Bioelect 4:103-120.

587. Hudson, R.J. et al. (1989) Wildlife Production Systems: Economic Utilization of Wild Ungulates. Cambridge Univ.

588. Hughes, N.F. (ed) (1973) Organisms and Continents Through Time. London, The Palaeontological Assoc.

589. Hull, D.L. (1988) Introduction. pp.1-10 in (eds) B.H. Weber et al., Entropy, Information and Evolution. Cambridge, Mass, MIT Press.

590. Hull, T.G. (1975) Diseases Transmitted from Animals to Man. 6th ed. Springfield, Ill, C Thomas.

591. Hunter, A.F., Aarssen, L.W. (1988) Plants Helping Plants: New Evidence Indicates that Beneficence is Important in Vegetation. BioScience 38:34-40.

592. Hunter, B.T. (1980) Fumigants and Solvents in Food Processing: Some Effects on Food and Health. J Appl Nutr 32(1&2):44-55.

593. Huntingford, F.A., Turner, A.K. (1987) Animal Conflict. NY, Chapman & Hall.

594. Hussein, H.A.S., Abdalla, M.M.F. (1979) Gamma-Ray and EMS-Induced Mutations in Vicia Faba L: Evaluation of Yield and Protein Traits of Mutants in the M4 and M5 Generations. pp.23-32 in FAO/IAEA/ Gesellschaft fur Strahlen- und Umweltforschung. Seed Protein Improvement in Cereals and Grain Legumes. Vol.2, 1978. Vienna, Intl Atomic Energy Agency.

595. Hutterman, J. (1978) Structure of Radicals from Nucleic Acid Constituents. pp.31-58 in (eds) J. Hutterman et al., Effects of Ionizing Radiation on DNA: Physical, Chemical and Biological Aspects. NY, Springer-Verlag.

596. Hutterman, J. et al. (eds) (1978) Effects of Ionizing Radiation on DNA: Physical, Chemical and Biological Aspects. NY, Springer-Verlag.

597. Iizuka, E., Kondo, J. (1979) Magnetic-Field Orientation of the Liquid Crystals of Polyribonucleotide Complexes. Mol Cryst Liq Cryst 51:285-294.

598. Impens, R. (1982) Atmospheric Pollution and Maintenance of Soil Fertility. pp.293-297 in (eds) S. Hill, P. Ott, Basic Technics in Ecological Farming. Basel, Switz, Birkhäuser Verlag.

599. Ingold, C.T. (1971) Fungal Spores, Their Liberation and Dispersal. Oxford, Clarendon Press.

600. Inman, R.E. et al. (1971) Soil: A Natural Sink for Carbon Monoxide. Science 172:1229-31.

601. International Atomic Energy Agency/FAO (1982) International Symposium on Agrochemicals: Fate in Food and the Environment Using Isotope Techniques. Vienna, Intl Atomic Energy Agency.

602. International Commission on Large Dams U.S. Committee. Committee on Environmental Effects (1978) Environmental Effects of Large Dams: Report. NY, Amer Soc Civil Eng.

603. Ip, C. (1983) Selenium-Mediated Inhibition of Mammary Carcinogenesis. Biol Trace Element Res 5(4/5):245-255.

604. Iskhakov, V.P. (1973) Question of Possible Effect of Solar Activity on Increase in Number of Schizophrenic Patients in the Population. Trudy Molodykh Uch Med Uzb 3:106-107.

605. Israel, H., Dolezalek, H. (1973) Atmospheric Electricity. Vol.2. Jerusalem, Israel Prog Sci Trans.

606. Iusupova, F.D., Iampol Skaia, I.B. (1987) Early Skin Changes in Persons Working With Pesticides. Vestnik Dermatolgii i Venerologii (10):55-57.

607. Ivanhoe, I. (1982) Coevolution of Human Brain and Paleolithic Culture in Northern Hemisphere: Relation to Geomagnetic Intensity. J Bioelect 1:13-57.

608. Iverson, W.P. (1972) Biological Corrosion. Adv Corrosion Sci Tech 2:1-42.

609. Iverson, W.P. (1974) Microbial Corrosion of Iron. pp.475-53 in (ed) J.B. Neilands, Microbial Iron Metabolsim. NY, Academic.

610. Iverson, W.P. (1981) Overview of the Anaerobic Corrosion of Underground Metallic Structures, Evidence for a New Mechanism. pp.33-52 in (ed) E. Escalante, Underground Corrosion. Pa, Amer Soc Testing & Materials.

611. Iverson, W.P. (1983) Anaerobic Corrosion Mechanisms. Corrosion 83, Paper no.243. Houston, NACE.

612. Jablonski, D. (1986) Causes and Consequences of Mass Extinctions: A Comparative Approach. pp.183-230 in (ed) D.K. Elliot. Dynamics of Extinction. NY, J Wiley.

613. Jacobs, M.M. (1983) Trace Element Inhibition of Carcinogenesis and Angiogenesis. Biol Trace Element Res 5(4/5):375-381.

614. Jacobs, M.M., Griffin, A.C. (1979) Effects of Selenium on Chemical Carcinogenesis: Comparative Effects of Antioxidants. Biol Trace Element Res 1(1):1-14.

615. Jacoff, F.S., et al. (1986) Source Assessment of Hexachlorobenzene from the Organic Chemical Manufacturing Industry. IARC Sci Pub (77):31-7.

616. Jaffe, L.F. Nuccitelli, R. (1977) Electrical Controls of Development. Ann Rev Biophys Bioeng 6:445-476.

617. Jahn, R.G. (ed) (1981) Role of Consciousness in the Physical World. AAAS. Boulder, Co, Westview Press.

618. Jakubczak, L.F. (1976) Behavioral Aspects of Nutrition and Longevity in Animals. pp.103-122 in (eds) M. Rockstein, M.L. Sussman, Nutrition, Longevity and Aging. NY, Academic.

619. Jamil, K., et al. (1987) Studies on Water Hyacinth as a Biological Filter for Treating Contaminants from Agriculture Wastes and Industrial Effluents. J Environ Sci Health 221B:103-112.

620. Jamison, V.C., et al. (1968) Soil and Water Research on a Claypan Soil. Tech Bull 1379, Agric Res Service, US Dept Agric, Wash, DC, USGPO.

621. Jansson, B. (1976) Proceedings of the Symposium on Selenium-Tellurium in the Environment, May 11-13.

622. Jarrett, R.J. (1979) Problems of Food Abundance. pp.131-172 in (ed) R.J. Jarrett, Nutrition and Disease. Baltimore, Univ Park Press.

623. Jarrett, R.J. (ed) (1979) Nutrition and Disease. Baltimore, Univ Park Press.

624. Jaynes, E.T. (1965) Gibbs vs Boltzmann Entropies. Amer J Phys 33:391-398.

625. Jeffers, J.N.R. (1986) Role of Ecosystem Theory in Upland Use and Management. pp.51-65 in (eds) G. Chisci, R.P.C. Morgan, Soil Erosion in the European Community. Impact of Changing Agriculture. Boston, Balkema.

626. Jewett, D.L., et al. (1985) Effects of Electric Lighting on Human Muscle Strength: Visible Spectrum and Low Frequency Electromagnetic Radiation. Ann NY Acad Sci 453:390-391.

627. Jeyaratnam, J., et al. (1987) Survey of Acute Pesticide Poisoning Among Agricultural Workers in Four Asian Countries. Bull WHO 65(4):521-7.

628. Johnson, A.R., et al. (1980) Biotin and the Sudden Infant Death Syndrome Nature 285:159-160.

629. Johnson, L. (1988) Thermodynamic Origin of Ecosystems: A Tale of Broken Symmetry. pp.75-106 in (eds) B.H. Weber et al., Entropy, Information and Evolution. Cambridge, Mass, MIT Press.

630. Jolivet, P. (1986) Insects and Plants: Parallel Evolution and Adaptations. NY, Brill Flora & Fauna.

631. Jones, D.M. (1984) Performance Effects. pp.155-184 in (eds) D.M. Jones, A.J. Chapman. Noise and Society. NY, J Wiley.

632. Jones, E.C., et al. (1984) Prolonged Anticoagulation in Rat Poisoning. JAMA 242:3005-7.

633. Jordan, W. (1985) Pest Management: Pursuing an Environmental Dream. EPA J 11:12-14.

634. Josephson, B.D. (1969) Weakly Coupled Superconductors. pp.423-447 in (ed) R.D. Parks, Superconductivity. Vol.1. NY, Marcel Dekker.

635. Jost, A. (1974) Mechanisms of Normal and Abnormal Sex Differentiation in Fetus. pp.116-135 in (ed) K.S. Moghissi. Birth Defects and Fetal Development. Springfield, Ill, Charles Thomas.

636. Joyce, C. (1988) Nature Helps Indonesia to Cut Its Pesticides Bill. New Sci 118(1617):35.

637. Joyce, G.F. (1989) RNA Evolution and the Origins of Life. Nature 338:217-224.

638. Jukes, T.H. (1987) Transitions, Transversions and the Molecular Evolutionary Clock. J Mol Evol 26:87-98.

639. Jung, C.G. (1952) Synchronicity: An Acausal Connecting Principle. Coll Works. Vol.8. NJ, Princeton: Bollingen Series.

640. Jung, C.G. (1975) Archetypes and the Collective Unconscious. Trans R.F.C. Hall. NJ, Princeton Univ.

641. Kahn, S.U. (1980) Pesticides in the Soil Environment. NY, Elsevier.

642. Kanja, L., et al. (1986) Organochlorine Pesticides in Human Milk from Different Areas of Kenya 1983-1985. J Toxicol Environ Health 19:449-64.

643. Kalmijn, Ad. J. (1979) Electromagnetic Guidance Systems in Fishes. pp.15-18 in (ed) T.S. Tenforde, Magnetic Field Effect on Biological Systems. NY, Plenum.

644. Kalmijn, Ad. J (1988) Electromagnetic Orientation: A Relativistic Approach. pp.23-46 in (eds) M.E. O'Connor, R.H. Lovely, Electromagnetic Fields and Neurobehavioral Function. NY, Alan Liss.

645. Karakaya, A.E., Ozalp, S. (1987) Organochlorine Pesticides in Human Adipose Tissue Collected in (Ankara, Turkey) 1984-1985. Bull Environ Contam Toxicol 38:941-945.

646. Karapina, T.N., et al. (1969) Relation Between Geomagnetic Activity and Course of Epilepsy. Proc 5th All-Union Congr Neuropathologists and Psychiatrists. Vol.2, Moscow.

647. Karsaevskaya, T.V. (1970) Social and Biological Factors Leading to Changes in Human Physical Development. Moscow, Meditsina.

648. Karson, C.N., et al. (1985) Schizophrenia: Altered Biological Responses to Light. Ann NY Acad Sci 453:392-393.

649. Katagiri, K., Nakajama, K. (1982) Tetraploid Induction by Gamma-Ray Irradiation in Milberry. in Induced Mutations in Vegetatively Propagated Plants II. FAO/IAEA Div Isotope & Radiation Appl Atomic Energy Food & Agric Develop, 1980. Vienna, Intl Atomic Energy Agency.

650. Katila, T., Varpula, T. (1983) Magnetic Fields of the Eye. pp.341-352 in (eds) S.J. Williamson, et al., Biomagnetism. NY, Plenum.

651. Kats, Yu, I. (1971) Study of Brachiopods in Light of the Problem of Planetary Periodism. pp.23-25 in Abstr Papers 2nd All-Union Conf Mesozoic & Cainozoic Brachiopods, Kharkov.

652. Kats, Yu, I., Bereznyakov, A.I. (1974) Geomagnetic Reversals: Rotational Causation and Correlation With Geological Processes and the Evolution of Organisms. pp.199-216 in The Cosmos and the Evolution of Organisms. ANSSSR, Moscow, Paleontolo Instit.

653. Kauffman, E.G. (1984) Toward a Synthetic Theory of Mass Extinction. Geol Soc Amer Abst Prog 16:555.

654. Kaufman, L. (1983) Perception and Event-Related Potentials and Fields. pp.385-398 in (eds) S.J. Williamson, et al., Biomagnetism. NY, Plenum.

655. Kaufman, R.H., et al. (1980) Upper Genital Tract Changes and Pregnancy Outcome in Offspring Exposed to Diethylstilberol. Amer J Obstet Gynecol 137:299.

656. Kavathas, P., et al. (1980) Gamma-Ray-Induced Loss of Expression in Lymphoblastoid Cells. Proc Natl Acad Sci (USA) 77:4251-4255.

657. Kavet, R.I., Banks, R.S. (1986) Emerging Issues in Extremely-Low-Frequency Electric and Magnetic Filed Health Research. Environ Res 39:386-404.

658. Keeton, W.T. (1979) Effects of Magnetic Fields on Avian Orientation. pp.18-20 in (ed) T.S. Tenforde, Magnetic Field Effect on Biological Systems. NY, Plenum.

659. Keller, G., et al. (1983) Multiple Microtektite Horizons in Upper Eocene Marine Sediments: No Evidence for Mass Extinctions. Science 221:150.

660. Kellogg III, E.W. (1984) Air Ions: Their Possible Biological Significance and Effects. J Bioelect 3:119-136.

661. Kennet, J.P. (ed) (1980) Magnetic Stratigraphy Sediments: A Memorial to Norman D. Watkins. Distr Academic Press, NY, Dowden, Hutchinson & Ross.

662. Kent, S. (1977) Do Free Radicals and Dietary Antioxidants Wage Intracellular War? Geriatrics 32(1):127-136.

663. Kerr, R.A. (1984) Periodic Impacts and Extinctions Reported. Science 223:1277-1279.

664. Kerxhall, J.S., et al. (1975) Effect of Ascorbic Acid on the Human Electroencephalgram. J Nutr 105:1356-1358.

665. Kety, S.S. (1975) Nutrition and Psychiatric Illness. pp.205-212 in (ed) G. Serban. Nutrition and Mental Functions. NY, Plenum.

666. Keusch, G.T. (1990) Vitamin A Supplements—Too Good to Be True. New Eng J Med 323:985-987.

667. Kevan, S.M. (1977) Month of Birth Versus Mental Development. pp.160-164 in (ed) S.W. Tromp, Progress in Biometerology. Amsterdam, Swets & Zeitlinger.

668. Keyes, K. (1982) Hundreth Monkey. NY, Vision Books.

669. Khodorkovski, R.I. (1971) Polonnikov Study of Extremely Weak Magnetic Reception in Fishes. p.72 in Questions of Fish Behavior, Kaliningrad.

670. Khvedelidze, M.A., et al. (1973) Bionic Aspects of Magnetobiological Effects. pp.196-201 in Problems of Bionics, Moscow, Nauka.

671. Kiely, M., et al. (1981) Zinc Status and Pregnancy Outcome. Lancet I (8225):893-894.

672. Kiester, A.R. (1971) Species Diversity of North American Amphibians and Reptiles. Syst Zool 20:127-137.

673. Killham, K., Wainwright, M. (1984) Chemical and Microbiological Changes in Soil Following Exposure to Heavy Atmospheric Pollution. Environ Poll 33A:121-131.

674. King, R.C. (1984) Dealing With Poisonings. Registered Nurse 47:45-48.

675. Kirschvink, J.L. (1983) Biogenic Ferrimagnetism: A New Biomagnetism. pp.501-531 in (eds) S.J. Williamson, et al., Biomagnetism. NY, Plenum.

676. Kirschvink, J.L., et al. (eds) (1985) Magnetite Biomineralization and Magnetoreception in Organisms: A New Magnetism. NY, Plenum.

677. Kitron, U. (1987) Malaria, Agriculture and Development Lessons from Past Campaigns. Intl J Health Serv 17:295-326.

678. Klevay, L.M. (1982) Magnesium, Calcium, Copper and Zinc in Meals: Correlations Related to the Epidemology of Ischemic Heart Disease. Biol Trace Element Res 4:95-104.

679. Kloke, A., et al. (1984) Contamination of Plants and Soils with Heavy Metals and Their Transport in Terrestrial Food Chains. pp.113-142 in (ed) J.O. Nriagu, Changing Metal Cycles and Human Health. NY, Springer-Verlag.

680. Klonowska, M.-T., Klonowski, W. (1985) Aging Processes and Enzymatic Proteins. J Bioelect 4:93-102.

681. Knoll, A.H. (1984) Patterns of Extinction in the Fossil Record of Vascular Plants. pp.21-68 in (eds) M.H. Nitecki, Extinctions. Univ Chicago Press.

682. Knowles, M., et al. (1981) Increased Bioelectric Potential Across Epithelia in Cytic Fibrosis. N Eng J Med 305:1489-95.

683. Kobrin, G. (1976) Corrosion by Microbiological Organisms in Natural Waters. Materials Performance 15(7):38-43.

684. Kodama, H., Ota, H. (1980) Transfer of Polychlorinated Biphenyls to Infants from Their Mothers. Arch Environ Health 35(2):95-100.

685. Koepf, H.H., et al. (1974) [Bio-Dynamic Agriculture] Biologische Landwirtschaft. Stuttgart, Germany, Verlag Eugen Ulmer. Trans c1976. Spring Valley, NY, Anthroposophic Press.

686. Kohnlein, W. (1978) Irradiation Effects on the Transforming and Transfecting Activity of DNA. pp.279-285 in (eds) J. Hutterman, et al. Effects of Ionizing Radiation on DNA. NY, Springer-Verlag.

687. Koilpakov, I.E., et al. (1987) Possibility of Early Detection of Disorders of the Respiratory System in Persons Who Work With Pesticides [in Russian with English Abstract]. Fiziologicheskii Zh 33(6):50-5.

688. König, H.L. (1974) Behavioral Changes in Human Subjects Associated With ELF Electric Fields. pp.81-100 in (ed) M.A. Persinger, ELF and VLF Electromagnetic Field Effects. NY, Plenum.

689. König, H.L. (1974) Signal Properties of ELF-VLF Fields. in (ed) M.A. Persinger, ELF and VLF Electromagnetic Fields Effects. NY, Plenum.

690. König, H.L. (1975) [Invisible World Around Us] Unsichtbare Umwelt. Munich Heinz, Mods Verlag.

691. Kornberg, A. et al. (1964) Enzymatic Synthesis of Deoxyribonucleic Acid. XVI. Oligonucleotides as Templates and the Mechanisms of Their Replication. Proc Natl Acad Sci (USA) 51:315-323.

692. Koval'skii, U.V., Pletneva, I.A. (1948) Biological Rhythms and Diurnal Periodism of Carbohydrate Function of Liver. Trudy Inst. A Kusherstfva i Ginekologii Amn SSSR 1:88-100.

693. Kozliuk, A.S., et al. (1987) Immunologic Status of Children Living in Regions With Various Intensities of Pesticide Use [English Abstract]. Gigiena i Sanitariia (6):26-28.

694. Krasil'nikov, N.A. (1958) Soil Microorganisms and Higher Plants. Trans Y. Halperin, Jerusalem, S. Monson [Available from Off Tech Serv, US Dept Comm, Wash, DC].

695. Krasniuk, E.P., et al. (1987) Effect of Pesticides on the Health of Workers Engaged in Raising Vegetables in Protected Soil [English Abstract]. Vrachebnoe Delo (6):105-8.

696. Krebs, C. (1985) Ecology: The Experimental Analysis of Distribution and Abundance. . NY, Harper & Row.

697. Kritchevsky, D., Story, J.A. (1977) Dietary Fiber and Cancer. pp.41-55 in (ed) M. Winick. Nutrition and Cancer. NY, J Wiley.

698. Krizaj, D. (1987) Electrical Stimulation: Its Effects on Growth and Ion Accumulation in *Lactuca Satival*. J Bioelect 6:129-136.

699. Krueger, A.P. (1957) Action of Air Ions on Bacteria. J Gen Physiol. Berkeley, Univ Calif.

700. Krueger, A.P. (1973) Are Negative Ions Good For You? New Sci 1:663.

701. Krueger, A.P., Reed, E.J. (1972) Effect of Air Ion Environment on Influenza in the Mouse. Intl J Biometeor 16:209-232.

702. Krueger, A.P., Reed, E.J. (1976) Biological Impact of Small Air Ions. Science 193:1209-13

703. Kryter, W.D. (1985) Effects of Noise on Man. NY, Academic.

704. Kucynska, I., Lachowski, A. (1987) Effect of Cattle Slurry on the Mineral Content of Pastures and Blood Serum in Dairy Cows. pp.357-360 in (eds) H.G. Van Der Meer, et al., Animal Manure on Grasslands and Fodder Crops. Dordrecht, Netherlands, Martinus Nihoff

705. Kuhn, T.S. (1981) A Function for Thought Experiments. pp.6-27 in (ed) I. Hacking, Scientific Revolutions. Oxford Univ.

706. Kumar, D. (1982) Trees Can Meet Human Needs. in (ed) S. Hill, Basic Technics in Ecological Farming. Basel, Switz, Birkhäuser Verlag.

707. Kunin, N. Yas, Sardonnikov, N.M. (1974) Cyclical Variation of the Magnetic Field and the Earth's Climate in the Phanerozoic. pp.61-82 in Cosmos and the Evolution of Organisms. AN SSSR, Moscow Paleontolo Instit.

708. Kursanov, A.L. (1976) Transport Assimilations in Plants. Moscow, Nauka.

709. Kursevich, N.V., Travkin, M.P. (1973) Effect of Weak Magnetic Fields on Radicle Growth and Respiration Rate of "Val'titskii" Barley Seedlings. pp.104-105 in Effect of Natural and Weak Artificial Magnetic Fields on Biological Objects. Belgorod.

710. Kurtén, B. (1973) Early Tertiary Land Mammals. pp.437-443 in (ed) A. Hallam, Atlas of Palaeobiogeography. NY, Elsevier.

711. Kurz, M.D. (1986) Cosmogenic Helium in a Terrestrial Igneous Rock. Nature 320:435-439.

712. Kutsky, R.J. (1981) Handbook of Vitamins, Minerals and Hormones. NY, Van Nostrand Reinhold.

713. Lakin, H.W. (1972) Selenium Accumulation in Soils and Its Absorption by Plants and Animals. Geol Soc Amer Spec Paper 140:45-54.

714. Lal, D. et al. (1987) Cosmogenic 10Be in Zaire Alluvial Diamonds: Implications for 3He Contents of Diamonds. Nature 328:139-141.

715. Landav, S.F., Drapkin, I. (1968) Ethnic Patterns of Criminal Homicide in Israel. Ford Foundation Report 5 F-8. Jerusalem, Hebrew Univ.

716. Landes, R.R., et al. (1977) Inquiry Into the Relation Between Water Hardness and the Frequency of Urolithiasis. pp.9-21 in (ed) M.S. Seelig, Nutritional Imbalances in Infant and Adult Disease. NY, S.P. Books Div Spectrum.

717. Lang, S.B. (1988) Bioelectric Pyroelectricity. pp.243-280 in (ed) A.A. Marino, Modern Bioelectricity. NY, Marcel Dekker.

718. Lang, S.B., Athenstaedt, II. (1977) Pyroelectricity and Induced-Pyroelectric Polarization in Leaves of the Palmlike Plant *Encephalartos Villosus*. Science 196:985-986.

719.   Langlois, R.G. et al. (1987) Evidence for Increased Somatic Cell Mutations at the Glycophorin A Locus in Atomic Bomb Survivors. Science 236:445-448.

720.   Lanzerotti, L. J., Gregori, G.P. (1986) Telluric Currents: The Natural Environment and Interactions with Man-Made Systems. pp.232-258 in Earth's Electrical Environment. Wash, DC, Natl Acad Press.

721.   Latin American Study Group Meeting on Induced Mutations and Plant Improvement (1972) Induced Mutations and Plant Improvement. Joint FAO/IAEA Div Atomic Energy Food & Agric. Vienna, Intl Atomic Energy Agency.

722.   Lautsevichus, L.Z., et al. (1977) Some Indexes of Solar Activity, Geomagnetic Disturbances and Cardiovascular Attacks. pp.211-212 in (eds) M.N. Gnevyshev, A.I. Ol', Effects of Solar Activity on the Earth's Atmosphere and Biosphere. Jerusalem, Israel Prog Sci Trans.

723.   Lavine, L.S., et al. (1972) Electric Enhancement of Bone Healing. Science 175:1118-1121.

724.   Lavine, L.S., et al. (1974) Clinical and Ultrastructural Investigations of Electrical Enhancement of Bone Healing. Ann N Y Acad Sci 238:552-563.

725.   Lawerence, J.S. (1977) Influence of Weather and Climate on Rheumatic Diseases. pp.83-88 in (ed) S.W. Tromp, Progress in Biometerology. Amsterdam, Swets & Zeitlinger.

726.   Layzer, D. (1988) Growth of Order in the Universe. pp.23-40 in (eds) B.H. Weber et al., Entropy, Information and Evolution. Cambridge, Mass, MIT Press.

727.   Leach, C.M. (1976) Electrostatic Theory to Explain Violent Spore Liberation by Drechslera Turcica and Other Fungi. Mycologia 68:63-86.

728.   Leach, C.M. (1980) Evidence for an Electrostatic Mechanism in Spore Discharge by Drechslera Turcica. Phytopathology 70:206-213.

729.   Leach, C.M. (1982) Active Sporangium Discharge by Peronspora Destructor. Phytopathology 72:881-885.

730.   Leach, C.M. (1987) Diurnal Electrical Potentials of Plant Leaves Under Natural Conditions. Environ Experi Bot 27:419-430.

731.   Leach, C.M., Apple, J.D. (1984) Leaf Surface Electrostatics: Behavior of Detached Leaves of Beans, Maize, and Other Plants Under Natural Conditions. Phytopathology 74:704-709.

732.   Leaf, A. (1986) New Perspective on the Medical Consequences of Nuclear War. New Eng J Med 315:905-912.

733.   Leaf, C.M. (1984) Leaf Surface Electrostatics: Response of Detached Leaves of Beans and Maize to Humidity and Red-Infrared Radiation Under Controlled Conditions. Phytopathology 76:695-701.

734.   Le Gall, J., et al. (1982) Hydrogenase and Other Iron-Sulphur Proteins from Sulphate-Reducing and Methane-Forming Bacteria. pp.177-248 in (ed) T.G. Spiro, Iron-Sulphur Proteins. NY, J Wiley.

735.   Legge, A.H., Krupa, S.V. (eds) (1986) Air Pollutants and Their Effects on the Terrestrial Ecosystem. NY, J Wiley.

736.   Lehmann, A.R. (1978) Repair Processes for Radiation-Induced DNA Damage. pp.312-334 in (eds) J. Hutterman, et al., Effects of Ionizing Radiation on DNA. NY, Springer-Verlag.

737.   Leiboici, M., Capetan-Bacskai, M. (1987) Ocular Burns Caused by Pesticides [English Abstract]. Revista de Chirurgie, Oncologie, Radiologie, Orl, Oftalmologie, Stomatologie. Seria: Oftalmologie 31(2):127-129.

738.   Lesi, F.E.A. (1978) Infant Mortality, Diet and Disease in Nigeria. Nigerian Med J 8:114-18

739. Lester, J.R. (1985) Reply to "Cancer Mortality and Air Force Bases: A Reevaluation." J Bioelect 4:129-131.

740. Lester, J.R., Moore, D.F. (1982) Cancer Incidence and Electromagnetic Radiation. J Bioelect 1:59-76.

741. Lester, J.R., Moore, D.F. (1982) Cancer Mortality and Air Force Bases. J Bioelect 1:77-82.

742. Levengood, W.C., Shinkle, W.P. (1960) Environmental Factors Influencing Progeny Yields in Drosophila. Science 132:34-35 [and] 133:115-116.

743. Levitsky, D. (ed) (1979) Malnutrition, Environment and Behavior. Ithaca, NY, Cornell U.

744. Levy, D.D. (1974) Pulsed Electrical Stimulation Technique for Inducing Bone Growth. Ann N Y Acad Sci 238:478-490.

745. Lewin, R. (1982) Can Genes Jump Between Eukaryotic Species? Science 217:42-43.

746. Lewin, R. (1985) Parkinson's Disease: An Environmental Cause? Science 229:257-8.

747. Lewin, R. (1988) Molecular Clocks Turn a Quarter Century. Science 239:561-563.

748. Lewin, R. (1988) DNA Clock Conflict Continues. Science 241:1756-1759.

749. Lewontin, R.C. (1974) Genetic Basis of Evolutionary Change. NY, Columbia Univ.

750. Liboff, A.R., Homer, L. (1983) Eddy Current Effects on DNA Synthesis at Geomagnetic Intensities. Bioelectromagnetic Soc 5th Ann Sci Sess, 3.

751. Liboff, A.R., et al. (1984) Time-Varying Magnetic Fields: Effect on DNA Synthesis. Science 223:818-820.

752. Liboff, A.R., Rinaldi, R.A. (1974) Electricity Mediated Growth Mechanisms in Living Systems. Ann N Y Acad Sci 238:1-593.

753. Liboff, R.L. (1979) Analysis of Stationary Magnetic Field Effects on Ionic Diffusion and Nerve Action Potentials. p.81 in (ed) T.S. TenForde, Magnetic Field Effects on Ionic Diffusion and Nerve Action Potentials. NY, Plenum.

754. Lichtenstein, E.P. (1973) Environmental Factors Affecting Penetration and Translocation of Insecticides from Soils into Crops. Qual Plant-Pl. Fds. Human Nutr 23:113-18.

755. Liddel, P.A., et al. (1986) Charge Separation and Energy Transfer in Carotenpyropheophorbide-Quinon Triads. J Amer Chem Soc 108:5350-5352.

756. Lieber, A.L. (1978) Human Aggression and the Lunar Synodic Cycle. J Clin Psychi 39:385-7.

757. Lieber, A.L. (1980) Lunar Effect: Biological Tides and Human Emotions. Garden City, NY, Anchor.

758. Lieber, A.L., Sherin, C.R. (1972) Homicides and the Lunar Cycle: Toward a Theory of Lunar Influence on Human Emotional Disturbance. Amer J Psych 129:101-6.

759. Lieberman, H.R., et al. (1985) Possible Behavioral Consequences of Light-Induced Changes in Melatonin Availability. Ann N Y Acad Sci 453:242-252.

760. Lima-de-Faria, A. (1983) Molecular Evolution and Organization of the Chromosome. NY, Elsevier.

761. Lindauer, M., Martin, H. (1968) [Body Orientation Under the Influence of the Geomagnetic Field] Schwereorientierug der Bienen unter dem Einfluss des Erdmagnetfeldes. Z Vgl Physiol 60(3):219.

762. Lipinski, B. (1984) Bioelectric Theory of Pathogenesis of Atherosclerosis. J Bioelect 3:177-191.

763. Lipton, M.A., et al. (1979) Vitamins, Megavitamin Therapy and the Nervous System. pp.183-264 in (ed) R.J. Wurtman, Nutrition and the Brain. Vol.3. NY, Raven.

764. Lipton, R.A., Klass, E.M. (1984) Human Ingestion of a Superwarfarin Rodenticide Resulting in a Prolonged Anticoagulant Effect. JAMA 252:3004-5.

765. Lisi, P. (1987) Cutaneous Pathology Caused by Pesticides. Giornale Italiano di Dermatologia e Venereologia 122:175-82.

766. Lisi, P., et al. (1986) A Test Series for Pesticide Dermatitis. Contact Dermatitis 15:266-9.

767. Lisi, P., et al. (1987) Irritation and Sensitization Potential of Pesticides. Contact Dermatitis 17:212-8.

768. Little, P., Martin, M.H. (1972) A Survey of Zinc, Lead and Cadmium in Soil and Natural Vegetation Around a Smelting Complex. Environ Poll(3):241-254.

769. Lockertz, W., et al. (1981) Organic Farming in the Corn Belt. Science 211:540-547.

770. Lohmann, K.J., Willows, A.O.D. (1987) Lunar-Modulated Geomagnetic Orientation by a Marine Mollusk. Science 235:331-334.

771. Lomsadze, M.S. (1971) Investigation of Electroelectric Properties of Plant Seeds in Relation to Their Vital Activity [Dissert Abstr]. Tbilisi State Pedagogic Instit.

772. Loper, D.E., et al. (1988) Model of Correlated Episodicity in Magnetic-Field Reversals, Climate and Mass Extinctions. J Geol 96:1-15.

773. Lopez, R., Coye, M.J. (1984) Insecticide Spraying in Enclosed Occupied Areas. JAMA 252:1782.

774. Lowe, L.E. (1978) Carbohydrates in the Soil. pp.63-93 in (eds) M. Schnitzer, S.U. Khan, Soil Organic Matter. NY, Elsevier.

775. Lowengart, R.A., et al. (1987) Childhood Leukemia and Parents' Occupational and Home Exposures. J Natl Cancer Instit 79(1):39-46.

776. Lowenhaupt, B. (1974) Thermodynamic Considerations of Bioelectric Potential. Ann N Y Acad Sci 238:214-217.

777. Lomsadze, M.S. (1971) Investigation of Magnetoelectric Properties of Plant Seeds in Relation to Their Vital Activity. [Dissert Abstr] Tbilisi State Pedagogic Instit.

778. Lovelock, J, (1995) Gaia, A New Look at Life on Earth. NY, Oxford Univ Press.

779. Lovtrup, S. (1983) Victims of Ambition: Comments on the Wiley and Brooks Approach to Evolution. Syst Zool 32:90-96.

780. Luben, R.A., et al. (1982) Effects of Electromagnetic Stimuli on Bone and Bone Cells *In Vitro*: Inhibition of Responses to Parathyroid Hormone by Low-Energy Low-Frequency Fields. Proc Natl Acad Sci 79:4180-4184.

781. Ludwig, W.W. (1974) Electric and Magnetic Field Strength in the Open and in Shielded Rooms in the ULF- to LF-Zone. pp.35-80 in (ed) M.A. Persinger, ELF and VLF Electromagnetic Field Effects. NY, Plenum.

782. Lund, E.J. (1977) Bioelectric Fields and Growth. Austin, Univ Texas Press.

783. Lunsford, C.A., et al. (1987) Uptake of Kenone by the Estuarine Bivalve Rannia Cuneata During the Dredging of Contaminated Sediments in the James River. Virginia Water Res 21:411-16.

784. Lynch, J.M. (1981) Promotion and Inhibition of Soil Stabilization by Microorganisms. J Gen Microbiol 126:371-5.

785. Mac Farlane, J.J., Roberts, E.A. (1968) Some Effects of Gamma Radiation on Washington Navel and Valencia Oranges. Australian J Exp Agric Animal Husband 8:625.

786. MacKenzie, D. (1986) Crayfish Pesticide Decimates Spanish Birds. New Sci 112(1530):24.

787. MacKenzie, D. (1987) Thousands Poisoned by Pesticide in Guyana. New Sci 113(1552):18.

788. MacKenzie, D. (1988) Mystery of Mussel Poison Deepens in Canada as the Chain of Death Spreads to Whales. New Sci 117(1597):30.

789. MacLean, T.D. (1987) Pesticides and Health. Lancet 2(8558):575.

790. Madden, P. (1987) Can Sustainable Agriculture Be Profitable? Environment 29:18-20+.

791. Mahaney, W.C., Ermuth F. (1975) Effects of Agriculture and Urbanization on the Natural Environment: A Study of Human Impact in Southern Ontario. Geogr Mono No 7, 1974. Dept Geol, Atkinson College, York Univ, Toronto, Ontario, Canada.

792. Malek, J., et al. (1962) Charateristics of the Daily Rhythm of Menstruation and Labor. Ann N Y Acad Sci 98:1042-1055.

793. Malhi, P.K., Grover, I.S. (1987) Genotoxic Effects of Some Organophosphorus Pesticides. II. *In Vivo* Chromosomal Aberration Bioassay in Bone Marrow Cells in Rat. Mutat Res 188(1):45-51.

794. Malik, I.A., et al. (1979) Evaluation of Radiation-Induced Mutant Lines of Mung Bean (*Vigna Radiata (l.) Wilcek*) for Grain Yield and Protein Content. p 445 in FAO/IAEA/Geseellschaft fur Strahlen- und Umweltforschung. Seed Protein Improvement in Cereals and Grain Legumes. Vol.2. 1978. Vienna, Intl Atomic Energy Agency.

795. Malin, S.R.C. (1979) Correlations Between Heart Attacks and Magnetic Activity. Nature 277:646-648.

796. Mangelsdorf, P.C. (1974) Corn—Origin, Evolution and Improvement. Cambridge, Mass, Harvard Univ.

797. Mann, J.I. (1979) A Prudent Diet for the Nation. J Human Nutr 33(1):57-63.

798. Manzoor, M., Runcie, J. (1976) Folate-Responsive Neuropathy: Report of 10 Cases. Brit Med J 1:1176-1178.

799. Mapes, G., et al. (1989) Evolution of Seed Dormancy. Nature 337:645-646.

800. Margulis, L. (1981) Symbiosis in Cell Evolution. SF, WH Freeman.

801. Marino, A.A.(1988) Environmental Electromagnetic Fields and Public Health. pp.965-1044 in (ed) A.A. Marino, Modern Bioelectricity. NY, Marcel Dekker.

802. Marino, A.A., Becker, R. (1970) Piezoelectric Effect and Growth Control in Bone. Nature 229:473-474.

803. Marino, A.A., et al. (1974) Electric Field Effects in Selected Biological Systems. Ann N Y Acad Sci 238:436-444.

804. Marti, L.R., et al. (1984) Screening for Organic Contamination of Groundwater: Ethylene Dibromide in Georgia Irrigation Wells. Environ Sci Tech 18:973-974.

805. Martin, J.P. (1966) Influence of Pesticides on Soil Microbes and Soil Properties. pp.95-108 in Pesticides and Their Effects on Soils and Water. ASA Spec Pub No 8 Soil Sci Soc Amer

806. Mason, M., et al. (1980) Nutrition and the Cell: The Inside Story. Chicago, Year Book Medical Publishers.

807. Matsumoto, H., et al. (1987) Average Daily Intake of Pesticides and Polychlorinated Biphenyls in Total Diet Samples in Osaka, Japan. Bull Environ Contam Toxicol 38:954-58

808. Matthews, W.H., et al. (eds) (1971) Man's Impact on the Climate. Cambridge, Mass, MIT Press.

809. Matthews, L.H. (1974) [Introduction to] Origin of Species by Charles Darwin. London, J.M. Dent & Sons.

810.   Matyushin, G.N. (1974) Role of Ionizing Radiation in Anthropogenesis. pp.276-292 in Cosmos and the Evolution of Organisms. AN SSSR, Moscow, Paleontol Inst.

811.   Maugh II, T.H. (1973) Trace Elements: A Growing Appreciation of Their Effects on Man. Science 181:253-254.

812.   Maugh II, T.H. (1978) Chemical Carcinogens: The Scientific Basis for Regulation. Science 202:1200-1205.

813.   Maugh II, T.H. (1983) Number of Organic Superconductors Grows. Science 222:606-607.

814.   Maw, M.G. (1961) Suppression of Oviposition Rate Scambus Buolianae in Fluctuating Electric Fields. Canad Entomol 93:602-604.

815.   Maxie, E.C., et al. (1964) Some Physiological Effects of Gamma Irradiation on Lemon Fruit. Radiation Botany 4:405.

816.   May, R.M. (1985) Evolution of Pesticide Resistence. Nature 315:12-13.

817.   McCaughey, J.H. (1985) Biomass Energy Storage in a Mixed Forest Canopy. pp.224-227 in Conference on Agriculture and Forest Meteorology (17th) & Biometeorology and Aerobiology (7th). Boston, Amer Meterolo Soc.

818.   McDaniel, T.A., Hajek, B.F. (1985) Soil Erosion Effects on Crop Productivity and Soil Properties in Alabama. pp.48-58 in (ed) D.K. McCool, Natl Symp Erosion and Soil Productivity. St Joseph, Amer Soc Agric Eng.

819.   McDanell, R.E., McLean, A.E. (1985) Role of Nutritional Status in Drug Metabolism and Toxicity. pp.321-356 in (ed) H. Sidransky, Nutritional Pathology. NY, Marcel Dekker.

820.   McDowell, L.L., et al. (1984) Methyl Parathion and EPN Washoff from Cotton Plants by Stimulated Rainfall. Environ Sci Tech 18:423-7.

821.   McKarness, R. (1990) Chemical Victims. Manchester, England. C Nichols.

822.   McLaren, D.J. (1984) Abrupt Extinctions. Terra Cognita 4:27-32.

823.   McLaren, D.J. (1985) Mass Extinction and Iridium Anomaly in the Upper Devonian of Western Australia: A Commentary. Geology 13:170-172.

824.   McLaren, D.J. (1986) Abrupt Extinctions. pp.37-48 in (ed) D.K. Elliot, Dynamics of Extinction. NY, Wiley.

825.   McLeod, K.J., et al. (1987) Frequency Dependence of Electric Field Modulation of Fibroblast Protein Synthesis. Science 236:1465-1469.

826.   McNeill, W.H. (1998) Plagues and Peoples. Garden City, NY, Anchor Press.

827.   Medeiros, D.M., Pellum, L.K. (1984) Elevation of Cadmium, Lead and Zinc in the Hair of Adult Black Female Hypertensives. Environ Contamin Toxicol 32:525-532.

828.   Medical World News (1979) Infertile Women Conceive After Vitamin B6 Therapy. Med World News 20(6):43.

829.   Medical World News (1979) Vitamin-A Cousin on Trial as Bladder-Carcinoma Inhibitor. Med World News 20(10):82.

830.   Medical World News (1980) If Vitamin A is Circulating It May Prevent Some Cancers. Med World News 21(26):13,19.

831.   Medical World News (1980) Multivitamins May Prevent Neural-Tube Defects. Med World News 21(19):20-21.

832.   Medical World News (1981) Breast-Cancer Patients Short on Vitamin C? Med World News 22(10):56.

833. Medical World News (1981) New Study Links Diet to Birth Defects. Med World News 22(3):12-13.

834. Medici, R. (1985) Behavioral Studies with Electromagnetic Fields. J Bioelect 4:527-552.

835. Medici, R. (1988) Behavioral Measures of Electromagnetic Field Effects. pp.557-587 in (ed) A.A. Marino, Modern Bioelectricity. NY, Marcel Dekker.

836. Mena, I. (1974) Role of Manganese in Human Disease. Ann Clin Lab Sci 4:487-491.

837. Menaker, W., Menaker, A. (1959) Lunar Periodicity in Human Reproduction: A Likely Unit of Biological Time. Amer J Obstetrics Gynecol 77:905-914.

838. Mengistu, M., Maru, M. (1979) Dry Beriberi: A Clinical Report from North-West Ethiopia. Ethiop Med J 17(2):29-32.

839. Merritt, Jr., C., et al. (1978) Effect of Radiation on Parameters on the Formation of Radiolysis Products in Meat. J Agric Food Chem 26:29.

840. Mertz, W. (1981) Essential Trace Elements. Science 213:1332-1338.

841. Messing, K., Bradley, W.E.C. (1985) In Vivo Mutant Frequency Rises Among Breast Cancer Patients After Exposure to High Doses of [Gamma]-Radiation. Mutat Res 152:107-112.

842. Michael, R.P., Zumpe, D. (1983) Sexual Violence in the United States and the Role of Season. Amer J Psychi 140:883-886.

843. Miller, A.L., et al. (1986) Endogenous Ion Current Traverse Growing Roots and Root Hairs of *Trifolium Repens*. Plant Cell Environ 9:79-83.

844. Miller, J.D.A. (1981) Metals. pp.149-202 in (ed) A.H. Rose, Microbial Biodegradation. Vol.6. London, Academic.

845. Miller, J.D.A., Tiller, A.K. (1970) Microbial Corrosion of Buried and Immersed Metals. pp.61-105 in (ed) J.D.A. Miller, Microbial Aspects of Metallurgy. NY, Elsevier.

846. Miller, M.F., Krusekopf, H.H. (1932) Influence of Cropping and Methods of Culture on Surface Runoff and Soil Erosion. Bull 177, Columbia, Missouri Agric Experimental Station.

847. Miller, P.R., Kickert, R.N. (1986) Gaseous Air Pollutants. pp.581-601 in (eds) A.H. Legge, S.V. Krupa, Air Pollutants and Their Effects on the Terrestrial Ecosystems. NY, J Wiley.

848. Miller, M.H. (1968) Santa Ana Winds of Crime. Prof Geog 20:23.

849. Mills, C.A. (1934) Suicides and Homicides in Their Relation to Weather Changes. Amer J Psych 91:669-77.

850. Milne, R. (1985) Missile That Kills Young Pheasants. New Sci 108:9.

851. Milne, R. (1986) Unions Point to Pesticide Perils. New Sci 112:18.

852. Minkh, A.A. (1961) Effect of Ionized Air on Work Capacity and Vitamin Metabolism. J Acad Med Sci, USSR. Trans US Dept Comm, Wash, DC, USGPO.

853. Mironov, A.T. (1948) An Electric Current in the Sea and the Effect of the Current on Fish [English Abstract]. Trudy Morskogo Gidrofiei. Inst AN SSSR 1:56-74.

854. Mitchel, H.H., et al. (1952) Relationship Between the Protein Content of Corn and the Nutritional Value of the Protein. J Nutr 48:461-76.

855. Mitchell, R.L. (1972) Trace Elements in Soils and Factors That Affect Their Availability. Geol Soc Amer Spec Paper 140:9-16.

856. Moghissi, K.S. (ed) (1974) Birth Defects and Fetal Development. Springfield, Ill, Charles Thomas.

857. Mondy, N.I., Ponnampalam, R. (1985) Effect of Magnesium Fertilizers on Total Glycoalkoloids and Nitrate-N in Katahdin Tubers. J Food Science 50:536-539.

858. Monod, J. (1972) Chance and Necessity. London, Collins.

859. Montalenti, G. (1974) From Aristole to Democritus via Darwin. in (eds) F.J. Ayala, T, Dobzhansky, Studies in the Philosophy of Biology. London, Macmillian.

860. Monyo, J.H. (1973) Breeding for High Protein Content and Quality of Rice by Nuclear Techniques. pp.149-151 in FAO/IAEA Division of Atomic Energy in Food & Agriculture & Gesellschaft fur Strahlen- und Umweltforschung, 1972. Vienna, Intl Atomic Energy Agency.

861. Moore, P.D. (1986) Plant Ecology: Pesticides and Pollination. Nature 321:854.

862. Moore, T., et al. (1987) Transmembrane Charge Transfer in Model Systems for Photosynthesis. pp.283-297 in (ed) V. Balzani, Supramolecular Photochemistry. Boston, Reidel.

863. Moore-Ede, M.C., et al. (1983) Circadian Timekeeping in Health and Disease. New Eng J Med 309:530-536.

864. Morgan, J.R., Greene, H.L. (1979) B Vitamins and Vitamin C in Human Nutrition. Amer J Dis Child 133:192-199.

865. Moriyama, H. (1970) Studies on X Agent. Tokyo, Igaku Shoin.

866. Moriyama, H. (1975) Challenge to Einstein's Theory of Relativity (Further Studies on X-Agent). Tokyo, Igaku Shoin.

867. Morris, S.C., Whittington, H.B. (1979) Animals of the Burgess Shale. Sci Amer 24(1):110-120.

868. Muirhead-Thomson, R.C. (1987) Pesticide Impact on Stream Fauna with Special Reference to Macroinvertebrates. NY, Cambridge Univ Press.

869. Murphy, J.G. (1982) Evolution, Mortality and the Meaning of Life. Totowa, NJ, Rowman & Littlefield.

870. Murphy, R., Harvey, C. (1985) Residues and Metabolites of Selected Persistent Halogenated Hydrocarbons Blood Specimens from a General Population Study. Environ Health Perspect 60:115-120.

871. Murr, L.E. (1963) Plant Growth Response in a Simulated Electric Field-Environment. Nature 200:490-491.

872. Murr, L.E. (1964) Mechanism of Plant-Cell Damage in an Electrostatic Field. Nature 201:1305-1306.

873. Murr, L.E. (1966) Biophysics of Plant Growth in a Reversed Electrostatic Field. A Comparison With Conventional Electrostatic and Electrokinetic Field Growth Responses. Intl J Biometeor 10:135-146.

874. Murr, L.E. (1966) Physiological Stimulation of Plants Using Delayed and Regulated Electric Field Environments. Intl J Biometeorol 10:147-153.

875. Murray, H.M., et al. (1984) Pulsed Electromagnetic Fields and Peripheral Nerve Regeneration in Cat. J Bioelect 3:19-32.

876. Myers, N. (1992) Primary Source: Tropical Forests and Our Future. NY, Norton.

877. Nag, M., Nandi, N. (1987) Inhibition of Monoamine Oxidase Activity by Some Organophosphate Pesticides. Indian J Exp Biol 25:567-568.

878. Nagai, N.Y., Moy, J.H. (1985) Quality of Gamma Irradiated California Valencia Oranges. J Food Sci 50:215-219.

879. Nair, I., et al. (1989) Biological Effects of Power Frequency Electric and Magnetic Fields. (May), Wash, DC, Office Tech Assess, USGPO.

880. Nambi, K.S.V., et al. (1985) Environmental Radioactivity and Thermoluminesence: A Review. J Environ Radioact 2:59-75.

881. Natelson, B.H., et al. (1985) Life in Constant Light Increases the Longevity of Cardiomyopathic Hamsters. Ann NY Acad Sci 453:397-398.

882. National Research Council (1975) Pest Control: An Assessment of Present and Alternative Technologies. Wash, DC, Natl Acad Sci.

883. National Research Council, US Committee on the Atmospheric Effects of Nuclear Explosions (1985) Effects on the Atmosphere of a Major Nuclear Exchange. Wash, DC, Natl Res Council, USGPO.

884. Ndiokwere, C.L. (1984) A Study of Heavy Metal Pollution from Motor Vehicle Emissions and Its Effect on Roadside Soil, Vegetation and Crops in Nigeria. Environ Poll 7B:35-42.

885. Needleman, H.L. (1984) Hazard to Health of Lead Expose at Low Dose. pp.311-322 in (ed) J.O. Nriagu, Changing Metal Cycles and Human Health. NY, Springer-Verlag.

886. Neel, J.V., et al. (1985) Delayed Biomedical Effects of the Bomb. Bull Atomic Sci 41(7):72-75.

887. Neve, J., et al. (1983) Erythrocyte and Plasma Trace Element Levels in Clinical Assessments—Zinc, Copper and Selenium in Normals and Patients with Down's Syndrome and Cystic Fibrosis. Biol Trace Element Res 5(2):75-79.

888. Newell, N.D. (1959) Nature of the Fossil Record. Proc Amer Phil Soc 103(2):264-285.

889. Nikitopoulou-Maratou, G., et al. (1981) Effect of Serotonin and Melatonin on the Electrophysiological Behavior of the Plasma Membrane. pp.299-315 in (eds) B.Haber, et al., Serotonin. NY, Plenum.

890. Nitecki, M.H. (ed) (1984) Extinctions. Univ Chicago Press.

891. Nitsch, K., Subertova, E. (1977) Preparation and Basic Electrical Properties of Cytosine Monohydrate Crystals. Czech J Phys 27B:1181-1186.

892. Nordenstrom, B. (1984) Biologically Closed Electric Currents: Activation of Vascular Interstital Closed Electric Circuit for Treatment of Inoperable Cancers. J Bioelect 3:137-53

893. Norton, L.A., et al. (1979) Adherence and DNA Synthesis Changes in Hard Tissue Cell Culture Produced by Electric Pertubation. p.443 in (ed) C.T. Brighton, et al., Electrical Properties of Bone and Cartilage. NY, Grune & Stratton.

894. Notkola, V.J., et al. (1987) Mortality Among Male Farmers in Finland During 1979-1983. Scand J Work Environ Health 13(2):124-128.

895. Novák, J., Valek, L. (1965) Attempt at Demonstrating the Effect of a Weak Magnetic Field on *Taxaxacum Officinale*. Biol Plantarum 7(6):469.

896. Noval, J.J., et al. (1976) Extremely Low Frequency Electric Field Induces Changes in Rate of Growth and Brain and Liver Enzymes of Rats, Compilation of Navy-Sponsored ELF Biomedical and Ecological Research Reports. Vol.3, AD Ao 35959.

897. Novikova, K.F., Ryvkin, B.A. (1977) Solar Activity and Cardiovascular Diseases. pp.184-200 in (eds) M.N. Gnevyshev, A.I. Ol', Effects of Solar Activity on the Earth's Atmosphere and Biosphere. Jerusalem, Israel Prog Sci Trans.

898. Nriagu, J.O. (1984) Introduction. pp.1-7 in (ed) J.O. Nriagu, Changing Metal Cycles and Human Health. NY, Springer-Verlag.

899. O'Brien, W.J., et al. (1984) Effects of Pulsing Electromagnetic Fields on Nerve Regeneration: Correlation of Electrophysiologic and Histochemical Parameters. J Bioelect 3:33-40.

900. O'Connor, M.E. (1988) Prenatal Microwave Exposure and Behavior. pp.265-288 in (eds) M.E. O'Connor, R.H. Lovely, Electromagnetic Fields and Neurobehavioral Function. NY, Alan Liss.

901. Odum, E.P. (1971) Fundamentals of Ecology. Philadelphia, Saunders.

902. Oelhaf, R.C. (1978) Organic Agriculture. NY, J Wiley.

903. Officer, C.B., et al. (1987) Late Cretaceous and Paroxysmal Cretaceous/Tertiary Extinctions. Nature 326:143-149.

904. O'Grady, R.T., Brooks, D.R. (1988) Teleology and Biology. pp.285-316 in (ed) B.H. Weber, Entropy, Information and Evolution. Cambridge, Mass, MIT Press.

905. Ohta, T. (1988) Multigene and Supergene Families. pp.41-65 in (eds) P.H. Harvey, L. Partridge. Oxford Surveys in Evolutionary Biology. NY, Oxford Univ.

906. O'Keefe, J.A. (1976) Tektites and Their Origin. NY, Elsevier.

907. Oldfield, J.E. (1972) Selenium Deficiency in Soils and Its Effect on Animal Health. Geol Soc Amer Spec Pap 140:57-63.

908. Olenchock, S.A., et al. (1987) Occupational Exposures to Airborne Endotoxins in Agriculture. Prog Clin Biol Res 231:475-487.

909. Olmsted III, J. (1988) Observations on Evolution. pp.243-262 in (eds) B.H. Weber, et al., Entropy, Information and Evolution. Cambridge, Mass, MIT Press.

910. Olsen, S.R. (1986) Role of Organic Matter and Ammonium in Producing High Corn Yields. pp.29-54 in (eds) Y. Chen, Y. Avnimelech, Role of Organic Matter in Modern Agriculture. Boston, Martinus Nijhoff.

911. Ordish, G. (1967) Biological Methods in Crop Pest Control. London, Constable & Company Ltd.

912. Ordóñez, L.A. (1977) Control of the Availability to the Brain of Folic Acid, Vitamin B12 and Choline. pp.205-245 in (eds) R.J. Wurtman, J.J. Wurtman, Nutrition and the Brain. Vol.1. NY, Raven.

913. Orskov, E.R. (1987) Role of Livestock in Africa: Are Livestock Occasionally Contributing to Famine? Proc Nutr Soc 46:301-308.

914. Orth, C.J., et al. (1984) Search for Iridium Abundance Anomalies at Two Late Cambrian Biomere Boundaries in Western Utah. Science 223:163-165.

915. Oskar, K.J., Hawkins, T.D. (1977) Microwave Alteration of the Blood-Brain Barrier System of Rats. Brain Res 126:281.

916. Ossenkopp, K.P., Ossenkopp, M.D. (1978) Self-Inflicted Injuries and the Lunar Cycle: A Preliminary Report. J Interdiscipl Cycle Res 4:337-48.

917. Ostrowski, K., et al. (1974) Accuracy, Sensitivity, and Specificity of Electron Spin Resonance Analysis of Mineral Constituents of Irradiated Tissues. Ann N Y Acad Sci 238:186-201.

918. Ott, J. (1988) Health and Light. Greenwich, Conn, Devin-Adair.

919. Pal, P.C., Creer, K.M. (1986) Geomagnetic Reversal Spurts and Episodes of Extraterrestrial Catastrophism. Nature 320:148-150.

920. Papatheofanis, F.J., et al. (1984) Intense Static Magnetic Field Induced Bone Growth *In Vivo*. J Bioelect 3:223-233.

921. Park, C.H., et al. (1980) Growth Suppression of Human Leukemia Cells *In Vitro* by L-Ascorbic Acid. Cancer Res 40:1062-1065.

922. Parodi, P.C., Nebreda, I.M. (1979) Protein and Yield of Six Wheat (*Triticum spp.*) Genotypes to Gamma Radiation. pp.201-209 in FAO/IAEA/Gesellschaft fur Strahlen- und Umweltforschung. Seed Protein Improvement in Cereals and Grain Legumes. Vol. 2, 1978. Vienna, Intl Atomic Energy Agency.

923. Parodi, P.C., Nebreda, I.M. (1984) Wheat and Triticale Breeding Using Gamma-Ray-Induced Variability. pp.3-15 in FAO/IAEA Division of Isotope & Radiation Applications of Atomic Energy for Food & Agricultural Development. Cereal Grain Improvement: 1982. Vienna, Intl Atomic Energy Agency.

924. Patterson, M.A., et al. (1984) Treatment of Drug, Alcohol and Nicotine Addiction by Neuroelectric Therapy: Analysis of Results Over 7 Years. J Bioelect 3:193-221.

925. Paul, C.R.C. (1988) Extinction and Survival in Echinoderms. pp.155-170 in (ed) G.P. Larwood, Extinction and Survival in the Fossil Record. Systematics Assoc Spec. Vol.34. Oxford, Clarendon.

926. Pavlovich, S.A., Sluvko, A.L. (1975) Effect of Shielding from Magnetic Field on Staphylococcus Aureus. p.56 in Proc 3rd All-Union Symp Effect of Magnetic Fields on Biological Objects. Kaliningrad State Univ.

927. Pavlyuchenko, V.M., et al. (1975) Gonadotropic Function of Adenohypophsis of Sables and Effect of Some Ecological Factors on It. pp.102-105 in Collection of Scientific Works of Moscow Veterinary Academy. Vol.80. Moscow.

928. Payne, B. (1983) New Device Which Detects and Measures an Energy Field Around the Human Body. Amer J Acupuncture 11(4):353-358.

929. Payne, B. (1984) Cycles of Peace, Sunspots and Geomagnetic Activity. Cycles 35:101-105.

930. Pearce, F. (1987) Pesticide Deaths: The Price of the Green Revolution. New Sci 114:30.

931. Pearce, N.E., et al. (1987) Non-Hodgkin's Lymphoma and Farming: An Expanded Case-Control Study. Intl J Cancer 39(2):155-61.

932. Pemadasa, M.A. (1981) Photocontrol of Stomatal Movements. Biol Rev 56:551-588.

933. Pereira, H.C. (1972) The Influence of Man on the Hydrological Cycle. In IASH, World Water Balance. Gentbrugge 3:553-569.

934. Perry, F.S., et al. (1981) Environmental Power Frequency Magnetic Fields and Suicide. Health Phys 41:267-277.

935. Persinger, M.A. (1974) Introduction. pp.1-8 in (ed) M.A. Persinger, ELF and VLF Electromagnetic Field Effects. NY, Plenum.

936. Persinger, M.A. (1975) Geophysical Models for Parapsychological Experiences. pp.1-2 in Psychoenergetic Systems. Vol.1. NY, Plenum.

937. Persinger, M.A. (1980) The Weather Matrix and Human Behavior. NY, Praeger.

938. Persinger, M.A. (1985) Classical Psychophysics and ELF Magnetic Field Detection. J Bioelect 4:577-584.

939. Persinger, M.A., et al. (1974) Psycho-Physiological Effects of ELF-EM. In (ed) M.A. Persinger, ELF and VLF Electromagnetic Field Effects. NY, Plenum.

940. Persinger, M.A., Krippner, S. (1989) Dream ESP Experiments and Geomagnetic Activity. J Amer Soc Psychical 83:101ff.

941. Peterson, I. (1984) Enforcing the Law: EPA's Toxics Problem. Sci News 126:36.

942. Peterson, I. (1985) Material Loss: Acid Rain is Leaving Its Mark on Buildings, Statues, Automobiles and Other Man-Made Structures. Sci News 128:154-6.

943. Petersson, L., Ehrenberg, A. (1985) Highly Sensitive Farday Balance for Magnetic Susceptiblity Studies of Dilute Protein Solutions. Rev Sci Instrum 56:575-580.

944. Pethig, R. (1982) Electronic Conduction in Biopolymers. pp.1-85 in (ed) B. Lipinski, Electronic Conduction and Mechanoelectrical Transduction in Biological Materials. NY, Marcel Dekker.

945. Phil, R.O., Parkes, M. (1977) Hair Element Content in Learning Disabled Children. Science 198:204-206.

946. Philla, A.A. (1979) Electrochemical Information Transfer and Its Possible Role in the Control of Cell Function. p.455 in (ed) C.T. Brighton, et al., Electrical Properties of Bone and Cartilage. NY, Grune & Stratton.

947. Phillips, J. (1986) Transferrin Receptors and Natural Killer Cell Lysis. A Study Using Colo 205 Cells Exposed to 60 Hz Electromagnetic Fields. Immunology Lett 13:295-300.

948. Philpott, W.H., Kalita, D. (1987) Brain Allergies. New Canaan, CT, Keats.

949. Piccardi, G. (1962) Chemical Basis of Medical Climatology. Springfield, Ill, C Thomas.

950. Pickard, B.G. (1973) Action Potentials in Higher Plants. Bot Rev 39:172-201.

951. Pickard, B.G., Van Sambeek, J.W. (1976) Mediation of Rapid Electrical, Metabolic, Transpirational, and Photosynthetic Changes by Factors Released from Wounds. I. Variation Potentials and Putative Action Potentials in Intact Plants. Canad J Bot 54:2642.

952. Pilla, A.A., Kaufman, J.J. (1984) Electromagnetic Modulation of Cell Function: Frequency Characteristics of Input Waveforms. J Bioelect 3:3-18.

953. Pimentel, D., Pimentel, M. (1996) Food, Energy and Society. Niwot, CO, Univ Press Colorado.

954. Pionka, H.B., Urban, J.B. (1985) Effect of Agricultural Land Use on Ground-Water Quality in a Small Pennsylvania Watershed. Ground Water 23:68-80.1

955. Pirozynski, K.A., Hawksworth, D.L. (eds)(1988) Coevolution of Fungi with Plants and Animals. San Diego, Academic.

956. Pittman, U.J. (1971) Biomagnetic Responses in Germinating Malting Barley. Can J Plant Sci 51:64-65.

957. Pittman, U.J. (1972) Biomagnetic Responses in Potatoes. Can J Plant Sci 52(5):727-733.

958. Playford, P.E., et al. (1984) Iridium Anomaly in the Upper Devonian of the Canning Basin, Western Australia. Science 226:437-439.

959. Plumstead, E.P. (1973) Late Palaeozoic Glossopteris Flora. pp.187-206 in (ed) A. Hallam, Atlas of Palaeobiogeography. NY, Elsevier.

960. Podzorou, N.V. (1970) Effect of Geomagnetic Field on Germination of Conifers. Izv Vyssh Uchelon Zaved Lesnoi Zh (5):155.

961. Pohl, H.A. (1977) Electroculture. J Biol Phys 5:3-23.

962. Pohl, H.A. (1984) Natural AC Electric Fields in and About Cells. pp.87-103 in (eds) W.R. Adey, A.F. Lawrence, Nonlinear Electrodynamics in Biological Systems. NY, Plenum.

963. Policy Research Project on Pesticide Regulation in Texas (1984) Pesticides and Worker Health in Texas: A Report. Austin, The Project (Univ Texas).

964. Popper, S.K. (1981) Rationality of Scientific Revolutions. pp.80-106 in (ed) I. Hacking. Scientific Revolutions. Oxford Univ.

965. Popkin, B.M., et al. (1986) Infant-Feeding Triad: Infant, Mother and Household. NY, Gordon & Breach.

966. Pool, R. (1989) Quantum Chaos: Enigma Wrapped in a Mystery. Science 243:893-895.

967. Pories, W.J., et al. (1979) Trace Element Profiles in Cancer Patients. Biol Trace Element Res 1:229-241.

968. Postel, S. (1984) Air Pollution, Acid Rain and the Future of Forests. Wash, DC, Worldwatch Institute.

969. Postgate, J.R. (1979) Sulphate-Reducing Bacteria. Cambridge Univ Press.

970. Prasad, A.S., Oberleas, D. (1974) Trace Elements in Aterial and Fetal Development. pp.33-68 in (ed) K.S. Moghissi, Birth Defects and Fetal Development. Springfield, Ill, C Thomas.

971. Preer, J.R., et al. (1984) Metals in Downtown Washington D.C. Gardens. Biol Trace Element Res 6(1):79-91.

972. Presman, A.S. (1971) Electromagnetic Fields in the Biosphere. Znaniya, Moscow.

973. Presman, A.S. (1972) [Acceleration and the Electromagnetic Field of the Biosphere] Akzeleration und Elektromagnetische Felder der Biosphäre. Moderne Medizin 2(4):224-28

974. Prien, Sr., E.L., Geroff, S.N. (1974) Magnesium Oxide-Pyridoxine Therapy for Recurrent Calcium Oxalate Calculi. J Urology 114:509-512.

975. Pritchard, J.A., Scott, D.E. (1974) Effect of Maternal Anemia on Fetal Growth and Development. pp.77-91 in (ed) K.S. Moghissi, Birth Defects and Fetal Development. Springfield, Ill, C Thomas.

976. Protasov, V.R., et al. (1975) Effect of Natural Electric Fields in Sea on Behavior and Distribution of Fish. Zool Zh 114:1098-1101.

977. Protopopov, V.V., Stakanov, V.D. (1970) Some Physiological Changes in Germinating Seeds and Seedlings of Trees Due to Electromagnetic Environmental Factors. p.24 in 3rd Ural Conf Physiology and Ecology of Woody Plants, UFA.

978. Purves, D. (1968) Trace Element Contamination of Soils in Urban Areas. 9th Intl Congr Soil Sci Trans (Adelaide) 2:351-355.

979. Puthoff, H.E., et al. (1981) Experimental Psi Research: Implications for Physics. pp.37-86 in (ed) R.G. Jahn, Role of Consciousness in the Physical World. AAAS. Boulder, Co, Westview Press.

980. Radaksie, V.D. (1976) The Threescore and Ten Life-span: Can It Be Prolonged? So African Med J 50:1189-1190.

981. Raisbeck, G.M., et al. (1985) Evidence for an Increase in Cosmogenic 10Be During a Geomagnetic Reversal. Nature 315:315-317.

982. Rajput, A.H., et al. (1987) Geography, Drinking Water Chemistry, Pesticides and Herbicides and the Etiology of Parkinson's Disease. Can J Neurol Sci 14(3 Suppl):414-8.

983. Rajput, A.H., Uitti, R.J. (1987) Paraquat and Parkinson's Disease. Neurology 37:1820-21.

984. Rakhno, P., et al. (1971) Population Dynamics of Soil Microorganisms and Nitrogen Compounds in the Soil. Valgus, Tallin.

985. Raloff, J. (1984) Picturing Pesticide Exposure Patterns. Sci News 125:246.

986. Raloff, J. (1985) AAAS: Tracing Disease to Trace Minerals. Sci News 127:357-358.

987. Raloff, J. (1985) Chinese Salted Fish Linked to Cancer. Sci News 127:404.

988. Raloff, J. (1985) Dirty Tricks: Plant Defense Backfires. Sci News 127:247.

989. Raloff, J. (1985) EPA Plans to Ban Carcinogen Daminozide. Sci News 128:149.

990. Raloff, J. (1985) Mutagens in the Air: They May Be a Gas. Sci News 127:166.

991. Raloff, J. (1985) Oxidized Lipids: A Key to Heart Disease. Sci News 127:278.

992. Raloff, J. (1987) Kid's Leukemia from Parents' Exposures. Sci News 132:38.

993. Ram, R.N., Sathyanesan, A.G. (1987) Histopathological Changes in Liver and Thyroid of the Teleost Fish *Channa Punctatus* (*Bloch*) in Response to Ammonium Sulfate Fertilizer Treatment. Ecotoxicol Environ Safety 13:185-190.

994. Randolph, T.G., Moss, R.W. (1989) An Alternative Approach to Allergies. Lippincott & Crowell.

995. Rands, M., Sotherton, N. (1985) Pesticides Threaten British Wildlife. New Sci 106:32.

996. Rao, A.R. (1985) Fertilizer Use Saves Fuels in India. Energy (Oxford, England) 10:989-91.

997. Rasekh, J.G. (1987) Marine Fish as Source of Protein Supplement in Meat. J Assoc Off Anal Chem 70:91-95.

998. Rashid, K.A., et al. (1987) Residues and Mutagenicity of Captan Applied to Apple Trees and Potential Human Exposure. J Environ Sci Health 22B:71-89.

999. Ratner, D., Eshel, E. (1986) Aerial Pesticide Spraying: An Environmental Hazard. JAMA 256:2516-17.

1000. Raup, D.M. (1979) Size of the Permo-Triassic Bottleneck and Its Evolutionary Implications. Science 206:217-220.

1001. Raup, D.M. (1984) Death of Species. pp.1-19 in (ed) M.D. Nitecki, Extinctions. Univ Chicago Press.

1002. Raup, D.M. (1984) Evolutionary Radiations and Extinctions. pp.5-14 in (eds) H.D. Holland, A.F. Trendall, Patterns of Change in Earth Evolution. NY, Spring-Verlag.

1003. Raup, D.M., et al. (1988) Testing for Periodicity of Extinction. Science 241:94-99.

1004. Raup, D.M., Sepkoski, J. (1984) Periodicity of Extinctions in the Geologic Past. Proc Natl Acad Sci 81:801-805.

1005. Raup, D.M., Sepkoski, J. (1986) Periodic Extinction of Families and Genera. Science 231:833-836.

1006. Ravitz, L.J. (1953) Electrodynamic Field Theory in Psychiatry. Southern Med J 46:650-60.

1007. Ravitz, L.J. (1962) History, Measurement and Application of Periodic Changes in the Electromagnetic Field in Health and Disease. Ann N Y Acad Sci 98:141-201.

1008. Rawal, B.D. (1978) Bactericidal Action of Ascorbic Acid in Pseudomonas Aeruginosa: Alteration of Cell Surface as a Possible Mechanism. Chemotherapy 24:166-171.

1009. Raymond, A., et al. (1985) Early Devonian Phytogeography. pp.129-167 in (ed) B.H. Tiffney. Geological Factors and the Evolution of Plants. New Haven, Conn, Yale Univ.

1010. Raymond, A., et al. (1985) Phytogeography and Paleoclimate of the Early Carboniferous. pp.169-220 in (ed) B.H. Tiffney, Geological Factors and the Evolution of Plants. New Haven, Conn, Yale Univ.

1011. Rayski, J. (1988) Interpretation of the Einstein-Podolsky-Rosen Effect in Terms of a Generalized Causality. pp.257-272 in (ed) F. Selleri. Quantum Mechanics Versus Local Realism. NY, Plenum.

1012. Reddy, K.R., et al. (1986) Effect of Soil Redox Conditions on Microbial Oxidation of Organic Matter. pp.117-156 in (eds) Y. Chen, Y. Avnimelech, Role of Organic Matter in Modern Agriculture. Boston, Martinus Nijhoff.

1013. Reed, B. (1977) Testimony Before the Select Committee on Nutrition and Human Needs. US Senate 95th Congr 1st Session 22 June.

1014. Reed, T. (1987) Auroville Forest: Reclaiming A Desert. Ecologist 17:203-204.

1015. Reed, J.P., et al. (1987) Characterization of Microorganisms in Soils Exhibiting Accelerated Pesticide Degradation. Bull Environ Contam Toxicol 39:776-82.

1016. Reganold, J.P., et al. (1987) Long-Term Effects of Organic and Conventional Farming on Soil Erosion. Nature 330:370-2.

1017. Reid, R.L., et al. (1987) Effects of Varying Zinc Concentrations on Quality of Alfalfa for Lambs. J Anim Sci 64:1735-1742.

1018. Rein, G., Dixey, R. (1984) Neurotransmitter Release Stimulated in a Clonal Nerve Cell Line by Low Intensity Pulsed Magnetic Fields. pp.79-86 in (eds) W.R. Adey, A.F. Lawrence, Nonlinear Electrodynamics in Biological Systems. NY, Plenum.

1019. Reisch, M. (1987) Fertilizer Based on Radioactive Wastes. Chem Eng News 65:5

1020. Reiter, R. (1952) [Biometeorological Index as Indicated in Total Admissions and Significant Prognosis] Biometeorologische Indikatoren von Groszraumiger und Prognostischer Bedeutung. Ber Dtsch Wetterdienstes U S Zone, Bad Kissingen 35:243.

1021. Reiter, R. (1952) [Traffic Accidents and Reaction Times Under Various Factors: Meteorology, Cosmic Rays and Atmospheric Electricity] Verkenhrsunfallziffer und Reaktionszeit Unterdem Einflusz Verschiedener Meteorologischer, Kosmischer und Luftelektrischer Faktoren. Meteorol Rundschau 5:14-17.

1022. Reiter, R. (1954) [Detection of a Biological Low-Frequency Electric Field] Nachweis der Biologischen Wirksamkeit Elektrisher Wechselfelder Niedriger Frequenz. Naturwissenschaten 41:22-23.

1023. Reiter, R. (1954) [Environmental Factors Influencing the Reaction Time of Healthy Humans] Umwelteinflusse auf die Reaktionszeit des Gesunden Menschen. Münch Med Wschr 17:479 [and] 18:526.

1024. Reiter, R. (1956) [Influence of the Weather Upon Daytime Industrial Accidents] Einflusz des Wetters auf die Häufigkeit von Unfällen im Untertagébetrieb. Arbeitschutz 3(Koln):133-134.

1025. Reiter, R. (1960) [Biometeorology and Electricity of the Atmosphere] Meteorobiologie und Elektrizität der Atmosphäre. Akad Verlagsgesllsch. Leipzig, Geest und Portig.

1026. Reiter, R.J. (1984) Neuroregulatory Aspects of the Pineal Gland in Reference to Reproduction. pp.112-121 in (eds) P. Pancheri, et al., Endorphins, Neuroregulators and Behavior in Human Reproduction. Amsterdam, Excerpta Medica.

1027. Repetto, M. (1987) Diagnosis of the Most Frequent Poisonings by Pesticides. J Toxicol Clin Experi 7(1):31-39.

1028. Rettig, B.A., et al. (1987) Incidence of Hospitalizations and Emergency Room Visits Resulting from Exposure to Chemicals Used in Agriculture. Nebr Med J 72:215-9.

1029. Reynolds, W.C., Perkins, H.C. (1977) Engineering Thermodynamics. 2nd ed. NY, McGraw-Hill.

1030. Rice-Evans, C. (ed) (1986) Free Radicals, Cell Damage and Disease. London, Richelieu Press.

1031. Richards, R.P., et al. (1987) Pesticides in Rainwater in the Northeastern United States. Nature 327:129-131.

1032. Rifkin, J. (1980) Entropy: A New World View. NY, Viking.

1033. Rimland, B.R., et al. (1978) Effects of High Doses of Vitamin B6 on Autistic Children: A Double-Blind Crossover Study. Amer J Psychi 135:472-475.

1034. Rimland, B.R., Larson, G.E. (1981) Nutritional and Ecological Approaches to the Production of Delinquency and Violence. J Appl Nutr 33:116-37.

1035. Rinaldi, R.A., Goodrich, J.D. (1982) Bone Electrical Conductance. J Bioelect 1:83-97.

1036. Rockstein, M., Sussman, M. (1976) Introduction: Food for Thought. pp.3-5 in (eds) M. Rockstein, M. Sussman, Nutrition, Longevity and Aging. NY, Academic.

1037. Rockstein, M., Sussman, M.L. (eds) (1976) Nutrition, Longevity and Aging. NY, Academic.

1038. Rodelsperger, K., et al. (1987) Potential Health Risks from the Use of Fibrous Mineral Absorption Granulates. Brit J Ind Med 44:337-43.

1039. Rodgers, C.D. (1985) Air Ionization Effects on Cardiovascular Parameters in Humans: A Review. J Bioelect 4:63-74.

1040. Rogers, J.M., et al. (1984) Plasma Glucose and Protein Concentrations in Rat Fetuses and Neonates Exposed to Calarmotogenic Doses of Mirex. Environ Res 34:155-61.

1041. Romer, A.S. (1973) Permian Reptiles. pp.159-168 in (ed) A. Hallam, Atlas of Palaeobiogeography. NY, Elsevier.

1042. Root, E.J., Longenecker, A. (1980) Brain Cell Alterations Resembling Senile Changes Induced by Deficiencies of Vitamin B-6 and/or Copper. Fed Proc 39:393.

1043. Ropper, A.H. (1979) Rational Approach to Dementia. Canad Med Assoc J 121:1175-1190.

1044. Rosenthal, R., Fode, K.L. (1963) Effect of Experimental Bias on the Performance of the Albino Rat. Behav Sci 8:183-189.

1045. Ross, M. (1962) Food in the Mental Hospital. J Amer Diet Assoc 40:318.

1046. Roukens De Lange, A. (1982) Evolution and Formative Causation. Parapsych J So Africa 3:84-105.

1047. Roukens De Lange, A. (1982) Matter, Life, Mind and Psi. Parapsych J So Africa 3:28-49.

1048. Rowland, A. (1984) Trace Metals Leave More than Trace Effects. Sci News 125:373.

1049. Rowley, B.A., et al. (1974) Influence of Electrical Current on an Infecting Microorganism in Wounds. Ann NY Acad Sci 238:543-551.

1050. Rozhdestvenskaya, E.D. (1972) Effect of Geomagnetic Disturbances on Vegetative Nervous System of Healthy People and Patients with Atherosclerosis. Abstr 4th Interregional Conf Ural Therapists, Sverdlousk.

1051. Rubes, J. (1987) Chromosomal Aberrations and Sister-Chromatid Exchanges in Swine. Mutat Res 191:105-9.

1052. Ruehle, J.L., Marx, D.H. (1979) Fiber, Food, Fuel and Fungal Symbionts. Science 206:419-422.

1053. Ruse, M.J. (1982) Darwinism Defended. Reading, Mass, Addison-Wesley.

1054. Russell, D.A. (1979) Enigma of the Extinction of the Dinosaurs. Rev Earth Planet Sci 7:163-182.

1055. Russell, E.W. (1973) Soil Conditions and Plant Growth. London, Longman Group Ltd.

1056. Russell, L.B., et al. (1976) Radiation-Induced Mutations at Mouse Hemoglobin Loci. Proc Natl Acad Sci (USA) 73:2843-6.

1057. Russell, L.B., Kelly, E.M. (1982) Mutation Frequencies in Male Mice and the Estimation of Genetic Hazards of Radiation in Men. Proc Natl Acad Sci (USA) 79:542-544.

1058. Rycroft, M.J. (1984) Interaction of Whistlers and Radiation Belt Electrons. Nature 312:698.

1059. Rycroft, M.J. (1987) Strange New Whistlers. Nature 327:368-369.

1060. Sadler, P.M. (1981) Sediment Accumulation Rates and the Completeness of Stratigraphic Sections. J Geol 89:569-584.

1061. Sagan, C. (1973) Ultraviolet Selection Pressure on the Earliest Organisms. J Theoretical Biol 39:195-200.

1062. Sagan, C., et al. (1979) Anthropogenic Albedo Changes and the Earth's Climate. Science 206:1363-1368.

1063. Sahlin, M., et al. (1987) Magnetic Interaction Between the Tyrosyl Free Radical and Antiferromagnetically Coupled Iron Center in Ribonucleotide Reductase. Biochemistry 26:5541-5548.

1064. Saito, T. (1936) Biological Changes in the Magnetic Field. Jap J Obstet Gynec 19:381.

1065. Sakamoto-Momiyama, M. (1977) Seasonality in Human Mortality. Univ Tokyo Press.

1066. Sakamoto-Momiyama, M., Katayama, K. (1971) Statistic Analysis of Seasonal Variation in Mortality. J Meterol Soc Jap 49:494-509.

1067. Sakula, A., et al. (1980) Vitamin A and Cancer. Lancet II (8202):1029.

1068. Salop, L.I. (1977) Relationship of Glaciations and Rapid Changes in Organic Life to Events in Outer Space. Intl Geol Rev 19:1271-1291.

1069. Salzman, C., Shader, R.I. (1978) Depression in the Elderly. II. Possible Drug Etiologies; Differential Diagnostic Criteria. J Amer Geriatrics Soc 26:303-308.

1070. Sampson, R.N. (1986) Herbicides and the Environment. American Forests 92:11ff.

1071. Samson, H.H. (1978) Sucrositis. So African Med J 54(15):590-591.

1072. Sanborn, W.B. (1950) Electrical "Bath" in Yellowstone. Nat Hist 59:258-259.

1073. Sanderson, B.J.S., et al. (1984) Mutations in Human Lymphocytes: Effect of X- and UV-Irradiation. Mutat Res 140:223-227.

1074. Sankaranarayanan, K. (1982) Genetic Effects of Ionizing Radiation in Multicellular Eukaroyotes and the Assessment of Genetic Radiation Hazards in Man. Amsterdam, Elsevier.

1075. Sanotskii IV (1987) [Current Problems of the Theory of Safe Levels of Chemicals.] Gigiena Truda i Professionalnye Zabolevaniia Aug (8):1-4.

1076. Sauberlich, H.E., et al. (1983) Comparative Nutritive Value of Corn of High and Low Protein Content for Growth in the Rat and Chick. J Nutr 51:623-35.

1077. Saunders, S., Hedlund, B.E. (1984) Electrostatic Modification of Protein Surface: Effect on Hemoglobin Ligation and Solubility. Biochemistry 23:1457-1461.

1078. Savage, R.J.G. (1988) Extinction and the Fossil Mammal Record. pp.319-335 in (ed) G.P. Larwood, Extinction and Survival in the Fossil Record. Systematics Assoc Special Vol. 34. Oxford, Clarendon.

1079. Schall, J.J., Pianka, E.R. (1978) Geographical Trends in Numbers of Species. Science 201:679-686.

1080. Schauss, A.G. (1978) Orthomolecular Treatment of Criminal Offenders. Berkeley, Ca, Michael Lesser, M.D.

1081. Schauss, A.G. (1979) A Critical Analysis of the Diets of Chronic Offenders: Part II. J Orthomol Psychi 8:222-226.

1082. Schauss, A.G. (1980) Diet, Crime and Delinquency. Berkeley, Ca, Parker House.

1083. Schauss, A.G., Simonsen, C. (1979) A Critical Analysis of the Diets of Chronic Juvenile Offenders (Part I). J Orthomol Psychi 8:149-157.

1084. Schindewolf, O.H. (1962) [Neocatastrophism?] Neokatastrophismus? Deutsche Geologishe Gesellschaft Zeitschrift 114:430-445.

1085. Schlesinger, K., et al. (1965) Genetics of Audiogenic Seizures. I. Relation to Brain Serotonin and Norepinephrine in Mice. Life Sci 4:2345-2351.

1086. Schneider, E.D. (1988) Thermodynamics, Ecological Succession and Natural Selection: A Common Thread. pp.107-138 in (eds) B.H. Weber, et al., Entropy, Information and Evolution. Cambridge, Mass, MIT Press.

1087. Schneider, F. (1958) Unusual Tropism of Feeder Roots in Sugar Beets and Its Possible Effect on Fertilizer Response. Can J Plant Sci 38:124.

1088. Schneider, S.H., Londer, R. (1984) Coevolution of Climate and Life. SF, Sierra Club.

1089. Schnitzer, M., Khan, S.U. (eds) (1978) Soil Organic Matter. NY, Elsevier.

1090. Schrodinger, E. (1944) What is Life? Cambridge Univ.

1091. Schua, L. (1952) [Research Concerning the Influence of Meteorological Elements Upon the Behavior of the Honeybee (*Apis Mellifera*)] Untersuchungen uber den Einfluz Meteorologischer Elemente auf das Verhalten der Honigbiene (*Apis Mellifera*). Z Vergl Physiol 34:258.

1092. Schua, L. (1954) [Does the Atmospheric Electric Field Effect Living Organisms?] Wirken Luftelektrische Felder auf Lebewesen? Umschau 15:468.

1093. Schua, L. (1954) [The Effect of the Atmospheric Electric Field Upon Animals] Die Wirkung von Luftelektrischen Feldern Auf Tiere. Verh Dtsch Zool G Tubingen 47:435-40.

1094. Schuphan, W. (1974) Nutritional Value of Crops as Influenced by Organic and Inorganic Fertilizer Treatments. Qual Plant Plant Foods Human Nutr 23(4):333-58.

1095. Schutte, K.H. (ed) (1959) Trace Element Problems in Nature: A Symposium. Univ Capetown, Dept Botany.

1096. Schwarzchild, B. (1987) Do Asteroid Impacts Trigger Geomagnetic Reversals? Phys Today 40(2):17-20.

1097. Scott, B.I.H. (1962) Electricity in Plants. Sci Amer 207:107-117.

1098. Scott, B.I.H. (1967) Electric Fields in Plants. Ann Rev Plant Physiol 18:409-418.

1099. Scott, B.I.H., Martin, D.W. (1962) Bioelectric Fields of Bean Roots and Their Relation to Salt Accumulation. Austral J Biol Sci 15:83-100.

1100. Scragg, R.K.R. (1978) Human Congenital Abnormalities and Their Relationship to Nutritional Related Environmental Factors in Rural South Australia. Proc Nutr Soc Austria 3:104.

1101. Scrimshaw, N.S. (1987) Phenomenon of Famine. Ann Rev Nutr 7:1-21.

1102. Sebastian, J., et al. (1987) Discovery of Cutinase-Producing *Pseudomonas sp.* Cohabiting with an Apparently Nitrogen-Fixing *Corynebacterium sp.* in the Phyllosphere. J Bacteriol 169:131-136.

1103. Segal, A.S., Magin, R.L. (1982) Microwaves and the Blood-Brain Barrier: A Review. J Bioelect 1:351-398.

1104. Segal, A.S., Magin, R.L. (1983) Reply to Comment On "Microwaves and the Blood-Brain Barrier." J Bioelect 2:89-92.

1105. Select Committee on Nutrition and Human Needs (1977) Nutrition and Mental Health, Wash, DC, USGPO.

1106. Select Committee on Nutrition and Human Needs (1980) Nutrition and Mental Health. Wash, DC, USGPO.

1107. Sellers, P.J., Lockwood, J.G. (1981) Numerical Simulation of the Effects of Changing Vegetation Type on Surface Hydroclimatology. Climatic Change 3:121-136.

1108. Senanayake, N., Karailiedde, L. (1987) Neurotoxic Effects of Organophosphorus Insecticides. New Eng J Med 316:761-763.

1109. Sepkoski, J.J., Raup, D.M. (1986) Periodicity in Marine Extinction Events. pp.3-36 in (ed) D.K. Elliot, Dynamics of Extinction. NY, J Wiley.

1110. Shacklette, H.T., et al. (1972) Distribution of Trace Elements in the Environment and the Occurrence of Heart Disease in Georgia. Geol Soc Amer Spec Paper 140:65-70.

1111. Shandala, M.G. (1988) Possible Physiological Mechanisms for Neurobehavioral Effects of Electromagnetic Exposure. pp. 367-376 in (eds) M.E. O'Connor, R.H. Lovely, Electromagnetic Fields and Neurobehavioral Function. NY, Alan Liss.

1112. Shandala, M.G., et al. (1988) Biological Effects of Power-Frequency Electric Fields in the Environment. pp.927-964 in (ed) A.A. Marino, Modern Bioelectricity. NY, Marcel Dekker.

1113. Shannon, C.E., Weaver, W. (1949) Mathematical Theory of Communication. Urbana, Univ Illinois.

1114. Shapely, D. (1977) Will Fertilizers Harm Ozone as Much as SST's? Science 195:658.

1115. Shapere, D. (1981) Meaning and Scientific Change. pp.28-59 in (ed) I. Hacking, Scientific Revolutions. Oxford Univ.

1116. Shapiro, I.M., Marecek, J. (1984) Dentine Lead Concentration as a Predictor of Neuropsychological Functioning in Inner-City Children. Biol Trace Element Res 6:69-78.

1117. Sharp, E.L. (1972) Relation of Air Ions to Air Pollution and Some Biological Effects. UK, Applied Sci.

1118. Sharp, P.M., et al. (1989) Chromosomal Location and Evolutionary Rate Variation in Enterobacterial Genes. Science 246:808-810.

1119. Sheard, C., Johnson, A.F. (1930) Effects of Infrared, Visible and Ultra-Violet Irradiation on Changes in Electrical Potentials and Currents in Plants. Science 71:246-248.

1120. Sheinkin, D., et al. (1987) Food Connection. NY, Bobbs-Merrill.

1121. Sheldrake, R. (1982) Morphic Resonance, Memory and Psychical Research. Parapsych J So Africa 3(2):70-76.

1122. Sheldrake, R. (1987) A New Science of Life: The Hypothesis of Formative Causation. London, Paladin [and] LA, CA, J.P. Tarcher.

1123. Shemi-zade, A.É., (1975) Increase in Natural Radioactivity of the Atmosphere Due to Geomagnetic Storms and Possible Biological Effect of this Phenomenon. pp.198-202 in Physicomathematical and Biological Problems of Effect of Electromagnetic Fields and Ionization of Air. Vol.1. Nauka, Moscow.

1124. Shemi-zade, A.É., Mambetov, R.U. (1974) Measurements of Natural Radioactivity of Air in Mountain Conditions. Atom Eng 36(1):61062.

1125. Shepparo, A.R. (1979) Magnetic Effects on Mammals. pp.33-37 in (ed) T.S. Tenforde, Magnetic Field Effect on Biological Systems. NY, Plenum.

1126. Shinbrot, M. (1987) Things Fall Apart. No One Doubts the Second Law, But No One's Proved It—Yet. Sciences 27(3):32-36.

1127. Sholto Douglas, J., Hart, R.A. (1984) Forest Farming: Towards a Solution to Problems of World Hunger and Conservation. London, Intermediate Technology.

1128. Shrivastava, P. (1987) Bhopal: Anatomy of a Crisis. Cambridge, Mass, Bassinger.

1129. Shul'ts, N.A. (1967) Effect of Solar Activity on Incidence of Functional Leucopenias and Relative Lymphocytoses. [Dissert Abstr]. USSR, Moscow, Acad Med Sci.

1130. Sibley, C.G., Ahlquist, J.E. (1986) Reconstructing Bird Phylogeny by Comparing DNAs. Sci Amer 254(2):82-92.

1131. Sidaway, G.H. (1969) Electrostatic Influence on Phytochrom-Mediated Photomorphogenesis. Intl J Biometeorol 13:219-230.

1132. Sidaway, G.H., Asprey, G.F. (1968) Influence of Electrostatic Fields on Plant Respiration. Intl J Biometeor 12(4):321-329.

1133. Silver, S. (1984) Bacterial Transformations of and Resistance to Heavy Metals. pp.199-224 in (ed) J.O. Nriagu, Changing Metal Cycles and Human Health. NY, Springer-Verlag.

1134. Simberloff, D. (1986) Are We on the Verge of a Mass Extinction in Tropical Rain Forests? pp.165-182 in (ed) D.K. Elliot, Dynamics of Extinction. NY, Wiley.

1135. Simic, M.G., et al. (eds)(1986) Mechanisms of DNA Damage and Repair. NY, Plenum.

1136. Simon, C. (1984) Abundant Ir Marks a Third Boundary. Sci News 125:213-214.

1137. Simonson, E., Keys, A. (1961) Research in Russia on Vitamins and Atherosclerosis. Circulation 24:1239-1248.

1138. Simpson, H.J., et al. (1977) Man and the Global Nitrogen Cycle Group Report. pp.253-274 in (ed) W. Stumm, Global Chemical Cycles and Their Alterations by Man. Berlin, Dahlem Konferenzen.

1139. Simpson, J.F. (1966) Evolutionary Pulsations and Geomagnetic Polarity. Geol Soc Amer Bull 77(2):197.

1140. Singh, S. (1984) Bioelectricity in Living Bone Under Ultrasound Exposure. J Bioelect 3:483-492.

1141. Sirsi, M. (1952) Antimicrobial Action of Vitamin C on M. Tuberculosis and Some Other Pathogenic Organisms. Indian J Med Sci (April):252-255.

1142. Sisken, B.F., et al. (1981) Effects of Direct and Indirectly Coupled Current, and Nerve Growth Factor on Nerve Regeneration *In Vitro*. pp.251-275 in (ed) R.O. Becker, Mechanisms of Growth Control. Springfield, Ill, C Thomas.

1143. Skaare, J.U., et al. (1988) Organochlorine Pesticides and Polychlorinated Biphenyls in Material Adipose Tissue, Blood, Milk and Cord Blood from Mothers and Their Infants Living in Norway. Arch Environ Contam Toxicol 17(1):55-63.

1144. Skrija, P. (1987) Investigations of the Fertilizer Value of Sheep Excrements Left on Pasture. pp.325-328 in (eds) H.G. Van Der Meer, et al., Animal Manure on Grasslands and Fodder Crops. Dordrecht, Netherlands, Martinus Nihoff.

1145. Smith, A.G., et al. (1973) Phanerozoic World Maps. pp.1-42 in (ed) N.F. Hughes, Organisms and Continents Through Time. London, Palaeontol Assoc.

1146. Smith, H.H. (ed) (1972) Evolution of Genetic Systems. NY, Gordon & Breach.

1147. Smith, H.W. (1973) Geophysical Effects on Pupal Weights of an Insect During Retrospective World Interval July 26 - August 1972. Report presented in Kyoto to IAGA Meeting. Agric Res Paper No 958, Dept Entomology, Univ Idaho.

1148. Smith, K.A., Van Dijk, T.A. (1987) Utilization of Phosphorus and Potassium from Animal Manures on Grassland and Forage Crops. pp.87-102 in (eds) H.G. Van Der Meer, et al., Animal Manure on Grasslands and Fodder Crops. Dordrecht, Netherlands, Martinus Nijoff.

1149. Smith, L.H., Leavitt, J.E. (1979) Nutrition, Pressures and Disciplines: Their Impact on Learning. Curriculum Bull 33(342):4-9.

1150. Smith, R.J. (1977) Tree Crops. Old Greenwich, Conn, Devin-Adair.

1151. Smith, S.D. (1988) Limb Regeneration pp.529-556 in (ed) A.A. Marino, Modern Bioelectricity. NY, Marcel Dekker.

1152. Smith, S.D., Feola, J.M. (1982) Pulsed Magnetic Field Modulation of LSA Tumors in Mice. J Bioelect 1:207-229.

1153. Smith, S.D., Pilla, A.A. (1981) Modulation of Newt Limb Regeneration by Electromagnetically Induced by Low Level Pulsating Current. pp.137-152 in (ed) R.O. Becker, Mechanisms of Growth Control. Springfield, Ill, C Thomas.

1154. Smithells, R.W., et al. (1980) Possible Prevention of Neural-Tube Defects by Periconceptional Vitamin Supplementation. Lancet I (8164):339-340.

1155. Smutkupt, S. (1973) Effects of Gamma Irradiation of Soybeans for Mutation Breeding. pp.255-261. in FAO/IAEA Division of Atomic Energy in Food & Agriculture & Gesellschaft fur Strahlen- and Umweltforschung, 1972. Vienna, Intl Atomic Energy Agency.

1156. Sobulo, R.A., Jaiyeola, K.E. (1977) Influence of Soil Organic Matter on Plant Nutrition in Western Nigeria. pp.105-116 in Soil Organic Matter Studies. Vol. I. Vienna, Intl Atomic Energy Agency.

1157. Sokoloff, L., et al. (1977) Cerebral Nutrition and Energy Metabolism. pp.87-140 in (eds) R.J. Wurtman, J.J. Wurtman, Nutrition and the Brain. Vol.1. NY, Raven.

1158. Solsky, R.L., Rechnitz, G.A. (1979) Antibody-Selective Membrane Electrodes. Science 204:1308-1309.

1159. Song, P-S. (1984) Primary Molecular Events in Aneural Cell Photoreceptors. pp.47-60 in (eds) G. Colombetti, et al., Sensory Perception and Transduction in Aneural Organisms. NY, Plenum.

1160. Sørensen, L.H. (1977) Factors Affecting the Biostability of Metabolic Materials in Soil. pp.3-14 in Soil Organic Matter Studies. Vol.II. Vienna, Intl Atomic Energy Agency.

1161. Sourkes, T.L. (1978) Nutrients and the Cofactors Required for Monoamine Synthesis in Nervous Tissue. pp.265-300 in (eds) R.J. Wurtman, J.J. Wurtman, Nutrition and the Brain. Vol.3. NY, Raven.

1162. Southern, W.E. (1988) Earth's Magnetic Field as a Navigational Cue. pp.35-74 in (ed) A.A. Marino, Modern Bioelectricity. NY, Marcel Dekker.

1163. Soyka, F., Allen, E. (1979) Ion Effect. NY, Bantam.

1164. Spadaro, J.A. (1982) Bioelectric Stimulation of Bone Formation: Methods, Models and Mechanisms. J Bioelect 1:99-128.

1165. Spanner, D.C. (1979) Electro-osmotic Theory of Phloem Transport: A Final Restatement. Plant Cell Environ 2:107-21.

1166. Sperry, R.W. (1964) Problems Outstanding in the Evolution of Brain Function. J Arthur Lecture. NY, Amer Mus Nat Hist.

1167. Sperry, R.W. (1972) Science and the Problem of Values. Perspect Biol Med 16:115-130.

1168. Sperry, R.W. (1980) Mind-Brain Interaction, Mentalism, Yes; Dualism, No. Neuroscience 5:195-206.

1169. Sperry, R.W. (1981) Changing Priorities. Annu Rev Neurosci 4:1-15.

1170. Speth, R.E., Yamamura, H.I. (1978) Sodium-Dependent High-Affinity Neuronal Choline Uptake. pp.129-139 in (eds) A. Barbeau, et al., Nutrition and the Brain. Vol.5. NY, Raven.

1171. Spittle, C.R. (1971) Artherosclerosis and Vitamin C. Lancet 1(7737):1280-1281.

1172. Spittle, C.R. (1972) Artherosclerosis and Vitamin C. Lancet 1(7754):798.

1173. Sporn, M.B. (1977) Vitamin A and its Analogs (Retinoids) in Cancer Prevention. pp.119-130 in (ed) M. Winick, Nutrition and Cancer. NY, J Wiley.

1174. Srikantia, S.G. (1986) Wholesomeness of Irradiated Wheat. MFI Bull (India) 7(3):9.

1175. Stacey, F.D. (1969) Physics of the Earth. NY, J Wiley.

1176. Stahr, A. (1986) Alden's Walden: Food for a Forest Dweller. Amer Forests 92:30-41.

1177. Stankowski, Jr., L.F., Hsie, A.W. (1986) Quantitative and Molecular Analyses of Radiation-Induced Mutation in AS52 Cells. Radiat Res 105:37-48.

1178. Stanley, N.F., Alpers, M.P. (1975) Man-Made Lakes and Human Health. NY, Academic.

1179. Stanley, S.M. (1979) Macroevolution: Pattern and Process. SF, WH Freeman.

1180. Stanley, S.M. (1982) New Evolutionary Timetable. NY, Basic.

1181. Stehli, F.G. (1968) Taxonomic Diversity Gradients in Pole Location: The Recent Model. pp.163-227 in (ed) E.T. Drake, Evolution and Environment. New Haven, Conn, Yale Univ Press.

1182. Stehr-Green, P.A., et al. (1986) Evaluation of Persons Exposed to Dairy Products Contaiminated with Heptachlor. JAMA 256:3350-1.

1183. Stein, G., et al. (1976) Relationship Between Mood Disturbances and Free and Total Plasma Tryptophan in Postpartum Women. Brit Med J 2(6033):457.

1184. Stephenson, M., Stickland, L.H. (1931) Hydrogenase. II. The Reduction of Sulphate to Sulphide by Molecular Hydrogen. Biochem J 25:215-220.

1185. Sternberg, M.J.E., et al. (1987) Prediction of Electrostatic Effects of Engineering of Protein Charges. Nature 330:86-88.

1186. Stillman, R.A. (1982) In Utero Exposure to Diethylstilberol. Adverse Effects on the Reproductive Performance in Male and Female Offspring. Amer J Obstet Gynecol 142:905.

1187. Stillman, R.J., Miller, L.C. (1984) Diethylstilberol Exposure in Utero and Endometriosis in Infertile Females. Fert Steril 41:369.

1188. Stone, J.F., et al. (1988) Relationships Between Clothing and Pesticide Poisoning—Symptoms Among Iowa Farmers. J Environ Health 50:210-15.

1189. Strahilevitz, M., et al. (1979) Air Pollutants and the Admission Rate of Psychiatric Patients. Amer J Psychi 136:205-206.

1190. Strauss, S.Y. (1987) Direct and Indirect Effects of Host-Plant Fertilization on an Insect Community. Ecology 68:1670-8.

1191. Sulman, F.G. (1971) Serotonin-Migraine in Climatic Heat Stress, its Prophylaxis and Treatment. Proc Intl Headache Symp. Elsinore, Denmark.

1192. Sulman, F.G. (1980) Effect of Air Ionization. Electric Fields, Atmospherics and Other Electric Phenomena on Man and Animal. Springfield, Ill, C Thomas.

1193. Sulman, F.G., et al. (1974) Ionometry of Hot, Dry Desert Winds (Sharav) and Application of Ionizing Treatment to Weather-Sensitive Patients. Intl J Biomet 18:393.

1194. Sultatos, L.G., Minor, L.D. (1987) Metabolic Activation of the Pesticide Azinphos-Methyl by Perfused Mouse Livers. Toxicol Appl Pharmacol 90(2):227-34.

1195. Sunderland, L.D. (1984) Darwin's Enigma: Fossils and Other Problems. Santee, Ca, Master Book.

1196. Sutta, T.V., et al. (1987) Concentration of Various Mineral Elements in the Placental Tissue of Women Engaged in the Production of Mineral Fertilizers. Gigiena Truda i Professionalnye Zabolevaniia (7):27-9.

1197. Sutton, C. (ed) (1985) Building the Universe. NY, Basil Blackwell.

1198. Sverlova, L.I. (1974) Effect of Magnetic Reversals on Evolution of Organic World. pp.340-342 in Cosmos and the Evolution of Organisms. AN SSSR, Moscow, Paleontolo Institute.

1199. Swanson, J.M., Kinsbourne, M. (1980) Food Dyes Impair Performance of Hyperactive Children on a Laboratory Learning Test. Science 207:1485-1487.

1200. Swift, R.S., Posner, A.M. (1977) Humification of Plant Materials. pp.171-182 in Soil Organic Matter Studies. Vol.I. Vienna, Intl Atomic Energy Agency.

1201. Symigielski, S. (1975) Acute Staphylcoccal Infections in Rabbits Irradiated With 3GHs Microwaves. Ann N Y Acad Sci 247:305.

1202. Symposium on the Effects of Neutron Irradiation Upon Cell Function (1974) Biological Effects of Neutron Irradiation; Proc Symp, Neuherberg, Munich, 22-26 Oct 1973. Vienna, Intl Atomic Energy Agency.

1203. Symposium on the Late Biological Effects of Ionizing Radiation (1978) Late Biological Effects of Ionizing Radiation: Proc Symp, Vienna, 13-17 March, 1978. Vienna, Intl Atomic Energy Agency.

1204. Sytin, A.G., Maksimov, A.A. (1972) Correlation of the Mobility of Rodents and Moles with Indices of Solar and Geomagnetic Activity. p.84 in Sun, Electricity and Life. Moscow, Izd MGU.

1205. Szmigielski, S., et al. (1988) Immunologic and Cancer-Related Aspects of Exposure to Low-Level Microwave and Radiofrequency Fields. pp.861-926 in (ed) A.A. Marino, Modern Bioelectricity. NY, Marcel Dekker.

1206. Tahir, M., et al. (1979) Seed Protein Improvement in Wheat and Grain Yield in Triticale Using Radiation-Induced Mutations. p.444 in FAO/IAEA/Gesellschaft fur Strahlen- und Umwltforschung. Seed Protein Improvement in Cereals and Grain Legumes. Vol.2, 1978. Vienna, Intl Atomic Energy Agency.

1207. Takayama, S., et al. (1989) Combination Effects of Forty Carcinogens Administered at Low Doses to Male Rats. Japanese J Cancer Res 80:732-736.

1208. Tamminga, C.A., Nutt, J.G. (1979) Cholinergic Influences on Affect. pp.409-416 in (eds) A. Barbeau, et al., Nutrition and the Brain. Vol.5. NY, Raven.

1209. Tanada, T. (1983) Photogeneration of a Bioelectric Field by Red and Far Red Irradiation in Soybean Hypocotyls. Plant Cell Environ 6:69-72.

1210. Tanner, J.A., Romero-Sierra, C. (1974) Beneficial and Harmful Accelerated Growth Induced by the Action of Nonionizing Radiation. Ann N Y Acad Sci 238:171-175.

1211. Tanner, J.A., Romero-Sierra, C. (1982) Effects of Chronic Exposure to Very Low Intensity Microwave Radiation on Domestic Fowl. J Bioelect 1:195-205.

1212. Tarakanova, G.A. (1971) Effect of Steady Magnetic Fields on Growth and Energy Metabolism of Plants. [Dissert Abstr] Moscow, Inst Plant Physiol, Acad Sci USSR.

1213. Tate, R.L. (1987) Soil Organic Matter: Biological and Ecological Effects. NY, J Wiley.

1214. Tatnall, R.E. (1981) Fundamentals of Bacteria Induced Corrosion. Materials Perform 20(9):32-38.

1215. Tatnall, R.E. (1981) Case Histories: Bacteria Induced Corrosion. Materials Perform 20(8):41-48.

1216. Taub, I.A., et al. (1976) Irradiated Food: Validity of Extrapolating Wholesomeness Data. J Food Sci 41:942.

1217. Taylor, L.J., Diespecker, D.D. (1972) Moon Phases and Suicide Attempts in Australia. Australian Psych Repts 31:110-20.

1218. Teoule, R., Cadet, J. (1978) Radiation-Induced Degradation of the Base Component in DNA and Related Substances—Final Products. pp.171-203 in (eds) J. Hutterman, et al., Effects of Ionizing Radiation on DNA. NY, Springer-Verlag.

1219. Tesch, F.W. (1970) [Homing Ability of Eels Following Impairment of the Sense of Smell After Adaptation or After Transplanting in an Estuary] Heimfindevermogen von Aalen nach Beeintractigung des Geruchssinnes nach Adaptation oder nach Verpflanzung in ein Astuar. Marine Biol 6(2):148.

1220. Thompson, D., Thompson, S. (1984) Farming Without Chemicals. EPA J 10:33-4.

1221. Thompson, J.R., et al. (1984) Effect of Dust on Photosynthesis and Its Significance for Roadside Plants. Environ Pollut 34A(2):171-190.

1222. Thomsen, D.E. (1984) Acid Rain Annual Report. Sci News 125:392.

1223. Thomsen, D.E. (1984) Iridium May Illuminate Mass Extinctions. Sci News 126:151.

1224. Thomsen, K.S. (1988) Marginalia. Anatomy of the Extinction Debate. Amer Sci 76:59-61.

1225. Thorington, L. (1985) Spectral Irradiance and Temporal Aspects of Natural and Artificial Light. Ann N Y Acad Sci 453:28-54.

1226. Thornton, I., Abrahams, P. (1984) Historical Records of Metal Pollution in the Environment. pp.7-26 in (ed) J.O. Nriagu, Changing Metal Cycles and Human Health. NY, Springer-Verlag.

1227. Tigerstedt, P.M.A., et al. (eds) (1985) Crop Physiology of Forest Trees. Proc Intl Conf Managing Forest Trees As Cultivated Plants, Finland, 23-28 July 1984. Finland, Univ Helsinki, Dept Plant Breeding.

1228. Tolgskaya, M., Gordon, F. (1973) Pathological Effects of Radio Waves. NY, Consultants Bureau.

1229. Tordoir, W.F., van Heemstra-Lequin, E.A.H. (eds) (1980) Field Worker Exposure During Pesticide Application. NY, Elsevier.

1230. Torrey, E.F., Peterson, M.R. (1976) Viral Hypothesis of Schizophrenia. Schizophrenia Bull 2:136-146.

1231. Townsend, M.G., et al. (1984) Assessment of Secondary Poisoning Hazard of Warfarin to Least Weasels. J Wildlife Mgnt 48:628-32.

1232. Toy, T.T. (1982) Accelerated Erosion: Process, Problems and Progress. Geology 10:524-9.

1233. Train, R.E. (1977) Environmental Cancer. Science 195:443.

1234. Tralau, H. (1973) Some Quaternary Plants. pp.499-504 in (ed) A. Hallam, Atlas of Palaeobiogeography. NY, Elsevier.

1235. Traut, H. (1978) Molecular Aspects of Mutagenesis Due to Ionizing Radiation. pp.335-347 in (eds) J. Hütterman, et al., Effects of Ionizing Radiation on DNA. NY, Springer-Verlag.

1236. Tregan, L. (1979) Survey of the Effect of Metals on the Immune Response. Biol Trace Element Res 1:141-8.

1237. Tribus, M. (1961) Thermostatics and Thermodynamics. Princeton, NJ, Van Nostrand-Reinhold.

1238. Tromp, S.W. (ed) (1963) Medical Biometerology. NY, Elsevier.

1239. Tromp, S.W. (1965) Influence of Weather and Climate on Blood Pressure, Blood Composition and Physico-Chemical State of the Blood of Blood Donors at Leiden, The Netherlands (Period 1953-1963). Monogr Ser, Biometeor Res Ctr, Leiden (7).

1240. Tromp, S.W. (1970) Seasonal and Yearly Fluctuations in Meteorologically Induced Electromagnetic Wave Patterns in the Atmosphere (Period 1956-1968) and Their Possible Biological Significance. A Review. Interdiscipl Cycle Res 1:193-199.

1241. Tromp, S.W. (1977) Influence of Weather and Climate on Cancer: General Considerations. pp.32-38 in (ed) S.W. Tromp, Progress in Biometerology. Vol.1, Part II, Period 1963-1975. Amsterdam, Swets & Zeitlinger.

1242. Tromp, S.W. (1977) Influence of Weather and Climate on Infectious Diseases in Man. pp.60-70 in (ed) S.W. Tromp, Progress in Biometerology. Vol.I, Part II, Period 1963-1975. Amsterdam, Swets & Zeitlinger.

1243. Tromp, S.W. (1977) Month of Birth and Proneness to Disease. pp.164-166 in (ed) S.W. Tromp, Progress in Biometeorology. Vol.1, No II, Period 1963-1975. Pathological Biometeorology. Amsterdam, Swets & Zeitlinger.

1244. Tromp, S.W. (1977) Regional Differences in the Effects of Weather and Climate on Asthma, Bronchitis and Allergic Diseases. pp.11-17 in (ed) S.W. Tromp, Progress in Biometeorology. Vol.1, No II, Period 1963-1975. Pathological Biometeorology. Amsterdam, Swets & Zeitlinger.

1245. Tromp, S.W. (1977) Various Aspects of Sociological and Psychological Biometerology. pp.296-306 in (ed) S.W. Tromp, Progress in Biometeorology. Amsterdam, Swets & Zeitlinger.

1246. Tromp, S.W. (1980) Biometerology: The Impact of the Weather and Climate on Humans and Their Environment (Animals and Plants). Philadelphia, Heyden.

1247. Tromp, S.W. (1981) New Evidence for Environmental and Extraterrestrial Effects on Organic Colloids on Earth. p.77 in Conf Agric & Forest Meteorol (15th) & Biometeorol (5th). Boston, Mass, Amer Meteorol Soc.

1248. Tromp, S.W., Faust, V. (1977) Influence of Weather and Climate on Mental Processes in General and Mental Diseases in Particular. pp.74-82 in (ed) S.W. Tromp, Progress in Biometerology. Vol.1, Part II, Period 1963-1975. Amsterdam, Swets & Zeitlinger.

1249. Trowell, H. C. (ed) (1975) Refined Carbohydrate Foods and Disease. NY, Academic.

1250. Trowell, H.C. (1978) Western Diseases, Western Diets and Fibre. East African Med J 55:283-289.

1251. Truppi, M.S. (1981) Estimation of Excess Human Mortality During Heat Waves and Cold Waves in the United States, 1962-1966, 1973-1977. pp.74-77 in Conf Agric & Forest Meteorol (15th)/Biometeorol (5th). Boston, Mass, Amer Meteorol Soc.

1252. Truswell, A.S. (1987) Food Irradiation. Brit Med J 294:1437-1438.

1253. Tsaplev, Yu.B., Zatsepina, G.N. (1980) Electrical Nature of the Spread of a Variable Potential in *Tradescantia*. Biophysics 25:723-728.

1254. Tschudy, R.H., Tschudy, B.D. (1986) Extinction and Survival of Plant Life Following the Cretaceous/Tertiary Boundary Event, Western Interior, North America. Geology 14:667-70

1255. Tudge, C. (1986) Two Cheers for Organic Farming. New Sci 112:64-5.

1256. Tupi, K., et al. (1987) Effects of Respiratory Morbidity on Occupational Activity Among Farmers. Eur J Respir Dis 152(suppl):206-11.

1257. Uffen, R.J. (1963) Influence of the Earth's Core on the Origin and Evolution of Life. Nature 198:143.

1258. Uhland, R.E. (1949) Crop Yields Lowered by Erosion. USDA-SCS-TEC Paper-75, USDA Soil Conserv Serv. Wash, DC, USGPO.

1259. United Nations Food and Agriculture Organization (1979) Trace Elements in Soils and Agriculture. Soils Bull No 17. UN, FAO.

1260. United Nations, Scientific Committee on the Effects of Atomic Radiation (1972) Ionizing Radiation: Levels and Effects; Report of the United Nations Scientific Committee on the Effects of Atomic Radiation to the General Assembly. NY, UN.

1261. Unwin, D.M. (1988) Extinction and Survival in Birds. pp. 215-318 in (ed) G.P. Larwood, Extinction and Survival in the Fossil Record. Systematics Spec. Vol.34. Oxford, Clarendon.

1262. Urey, H.C. (1973) Cometary Collisions and Geological Periods. Nature 242:32-33.

1263. Uvarov, B.P. (1931) Insects and Climate. Trans Entomol Soc Lond 79:1-247.

1264. van Andel, T.H. (1981) Consider the Incompleteness of the Geological Record. Nature 294:397-398.

1265. Van De Vorst, A., Westhof, E. (1978) Structure and Electronic Properties of DNA. pp.3-20 in (eds) J. Hutterman, Effects of Ionizing Radiation on DNA. NY, Springer-Verlag.

1266. Van Der Meer, H.G., et al. (1987) Animal Manure on Grassland and Fodder Crops. Dordrecht, Netherlands, Martinus Nihoff.

1267. Van Dyke, J.H., Halpern, M.H. (1965) Observations on Selected Life Processes in Null Magnetic Field. Anat Rec 151(3):480.

1268. Van Sambeck, J.W., et al. (1976) Mediation of Rapid Electrical, Metabolic, Transpirational and Photosynthetic Changes by Factors from Wounds. II. Mediation of the Variation Potential by Ricca's Factor. Canad J Bot 54:2651.

1269. Van Strum, C. (1979) Herbicides: A Faustian Bargain. Co Evolutionary Q 21(Spr):22-25.

1270. Van Valen, L.M. (1983) How Pervasive is Coevolution? pp.1-19 in (ed) W.H. Nitecki, Coevolution. Univ Chicago.

1271. Varela, G., Navarro, M.P. (1987) Influence of Pesticides on the Utilization of Food. Bibliotheca Nutritio Dieta (41):40-54.

1272. Vasilik, P.V. (1972) Western Drift of Geomagnetic Field Variations and Some Features of the Manifestation of Acceleration in the Past. Kiev, Abstr Papers 21st Ukranian Republic Sci Tech Conf.

1273. Vasilik, P.V. (1973) Changes in Physical Development of Contemporary Man and Their Relationship to Geographic Characteristics of Geomagnetic Field. pp.76-78 in Effect of Natural and Weak Artificial Magnetic Fields on Biological Objects, Belgorod.

1274. Vaughan, C. (1988) Disarming Farming's Chemical Warriors. Sci News 134:120-121.

1275. Vawter, L., Brown, W.M. (1986) Nuclear and Mitochondrial DNA Comparisons Reveal Extreme Rate Variation in the Molecular Clock. Science 234:194-196.

1276. Verma, S.R., et al. (1983) Pesticide-Induced Dysfunction in Carbohydrate Metabolism in Three Freshwater Fishes. Environ Res 32:127-33.

1277. Vijayalaxmi, Evans, H.J. (1984) Measurement of Spontaneous and X-Irradiation-Induced 6-Thioguanine-Resistant Human Blood Lymphocytes Using a T-cell Cloning Technique. Mutat Res 125:87-94.

1278. Villee, C.A. (1974) Endocrine Regulation of Pregnancy and Fetal Development. pp.106-115 in K.S. Moghissi, Birth Defects and Fetal Development. Springfield, ILL, C Thomas.

1279. Vinogradova, L.I., et al. (1975) Polarity of Interplanetary Magnetic Field Sectors and Frequency of Occurrence of Vegetative Vascular Paroysms Due to Disturbance of Activity of Hypothalamic Brain Structures. Reprint No 17(132) Moscow, Instit Terrestrial Magnetism, the Ionosphere, & Radiowave Propagation Acad Sci USSR.

1280. Vitousek, P.M., et al. (1979) Nitrate Losses from Disturbed Ecosystems. Science 204:469-74

1281. Voitinskii, E. Ya, et al. (1974) Effect of a Weak Magnetic Field on Electrical Activity of Brain. pp.21-22 in Magnetic Field in Medicine. Frunze, Kirgizskii Gos Med Inst.

1282. von Deschwanden Jr., P.L. (1977) Influence of Weather and Climate on Foehn and Related Diseases. pp.142-154 in (ed) S.W. Tromp, Progress in Biometerology. Vol.1, Part II, Period 1963-1975. Amsterdam, Swets & Zeitlinger.

1283. Voss, H.D., et al. (1984) Lightning-Induced Electron Precipitation. Nature 312:740-742.

1284. Vouk, V.B. (1976) WHO-NIEHS Symposium on Potential Health Hazards from Technological Developments in Plastic and Synthetic Rubber Industries. March 1-3. N.C., Research Training Park.

1285. Wadman, W.P., et al. (1987) Value of Animal Manures: Changes in Perception. pp.1-16 in (eds) H.G. Van Der Meer, et al., Animal Manure on Grasslands and Fodder Crops. Dordrecht, Netherlands, Martinus Nihoff.

1286. Wagner, G.C., et al. (1985) Fractal Models of Protein Structure, Dynamics and Magnetic Relaxation. J Amer Chem Soc 107:5589-5594.

1287. Wagner, O.E. (1988) Wave Behavior in Plant Tissues. Northwest Sci 62(5):263-275.

1288. Wagner, O.E. (1989) W-Waves and Plant Communication. Northwest Sci 63(3):119-131.

1289. Wagner, O.E. (1990) W-Waves and Plant Spacings. Northwest Sci 64(1):28-38.

1290. Wainwright, M. (1980) Effect of Exposure to Atmospheric Pollution on Microbial Activity in Soil. Pl Soil 55:199-204.

1291. Wakerley, D.S. (1979) Microbial Corrosion in UK Industry; A Preliminary Survey of the Problem. Chem Industry 19:657-8.

1292. Walcott, C., et al. (1988) Homing of Magnetized and Demagnetized Pigeons. J Exp Biol 134:27-41.

1293. Waldern, C., et al. (1979) Measurement of Mutagenesis in Mammalian Cells. Proc Natl Acad Sci (USA) 76:1358.

1294. Waldhauser, F., Dietzel, M. (1985) Daily and Annual Rhythms in Human Melatonin Secretion: Role in Puberty Control. Ann NY Acad Sci 453:205-214.

1295. Walker, C.F., et al. (1982) Effects of High-Intensity 60 Hz Fields on Bone Growth. J Bioelect 1:339-349.

1296. Walters, A.H. (1985) Nitrates in Food: Another Government Cover-Up? Ecologist 15:189-194.

1297. Wang, Y.-M., et al. (1980) Effect of Vitamin E Against Adriamycin-Induced Toxicity in Rabbits. Cancer Res 49:1022-1027.

1298. Wark, K. (1977) Thermodynamics. NY, McGraw-Hill.

1299. Warnke, U. (1984) Avian Flight Formation with the Aid of Electromagnetic Forces: A New Theory for the Formation Alignment of Migrating Birds. J Bioelect 3:493-508.

1300. Warters, R.L., Childers, T.J. (1982) Radiation-Induced Tyhmine Base Damage in Replicating Chromatin. Radiat Res 90:564-574.

1301. Watkin, D.M. (1976) Biochemical Impact of Nutrition on the Aging Process. in (eds) M. Rockstein, M.L. Sussman, Nutrition, Longevity and Aging. NY, Academic.

1302. Watkin, D.M. (1976) Role of Nutrition in Human Aging. pp. 29-46 in (eds) M. Rockstein, M.L. Sussman, Nutrition, Longevity and Aging. NY, Academic.

1303. Watson, A.K. (1982) Biological Control of Weeds in Pastures in Canada. pp.178-181 in (ed) S.Hill, Basic Technics in Ecological Farming. Basel, Switz, Birhäuser Verlag.

1304. Watson, J. (1979) Electrical Stimulation of Bone Healing. Proc IEEE 67:1339-1352.

1305. Watt, J.B.A. (1885) Electrical Phenomenon. Nature 32:316.

1306. Webb, H., et al. (1961) Organismic Responses to Differences in Weak Horizontal Electrostatic Fields. Biol Bull 121:413.

1307. Webb, T., Lang, T. (1987) Food Irradiation: The Facts. Wellingsborough, Northants, Thorsons.

1308. Weber, B.H., et al. (eds)(1988) Entropy, Information and Evolution. Cambridge, Mass, MIT Press.

1309. Webster, J.C. (1984) Noise and Communication. pp.185-220 in (eds) D.M. Jones, A. Chapman, Noise and Society. NY, Wiley.

1310. Weidensaul, T.C., McClenahen, J.R. (1986) Soil-Air Pollutant Interactions. pp.398-414 in (eds) A.H. Legge, S.V. Krupa, Air Pollutants and Their Effects on the Terrestrial Ecosystems. NY, J Wiley.

1311. Weisburd, S. (1984) DNA Helix Found to Oscillate in Resonance with Microwaves. Sci News 125:248.

1312. Weisburd, S. (1987) Electric Life of Plants Gives Fungal Spores a Charge. Sci News 132(4):53.

1313. Weisburger, J. (1979) Nutritional Factors That May Be Involved in Cancer of the Bladder. Nutr Cancer 1(2):74-81.

1314. Weiss, B., et al. (1980) Behavioral Responses to Artificial Food Colors. Science 207:1487-9

1315. Weiss, H.V., et al. (1971) Mercury in a Greenland Ice Sheet: Evidence of Recent Input by Man. Science 174:692-694.

1316. Weiss, R. (1987) Gamma-Ray Gourmet: Scientists Cook Up Tests for Irradiated Food. Sci News 132:398-399.

1317. Wellington, W.G. (1957) Synoptic Approach to Studies of Insects and Climate. Ann Rev Entomol 2:143-162.

1318. Wenz, C. (1984) Pesticides: New Chemicals Under Fire. Nature 309:741.

1319. Wesley, A. (1973) Jurassic Plants. pp.329-338 in (ed) A. Hallam, Atlas of Palaeobiogeography. NY, Elsevier.

1320. West, S. (1980) Acid from Heaven. Sci News 117:76-78.

1321. Wever, R. (1973) Human Circadian Rhythms Under the Influence of Weak Electric Fields and the Different Aspects of These Studies. Intl J Biometerol 17:227-232.

1322. Wever, R. (1974) ELF-Effects on Human Circadian Rhythms. pp.101-144 in (ed) M.A. Persinger, ELF and VLF Electromagnetic Field Effects. NY, Plenum.

1323. Wheaton, F.W. (1970) Influence of Electrical Energy on Plants: A Review. Report No 4262. Maryland Agricultural Experimental Station, Univ Maryland.

1324. White, F.M., et al. (1988) Chemicals, Birth Defects and Stillbirths in New Brunswick: Associations with Agricultural Activity. Canadian Med Assoc J 138(2):117-124.

1325. White, L.F., et al. (1987) Health Effects from Indoor Air Pollution: Case Studies. J Community Health 12:147-155.

1326. Whitham, T.G. (1989) Plant Hybrid Zones as Sinks for Pests. Science 244:1490-1496.

1327. Whitmore, T.C. (ed)(1981) Wallace's Line and Plate Tectonics. Oxford, Clarendon Press.

1328. Wicken, J.S. (1980) Thermodynamics, Evolution and Emergence: Ingredients for a New Synthesis. pp.139-172 in (eds) B.H. Weber, et al., Entropy, Information and Evolution. Cambridge, Mass, MIT Press.

1329. Wiklander, L. (1973/74) Acidification of Soil by Acid Percipitation. Grundforbattring 26:155-164.

1330. Wilde, P., et al. (1986) Iridium Abundances Across the Ordovician-Silurian Stratotype. Science 233:339-341.

1331. Wiley, E.O. (1988) Entropy and Evolution. pp.173-188 in (eds) B.H. Weber, et al., Entropy, Information and Evolution. Cambridge, Mass, MIT Press.

1332. Wiley, Jr.,J.P. (1981) Phenomena, Comments and Notes. Smithsonian 11(11):30-38.

1333. Wilford, J.N. (1984) Study Hints Extinctions Strike in Set Intervals. Cycles 35:53-54.

1334. Wilhelmi, Th. (1963) [Can the Forest Tree's Growth Reaction to Thunder Storms Be Understood?] Losen Gewitter Wachstumsreaktionen Bei den Waldbaumen Aus? Wetter und Leben 15:93-97.

1335. Wilkinson, S.R., et al. (1987) Relation of Soil and Plant Magnesium to Nutrition of Animals and Man. Magnesium 6(2):74-90.

1336. Williams, G.C. (1974) Adaptation and Natural Selection. NJ, Princeton Univ.

1337. Williams, R.J. (1998) Biochemical Individuality. New Canaan, CT, Keats Pubs.

1338. Williams, R.J., Rimland, B. (1977) Individuality. in Encyclopedia of Psychiatry, Neurology and Psychoanalysis. NY, Van Nostrand-Reinhold.

1339. Winick, M. (ed) (1977) Nutrition and Cancer. NY, J Wiley.

1340. Winston, F. (1973) Oral Contraceptives, Pyridoxine and Depression. Amer J Psychi 130:1217-1221.

1341. Wiszkowska, H.K., Trena, M.A. (1988) Effect of Malathion on RNA Polymarase Activity of Cell Nuclei and Transcription Products in Lymphocyte Culture. Environ Res 41:372-7.

1342. Wittes, R.E. (1985) Vitamin C and Cancer. New Eng J Med 312:178-179.

1343. Woese, C. (1965) On the Origin of the Genetic Code. Proc Natl Acad Sci (USA) 54:1546-52

1344. Wolff, I.A., Wasserman, A.E. (1972) Nitrates, Nitrites and Nitrosamines. Science 177:15-19.

1345. Wood, J.M., Goldberg, E.D. (1977) Impact of Metals on the Biosphere. pp.137-154 in (ed) W. Stumm, Global Chemical Cycles and Their Alterations by Man. Berlin, Dahlem Konferenzen.

1346. Woodell, G.M. (1967) Radiation and the Patterns of Nature. Science 156:461-470.

1347. World Health Organization (1980) Environmental Health Criteria 12—Noise. Geneva, WHO.

1348. Wunderlich, R.C. (1982) Sugar and Your Health. St Petersburg, Fl, Good Health Pubs.

1349. Wurtman, R.J., et al. (1977-89) Nutrition and the Brain. Volumes 1-8. NY, Raven.

1350. Wynder, E.L., Reddy, B.S. (1977) Diet and Cancer of the Colon. pp.59-69 in (ed) M. Mimick, Nutrition and Cancer, NY, J Wiley.

1351. Xu, D.-Y., et al. (1985) Abundance Variation of Iridium and Trace Elements at the Permian/Triassic Boundary at Shangsi in China. Nature 314:154-156.

1352. Yaryura-Tobias, J. (1975) Violent Behavior, Brain Dysrhythmia and Glucose Dysfunction. J Orthomol Psychi 4(3):182-188.

1353. Young, V.R. (1976) Protein Metabolism and Needs in Elderly People. pp.65-71 in (eds) M. Rockstein, M.L. Sussman, Nutrition, Longevity and Aging. NY, Academic.

1354. Zahm, S.H., et al. (1987) National Bladder Cancer Society: Employment in the Chemical Industry. J Natl Cancer Instit 79:217-22.

1355. Zapata, G.N., et al. (1987) Clastogenic Changes in the Chromosomes of a Population of Individuals Occupationally Exposed to Different Pesticides. Salud Publica de Mexico 29:506-11.

1356. Zelles, L., et al. (1985) Side Effects of Some Pesticides on Non-Target Soil Microorganisms. J Environ Sci Health 20B:457-488.

1357. Zimmerman, R.L. (1972) Relative Effects of Alpha and Beta Radiation. Radiation Effects 14:81-92.

1358. Zimmerman, R.L. (1982) Review Article: Piezoelectricity and Biological Materials. J Bioelect 1:265-287.

1359. Zinke, P.J. (1977) Man's Activities and Their Effect Upon Limiting Nutrients for Primary Productivity in Marine and Terrestrial Ecosystems. pp.89-98 in (ed) W. Stumm, Global Chemical Cycles and Their Alterations by Man. Berlin, Dahlem Konferenzen.

1360. Zon, J.R., TiTien, H. (1988) Electronic Properties of Natural and Modeled Bilayer Membranes. pp.181-242 in (ed) A.A. Marino, Modern Bioelectricity. NY, Marcel Dekker.

1361. Zuckerkandl, E. (1975) Appearance of New Structures in Proteins During Evolution. Sci Mon 64:481-495.

Tome Three

1.   Abe, K., Kanamori, H.J. (1979) Temporal Variation of the Activity of Intermediate and Deep Focus Earthquakes. J Geophys Res 84:3589-3595.

2.   Ager, D.V. (1976) Nature of the Fossil Record. Proc Geol Assoc 87:131-160.

3.   Ahlfeld, F. (1965) Bolivia. Geol Soc Amer Mem 65:171-186.

4.   Alexopoulos, J.S., et al. (1988) Microscopic Lamellar Deformation Features in Quartz: Discriminative Characteristics of Shock-Generated Varieties. Geology 16:796-799.

5.   Alvarez, L.W., et al. (1980) Extraterrestrial Cause for the Cretaceous-Tertiary Extinction. Science 208:1095-1108.

6.   Alvarez, W. (1986) Toward a Theory of Impact Crises. Eos 67:649-658.

7. Alvarez, W., et al. (1982) Current Status of the Impact Theory for the Terminal Cretaceous Extinction. Geol Soc Amer Spec Paper (190):305-316.

8. Alvarez, W., et al. (1982) Iridium Anomaly Approximately Synchronous with Terminal Eocene Extinction. Science 216:886-888.

9. Alvarez, W., et al. (1984) Impact Theory of Mass Extinctions and the Invertebrate Fossil Record. Science 223:1135-1141.

10. Alvarez, W., et al. (1984) End of the Cretaceous: Sharp Boundary or Gradual Transition? Science 223:1183-1186.

11. Alvarez, W., Muller, R.A. (1984) Evidence from Crater Ages for Periodic Impacts on Earth. Nature 308:718-720.

12. Anders, E., et al. (1986) Cretaceous Extinctions and Wildfires. Science 234:263-264.

13. Anderson, D.L. (1984) The Earth as a Planet: Paradigms and Paradoxes. Science 223:347- 355.

14. Anderson, I. (1987) Chinese Unearth a Dinosaur's Graveyard. New Sci 116(1586):28-29.

15. Anderson, T.A., Schmiot, V.A. (1983) Evolution of Middle America and the Gulf of Mexico—Caribbean Sea Region During Mesozoic Time. Geol Soc Amer Bull 94:941-966.

16. Anonymous (1912) Curious Lightning in the Andes. Sci Amer 106:464.

17. Anonymous (1971) Chandler Wobble, Earthquake Correlations. Nature 229:227.

18. Anonymous (1978) Evidence Builds Up for Magnetic Changes from Crater-Forming Impacts. New Sci 79(1119):685.

19. Anonymous (1983) New Data Cast Doubt on Dinosaurs' Death (Mass Extinction Theory and Dinosaurs' Disappearance Debated). Earth Science 36(4):90.

20. Anonymous (1987) Slide Sets Show Earthquake, Tsunami, and Volcano Damage. Earth Science 40:7.

21. Anonymous (1988) Earthquake Waves Give "Outrageous" Result. New Sci 117 (1600):38.

22. Archibald, J.D. (1982) Study of Mammalia and Geology Across the Cretaceous-Tertiary Boundary in Garfield County, Montana. Univ Calif Pub Geol Sci 122:1-288.

23. Archibald, J.D. (1984) Bug Creek Anthills (BCA), Montana: Faunal Evidence for Cretaceous Age and Non-Catastrophic Extinctions. Geol Soc Amer Abstr Prog 16:432.

24. Archibald, J.D., Clemens, W.A. (1982) Late Cretaceous Extinctions. Amer Sci 70:377-385.

25. Argyle, E. (1986) Cretaceous Extinctions and Wildfires. Science 234:261.

26. Arthur, M.A., et al. (1985) Variations in the Global Carbon Cycle During the Cretaceous Related to Climate, Volcanism and Changes in Atmospheric CO_2. pp.504-529 in (eds) E.T. Sundquist, W.A. Broecker. Carbon Cycle and Atmospheric CO_2: Natural Variations Archean to Present. Wash, Amer Geophys U.

27. Asama, K. (1960) Evolution of the Leaf Forms Through the Ages Explained by the Successive Retardation and Neoteny: Tohoku Univ Sci Rpt, 2nd Ser. Geology 4:252-280.

28. Asama, K. (1962) Evolution of Shansi Flora and Origin of Simple Leaf: Tohoku Univ Sci Rpt, 2nd Ser. Geology 5:247-73.

29. Asama, K. (1975) Evolutionary Biology in Plants. IV. The Origin of the Angiosperms. Tokyo, Sanseido Co.

30. Asaro, F., et al. (1982) Geochemical Anomalies Near the Eocene/Oligocene and Permian/Triassic Boundaries. Geol Soc Amer Spec Paper (190):517-528.

31. Askin, R.A. (1988) Palynological Record Across the Cretaceous/Tertiary Transition on Seymour Island, Antarctica. Geol Soc Amer Mem 169:155-162.

32.   Audley-Charles, M.G., et al. (1981) Continental Movements in the Mesozoic and Cenozoic. pp.9-23 in (ed) T.C. Whitmore, Wallace's Line and Plate Tectonics. Oxford, Clarendon.

33.   Audley-Charles, M.G. (1981) Geological History of the Region of Wallace's Line. pp.24-35 in (ed) T.C. Whitmore, Wallace's Line and Plate Tectonics. Oxford, Clarendon.

34.   Austin, R.B., et al. (1986) Molecular Biology and Crop Improvement. NY, Cambridge Univ.

35.   Axelrod, D.I. (1959) Poleward Migration of the Early Angiosperm Flora. Science 130:203-7

36.   Axelrod, D.I. (1960) Evolution of Flowering Plants. pp.227-306 in (ed) S. Tax, The Evolution of Life. Univ Chicago.

37.   Axelrod, D.I. (1970) Mesozoic Paleogeography and Early Angiosperm History. Bot Rev 36:277-319.

38.   Axelrod, D.I. (1974) Revolutions in the Plant World. Geophytology 4:1-6.

39.   Axelrod, D.I. (1976) Roles of Plate Tectonics in Angiosperm History. pp.435-448 in (eds) J. Gray, A.J. Boucot, Historical Biogeography, Plate Tectonics and the Changing Environment. Proc 37th Ann Biol Colloq. Oregon State Univ.

40.   Axelrod, D.I. (1981) Role of Volcanism in Climate Evolution. Geol Soc Amer Spec Paper (185):1-32

41.   Axelrod, D.I., Bailey, H.P. (1968) Cretaceous Dinosaur Extinction. Evolution 22:595-611.

42.   Bachmann, E. (1965) Wer hat Himmel und Erde gemessen. Zurich, Buechergilde Gutenberg.

43.   Bailey, M.E., et al. (1987) Can Episodic Comet Showers Explain the 30-Myr Cyclicity in the Terrestrial Record? Mon Not Roy Astron Soc 227:863-885.

44.   Barnes, V.E. (1940) North American Tektites. Univ Texas Pub (3945):477-582.

45.   Barron, E.J., Washington, W.M. (1982) Cretaceous Climate: A Comparison of Atmospheric Simulations with the Geologic Record. Palaeogeogr Palaeoclim Palaeoec 40:103-134.

46.   Bauer, E., Gilmore, F.R. (1975) Effect of Atmospheric Nuclear Explosions on Total Ozone. Rev Geophys Space Phys 13:451-458.

47.   Beets, D.J., et al. (1984) Magmatic Rock Seriess and High-Pressure Metamorphism as Constraints on the Tectonic History of the Southern Caribbean. Geol Soc Amer Mem 162:95-130.

48.   Béland, P. (1977) Models for the Collapse of Terrestrial Communities of Large Vertebrates. Syllogeus 12:25-37.

49.   Béland, P., et al. (1977) Chains of Events Leading to Mass Extinctions: Two Synopses. pp.155-159 in (eds) K-Tec Group, Cretaceous-Tertiary Extinctions and Possible Terrestrial and Extraterrestrial Causes. Proc Workshop, Ottawa, Canada 16-17 Nov 1976. Ottawa, Natl Museums of Canada.

50.   Bell, R.T., et al. (1976) Uranium-Bearing Bone Occurrences. Geol Surv Canada Pap 76A:339-340.

51.   Beloussov, V.V. (1970) Against the Hypothesis of Ocean-Floor Spreading. Tectonophysics 9:489-511.

52.   Ben-Menahem, A., Israel, M. (1970) Effects of Major Seismic Events on the Rotation of the Earth. J Geophys 19:367-393.

53.   Benson, R.H. (1984) Phanerozoic "Crisis" as Viewed from the Miocene. pp.437-446 in (eds) W.A. Berggren, J.A. Van Couvering, Catastrophes and Earth History. NJ, Princeton.

54.   Bentley, C.R. (1979) No Giant Meteor Crater in Wilkes Land, Antarctica. J Geophys Res 84:5681-5682.

55. Bentor, Y.K. (1986) New Approach to the Problem of Tektite Genesis. Earth Planet Sci Lett 77:1-13.

56. Berggren, W.A., Van Couvering, J.A. (1974) Late Neogene. Palaeogeogr Palaeoclimatol Palaeoecol 16:1-216.

57. Bernal, J. (1967) Origin of Life. London, Weidenfeld & Nicolson.

58. Berry, E.W. (1911) Lower Cretaceous Floras of the World. pp.99-151 in Maryland Geological Survey. Vol.I. Lower Cretaceous. Baltimore, John Hopkins.

59. Besse, J., Courtillot, V. (1986) Master Apparent Polar Wander Path for Africa, Eurasia, India and North America: Consequences for True Polar Wander. Eos 67:925.

60. Besse, J., et al. (1986) Deccan Traps (India) and Cretaceious-Tertiary Boundary Events. pp.365-370 in (eds) O.H. Walliser, et al., Global Bio-Events, A Critical Approach. Lecture Notes in Earth Sciences, No 8. Berlin, Springer-Verlag.

61. Betzer, P.R., et al. (1988) Long-Range Transport of Giant Aerosol Particles. Nature 336:568-571.

62. Bickle, M. (1986) Plate Tectonics: Global Thermal Histories. Nature 319:13-14.

63. Bird, J. (ed) (1980) Plate Tectonics; Selected Papers. Wash, DC, Amer Geophys U.

64. Birkelund, T., et al. (1979) Cretaceous-Tertiary Boundary Events. Vol.2. Denmark, Univ Copenhagen.

65. Birkelund, T., Hakansson, E. (1982) Terminal Cretaceous Extinction in Boreal Shelf Seas— A Multicausal Event. Geol Soc Amer Spec Pap (190):373-384.

66. Black, D.I. (1967) Cosmic Ray Events and Faunal Extinctions at Geomagnetic Field Reversals. Earth Planet Sci Lett 3:225-236.

67. Blake, M.C., et al. (1974) Active Continental Margins; Contrasts Between California and New Zealand. pp.853-872 in (eds) C.A. Burk, C.L. Drake, Geology of Continental Margins. NY, Springer-Verlag.

68. Boersma, A. (1984) Campanian Through Paleocene Paleotemperature and Carbon Isotope Sequence and the Cretaceous-Tertiary Boundary in the Atlantic Ocean. pp.247-277 in (eds) W.A. Berggren, J.A. Van Couvering, Catastrophes and Earth History. NJ, Princeton Univ.

69. Boersma, A., et al. (1979) Carbon and Oxygen Isotope Records at DSDP Site 384 (North Atlantic) and Some Paleocene Paleotemperatures and Carbon Isotope Variations in the Atlantic Ocean. Initial Rpts DSDP 43:695-718.

70. Boersma, A., Shackleton, N. (1979) Some Oxygen and Carbon Isotope Variations Across the Cretaceous/Tertiary Boundary in the Atlantic Ocean. pp.50-53 in (eds) W.K. Christensen, T. Birkelund, Cretaceous-Tertiary Boundary Events. Vol.2, Symp Proc, Copen Hagen Univ.

71. Bohor, B.F., et al. (1984) Mineralogic Evidence for an Impact Event at the Cretaceous-Tertiary Boundary. Science 224:867-869.

72. Bohor, B.F., et al. (1985) Search for Shock-Metamorposed Quartz at the K-T Boundary. Lunar Planet Sci Conf Abstr 16:79-80.

73. Bohor, B.F., et al. (1987) Shocked Quartz in the Cretaceous-Tertiary Boundary Clays: Evidence for a Global Distribution. Science 236:7055-709.

74. Bohor, B.F., et al. (1987) Dinosaurs, Spherules and the "Magic" Layer: A New K-T Boundary Clay Site in Wyoming. Geology 15:896-899.

75.   Bohor, B.F., Foord, E.E. (1987) Magnesioferrite from a Nonmarine K-T Boundary Clay in Wyoming. pp.101-102 in (abs) Lunar and Planetary Science 18. Houston, Lunar & Planetary Instit.

76.   Bohor, B.F., Seitz, R. (1990) Cuban K/T Catastrophe. Nature 344:593.

77.   Bond, G.C. (1972) A Late Paleozoic Volcanic Arc in the Eastern Alaska Range, Alaska. J Geol 81:557-575.

78.   Bond, G. (1976) Evidence for Continental Subsidence in North America During the Late Cretaceous Global Submergence. Geology 4:557-560.

79.   Bonte, P. et al. (1984) Iridium Rich Layer at the Cretaceous/Tertiary Boundary in the Bidart Section (Southern France). Geophys Res Lett 11:473-476.

80.   Borella, P.E. (1984) Sedimentology, Petrology, And Cyclic Sedimentation Patterns, Walvis Ridge Transect, Leg 74, D.S.D.P. Initial Rpts DSDP 74:645-662.

81.   Borsa, F., Tognetti, V. (eds) (1988) Magnetic Properties of Matter. Teaneck, NJ, World Sci.

82.   Boucot, A.J. (1988) Periodic Extinctions within the Cenozoic. Nature 331:395-396.

83.   Bourgeois, J., et al. (1988) Tsunami Deposit at the Cretaceous-Tertiary Boundary in Texas. Science 241:567-70.

84.   Boyajian, G.F. (1986) Phanerozoic Trends in Background Extinction: Consequences of an Aging Fauna. Geology 14:955-958.

85.   Brandt, E.H. (1989) Levitation in Physics. Science 243:349-354.

86.   Bramlette, M.N. (1965) Massive Extinctions in Biota at the End of Mesozoic Time. Science 148:1696-1699.

87.   Brenneke, J.C., Anderson, T.F. (1977) Carbon Isotope Variations in Pelagic Carbonates. Eos 58:415.

88.   Brenner, G.J. (1976) Middle Cretaceous Floral Provinces and Early Migrations of Angiosperms. pp.23-47 in (ed) C.B. Beck, Origin and Early Evolution of Angiosperms. NY, Columbia Univ.

89.   Brett, R. (1967) Metallic Spherules in Impactite and Tektite Glass. Amer Minerol 52:721-733.

90.   Briggs, J.C. (1987) Biogeography and Plate Tectonics. NY, Elsevier.

91.   Broertjes, C., Van Harten, A.M. (1978) Application of Mutation Breeding Methods in the Improvement of Vegetatively Propagated Crops. NY, Elsevier.

92.   Brooks, R.B., et al. (1984) Weathered Spheroids in a Cretaceous/Tertiary Boundary Shale at Woodside Creek, New Zealand. Geology 13:738-740.

93.   Brouwers, E.M., et al. (1987) Dinosaurs on the North Slope, Alaska: High Latitude, Latest Cretaceous Environments. Science 237:1608-1610.

94.   Brumsack, H.-J. (1986) Trace Metal Accumulation in Black Shales from the Cenomanian/Turonian Boundary Event. pp.337-344 in (eds) O.H. Walliser, et al., Global Bio-Events, A Critical Approach. Lecture Notes in Earth Sciences, No.8. NY, Springer-Verlag.

95.   Buffetaut, E. (1984) Selective Extinctions and Terminal Cretaceous Events. Nature 310:276.

96.   Buffetaut, E. (1990) Vertebrate Extinctions and Survival Across the Cretaceous-Tertiary Boundary. Tectonophysics 171:337-345.

97.   Bullard, E.C. (1968) Reversals of the Earth's Magnetic Field: The Bakerian Lecture 1967. Phil Trans Roy Soc Lond 263A:481-524.

98. Bullard, E.C. (1975) Overview of Plate Tectonics. pp.5-52 in (eds) A.G. Fischer, S. Judson, Petroleum and Global Tectonics. NJ, Princeton Univ.

99. Bullock, W.M., et al. (1911) Upper Cretaceous Floras of the World. in Maryland Geological Survey. Vol.II, The Upper Cretaceous. Baltimore, John Hopkins.

100. Burchfiel, B.C., Davis, G.A. (1972) Structural Framework and Evolution of the Southern Part of the Cordilleran Orogen, Western United States. Amer J Sci 272:97-118.

101. Burke, K., et al. (1984) Caribbean Tectonics and Relative Plate Motions. Geol Soc Amer Mem 162:31-64.

102. Burke, W.H., et al. (1982) Variation of Seawater 87Sr/86Sr Throughout Phanerozoic Time. Geology 10:516-519.

103. Burrett, C.F. (1974) Plate Tectonics and the Fusion of Asia. Earth Planet Sci Lett 21:181-9.

104. Campsie, J., et al. (1984) Episodic Volcanism and Evolutionary Crises. Eos 65:796-800.

105. Carey, S.W. (1981) Expanding Earth: A Symposium. Australia, Earth Resources Foundations, Univ Sydney.

106. Carpenter, K. (1983) Evidence Suggesting Gradual Extinction of Latest Cretaceous Dinosaurs. Die Naturwissenschaften 70:611-612.

107. Carr, J.B. (1968) When Did Mantle Convection Cease? Tectonophysics 6:413-424.

108. Carter, N.L., et al. (1986) Dynamic Deformation of Volcanic Ejecta from the Toba Caldera: Possible Relevance to Cretaceous/Tertiary Boundary Phenomena. Geology 14:380-383.

109. Carter, N.L., et al. (1990) Dynamic Deformation of Quartz and Feldspar: Clues to Causes of Some Natural Crises. Tectonophysics 171:373-391.

110. Chadwick, K.H., Leenhouts, H.P. (1981) Molecular Theory of Radiation Biology. NY, Springer-Verlag.

111. Chamley, H., et al. (1984) Paleoenvironmental History of the Walvis Ridge at the Cretaceous-Tertiary Transition, from Minerological and Geochemical Investigations. pp.685-695 in (eds) T.C. Moore, et al., Initial Rpts DSDP, Leg 74. Wash, DC, USGPO.

112. Chao, B.F., Gross, R.S. (1987) Changes in the Earth's Rotation and Low-Degree Gravitational Field Induced by Earthquakes. Geophys J Roy Astron Soc 91:569-596.

113. Chapman, D.R., Larson, H.K. (1963) On the Lunar Origin of Tektites. J Geophys Res 68:4305-4358.

114. Chase , C.G. (1979) Interpretation of the Geoid: Subduction and Lower Mantle Convection. (abstr) p.348, Intl Union Geod Geophys Gen Assem, No.17.

115. Cheng, L-H., et al. (1974) Chinese Fossil Plants. Vol.1 [in Chinese]. Peking, Academia Sinica.

116. Chinnery, M.A., Landers, T.E. (1975) Evidence for Earthquake Triggering Stress. Nature 258:490-493.

117. Chikov, B.M. (1972) Chief Characteristics of Structure of Median Massifs of the Circum-Pacific Orogenic Belt. Pacific Geol 5:81-87.

118. Chouet, B., et al. (1974) Photoballistics of Volcanic Jet Activity at Stromboli, Italy. J Geophys Res 79:4961-4976.

119. Churkin, Jr., M. (1970) Fold Belts of Alaska and Siberia and Drift Between North America and Asia. pp.G1-G17 in (eds) W.L. Adkinson, M.M. Brosgé, Proceedsings of the Geological Seminar on the North Slope of Alaska. Bull Amer Assoc Petroleum Geologists. Pacific Section.

120.   Churkin, Jr., M. (1972) Western Boundary of the North America Continental Plate in Asia. Geol Soc Amer Bull 83:1027-1036.

121.   Churkin, Jr., M., Eberlein, G.D. (1977) Ancient Borderland Terranes of the North American Cordillera: Correlation and Microplate Tectonics. Geol Soc Amer Bull 88:769-786.

122.   Cifelli, R., et al. (1966) Cemented Foraminiferal Oozes from the Mid-Atlantic Ridge. Nature 209:32-34.

123.   Cisowski, S.M. (1988) Analogues for Magnetic Microspherules Associated with the K/T and Upper Eocene Extinction Events. pp.186-187 in (abs) Lunar and Planetary Science 19. Houston, Lunar & Planetary Instit.

124.   Cisowski, S.M. (1988) Paleomagnetism of Manson Structure Cores Inconsistent with a K/T Link. pp.188-189 in (abs) Lunar and Planetary Science 19. Houston, Lunar & Planetary Instit.

125.   Cisowski, S.M. (1990) Critical Review of the Case for, and Against, Extraterrestrial Impact at the K/T Boundary. Surv Geophys 11:55-131.

126.   Cisowski, S.M., Fuller, M. (1986) Cretaceous Extinctions and Wildfires. Science 234:261-2.

127.   Clark, D.L., et al. (1986) Conodont Survival and Low Iridium Abundances Across the Permian-Triassic Boundary in South China. Science 233:984-986.

128.   Clemens, W.A. (1986) Evolution of the Terrestrial Vertebrate Fauna During the Cretaceous-Tertiary Transition. pp.63-86 in (ed) D.K. Elliott, Dynamics of Extinction. NY, J Wiley.

129.   Clube, S.V., Napier, B. (1982) Cosmic Serpent; A Catastrophist View of Earth History. NY, Universe Books.

130.   Clube, S.V., Napier, B. (1984) Terrestrial Catastrophism—Nemesis or Galaxy. Nature 311:635-636.

131.   Cobbing, E.J. (1976) Geosynclinal Pair at the Continental Margin of Peru. Tectonophysics 36:157-165.

132.   Cohen, A.J. (1963) Asteroid- or Comet-Impact Hypothesis of Tektite Origin: The Moldavite-Strewn Field. pp.189-211 in J.A. O'Keefe (ed) Tektites. Univ Chicago.

133.   Colbert, E.H. (1965) Age of Reptiles. NY, WW Norton.

134.   Condie, K.C. (ed)(1982) Plate Tectonics and Crustal Evolution. NY, Pergamon.

135.   Coode, A.M. (1967) Spherical Harmonic Analysis of Major Tectonic Features. p 489 in (ed) S.K. Runcorn. Mantles of the Earth and Terrestrial Planets. London, Interscience Wiley.

136.   Cooper, H.F. (1977) Summary of Explosion Cratering Phenomena Relevant to Meteor Impact Events. pp.11-14 in (eds) D.J. Roddy et al., Impact and Explosion Cratering. NY, Pergammon.

137.   Cooper, M.R. (1977) Eustacy During the Cretaceous: Its Implications and Importance. Palaeogeogr Palaeoclimatol Palaeoecol 22:1-60.

138.   Corliss, W.R. (comp) (1983) Earthquakes, Tides, Unidentified Sounds and Related Phenomena. A Catalog of Geophysical Anomalies. Glen Arm, Md, Sourcebook Project.

139.   Courtillot, V. (1987) Magnetic Field Reversals, Polar Wander and Core-Mantle Coupling. Science 237:1140-1147.

140.   Courtillot, V., Cizowski, S. (1987) Cretaceous-Tertiary Boundary Events: External or Internal Causes. Eos 68:193.

141.   Courtillot, V., et al. (1986) Deccan Flood Basalts at the Cretaceous/Tertiary Boundary? Earth Planet Sci Lett 80:361-374.

142. Courtillot, V., et al. (1988) Deccan Flood Basalts and the Cretaceous/Tertiary Boundary. Nature 333:843-846.

143. Courtillot, V., Vink, V.C. (1983) How Continents Break Up. Sci Amer 249(1):42.

144. Cox, A. (1973) Plate Tectonics and Geology. SF, WH Freeman.

145. Crane, P.R., Ligard, S. (1989) Angiosperm Diversification and Paleolatitudinal Gradients in Cretaceous Floristic Diversity. Science 246:675-678.

146. Crawford, A.R. (1963) Large Ring Structures in a South Australian Precambrian Volcanic Complex. Nature 197:140-2.

147. Creber, G.T., Chaloner, W.G. (1985) Tree Growth in the Mesozoic and Early Tertiary and the Reconstruction of Palaeoclimates. Palaeogeogr Palaeoclimatol Palaeoecol 52:35-60.

148. Cristi, J.M. (1956) Chile. Geol Soc Amer Mem 65:187-214.

149. Crocket, J.A. et al. (1988) Distribution of Noble Metals Across the Cretaceous/Tertiary Boundary at Gubbio, Italy: Iridium Variation as a Constraint on the Duration and Nature of Cretaceous/Tertiary Boundary Events. Geology 16:77-80.

150. Crough, S.T. (1979) Geiod Anomalies Across Fracture Zones and the Thickness of the Lithosphere. Earth Planet Sci Lett 44:224-230.

151. Crough, S.T., Judy, D.M. (1980) Subducted Lithosphere, Hotspots and the Geiod. Earth Planet Sci Lett 48:15-22.

152. Croxton, C.A. (1980) Aquilapolleniates from the Late Cretaceous-Paleocene (?) of Central West Greenland. Rapports Gronlands Geologiske Undersogelse 101:5-27.

153. Crutzen, P.J. et al. (1975) Solar Proton Events: Stratospheric Sources of Nitric Oxide. Science 189:457-59.

154. Currie, P.J. (1989) Long-distance Dinosaurs: Annual Migrations May Have Taken Some Dinosaurs to the Arctic—and Beyond. Nat Hist (6):60-65..

155. Cutchis, P.(1974) Stratospheric Ozone Depletion and Solar Ultraviolet Radiation on Earth. Science 184:13-19.

156. Cuttitta, F., et al. (1972) New Data on Selected Ivory Coast Tektites. Geochim Cosmochim Acta 36:1297-1300.

157. Cuvier, G. (1828) [Essays on Revolutionary Changes of the Earth's Surface and Final Species Variations in the Animal Kingdom]. Dicours sur les Revolutions de la Surface du Globe, et sur les Changements qu'elles ont Produits dans le Regne Animal. Paris, Schubart & Heideloff [and] Didot Freres, c1867.

158. Dalziel, I.W.D., et al. (1974) Fossil Marginal Basin in the Southern Andes. Nature 250:291-4.

159. Damon, P.E. (1971) Relationship Between Late Cenozoic Volcanism and Tectonism and Orogenic-Epeirogenic Periodicity. pp.15-36 in (ed) K.Turekian, Late Cenozoic Glacial Ages. New Haven, Conn, Yale Univ.

160. Darwin, C. (1872) Origin of the Species. NY, Collier,c1962.

161. Davies, P.A., Runcorn, S.K. (eds)(1980) Mechanics of Continental Drift and Plate Tectonics. NY, Academic.

162. Davidson, C.F., Atkin, D. (1953) On the Occurrence of Uranium in Phosphate Rock. Geol Cong, Algiers, Comptes Rendus, Sec 11, Fasc 11:13-31.

163. Davis, M., et al. (1984) Extinction of Species by Periodic Comet Showers. Nature 308:715-7

164. Davis, P.M., et al. (1973) Kilauea Volcano, Hawaii: A Search for the Volcano-Magnetic Effect. Science 180:73.

165. Denton, M. (1985) Evolution: A Theory in Crisis. London, Burnett.

166. DePaolo, D.J., et al. (1983) Rb-Sr, Sm-Nd, K-Ca, O, and H Isotope Study of Cretaceous-Tertiary Boundary Sediments, Caravaca, Spain: Evidence for an Oceanic Impact Site. Earth Planet Sci Lett 64:356-373.

167. Derry, D.R. (1980) World Atlas of Geology and Mineral Deposits. NY, Wiley.

168. Desmond, A.J. (1975) Hot-Blooded Dinosaurs; A Revolution in Palaeontology. NY, Dial Press/James Wade.

169. Devine, J.D., et al. (1984) Estimates of Sulfur and Chlorine Yield to the Atmosphere from Volcanic Eruptions and Potential Climatic Effects. J Geophys Res 89:6309.

170. D'Hondt, A.V. (1983) Campanian and Maastrichtan Inoceramids: A Review. Zitteliana 10:689-701.

171. D'Hondt, S., Keller, G. (1985) Late Cretaceous Stepwise Mass Extinction of Planktonic Foraminifera. Geol Soc Amer Abstr Prog 17:557-558.

172. Dickinson, W.R. (1971) Plate Tectonic Models of Geosynclines. Earth Planet Sci Lett 10:165-174.

173. Dickinson, W.R. (1973) Reconstruction of Past Arc-Trench Systems from Petrotectonic Assembleages in the Island Arcs of the Western Pacific. pp.569-601 in (eds) P.J. Coleman, Western Pacific: Island Arcs, Marginal Seas, Geochemistry. NY, Crane, Russak.

174. Dickinson, W.R., Yarborough, H. (1976) Plate Tectonics and Hydrocarbons. Tulsa, Okla, AAPG Dept Ed Activities.

175. Diggle, W.R., Saxon, J. (1965) Unusually Radioactive Fossil Fish from Thurso, Scotland. Nature 208:400.

176. Donovan, A.D., Vail, P.R. (1986) Sequence Stratigraphy of the K-T Boundary in Alabama: A Non-catastrophic Alternative (abst) Geol Soc Amer Ann Mtg 587.

177. Donovan, D.T., Jones, E.J.W. (1979) Causes of Worldwide Changes in Sea Level. J Geol Soc 136:187-192.

178. Donn, W.L. (1982) Enigma of High-Latitude Paleoclimate. Palaeogeogr Palaeoclimatol Palaeoecol 40:199-212.

179. Dorman, F.H. (1968) Some Australian Oxygen Isotope Temperature and a Theory for a 30-Million-Year Cycle. J Geol 76:297-313.

180. Dott, R.H., Batten, R.L. (1976) Evolution of the Earth. NY, Mc Graw-Hill.

181. Doyle, J.A., Hickey, L.J. (1976) Pollen and Leaves from the Mid-Cretaceous Potomac Group and their Bearing on Early Angiosperm Evolution. pp.139-206 in (ed) C.B. Beck, Origin and Early Evolution of Angiosperms. NY, Columbia Univ.

182. Drake, C.L., Officer, C.B. (1983) Cretaceous-Tertiary Transtion. Science 219:1383-1390.

183. Dransfield, J. (1981) Palms and Wallace's Line. pp.43-56 in (ed) T.C. Whitmore, Wallace's Line and Plate Tectonics. Oxford, Clarendon.

184. Dubinin, N.P. (ed)(1969) Current Problems of Radiation Genetics. [in Russian] Moscow.

185. Dubrov, A.P. (1978) Geomagnetic Field and Life: Geomagnetobiology. NY, Plenum.

186. Dugle, J.R., El-Lankany, M.H. (1971) Check List of the Plants of the Whiteshell Area, Manitoba Including a Summary of Their Published Radiosensitivities. Pub AECL-3678, Pinawa, Manitoba, Whiteshell Nuclear Res Establ.

187. Duncan, R.A., Hargraves, R.B. (1984) Plate Tectonic Evolution of the Caribbean Region in the Mantle Reference Frame. Geol Soc Amer Memiors 162:81-94.

188. Durham, J.W. (1985) Movement of the Caribbean Plate and its Importance for Biogeography in the Caribbean. Geology 13:123-125.

189. Durrani, S.A., Khan, H.A. (1971) Ivory Coast Microtektites Fission Track Age and Geomagnetic Reversals. Nature 232:320-323.

190. Dziewonski, A.M., Woodhouse, J.H. (1987) Global Images of the Earth's Interior. Science 236:37-48.

191. Ehling, P.R. (1986) Extinction: What is Happening Now and What Needs to be Done. pp.157-164 in (ed) D.K. Elliott. Dynamics of Extinction. NY, Wiley.

192. Ehling, U.H. (1984) Methods to Estimate the Genetic Risk. pp.292-318 in (ed) G. Obe, Mutations in Man. NY, Springer-Verlag.

193. Ehling, U.H., Randolph, M.L. (1962) Skeletal Abnormalities in the F1 Generation of Mice Exposed to Ionizing Radiation. Genetics 47:1543-1555.

194. Eicher, D. (1976) Geologic Time. Englewood Cliffs, NJ, Prentice-Hall.

195. Eicher, D.L., et al. (1984) History of the Earth's Crust. Englewood Cliffs, NJ, Prentice-Hall.

196. Einarsson, T. (1968) Submarine Ridges as an Effect of Stress Fields. J Geophys Res 73:7561-7576.

197. Einarsson, T. (1975) Several Problems in Radiometric Dating. Jokull 25:15-33.

198. Einarsson, T. (1976) Upper Pleistocene Volcanism and Tectonism in the Southern Median Active Zone of Iceland. Problems in Geology and Geophysics, Part I, Soc Sci Iceland. Greinar V:119-159.

199. Einarsson, T. (1977) Palaeomagnetism and the Possible Effect of Depth of Burial and Non-Hydrostatic Stresses. Greinar VI:66-74.

200. Ekdale, A.A., Bromley, R.G. (1984) Sedimentology Ichnology of the Cretaceous-Tertiary Boundary in Denmark: Implications for the Causes of the Terminal Cretaceous Extinction. J Sedimentary Petrol 54:681-703.

201. Elderfield, H. (1986) Strontium Isotope Stratigraphy. Palaeogeogr Palaeoclimatol Palaeoecol 57:71-90.

202. Emiliani, C. (1980) Death and Renovation at the End of the Mesozoic. Eos 61:505-506.

203. Eos (1985) Transactions. Amer Geophys U p.441.

204. Erben, H.K. (1972) [Ultrastructure and Thickness of Fossil Egg Shells.] Ultrastrukturn und Dicke der Wand Pathologischer Eischalen. Akad Wiss Lit Mainz Abh Math-Natuwiss Kl 6:193-216.

205. Erben, H.K., et al (1979) Paleobiological and Isotopic Studies of Eggshells from a Declining Dinosaur Species. Paleobiology 5:380-414.

206. Espenshade, E.B., Morrison, J.L. (1975) Goode's World Atlas. Chicago, Rand McNally.

207. Evans, A.E., et al. (1973) Potassium-Argon Age Determinations on Some British Tertiary Igneous Rocks. J Geol Soc London 129:419.

208. Ewing, J. et al. (1966) Ages of Horizon A and the Oldest Atlantic Sediments. Science 154:1125-1132.

209. Ewing, J., Ewing, M. (1966) Sediment Distribution on the Mid-Ocean Ridges with Respect to Spreading of the Seafloor. Science 156:1590-1592.

210. Ewing, M. et al. (1966) Crustal Structure of the Mid-Ocean Ridge. J Geophys Res 71:1611-1636.

211. Farabee, M.J. (1983) Stratigraphic Palynology of the Lower Part of the Lance Formation (Maestrichtian) of Wyoming. Thesis Dissertation (M.S.). Tempe, Ariz State.

212. Feist, M. (coordinator) (1986) Bio-Events in the Continental Realm during the Cretaceous-Tertiary Transition: A Multidiscliplinary Approach. pp.411-418 in (eds) O.H. Wallister, et al. Global Bio-Events, A Critical Approach. Lecture Notes in Earth Sciences, No.8. NY, Springer-Verlag.

213. Feldman, P. (1977) Astronomical Evidence Bearing on the Supernova Hypothesis for the Mass Extinctions at the End of the Cretaceous. pp.125-136 in (eds) K-Tec Group, Cretaceous-Tertiary Extinctions and Possible Terrestrial and Extraterrestrial Causes. Proc Workshop Ottawa, Canada 16-17 Nov 1976. Ottawa, Natl Museums of Canada.

214. Fischer, A.G. (1981) Climatic Oscillations in the Biosphere. pp.103-131 in (ed) M.H. Nitecki, Biotic Crises in Ecological and Evolutionary Time. NY, Academic.

215. Fischer, A.G. (1982) Carbon Cycle—Controls on Atmospheric COO2 and Climate in the Geological Past. pp.55-67 in Climate in Earth History. Wash, DC, Natl Acad.

216. Fischer, A.G., Arthur, M.A. (1977) Secular Variations in the Pelagic Realm. pp.19-50 in (eds) H.E. Cook, P. Enos, Deep Water Carbonate Environments. SEPM Spec Pub 25.

217. Flemming, R.F. (1988) Palynology of the Cretaceous-Tertiary Boundary in the Raton Basin: Implications for Development of Tertiary Flora. p.50 in Abstr 7th Intl Palynological Congr, Brisbane.

218. Flessa, K.W. (1980) Biological Effects of Plate Tectonics and Continental Drift. BioScience 30:518-523.

219. Flessa, K.W., et al. (1986) Causes and Consequences of Extinction. in (eds) D.M. Raup, D. Jablonski, Pattern and Process in the History of Life. NY, Springer-Verlag.

220. Foster, J.H. (1977) Geomagnetic Field and the Cretaceous-Tertiary Extinctions. pp.63-74 in (eds) K-Tec Group, Cretaceous-Tertiary Extinctions and Possible Terrestrial and Extraterrestrial Causes. Proc Workshop Ottawa, Canada 16-17 Nov 1976. Ottawa, Natl Museums of Canada.

221. Fraser-Smith, A.C., et al. (1990) Low-Frequency Magnetic Field Measurements Near the Epicenter of the Mg 7.1 Loma Prieta Earthquake. Geophys Res Lett 17:1465-1468.

222. French, B.M. (1985) Cretaceous-Tertiary Extinctions Alternative Models. Science 230:1293-1294.

223. Friis, E.M., et al. (1987) Introduction to Angiosperms. pp.1-16 in (eds) E.M. Friis et al., Origins of Angiosperms and Their Biological Consequences. NY, Cambridge Univ.

224. Friis, E.M., Crepet, W.L. (1987) Time of Appearance of Floral Features. pp.145-179. in (eds) E.M. Friis et al., Origins of Angiosperms and Their Biological Consequences. NY, Cambridge Univ.

225. Fuller, M., et al. (1979) Tectonomagnetics and Local Geomagnetic Field Variations. Proc IAGA/IAMAP Joint Assembly, Aug 1977, Seattle, Wash. Tokyo, Center Acad Pubs Japan, Japan Sci Soc.

226. Fujinawa, Y., Takahashi, K. (1990) Emission of Electromagnetic Radiation Preceeding the Ito Seismic Swarm of 1989. Nature 347:376-378.

227. Fujita, K. (1978) Pre-Cenozoic Tectonic Evolution of Northeast Siberia. J Geol 86:159-172.

228. Ganapathy, R. (1980) Major Meteorite Impact on the Earth 65 Million Years Ago: Evidence from the Cretaceous-Tertiary Boundary Clay. Science 209:921-923.

229. Ganapathy, R. (1982) Evidence for a Major Meteorite Impact on the Earth 34 Million Years Ago: Implication for Eocene Extinctions. Science 216:885-888.

230. Garner, R.J. (1972) Transfer of Radioactive Materials from the Terrestrial Environment to Animals and Man. Cleveland, Ohio, CRC Press.

231. Gartner, S. (1977) Nannofossils and Biostratigraphy: An Overview. Earth Sci Rev 13:227-50.

232. Gartner, S., Mc Guirk, J.P. (1979) Terminal Cretaceous Extinction: Scenario for a Catastrophe. Science 206:1272-6.

233. Gastil, G. (1960) Distribution of Mineral Dates in Time and Space. Amer J Sci 258:1-35.

234. Gee, H. (1988) Cretaceous Unity and Diversity. Nature 322:487.

235. George, W. (1981) Wallace and His Line. pp.3-8 in (ed) T.C. Whitmore, Wallace's Line and Plate Tectonics. Oxford, Clarendon.

236. Ghosh, S.K. (1984) Late Cretaceous Condensed Sequences, Venezuelan Andes. Geol Soc Amer Memiors 162:317-324.

237. Giardini, D., Woodhouse, J.H. (1986) Horizontal Shear in the Mantle Beneath the Tonga Arc. Nature 319:551-555.

238. Gilmore, F.R. (1975) Production of Nitrogen Oxides by Low-Altitude Nuclear Explosions. J Geophys Res 80:4553.

239. Gilmore, J.S., et al. (1984) Trace Elements Patterns at a Nonmarine Cretaceous-Tertiary Boundary, Western Interior. Nature 307:224-228.

240. Gilmour, I., Anders, E. (1989) Cretaceous-Tertiary Boundary Event: Evidnece for a Short Time Scale. Geochim Cosmochim Acta 53:503-511.

241. Gilmour, I., Geunther, F. (1988) Global Cretaceous-Tertiary Fire: Biomass or Fossil Carbon? (abst) pp.60-61 in Global Catastrophes in Earth History: An Interdisciplinary Conference on Impacts, Volcanism and Mass Mortality. Houston, Lunar & Planetary Instit.

242. Glass, B.P. (1982) Possible Correlations Between Tektite Events and Climate Changes. Geol Soc Amer Spec Pap 190:251-256.

243. Glass, B.P. (1990) Tektites and Microtektites: Key Facts and Inferences. Tectonophysics 171:393-404.

244. Glass, B.P., Barlow, R.A. (1979) Mineral Inclusions in Muong Nong-type Indochinites: Implications Concerning Parent Material and Process Formation. Meteoritics 14:55-67.

245. Glass, B.P., Crosbie, J.R. (1982) Age of Eocene/ Oligocene Boundary Based on Extropolation from North American Microtektite Layer. Bull Amer Assoc Petrol Geol 66:471-476.

246. Glass, B.P., et al. (1979) Australian, Ivory Coast and North American Tektite Strewnfields: Size, Mass and Correlation with Geomagnetic Reversals and Other Earth Events. Proc Lunar Planet Sci Conf 10th:2535-2545.

247. Glass, B.P., et al. (1979) Late Eocene North American Microtektites and Clinopyroxene-Bearing Spherules. Proc Lunar Planet Sci Conf 16D:175-196.

248. Glass, B.P., et al. (1986) Further Evidence for the Impact Origin of Tektites. Meteoritics 21:369-370.

249. Glass, B.P., Heezen, B.C. (1967) Tektites and Geomagnetic Reversals. Nature 214:372.

250. Glass, B.P., Heezen, B.C. (1967) Tektites and Geomagnetic Reversals. Sci Amer 217:32-38.

251. Glass, B.P., Zwart, M.J. (1977) North American Microtektites, Radiolarian Extinctions and the Age of the Eocene-Oligocene Boundary. pp.553-568 in (ed) F.M. Swain, Stratigraphic Micropaleontology of Atlantic Basin and Borderlands. NY, Elsevier.

252. Glass, B.P., Zwart, M.J. (1979) North American Microtektites in Deep Sea Drilling Project Cores from the Carribbean Sea and Gulf of Mexico. Geol Soc Amer Bull 90:595-602.

253. Godby, E.A. et al. (1968) Aeromagnetic Profiles Across the Reykjanes Ridge Southwest of Iceland. J Geophys Res 73:7637-7649.

254. Gokhberg, M.G., et al. (1982) Experimental Measurement of Electromagnetic Emissions Possibly Related to Earthquakes in Japan. J Geophys Res 87B:7824.

255. Golding, G.B. (1987) Nonrandom Patterns of Mutation are Reflected in Evolutionary Divergence and May Cause Some of the Unusual Patterns Observed in Sequences. in (ed) E. Loeschcke, Genetic Constraints on Adaptive Evolution. NY, Springer-Verlag.

256. Goldschmidt, R.B. (1940) Material Basis of Evolution. New Haven, Yale Univ.

257. Goldreich. P. Toomre, A. (1969) Some Remarks on Polar Wandering. J Geophys Res 74:2555-2567.

258. Golenkin, M.I. (1947) Victors in the Struggle for Existence in the Plant World. [in Russian] Moscow.

259. Gorgoni, C., et al. (1988) Radon and Helium Anomalies in Mud Volcanoes from Northern Apennines (Italy): A Tool for Earthquake Prediction. Geochem J 22:265-274.

260. Gose, W.A., Swartz, D.G. (1977) Paleomagnetic Results from Cretaceous Sediments in Honduras: Tectonic Implications. Geology 5:505-515.

261. Gostin, V.A., et al. (1989) Iridium Anomaly from the Acraman Impact Ejecta Horizon: Impacts Can Produce Sedimentary Iridium Peaks. Nature 340:542-544.

262. Gottlieb, L.D. (1984) Genetics and Morphological Evolution in Plants. Amer Naturalist 123:681-709.

263. Gottlieb, L.D., Ford, V.S. (1987) Genetic and Developmental Studies of the Absence of Ray Florets in *Layia Discoidea*. pp.2-17 in (eds) H. Thomas, D. Grierson, Developmental Mutants in Higher Plants. NY, Cambridge Univ.

264. Gottlieb, L.D., Jain, S.K. (1988) Plant Evolution Biology. NY, Chapman & Hall.

265. Gould, S.J. (1977) Ontogeny and Phylogeny. Cambridge, Mass, Belknap Press Harvard.

266. Gould, S.J. (1977) Return of the Hopeful Monsters. Nat Hist 86(6):22-30.

267. Graup, G. (1988) Mineralogy and Phase-Chemistry of the Cretaceous/Tertiary Section in the Lattengbirge, Bavarian Alps (abs) p.65 in Global Catastrophes in Earth History: An Interdisciplinary Conference on Impacts and Mass Motality. Houston, Lunar & Planetary Instit.

268. Grieve, R.A.F., et al. (1985) Periodic Cometary Impacts and the Terrestrial Cratering Record. Eos 66:813.

269. Grieve, R.A.F., Robertson, P.B. (1979) Terrestrial Cratering Record, I. Current Status of Observations. Icarus 38:212-29.

270. Grieve, R.A.F., Sharpton, V.L. (1986) K/T Impact Event: Some Implications from the Evidence. (abs) pp.289-290 in Lunar and Planetary Science 17, Houston, Lunar & Planetary Instit.

271. Grigoryev, A.I., et al. (1989) Parametric Instablility of Water Drops in an Electric Field as a Possible Mechanism for Luminous Phenomena Accompanying Earthquakes. Phys Earth Planet Interiors 57:139-143.

272. Gross, M.G. (1980) Oceanography. Columbus, C Merrill.

273. Gross, S. (1977) Mineralogy of the Hatrurim Formation. Israel Geol Surv Israel Bull 70:8-9.

274. Grzimek, H.C. (ed)(1976) Grzimek's Encyclopedia of Evolution. NY, Van Nostrand Reinhold.

275. Gurley, L.R., et al. (1991) Proteins in the Fossil Bone of the Dinosaur, Seismosaurus. J Protein Chem 10:75-102.

276. Gurnis, M. (1988) Large-scale Mantle Convection and the Aggregation and Dispersal of Supercontinents. Nature 322:696-699.

277. Habicht, J.A. (1979) Paleoclimate, Paleomagnetism, and Continental Drift. Tulsa, Oklahoma, Amer Assoc Petro Geol.

278. Hall, J.W., Norton, N.J. (1967) Palynological Evidence of Floristic Changes Across the Cretaceous-Tertiary Boundary in Eastern Montana. Palaeogeogr Palaeoclimatol Palaeoecol 3:121-131.

279. Hallam, A. (1984) Asteriods and Extinctions—No Cause for Concern. New Sci 104(1429):30-33.

280. Hallam, A. (1984) Causes of Mass Extinctions. Nature 308:686-687.

281. Hallam. A. (1984) Pre-Quarternary Sea-Level Changes. Ann Rev Earth Planet Sci 12:205.

282. Hallam, A. (1985) A Review of Mesozoic Climates. J Geol Soc 142:433.

283. Hallam, A. (1987) End-Cretaceous Mass Extinction Event: Argument for Terrestrial Causation. Science 238:1237-1242.

284. Hallam, A. (1989) The Case for Sea-level Change as a Dominant Causal Factors in Mass Extinction of Marine Inverterbrates. Phil Trans Roy Soc London 325B:437-455.

285. Hallam, A., Perch-Nielsen, K. (1990) The Biotic Record of Events in the Marine Realm at the End of the Creatceous: Calcareous, Siliceous and Organic-walled Microfossils and Microinvertebrates. Tectonophysics 171:347-357.

286. Hamilton, E.I., Farquhar, P.M. (1968) Radiometric Dating for Geologists. NY, Wiley Interscience.

287. Hamilton, W. (1969) Mesozoic California and the Underflow of the Pacific Mantle. Geol Soc Amer Bull 80:2409-2430.

288. Hamilton, W. (1970) Uralides and the Motion of the Russian and Siberian Platform. Geol Soc Amer Bull 81:2553-2576.

289. Hancock, J.M. (1967) Some Cretaceous-Tertiary Marine Faunal Changes. pp.91-104 in (eds) W.B. Harland et al., Fossil Record. Special Pub No. 2. Geol Soc London.

290. Hansen, H.J., et al. (1986) Cretaceous-Tertiary Boundary Spherules from Denmark, New Zealand and Spain. Bull Geol Soc Denmark 35:75-82.

291. Hansen, H.J., et al. (1987) Iridium-Bearing Carbon Black at the Cretaceous-Tertiary Boundary. Bull Geol Soc Denmark 36:305-314.

292. Haq, B.U., et al. (1987) Chronology of Fluctuating Sea Levels Since the Triassic (250 Million Years Ago to Present). Science 235:1156-1167.

293. Harris, W.B., et al. (1986) Cretaceous-Tertiary Boundary on the Cape Fear Arch, North Carolina, U.S.A. Cret Res 7:1-17.

294 Harriss, R.C., et al. (1973) Palladium, Iridium and Gold in Deep-sea Manganese Nodules. Geochim Cosmochim Acta 32:1049-1056.

295. Hast, N. (1969) State of Stress in the Upper Part of the Earth's Crust. Tectonophysics 8:169-210.

296. Hatfield, G.B., Camp, M.J. (1970) Mass Extinction Correlated with Periodic Galactic Events. Geol Soc Amer Bull 81:911-4.

297. Hattori, I., Hirooka, K. (1979) Paleomagnetic Results from Permian Greeenstones in Central Japan and Their Geologic Significance. Tectonophysics 57:211-235.

298. Haq, B.U. (1987) Chronology of Fluctuating Sea Levels Since the Triassic. Science 235:1156-1167.

299. Hawkins, G.S. (1960) Tektites and the Earth. Nature 185:300-301.

300. Hays, J.D. (1971) Faunal Extinctions and Reversals of the Earth's Magnetic Field. Geol Soc Amer Bull 82:2433.

301. Hays, J.D., Pitman, III, W.C. (1973) Lithosphere Plate Motion, Sea Level Changes and Climatic Ecological Consequences. Nature 246:18-22.

302. Haxby, W.F., Turcotte, D.L. (1978) On Isostatic Geoid Anomalies. J Geophys Res 83:5473-5478.

303. Heath, D.F. et al. (1977) Solar Proton Event: Influence on Stratospheric Ozone. Science 197:886-889.

304. Heezen, B.C. et al. (1953) Trans-Atlantic Profile of Total Magnetic Intensity and Topography, Dakar to Barbados. Deep-Sea Res 1:25-33.

305. Heezen, B.C., Hollister, C.D. (1971) Face of the Deep. NY, Oxford Univ.

306. Heirtzler, J.R. et al. (1968) Marine Magnetic Anomalies, Geomagnetic Field Reversals and Motions of the Ocean Floor and Continents. J Geophys Res 73:2119-2136.

307. Heller, J., Teixeria-Pinto, A. (1959) New Physical Method of Creating Chromosomal Aberrations. Nature 183:905-906.

308. Hemenway, D. (1984) Dinosaur Skulls Tell Tales. Earth Sci 17(3):20-21.

309. Henshaw, P.S. (1963) Radiation Effects and Peaceful Uses of Atomic Energy in Animal Science Radiation and Biologic Capability. NY, Radioecology.

310. Herring, J.R. (1985) Charcoal Fluxes into Sediments of the North Pacific Ocean: the Cenozoic Record of Burning. pp.419-442 in E.T. Sunquist, W.S. Broecker. Carbon Cycle and Atmospheric CO2: Natural Variations Archean to Present. Wash, DC, Amer Geophys Union.

311. Hess, H.H., Poldervaart, A. (eds)(1967-8) Basalts: The Poldervaart Treatise on Rocks of Basaltic Composition. NY, Wiley Interscience.

312. Hess, J. et al. (1986) Evolution of the Ratio of Strontium-87 to Strontium-86 in Seawater from Cretaceous to Present. Science 231:979-984.

313. Hickey, L.J. (1977) Changes in Angiosperm Flora Across the Cretaceous-Paleocene Boundary. J Paleontol 51(sup 2):14-5.

314. Hickey, L.J. (1981) Land Plant Evidence Compatible with Gradual, Not Catastrophic, Change at the End of the Cretaceous. Nature 292:529-531.

315. Hickey, L.J. (1984) Changes in the Angiosperm Flora Across the Cretaceous-Tertiary Boundary. pp.279-313 in (eds) W.A. Berggren, J.A. Van Couvering, Catastrophes and Earth History. NJ, Prnceton Univ.

316. Hickey, L.J. (1985) Once-Fertile Wasteland. Sciences 25(1):43-46.

317. Hickey, L.J., et al. (1983) Arctic Terrestrial Biota: Paleomagnetic Evidence of Age Disparity with Mid-Northern Latitudes During the late Cretaceous and early Tertiary. Science 221:1153-1156.

318. Higdon, J.C., Lingenfelter, R.E. (1973) Sea Sediments, Cosmic Rays and Pulsars. Nature 246:403-405.

319. Hildebrand, A.R., Boynton, W.V. (1988) Impact Wave Deposits Provide a New Constraint on the Location of the K/T Boundary Impact. pp.76-77 in Global Catastrophes in Earth History: An Interdisciplinary Conference on Impacts, Volcanism and Mass Mortality. Houston, Lunar & Planetary Instit.

320. Hildebrand, A.R., et al. (1984) Kilauea Volcano Aerosols: Evidence in Siderophile Element Abundances for Impact-induced Oceanic Volcanism at the K/T Boundary (abs) p.239 in Lunar and Planetary Science 19. Houston, Lunar & Planet Instit.

321. Hildebrand, A.R., Boynton, W.V. (1988) Provenance of the K/T Boundary Layers. pp.78-79 in Global Catastrophes in Earth History: An Interdisciplinary Conference on Impacts, Volcanism and Mass Mortality. Houston, Lunar & Planet Inst.

322. Hildebrand, A.R., Boynton, W.V. (1990) Proximal Cretaceous-Tertiary Boundary Impact Deposits in the Caribbean. Science 248:843-846.

323. Hill, M. (1987) Underwater Earthquakes Create Devastating Sea Waves that Can Kill Thousands and Wash Away Towns. Earth Sci 40:38.

324. Hillhouse, J.W. (1977) Paleomagnetism of the Traissic Nikolai Greenstone, Mc Carthy Quadrangle, Alaska. Canadian J Earth Sci 14:2578-2592.

325. Hirschberg, J., et al. (1967) Long Period Geomagnetic Fluctuations After the March 1964 Alaska Earthquake. Eos 48:80.

326. Hirschberg, J., et al. (1967) Long Period Geomagnetic Fluctuations After the March 1964 Alaska Earthquake. Earth Planet Sci Lett 3:426.

327. Hoffman, A. (1985) Patterns of Family Extinction and Geological Timescale. Nature 315:659-662.

328. Hoffman, K.A. (1988) Ancient Magnetic Reversals: Clues to the Geodynamo. Sci Amer 258:76-83.

329. Horner, J.R., Weishampel, D.B. (1988) Comparative Embryological Study of Two Orinithiscian Dinosaurs. Nature 332:256-257.

330. House, M.R. (1985) Correlation of Mid-Palaeozoic Ammonoid Evolutionary Events with Global Sedimentary Perturbations. Nature 313:17-22.

331. Hsu, K.J. (1971) Franciscan Melange as a Model for Eugeosynclinical Sedimentation and Underthrusting Tectonics. J Geophys Res 76:1162-1170.

332. Hsu, K.J. (1971) Origin of the Alps and Western Mediterranean. Nature 233:44-48.

333. Hsu, K.J. (1980) Terrestrial Catastrophe Caused by Cometary Impact at the End of the Cretaceous. Nature 285:201-203.

334. Hsu, K.J. (1986) Darwin's Three Mistakes. Geology 14:532-4.

335. Hsu, K J., McKenzie, J.A. (1985) A "Strangelove" Ocean in the Earliest Tertiary. pp.487-492. in (eds) E.T. Sunquist, W.S. Broecker. Carbon Cycle and Atmospheric CO2: Natural Variations Archean to Present. Wash, DC, Amer Geophys U.

336. Huffman, A.R., et al. (1988) Iridium, Shocked Minerals and Trace Elements Across the Creatceous/Tertiary Boundary at Maud Rise, Weddell Sea and Walvis Ridge. South Atlantic Ocean pp.81-82 in Global Catastrophes in Earth History: An Interdisciplinary Confernece on Impacts, Volcanism and Mass Mortality. Houston, Lunar & Planetary Instit.

337. Hughes, N.F. (1976) Palaeobiology of Angiosperm Origins. Mass, Cambridge Univ.

338. Hughes, N.F. (1976) Paleobiology of Angiosperms: Problems of Mesozoic Seed-Plant Evolution. Mass, Cambridge Univ.

339. Hulett, L.D., et al. (1980) Chemical Species in Fly-Ash from Coal-Burning Power Plants. Science 210:1356.

340. Hurley, P.M. (1971) Possible Inclusion of Korea, Central and Western China, and India in Gondwanaland. Eos 52:356.

341. Hurley, N.F., Van der Voo, R. (1990) Magnetostratigraphy of the Late Devonian Iridium Anomaly and Impact Hypotheses. Geology 18:291-294.

342. Hut, P., et al. (1987) Comet Showers as a Cause of Mass Extinctions. Nature 329:118-126.

343. Hutchinson, J.H., Archibald, J.D. (1986) Diversity of Turtles Across the Cretaceous/ Teritary Boundary in Northeastern Montana. Palaeogeogr Palaeoclimatol Palaecol 55:1-22.

344. Huttermann, J., et al. (eds)(1978) Effects of Ionizing Radiation on DNA: Physical, Chemical and Biological Aspects. NY, Springer-Verlag.

345. Igarashi, C., Wakita, H. (1990) Groundwater Radon Anomalies Associated with Earthquakes. Tectonophysics 180:237-254.

346. International Atomic Energy Agency (1969) Induced Mutations in Plants. Proc Symp Nature, Induction and Utilization of Mutations in Plants. Joint IAEA/FAO, Pullman, Wash 14-18 July 1969. Vienna, Intl Atomic Energy Agency.

347. International Atomic Energy Agency (1984) Selection in Mutation Breeding Proc Consultants Meeting. Joint FAO/IAEA Div Isotope & Radiation Appl Atomic Energy for Food & Agric Develop, Vienna, 21-25 June 1982. Intl Atomic Energy Agency.

348. International Atomic Energy Agency (1986) Nuclear Techniques and *In Vitro* Culture for Plant Improvement. Proc Symp Joint IAEA/FAO, Vienna 19-23 Aug 1985. Intl Atomic Energy Agency.

349. Irving, E., Robertson, W.A. (1969) Test for Polar Wandering and Some Pssible Implications. J Geophys Res 74:1026-1036.

350. Isaacson. P.E. (1975) Evidence for a Western Extracontinental Land Source during the Devonian Period in the Central Andes. Geol Soc Amer Bull 86:39-46.

351. Ivanova, Ye. A. (1955) Relationship Between Stages of Evolution of Organic Life and Stages of Evolution of the Earth's Crust. [in Russian] AN SSSR Doklady 105(1).

352. Izett, G.A. (1987) Authigenic "Spherules" in K-T Boundary Sediments at Carvaca, Spain and Raton Basin, Colorado and New Mexico May Not be Impact Derived. Geol Soc Amer Bull 99:78-86.

353. Izett, G.A. (1987) Cretaceous-Tertiary (K-T) Boundary Interval, Raton Basin, Colorado and New Mexico, and its Content of Shocked-Metamorphosed Minerals: Implications Concerning the K-T Boundary Impact-Extinction Theory. US Geol Survey Open-File Rept 87-606.

354. Izett, G.A., Pillmore, C.L. (1985) Abrupt Appearance of Shocked Quartz at the Cretaceous-Tertiary Boundary, Raton Basin, Colorado and New Mexico. Geol Soc Amer Abstr Prog 17:617.

355. Jablonski, D. (1986) Background and Mass Extinctions: The Alteration of Macroevolutionary Regimes. Science 231:129-133.

356. Jacobs, J.A. (1984) Reversals of the Earth's Magnetic Field. Bristol, Adam Hilger Ltd.

357. Jakubovics, J.P. (1987) Magnetism and Magnetic Materials. Brookfield, Vt, Insit Metals.

358. Jarzen, D.M. (1977) Angiosperm Pollen as Indicators of Cretaceous-Tertiary Environments. Syllogeus 12:39-49.

359. Jaworowski, Z., Pensko, J. (1967) Unusually Radioactive Bones from Mongolia. Nature 214:161-163.

360. Jeletzky, J.A. (1978) Causes of Cretaceous Oscillations of Sea Level in Western and Arctic Canada and Some General Geotectonic Implications. Geol Surv Can Paper 77:18.

361. Jerzykiewicz, T., Sweet, A.R. (1986) Cretaceous-Tertiary Boundary in the Central Alberta Foothills. I: Stratigraphy. Canadian J Earth Sci 23:1356-1374.

362. Jiang, M.J., Gartner, S. (1986) Calcareous Nannofossil Succession Across the Cretaceous/Tertiary Boundary in East-Central Texas. Micropaleontology 32:232-255.

363. Johnston, H.S. et al. (1973) Effects of Nuclear Explosions on Stratospheric Nitric Oxide and Ozone. J Geophys Res 78:6107.

364. Johnston, M.J.S. (1978) Tectomagnetic Effects. Earthq Info Bull 10:82.

365. Johnston, M.J.S. (1989) Review of Magnetic and Electric Field Effects Near Active Faults and Volcanoes in the U.S.A. Phys Earth Planet Interiors 57:47-63.

366. Johnston, M.J.S., Mueller, R.J. (1987) Seismomagnetic Observation during the 8 July 1986 Magnitude 5.9 North Palm Springs Earthquake. Science 237:1201-1203.

367. Johnston, M.J.S., et al. (1976) Tectonomagnetic Experiments and Observations in the Western U.S.A. J Geomag Geoelect 28:85-97.

368. Jones, D.L., et al. (1977) Wrangellia—A Displaced Terrane in Northwestern North America. Can J Earth Sci 14:2565-77.

369. Jones, D.S., et al. (1987) Biotic, Geochemical and Paleomagnetic Changes Across the Cretaceous/Tertiary Boundary at Braggs, Alabama. Geology 15:311-315.

370. Jones, E.M., Sanford, M.T. (1977) Numerical Simulations of a Very Large Explosion at the Earth's Surface with Possible Applications to Tektites. p.1009 in (eds) D.J. Roddy, et al., Impact and Explosion Cratering. NY, Pergamon.

371. Jones, E.M., Kodis, J.W. (1982) Atmospheric Effects of Large-Body Impacts: The First Few Minutes. Geol Soc Amer Spec Paper 190:175-186.

372. Kanamori, H. (1977) Energy Relaese in Great Earthquakes. J Geophys Res 82:2981-2987.

373. Karig, D.E. (1971) Origin and Development of Marginal Basins in the Western Pacific. J Geophys Res 76:2545-2561.

374. Kastner, M., et al. (1984) Precursor of the Cretaceous-Tertiary Boundary Clays at Stevns Klint, Denmark and DSDP Hole 465A. Science 226:137-143.

375. Kauffman, E.G. (1977) Cretaceous Extinction and Collapse of Marine Trophic Structure. J Paleontol 51(suppl 2):16

376. Kauffman, E.G. (1984) Fabric of Cretaceous Marine Extinctions. pp.151-246 in (eds) W.A. Berggren, J.A. VanCouvering, Catastrophes and Earth History. NY, Princeton.

377. Kauffman, E.G. (1984) Toward a Synthetic Theory of Mass Extinction. Geol Soc Amer Abst Prog 16:555.

378. Kauffman, E.G. (1985) Cretaceous Evolution of the Western Interior Basin of the United States. pp.IV-XIII in (eds) L.M. Pratt, et al., Fine-Grained Deposits and Biofacies of the Cretaceous Western Interior Seaway: Evidence of Cyclic Sedimentary Processes, IV-XI. Tulsa, Soc Econ Paleontol Mineralogy.

379. Kauffman, E.G. (1986) High Resolution Event Stratigraphy: Regional and Global Cretaceous Bio-Events. pp.279-336 in (eds) O.II. Walliser, et al., Global Bio-Events, A Critical Approach. Lecture Notes in Earth Sciences, No 8. Berlin, Springer-Verlag.

380. Kawai, N. et al. (1969) Palaeomagnetic and Potassium-Argon Age Information Supporting Cretaceous-Tertiary Hypothetic Bend of Main Island Japan. Palaeogeogr Palaeoclim Palaeoecol 6:277-282.

381. Keany, J., Kennet, J.P. (1972) Pliocene-Early Pleistocene Paleoclimatic History Record in Antarctic-Subantarctic Deep-Sea Cores. Deep-Sea Res 19:529.

382. Keller, G. (1983) Biochronolgy and Paleoclimatic Implications of Middle Eocene to Oligocene Planktonic Forminiferal Faunas. Marine Micropaleont 7:463-486.

383. Keller, G. (1985) Eocene and Oligocene Stratigraphy Uncomformities in the Gulf of Mexico and Gulf Coast. J Paleont 59:882-903.

384. Keller, G. (1986) Stepwise Mass Extinctions and Impact Events: Late Eocene to Early Oligocene. Marine Micropaleontol 10:267-293.

385. Keller, G. (1988) Biotic Turnover in Benthic Foraminifera Across the Cretaceous/Tertiary Boundary at El Kef, Tunisia. Palaeogeogr Palaeoclimat Palaeoecol 66:153-171.

386. Keller, G. (1988) Extinction, Survivorship and Evolution of Planktonic Foraminifera Across the Cretaceous/Tertiary Boundary at El Kef, Tunisia. Mar Micropaleontol 13:239-263.

387. Keller, G., et al. (1983) Multiple Microtektities in Upper Eocene Marine Sediments: No Evidence of Mass Extinctions. Science 221:150-152.

388. Keller, G., Lindinger, M. (1989) Stable Istope, TOC and $CaCO_3$ Record Across the Cretaceous/Tertiary Boundary at El Kef, Tunisia. Palaeogeogr Palaeoclimat Palaeoecol 73:243-265.

389. Keller, G.V., et al. (1972) Magnetic Noise Preceeding the August 1971 Summit Eruption of Kilauea Volcano. Science 175:1457.

390. Kelley, M.C., et al. (1985) Large Amplitude Thermospheric Oscillations Induced by an Earthquake. Geophys Res Lett 12:577-580.

391. Kennett, J.P. (1982) Marine Geology. NY, Prentice Hall.

392. Kennett, J.P. (1986) Geologic Record of Explosive Volcanism. pp.38-72 in Abstract Volume, Norman D. Watkins Symp Environmental Impact of Volcanism.

393. Kent, D.V. (1977) Estimate of the Duration of the Faunal Change at the Cretaceous-Tertiary Boundary. Geology 5:769-771.

394. Kent, P.E. (1977) Mesozoic Development of Aseismic Continental Margins. J Geol Soc London 134:1-18.

395. Kerr, R.A. (1980) Asteriod Theory of Extinctions Strengthened. Science 210:514-517.

396. Kerr, R.A. (1982) Extinctions: Iridium and Who Went When. Science 215:389.

397. Kerr, R.A. (1984) How to Make a Warm Cretaceous Climate. Science 223:677-678.

398. Kerr, R.A. (1984) Periodic Impacts and Extinctions Reported. Science 223:1277-1279.

399. Kerr, R.A. (1985) Periodic Extinctions and Impacts Challenged. Sceince 227:1451-1453.

400. Kerr, R.A. (1987) Asteriod Impact Gets More Support. Science 236:666-668.

401. Kerr, R.A. (1987) Beyond the K-T Boundary. Science 236:667.

402. Kerr, R.A. (1987) Do Tectonic Plates Drive Themselves? Science 236:1426.

403. Kerr, R.A. (1987) Impact Cratering Looks Clustered, Not Periodic. Science 236:1426-1427.

404. Kerr, R.A. (1988) Was There a Prelude to the Dinosaurs' Demise? Science 239:729-730.

405. Khain, V.E., Seslavinsky, K.B. (1973) Some Basic Problems of Structure and Tectonic History of the North-Western Segment of the Pacific Mobile Belt. pp.389-414 in (eds) P.J. Coleman, The Western Pacific; Island Arcs, Marginal Seas, Geochemistry. NY, Crane, Russak.

406. Khan, H.R. (1987) Ferromagnetism. pp.340-348 in (ed) R.A. Meyer, Encyclopedia of Physics, Science and Technology. NY, Academic.

407. Kieffer, S.W. (1971) Shock Metamorphism of the Coconino Sandstone at Meteor Crater, Arizona. J Geophys Res 76:5449-5473.

408. Kielan-Jaworowska, Z., Barbold, R. (1972) Results of the Polish-Mongolian Palaeontological Expedition. Part IV. Palaeont Polonica 27:5-13.

409. Kimura, T. (1973) Old "Inner" Arc and its Deformation in Japan. pp.255-273 in (ed) P.J. Coleman. The Western Pacific; Island Arcs, Marginal Seas, Geochemistry. NY, Crane, Russak.

410. Kimura, T. (1974) Ancient Continental Margin of Japan. pp.817-829 in (eds) C.A. Burk, C.L. Drake, The Geology of Continental Margins. NY, Springer-Verlag.

411. King, L.C. (1983) Wandering Continents and Spreading Sea Floors on an Expanding Earth. NY, Wiley.

412. Kirillov, M. (1970) Why Did the Saurians Die Out? [in Russian] Khimiya i Zhazn (3).

413. Kistler, R.W., et al. (1971) Sierra Nevada Plutonic Cycle: Part I, Origin of Composite Granitic Batholiths. Geol Soc Amer Bull 82:853-868.

414. Kitchell, J.A., et al. (1986) Biological Selectivity of Extinction: A Link Between Background and Mass Extinction. Palaios 1:504-511.

415. Kitchell, J.A., Pena, D. (1984) Periodicity of Extinctions in the Geologic Past: Deterministic Versus Stochastic Explanations. Science 226:689-692.

416. Klekowski, E.J., Jr., et al. (1985) Shoot Apical Meristems and Mutation: Stratified Meristems and Angiosperm Evolution. Amer J Bot 72:1788-1800.

417. Klusek, C.S. (1987) Strontium-90 in Food and Bone from Fallout. J Environ Quality 16(3):195-199

418. Knobloch, E. (1986) Palaeofloristic and Palaeoclimatic Changes in the Cretaceous and Teritary Periods (Facts, Problems and Tasks). pp.371-374 in (eds) O.H. Walliser, et al., Global Bio-Events, A Critical Approach. Lecture Notes in Earth Sciences, No. 8. NY, Springer-Verlag.

419. Knoll, A.H. (1984) Patterns of Extinction in the Fossil Record of Vascular Plants. pp.21-68 in (eds) M.H. Nitecki, Extinctions. Univ Chicago.

420. Koch, B.E. (1959) Contribution to the Stratigraphy of the Non-Marine Tertiary Deposits on the South Coast of the Nugssuaq Peninsula, Northwest Greenland. Medd Gronland 162(1):1-100.

421. Koch, B.E. (1964) Review of Fossil Floras and Non-Marine Deposits od West Greenland. Geol Soc Amer Bull 75:535-548.

422. Koeberl, C. (1986) Geochemistry of Tektites and Impact Glasses. Ann Rev Earth Planet Sci 14:323-350.

423. Koeberl, C., et al. (1988) Kara and Ust-Kara Impact Structures (USSR) and Their Relevance to the K/T Boundary Event. Earth Planet Sci Lett 92:95-96.

424. Krasovskiy, V.I., Shklovskiy, I.S. (1957) Possible Effect of Supernova Explosions on the Evolution of Life on Earth. [in Russian] AN SSSR Doklady 116(2).

425. Krassilov, V.A. (1975) Ancestors of the Angiosperms. pp.76-106 (ed) N.N. Vorontsova, Contemporary Problems of Evolution, 4. Vladivostok, Akad Nauk SSSR.

426. Krassilov, V.A. (1975) Climatic Changes in Eastern Asia as Indicated by Fossil Floras II. Palaeogeogr Palaeoclimatol Palaeoecol 17:157-172.

427. Krassilov, V.A. (1975) Dirhopalostachyaceae—A New Family of Proangiosperms and its Bearing on Angiosperm Ancestry. Palaeontographica, Abt B 153:100-110.

428. Krishnamurthy, R., et al. (1986) Paleoclimatic Changes Deduced from 13C/12C and C/N Ratios of Karewa Lake Sediments. Nature 323:150-152.

429. K-TEC Group (1977) Cretaceous-Tertiary Extinctions and Possible Terrestrial and Extraterrestrial Causes. Proc Workshop Ottawa, Canada, 16-17 Nov, 1976. Ottawa, Natl Museums Canada.

430. Kucha, H. (1981) Precious Metal Alloys and Organic Matter in the Zechstein Copper Deposits Polland. TMPM Tschermaks Mineralogische und Petrographische Mitteilungen 28:1-16.

431. Kuhn, T.S. (1967) The Structure of Scientific Revolutions. Univ Chicago.

432. Kunk, M.J. et al. (1989) 40Ar-39Ar Dating of the Manson Impact Structure: A Cretaceous-Tertiary Boundary Crater Candidate. Science 244:1565-1568.

433. Kyte, F.T., Brownlee, D.E. (1985) Unmelted Meteoritic Debris in the Late Pliocene Iridium Anomaly: Evidence for the Ocean Impact of a Nonchondritic Asteriod. Geochim Cosmochim Acta 49:1095-1108.

434. Kyte, F.T., Smit, J. (1986) Regional Variations in Spinel Composition: An Important Key to the Cretaceous/Tertiary Event. Geology 14:485-487.

435. Kyte, F.T., Wasson, J.T. (1982) Geochemical Constraints on the Nature of Large Accretionary Events. Geol Soc Amer Special Paper 190:235-242.

436. Kyte, F.T., et al. (1980) Siderophile-Enriched Sediments from the Cretaceous-Tertiary Boundary. Nature 288:651-656.

437. Kyte, F.T., et al. (1981) High Noble Metal Concentrations in a Late Pliocene Sediment. Nature 292:417-420.

438. Kyte, F.T., et al. (1985) Siderophile-Enriched Sediments from the Cretaceous-Tertiary Boundary Sediments from Carvaca, Spain. Earth Planet Sci Lett 73:183-195.

439. La Brecque, J.L., et al. (1977) Revised Magnetic Polarity Time Scale for Late Cretaceous and Cenozoic Time. Geology 5:330-335.

440. Ladd, J.W., et al (1984) Seismic Reflection Across the Southern Margin of the Caribbean. Geol Soc Amer Memoirs 162:153-160.

441. Lal, D., Peters, B. (1967) [Handbook of Physics] Handbuch der Physik 46/2, Cosmic Rays 2:551. Berlin, Springer-Verlag.

442. Lambeck, K., et al. (1984) State of Stress within the Australian Continent. Annales Geophysicae 2:723-742.

443. Landis, C.A., Bishop, D.G. (1972) Plate Tectonics and Regional Stratigraphic-Metamorphic Relations in the Southern Part of the New Zealand Geosyncline. Geol Soc Amer Bull 83:2267-2284.

444. Lapedes, D.N. (1978) Mc Graw-Hill Encyclopedia of the Geological Sciences. NY, Mc Graw-Hill.

445. Laster,H. (1968) Cosmic Rays from Nearby Supernovae: Biological Effects. Science 160:1138.

446. Lehman, T.M. (1990) Paleosols and the Cretaceous-Tertiary Transition in the Big Bend Region of Texas. Geology 18:362-364.

447. Leonov, G.P. (1973) Principles of Stratigraphy [in Russian] Vol.1. Moscow, Izd-vo Mosk Univ.

448. Lerbekmo, J.F., et al. (1987) Relationship Between the Iridium Anomaly and Palynological Floral Events at Three Cretaceous-Tertiary Boundary Localities in Western Canada. Geol Soc Amer Bull 99:325-330.

449. Lewin, R. (1983) Extinctions and the History of Life. Science 221:935-937.

450. Lin, J., et al. (1985) Preliminary Phanerozoic Polar Wander Paths for the North and South China Blocks. Nature 313:444-449.

451. Lindley, D. (1988) Is the Earth Alive or Dead? Nature 332:483-484.

452. Lindsay, E.H., et al. (1978) Biostratigraphy and Magnetostratigraphy of Paleocene Terrestrial Deposits, San Juan Basin, New Mexico. Geology 6:425-429.

453. Lisowski, M., et al. (1990) Possible Geodetic Anomaly Observed Prior to the Loma Prieta, California, Earthquake. Geophys Res Lett 17:2011-2014.

454. Lockley, M.G., et al. (1986) North America's Largest Dinosaur Trackway Site: Implications for Morrison Formation Paleoecology. Geol Soc Amer Bull 97:1163-1176.

455. Lombardi, S., Reimer, G.M. (1990) Radon and Helium in Soil Gases in the Phlegraean Fields, Central Italy. Geophys Res Lett 17:849-852.

456. Loper, D.E., et al. (1988) Model of Correlated Episodicity in Magnetic-Field Reversals, Climate and Mass Extinctions. J Geol 96:1-15.

457. Loper, D.E., Mc Cartney, K. (1986) Mantle Plumes and the Periodicity of Magnetic Field Reversals. Geophys Res Lett 13:1525-1528.

458. Lowenstam, H.A., Epstein, S. (1954) Paleotemperatures of the Post-Aptian Cretaceous as Determined by the Oxygen Isotope Method. J Geol 62:207-248.

459. Lowrie, W., et al (1980) A Review of Magnetic Stratigraphy Investigations in Cretaceous Pelagic Carbonate Rocks. J Geol Res 85B:3597-3605.

460. Luck, J.M., Turekin, K.K. (1983) Osmium-187/Osmium-186 in Manganese Nodules and the Cretaceous-Tertiary Boundary. Science 222:613-615.

461. Lyell, C. (1830-33) Principles of Geology. NY, S-H Service Agency [Fascimilie Reprint of 1st Edition, c1970].

462. MacDonald, K.C. (1988) Cracks in the Pacific Plate and Mantle Convection. Nature 331:395.

463. MacDougall, J.D. (1988) Seawater Strontium Isotopes, Acid Rain and the Cretaceous-Tertiary Boundary. Science 239:485-487.

464. Maddox, J. (1988) Earthquakes and the Earth's Rotation. Nature 332:11.

465. Magaritz, M. (1989) 13C Minima Folow Extinction Events: A Clue to Faunal Radiation. Geology 17:337-340.

466. Mapes, G., et al. (1989) Evolution of Seed Dormancy. Nature 337:645-646.

467. Mal'kov, B.A. (1978) Global Epochs of Kimberlite Volcanism in the Phanerozoic. Doklady Earth Sci Sections 42:113-115 [English Trans] or 1152-1154 [in Russian] Nauk SSSR, Doklady Akademii.

468. Mansinha, L., Smylie, D.E. (1967) Effect of Earthquakes on the Chandler Wobble and the Secular Polar Shift. J Geophys Res 72:4731-4743.

469. Mantura, A.J. (1972) Geophysical Illusions of Continental Drift. Amer Assoc Petrol Geol Bull 56:1152-1156 & 2451-5.

470. Margolis, S.V., et al. (1977) Cenozoic and Late Mesozoic Paleooceanographic and Paleoglacial History Recorded in Circum-Antarctic Deep-Sea Sediments. Marine Geol 25:131-47.

471.　Margolis, S.V., Kennett, J.P. (1970) Antarctic Glaciation during the Tertiary Recorded in Sub-Antarctic Deep-Sea Cores. Science 170:1085-1087.

472.　Margulis, L., et al. (1976) Reassessment of Roles of Oxygen and Ultraviolet Light in Precambrian Evolution. Nature 264:620-624.

473.　Marks, J.G. (1965) Ecuador; Pacific Coast Geologic Province. Geol Soc Amer Mem 65:277-291.

474.　Markson, R., Nelson, R. (1971) Mountain-Peak Potential-Gradient Measurements and the Andes Glow. Weather 25:350.

475.　Marino, A.A., et al. (1974) Electric Field Effects in Selected Biological Systems. Ann NY Acad Sci 238:436-444.

476.　Martin, W, et al. (1989) Molecular Evidence for Pre-Cretaceous Angiosperm Origins. Nature 339:46-48.

477.　Matese, J.J., Whitmire, D.P. (1986) Planet-X and the Origins of the Shower and Steady-State Flux of Short-Period Comets. Icarus 65:37-50.

478.　Mattson, P.H. (1984) Caribbean Structural Breaks and Plate Movements. Geol Soc Amer Memoirs 162:131-152.

479.　Masaitis, V.L. (1975) Astroblemes in the Soviet Union. [in Russian] Soviet Geol 11:52-64.

480.　Matsumoto, T. (1967) Fundamental Problems in the Circum-Pacific Orogenesis. Tectonophysics 4:595-613.

481.　Matsumoto, T. (1977) Timing of Geological Events in the Circum-Pacific Region. Canadian J Earth Sci 14:551-561.

482.　Mayers, I.R., Worsley, T.R. (1973) Statistical Recognition of Late Cretaceous Cyclic Sedimentation by Means of Calcareous Nannofossil Population Studies. Palaeogeogr Palaeoclimat Palaeoecol 13:81-90.

483.　McCartney, K., Nienstedst, J. (1986) Cretaceous/Tertiary Controversy Reconsidered. J Geol Ed 34:90-94.

484.　McCrea, W.H. (1981) Long Time-Scale Fluctions in the Evolution of the Earth. Proc Roy Soc London 375A:1-41.

485.　McElhinny, M.W. (1973) Palaeomagnetism and Plate Tectonics of Eastern Asia. pp.407-414 in (ed) P.J. Coleman, Western Pacific: Island Arcs, Marginal Seas, Geochemistry. NY, Crane, Russak.

486.　McElhinny, M.W. (1973) Palaeomagnetism and Plate Tectonics. Cambridge Univ.

487.　McHone, J.F. et al. (1989) Stishovite at the Cretaceous-Tertiary Boundary, Raton, New Mexico. Science 243:1182-4.

488.　McKinney, M.L. (1987) Taxonomic Selectivity and Continous Variation in Mass and Background Extinctions of Marine Taxa. Nature 325:143-145.

489.　McLaren, D.J. (1970) Time, Life and Boundaries. J Paleontol 44:801-815.

490.　McLaren, D.J. (1983) Bolides and Biostratigraphy. Geol Soc Amer Bull 94:313-324.

491.　McLaren, D.J. (1984) Abrupt Extinctions. Terra Cognita 4:27-32.

492.　McLaren, D.J. (1985) Mass Extinction and Iridium Anomaly in the Upper Devonian of Western Australia: A Commentary. Geology 13:170-172.

493.　McLaren, D.J. (1986) Abrupt Extinctions. pp.37-48 in (ed) D.K. Elliot. Dynamics of Extinction. NY, Wiley.

494. McLean, D.M. (1978) A Terminal Mesozoic "Greenhouse": Lessons from the Past. Science 101:401-406.

495. McLean, D.M. (1980) Terminal Cretaceous Catastrophe. Nature 287:260.

496. McLean, D.M. (1983) Mantle Degassing, Williams-Riley "Pump" Disruption. Carbon Cycle Perturbation in the K-T Transition. Eos 64:245.

497. McLean, D.M. (1985) Deccan Traps Mantle Degassing in the Terminal Cretaceous Marine Extinctions. Cret Res 6:235-259.

498. Melosh, H.J., et al. (1990) Ignition of Global Wildfires at the Creatceous/Tertiary Boundary. Nature 343:251-254.

499. Mercer, J.H. (1968) West Antarctic Ice Sheet. Intl Assoc Sci Hydrol Pub 79:217-225.

500. Mercer, J.H. (1978) West Antarctic Ice Sheet and CO_2 Greenhouse Effect: A Threat of Disaster. Nature 271:321-5.

501. Merill, R.T., McElhinny, M.W. (1983) Earth's Magnetic Field: Its History, Origin and Planetary Perscpective. NY, Academic.

502. Merrill, J.R., et al. (1960) Sedimentary Geochemistry of the Beryllium Isotopes. Geochim Cosmochim Acta 18:108.

503. Meyerhoff, A.A., Meyerhoff, H.A. (1972) New Global Tectonics: Major Inconsistencies. AAPG Bull 56:269-336.

504. Meyers, P.A., Simonett, B.R.T. (1989) Global Comparisons Matter in Sediments Across the Cretaceous/Tertiary Boundary. Adv Organic Geochem 16:641-648.

505. Minato, M. (1968) Basement Complex and Paleozoic Orogeny in Japan. Pacific Geol 1:85-95.

506. Mitra, A.P. (1974) Ionospheric Effects of Solar Flares. Boston, Reidel.

507. Mizutani, H., et al. (1976) Electrokinetic Phenomena Associated with Earthquakes. Geophys Res Lett 3:365-368.

508. Mogi, K. (1974) Active Periods in the World's Cheif Seismic Belts. Tectonophysics 22:265-82.

509. Monastersky, R. (1987) Quake Prediction: Magnetic Signals? Sci News 132:167.

510. Monastersky, R. (1988) Whole-Earth Syndrome. Sci News 133:378-380.

511. Monger, J.W.H., Ross, C.A. (1971) Diatribution of Fusulinaceans in the Western Canadian Cordillera. Canadian J Earth Sci 8:259-278.

512. Montanari, A., et al. (1983) Spheriods at the Cretaceous-Tertiary Boundary are Altered Impact Droplets of Basaltic Composition. Geology 11:668-671.

513. Moorbath, S.H., et al. (1968) K-Ar Dates of the Oldest Exposed Rocks in Iceland. Earth Planet Sci Lett 4:197-205.

514. Moore, G.W. (1964) Magnetic Disturbances Preceeding the 1964 Alaska Earthquake. Nature 203:508.

515. Moore, T.C., et al. (1978) Cenozoic Hiatuses in Pelagic Sediments. Micropaleontology 24:113-138.

516. Morel, P., Irving, E. (1981) Evolution of the Continents. Geos (Spr):2-6.

517. Mount, J.F., et al. (1986) Carbon and Oxygen Isotope Stratigraphy of the Upper Maestrichtian, Zumaya, Spain: a Record of Oceanographic and Biologic Changes at the End of the Cretaceous Period. Palaios 1:87-92.

518. Mueller, R.J., Johnston, M.J.S. (1990) Seismomagnetic Effect Genertated by the October 18, 1989, Ms 7.1 Loma Prieta, California, Earthquake. Geophys Res Lett 17:1777-1780.

519. Muller, R.A. (1985) Evidence for a Solar Companion. pp.233-243 in (ed) M.D. Papagiannis, Search for Extraterrestrial Life: Recent Developments. Boston, Mass, D Reidel.

520. Muller, R.A., Morris, D.E. (1986) Geomagnetic Reversals from Impacts on the Earth. Geophys Res Lett 13:1177-80.

521. Mutter, J.C., et al. (1985) Breakup Between Australia and Antarctica: A Brief Review in Light of New Evidence. Tectonophysics 114:255-279.

522. Myers, A.T., Hamilton, J.C. (1961) Rhenium in Plant Samples from the Colorado Plateau. US Geol Survey Res B286-B288.

523. Myerson, R.J. (1970) Long-Term Evidence for the Association of Earthquakes with the Excitation of the Chandler Wobble. J Geophys Res 75:6612-6617.

524. Nafe, J.E., Drake, C.L. (1968) Floor of the North Atlantic—Summary of Geophysical Data. Amer Assoc Petrol Geol Mem 12:59-87.

525. Nagappa, Y. (1960) Cretaceous-Tertiary Boundary in the India-Pakistan Subcontinent. Intl Geol Congr 21st Copenhagen Report, Pt 5:41-49.

526. Naidin, D.P. (1976) Epeirogeny and Eustacy. Vestn Mosk Univ Geol (2):3-16.

527. Naidin, D.P. (1987) Cretaceous-Tertiary Boundary in Mangyshlak, U.S.S.R. Geol Mag 1:13-19.

528. Nambi, K.S.V., et al. (1985) Environmental Radioactivity and Thermoluminesence: A Review. J Environ Radioactiv 2:59-75.

529. Namias, J. (1988) Similarity of Anomalous Sea Level Pressure Fields During the July 1986 and September 1987 Southern California Quakes—Accidental or Indicative? Geophys Res Lett 15:350-352.

530. Naslund, H.R., et al. (1986) Microspherules in Upper Creataceous and Lower Tertiary Clay Layers at Gubbio Italy. Geology 14:923-926.

531. National Research Council, U.S. Commitee on the Atmospheric Effects of Nuclear Explosions (1985) Effects on the Atmosphere of a Major Nuclear Exchange. Wash, DC, Natl Res Council, USGPO.

532. Negi, J.G., Tiwari, R.K. (1983) Matching Long Term Periodicities of Geomagnetic Reversals and Galactic Motions of the Solar System. Geophys Res Lett 10:713-716.

533. Nelson, S.O. (1965) Electromagnetic Radiation Effects on Seeds. pp.60-63 in E.M. Radiation in Agriculture. Conf Proc Illuminating Engineering Soc & Amer Soc Agric Engineers, Oct. Roanoke, Virginia.

534. Newell, N.D. (1962) Palaeontologic Gaps and Geochronology. J Palaeontol 36:592-610.

535. Newell, N.D. (1967) Revolutions in the History of Life. Geol Soc Amer Spec Paper 89.

536. Ngo, H.H., et al. (1985) Nd and Sr Isotopic Compostions of Tektite Material from Barbados and Their Relationship to North American Tektities. Geochim Cosmochim Acta 49:1479-85.

537. Nicholaysen, L.O. (1987) Tektites: Ejecta from Massive Cratering Events, Caused by Periodic Escape and Detonation of Deep Mantle Fluids. in Intl Workshop Cryptoexplosions and Catastrophes in the Geol Record (Parys). Sect. N3, 15.

538. Nichols, D.J., et al. (1986) Palynological and Iridium Anomalies at the Creataceous-Tertiary Boundary, Southeastern Saskatchewan. Science 231:714-717.

539. Nier, A.O.C., et al. (1975) Long-Term Worldwide Effects of Multiple Nuclear-Weapons Detonations. Wash, DC, Natl Acad Sci.

540. Norman, D.B. (1988) Embryo in Dinosaur Nests. Nature 332:202-203.

541. Norris, G., et al. (1975) Evolution of the Cretaceous Palynoflora in Western Canada. Geol Assoc Canada Spec Paper 13:333-364.

542. Nur, A., Ben-Avraham, Z. (1977) Lost Pacifica Continent. Nature 270:41-43.

543. Officer, C.B., Drake, C.L. (1983) Cretaceous-Tertiary Transition. Science 219:1383-1390.

544. Officer, C.B., Drake, C.L. (1985) Terminal Cretaceous Environmental Events. Science 227:1161-1167.

545. Officer, C.B., Ekdale, A.A. (1986) Cretaceous Extinctions and Wildfires. Science 234:262-3.

546. Officer, C.B., et al. (1987) Late Cretaceous and Paroxysmal Cretaceous/Tertiary Extinctions. Nature 326:143-149.

547. Ogren, J.A. (1982) Deposition of Particulate Elemental Carbon from the Atmosphere. pp.370-391 in (eds) G.T. Wolf, R.L. Klimisch, Particulate Carbon: Atmospheric Life Cycle. NY, Plenum.

548. O'Connell, R.J., Dziewonski, A.M. (1976) Excitation of the Chandler Wobble by Large Earthquakes. Nature 262:259-262.

549. O'Keefe, J.A. (1978) The Tektite Problem. Sci Amer 239(2):116-125.

550. O'Keefe, J.A. (1980) Comments on "Chemical Relationship Among Irghizites, Zhamanshinites, Australasian Tektites and Henbury Impact Glass." Geochim Cosmochim Acta 44:2151-2152.

551. O'Keefe, J.A. (1985) Terminal Createcous Event: Circumterrestrial Rings of Tektite Glass Particles? Cretaceous Res 6:261-269.

552. O'Keefe, J.A. (1985) Coming Revolution in Planetology. Eos 66:89.

553. O'Keefe, J.A. (1987) Zhamanshin and Aouelloul: Craters Produced by Impact of Tektite-like Glasses? Meteoritics 22:219-228.

554. O'Keefe, J.D., Ahrens, T.J. (1982) Interaction of the Cretaceous-Tertiary Bolide with the Atmosphere, Ocean and Solid Earth. Geol Soc Amer Spec Paper 190:103-120.

555. Ollier, O.D. (1981) Tectonics and Landforms. NY, Longman.

556. Olsen, P.E., et al. (1987) New Early Jurassic Tetrapod Assemblages Constrain Triassic-Jurassic Tetrapod Extinction Event. Science 237:1025-1029.

557. Olson, P. (1987) Plate Tectonics: Drifting Mantle Hotspots. Nature 327:559.

558. Olson, P., et al. (1987) Plume Formation in the D-Layer and Roughness of the Core-Mantle Boundary. Nature 327:409.

559. Orth, C.J., et al. (1981) Iridium Abundance Anomaly at the Palynological Cretaceous-Tertiary Boundary in Northern New Mexico. Science 214:1341-1343.

560. Orth, J.S., et al. (1984) A Search for Iridium Abundance Anomalies at Two Late Cambrian Biomere Boundaries in Western Utah. Science 223:163-165.

561. Orth, J.S., et al. (1988) Siderophile Abundance Maxima at Upper Cenomanian Marine Invertebrate Extinction Horizon: Western Interior, North America. Oral Presentation at 19th Lunar & Planetary Sci Conf, Houston.

562. Orth, J.S., et al. (1988) Iridium Abundance Maxima in the Upper Cenomanian Extinction Interval. Geophys Res Lett 15:346-349.

563. Osborn, H.F. (1911) Dinosaur Mummy. Amer Mus J 2:7-11.

564. Ostram, J.H. (1980) Evidence for Endothermy. pp.15-54 in (eds) R.D.K. Thaomas, E.C. Olson, A Cold Look at the Warm-Blooded Dinosaurs. Boulder, Colo, Westview.

565. Ostrom, P.H., et al. (1990) Geochemical Characterization of High Molecular Weight Material Isolated from Late Cretaceous Fossils. Organic Geochem 16:1139-1144.

566. Ouellet, M. (1990) Earthquake Lights and Seismicity. Nature 348:492.

567. Owen, H.G. (1983) Atlas of Continental Displacement: 200 Million Years to the Present. NY, Cambridge Univ.

568. Owen, M.R., Anders, M.H. (1988) Evidence from Cathodluminescence for Non-Volcanic Origin of Shocked Quartz at the Cretaceous/Tertiary Boundary. Nature 334:145-147.

569. Oxburgh, E.R., O'Nions, R.K. (1987) Helium Loss, Tectonics and the Terrestrial Heat Budget. Science 237:1583-1587.

570. Pacey, N.R. (1984) Bentonities in Chalk of Central Eastern England and their Relation to the Opening of the Northeast Atlantic. Earth Planet Sci Lett 67:48-60.

571. Packer, D.R., Stone, D.B. (1974) Paleomagnetism of Jurassic Rocks from Southwestern Alaska and Their Tectonic Implication. Canadian J Earth Sci 11:976-997.

572. Padian, K., Clemens, W.A. (1985) Terrestrial Vetebrate Diversity: Episodes and Insights. pp.41-96 in (ed) J.W. Valentine, Phanerozoic Diversity Patterns. NJ, Princeton.

573. Padian, K., et al. (1984) Possible Influences of Sudden Events on Biological Radiation and Extinctions. pp.77-102 in (eds) H.D. Holland, A.F. Trendall, Patterns of Change in Earth Evolution. Berlin, Springer-Verlag.

574. Pal, P.C., Creer, K.M. (1986) Geomagnetic Reversal Spurts and Episodes of Extraterrestrial Catastrophism. Nature 320:148-150.

575. Pal, D.K., et al. (1982) Beryllium-10 in Australasian Tektites: Evidence for a Sedimentary Precursor. Science 218:787-789.

576. Pantic, N. (1986) Global Tertiary Climatic Changes, Paleophytogeography and Phytostratigraphy. pp.419-428 in (eds) O.H. Walliser, et al. Global Bio-Events: A Critical Approach. Lecture Notes in Earth Sciences, No 8. Berlin, Springer-Verlag.

577. Parrish, J.D., Spicer, R.A. (1988) Late Cretaceous Terrestrial Vegetation: A Near-Polar Temperature Curve. Geology 16:22-25.

578. Patwardham, A.M., Ahluwalia, A.D. (1971) Discovery of Pre-Tertiary Fossils Indigenous to the Lower Himalayan Basin. Nature 230:451-452.

579. Phillips, N.D. (1967) Magnetic Anomalies Over the Mid-Atlantic Ridge Near 27o N. Science 157:920-923.

580. Pillmore, C.L., et al. (1984) Geologic Framework of Nonmarine Cretaceous-Tertiary Boundary Sites, Raton Basin, New Mexico and Colorado. Science 223:1180-1183.

581. Pines, D., Shaham, J. (1973) Seismic Activity, Polar Tides and the Chandler Wobble. Nature 245:77.

582. Pitman, W.C. (1978) Relationship Between Eustacy and Stratigraphy Sequences of Passive Margins. Bull Geol Soc Amer 89:1389-1403.

583. Pitman, W., Hays, J. (1973) Upper Cretaceous Spreading Rates and the Great Transgression. Geol Soc Amer Abst 5:768.

584. Pollack, J.B., et al. (1982) Environmental Effects of an Impact Generated Dust Cloud: Implications for the Cretaceous-Tertiary Extinctions. Science 219:287-289.

585. Pollastro, R.M., Pillmore, C.L. (1987) Minerlogy and Petrology of the Cretaceous-Tertiary Boundary Clay Bed and Adjacent Clay-Rich Rocks, Raton Basin, New Mexico and Colorado. J Sedimentary Petrol 57:456-466.

586. Pomerol, C. (1982) Cenozoic Era—Tertiary and Quarternary. NY, Wiley..

587. Poplavko, E.M., et al. (1974) On the Concentration of Rhenium in Petroleum, Petroleum Bitumens and Oil Shales. Geochem Intl 969-972.

588. Preisinger, A., et al. (1986) Cretaceous/Teritary Boundary in the Gosau Basin, Austria. Nature 322:794-799.

589. Press, F. Briggs, P. (1975) Chandler Wobble, Earthquakes, Rotation and Geomagnetic Changes. Nature 256:270.

590. Prinn, R.G. (1985) Impacts, Acid Rain and Biospheric Traumas. Eos 66:813.

591. Raisbeck, G.M. et al. (1985) Evidence for an Increase in Cosmogenic 10Be During a Geomagnetic Reversal. Nature 315:315-317.

592. Rampino, M.R. (1987) Impact Cratering and Flood Basalt Volcanism. Nature 327:468.

593. Rampino, M.R. (1987) Worlds in Collusion: How Galactic Disturbances Trigger Earthly Cycles. Sciences 27(4):30-37 [and] Cycles 39:23-27 (1988).

594. Rampino, M.R., Stothers, R.B. (1984) Terrestrial Mass Extinctions, Cometary Impacts and the Sun's Motion Perpendicular to the Galactic Plane. Nature 308:709-712.

595. Rampino, M.R., Stothers, R.B. (1984) Geological Rhythms and Cometary Impacts. Science 226:1427-1431.

596. Rao, B.S.N. (1987) Radionuclides in Foods Due to Radioactive Fallout and Their Biological Hazards. NFI Bull (India) 8(3):6-7.

597. Raup, D.M. (1979) Size of the Permo-Triassic Bottleneck and its Evolutionary Implications. Science 206:217.

598. Raup, D.M. (1985) Magnetic Reversals and Mass Extinctions. Nature 314:341-343.

599. Raup, D.M. (1985) Rise and Fall of Periodicity. Nature 317:384-385.

600. Raup, D.M. (1989) The Case For Extraterrestrial Causes of Extinction. Phil Trans Roy Soc London 325B:421-435.

601. Raup, D.M., et al. (1988) Testing for Periodicity of Extinction. Science 241:94-99.

602. Raup, D.M., Sepkoski, J. (1984) Periodicity of Extinctions in the Geologic Past. Proc Natl Acad Sci 81:801-805.

603. Raup, D.M., Sepkoski, J. (1986) Periodic Extinction of Families and Genera. Science 231:833-836.

604. Rea, D.K., Vallier, T.L. (1983) Two Cretaceous Volcanic Episodes in the Western Pacific Ocean. Geol Soc Bull 94:1430-1437.

605. Reid, G.C. (1977) Stratospheric Aeronomy and the Cretaceous-Tertiary Extinctions. Syllogeus 12:75-88.

606. Reid, G.C. (1977) Stratospheric Aeronomy and the Cretaceous-Tertiary Extinctions, pp.75-89 in (eds) K-Tec Group, Cretaceous-Tertiary Extinctions and Possible Terrestrial and Extraterrestrial Causes. Proc Workshop Ottawa, Canada 16-17 Nov 1976. Ottawa, Natl Museums of Canada.

607. Reid, G.C., et al. (1976) Influences of Ancient Solar-Proton Events on the Evolution of Life. Nature 259:177-179.

608. Reid, G.C., et al. (1978) Effects of Intense Stratospheric Ionization Events. Nature 257:489-92.

609. Reimer, G.M. (1985) Prediction of Central California Earthquakes from Soil-Helium Fluctuations. Pure Appl Geophys 122:369-375.

610. Reimer, G.M. (1990) Helium Increase. Nature 347:342.

611.   Research Group of the Tanba Belt (1971) Paleozoic System in the Tanba Belt (part 2) [in Japanese, English Abst]. J Assoc Geol Colloquim Japan 25:211-218.

612.   Reyment, R.A., Morner, N.A. (1977) Cretaceous Transgressions and Regressions Exemplified by the South Atlantic. Paleontol Soc Japan Sec Paper 21:248-261.

613.   Rice, A. (1987) Shocked Minerals at the K/T Boundary: Explosive Volcanism as a Source. Phys Earth Planet Interiors 48:167-174.

614.   Rich, P.V., et al. (1986) Significant Correlation Between Fluctuations in Seafloor Spreading Rates and Evolutionary Pulsations. Paleoceanography 1:83-95.

615.   Rich, P.V., et al. (1988) Evidence for Low Temperatures and Biologic Diversity in Cretaceous High Latitudes of Australia. Science 242:1403-1406.

616.   Richards, H.G. (1974) Tectonic Evolution of Alaska. Bull Amer Assoc Petro Geol 58:79-105

617.   Richardson, R.M.S., et al. (1979) Tectonic Stress in Plates. Rev Geophys Space Phys 17:981-1019.

618.   Rigby, J.K., Jr., Sloan, R.E. (1985) Dinosaur Decline and Eventual Extinction Near the Cretaceous/Tertiary Boundary, Hell Creek Fm., Mt. Geol Soc Amer Abstr Prog 17:700.

619.   Robert, C., H. Chamley (1990) Paleoenvironmental Significance of Clay Mineral Associations at the Cretaceous-Tertiary Passage. Palaeogeogr Palaeoclim Palaeoecol 79:205-219.

620.   Robertson, P.B., et al. (1968) Deformation in Rock-Forming Minerals from Canadian Craters. pp.433-452 in (ed) N.M. Short, Shock Metamorphism of Natural Materials. Baltimore, Md, Mono Book.

621.   Rocchia, R., et al. (1984) [Essay Evaluating the K-T Transition for the Evolution of the Iridium Anomaly: Implications for the Cause of the Mass Extinction.] Eassai d'Evaluation de la Transition C-T par l'Evolution de l'Anomalie en Iridium: Implications dans la Recherche de la Cause de la Crise Biologique. Bull Soc Geol France 26:1193-1202.

622.   Rocchia, R., et al. (1987) [Comparsion of the Distribution of Iridium Observations at the Cretaceous-Tertiary Boundary from Various European Sites]. Comparison des Distribution de l'Iridium Observees a la Limite Cretace-Tertiaire dans Divers Sites Europeens. Memiors Soc Geol France 150:95-103.

623.   Ross, C.A. (1976) Paleobiogeography. NY, Hutchinson & Ross.

624.   Ross, D.A. (1982) Introduction to Oceanography. NJ, Prentice-Hall.

625.   Roy, J.-R. (1977) Variations of the Luminosity of the Sun and "Super" Solar Flares: Possible Causes of Extinctions. pp.89-110 in (eds) K-Tec Group, Cretaceous-Tertiary Extinctions and Possible Terrestrial and Extraterrestrial Causes. Proc Workshop Ottawa, Canada 16-17 Nov 1976. Ottawa, Natl Museums of Canada.

626.   Roy, J.-R. (1977) Variations of the Luminosity of the Sun and "Super" Solar Flares: Possible Causes of Extinctions. Syllogeus 12:89-110.

627.   Runcorn, S.K. (1972) Flow in the Mantle Inferred from the Low Degree Harmonics of the Geopotential. Geophys J Roy Astron Soc 14:375.

628.   Russell, D.A. (1971) Disappearence of the Dinosaurs. Can Geogr J 83:204-215.

629.   Russell, D.A. (1973) Environment of Canadian Dinosaurs. Can Geogr J 87:4-11.

630.   Russell, D.A. (1977) Biotic Crisis at the End of the Cretaceous Period. pp.11-24 in (eds) K-Tec Group, Cretaceous-Tertiary Extinctions and Posible Terrestrial and Extraterrestrial Causes. Proc Workshop Ottawa, Canada 16-17 Nov 1976. Ottawa, Natl Museums Canada.

631. Russell, D.A. (1979) Enigma of the Extinction of the Dinosaurs. Ann Rev Earth Planet Sci 7:163-182.

632. Russell, D.A. (1982) Mass Extinctions of the Late Mesozoic. Sci Amer 246(1):58-65.

633. Russell, D.A. (1984) Terminal Cretaceous Extinctions of Large Reptiles. pp.373-384 in (eds) W.A. Berggren, J.A.VanCouvering, Catastrophes and Earth History. NJ, Princeton.

634. Russell, D.A., Singh, C. (1978) Cretaceous-Tertiary Boundary in Southcentral Alberta—A Reappraisal Based on Dinosaurian and Microfloral Extinctions. Can J Earth Sci 15:284-292.

635. Russell, D.A., Tucker, W. (1971) Supernovae and the Extinction of the Dinosaurs. Nature 229:553-554.

636. Rutland, R.W.R. (1976) Orogenic Evolution of Australia. Earth Sci Rev 12:161-196.

637. Ruzhentsev, V.Ye. (1957) Expansions and Crises in the History of the Ammonoids. [in Russian]. AN SSSR Doklady 115(4).

638. Sager, W.W., Bleil, U. (1987) Latitudinal Shift of Pacific Hotspots During the Late Cretaceous and Early Tertiary. Nature 326:488-490.

639. Sager, W.W., Pringle, M.S. (1988) Mid-Cretaceous to Early Tertiary Apparent Polar Wander Path of the Pacific Plate. J Geophys Res 93B:11753-11771.

640. Saito, T., et al. (1966) Tertiary Sediment from the Mid-Atlantic Ridge. Science 151:1075-79

641. Saito, T., et al. (1986) End-Cretaceous Devastation of Terrestrial Flora in the Boreal Far East. Nature 323:253-5.

642. Saito, T., Van Donk, J. (1974) Oxygen and Carbon Isotope Measurements of Late Cretaceous and Early Tertiary Foraminifera. Micropaleontology 20(2):152-177.

643. Salop, L.I. (1964) Geology of Baykal Mountain Region. Vol. 1, Stratigraphy [in Russian]. Moscow.

644. Salop, L.I. (1973) Precambrian Tillites and the Great Glaciations. [in Russian]. Mosk Obshch Ispytat Prirody Byull Otd Geol 48(6).

645. Salop, L.I. (1977) Relationship of Glaciations and Rapid Changes in Organic Life to Events in Outer Space. Intl Geol Rev 19:1271-1291.

646. Saltzman, E.S., Barron, E.J. (1982) Deep Circulation in the Late Cretaceous: Oxygen Isotope Paleotemperatures from *Inoceramus* Remains in D.S.D.P. Cores. Palaeogeogr Palaeoclim Palaeoecol 40:167-181.

647. Sankaranarayanan, K. (1982) Genetic Effects of Ionizing Radiation in Multicellular Eukaryotes and the Assessment of Genetic Radiation Hazards in Man. NY, Elsevier.

648. Schindewolf, O.H. (1954) [About the Possible Causes and Extent of Conflagrations in Historical Geology] Uber Die Moglichen Ursachen der Grossen Erdgeschichtlichen Faunenschnitte. Neues Jahrb Geol u Palaontol Abh (10).

649. Schindewolf, O.H. (1958) [Proclaiming the Magnitude of Conflagrations and Their Causes in Historical Geology]. Zur Aussprache Uber Die Grossen Erdgeschichtlichen Feunenschnitte und Ihre Verursachung. Neues Jahrb Geol u Palaontol (6).

650. Schindewolf, O.H. (1962) [Neocatastrophism?] Neokatastropismus? Zeit Deutsch Geol Gesselschaft 114(2).

651. Schlanger, S.O., et al. (1981) Volcanism and Vertical Tectonics in the Pacific Basin Related to Global Cretaceous Transgressions. Earth Planet Sci Lett 52:435-449.

652. Schmitz, B. (1985) Metal Precipitation in the Cretaceous-Tertiary Boundary Clay at Stevns Klint, Denmark. Geochim Cosmochim Acta 49:2361-2370.

653. Schmitz, B. (1988) Origin of Microlayering in Worldwide Distributed Ir-Rich Marine Cretaceous/Tertiary Boundary Clays. Geology 16:1068-1072.

654. Schmitz, B., et al. (1988) Iridium, Sulfur Isotopes and Rare Earth Elements in the Cretaceous/Tertiary Boundary Clay at Stevns Klint, Denmark. Geochim Cosmochim Acta 52:229-236.

655. Schneider, D.A., Kent, D.V. (1990) Ivory Coast Tektites and Geomagnetic Reversals. Geophys Res Lett 17:163-166.

656. Schopf, T.J.M. (1981) Evidence from Findings of Molecular Biology with Regard to the Rapidity of Genomic Change: Implications for Species Durations. pp.135-192 in (ed) K.J. Niklas, Paleobotany, Paleoecology and Evolution. Vol.1. NY, Praeger.

657. Schuchert, C. (1955) Atlas Paleographic Maps of North America. NY, Wiley.

658. Schwartz, R.D., James, P.B. (1984) Periodic Mass Extinction and Sun's Oscillation About the Galactic Plane. Nature 308:712-713.

659. Schwarzchild, B. (1987) Do Asteriod Impacts Trigger Geomagnetic Reversals? Phys Today 40(2):17-20.

660. Schweickert, R.A. (1976) Early Mesozoic Rifting and Fragmentation of the Cordilleran Orogen in the Western U.S.A. Nature 260:586-591.

661. Schwiderski, E.W. (1968) Mantle Convection and Crustal Tectonics Inferred from a Satellites' Orbit: A Different View of Sea-Floor Spreading. J Geophys Res 73:2828-2833.

662. Sepkoski, J.J., Jr. (1982) A Compendum of Fossil Marine Families. Contributions in Biology and Geology, No 51, Milwaukee Public Museum.

663. Sepkoski, J.J., Jr. (1986) Global Bioevents and the Question of Periodicity. pp.47-62 in (eds) O.H. Walliser, et al, Global Bio-Events: A Crtical Approach. Lecture Notes in Earth Sciences, No 8. NY, Springer-Verlag.

664. Sepkoski, J.J., Jr. (1986) Phanerozoic Overview of Mass Extinction. in (eds) D.M. Raup, D. Jablonski, Patterns and Process in the History of Life. NY, Springer-Verlag.

665. Sepkoski, J.J., Jr., Raup, D.M. (1986) Periodicity in Marine Extinction Events. pp.3-36 in (ed) E.K. Elliott, Dynamics of Extinction. NY, Wiley.

666. Seyfert, C.K., Sirkin. L.A. (1979) Earth History and Plate Tectonics: an Introduction to Historical Geology. NY, Harper & Row.

667. Shackleton, N.J., Hall, M.A. (1984) Carbon Isotope Data from Leg 74 Sediments. pp.613-619 in (eds) Init Rpts DSDP. Vol. 74. Wash, DC, USGPO.

668. Shackleton, N.J., Kennett, J. (1975) Paleotemperature History of the Cenozoic and the Initiation of Antarctic Glaciation: Oxygen and Carbon Isotope Analyses in DSDP Sites 277, 279 and 2821. pp.743-755. in Int Rpts DSDP, 29. Wash, DC, USGPO.

669. Shagam, R., et al. (1984) Tectonic Implications of Cretaceous-Pliocene Fission-Track Ages from the Rocks of the Circum-Maracaibo Basin Region of Western Venezuela and Eastern Colombia. Geol Soc Amer Memiors 162:385-412.

670. Sharpton, V.L., et al. (1988) K-T Impact(s): Continental, Oceanic or Both? pp.172-173. in Global Catastrophes in Earth History: An Interdisciplinary Confernece on Impacts, Volcanism and Mass Mortality. Houston, Lunar & Planetary Instit.

671. Shaw, H.F., Wasserburg, G.J. (1982) Age and Provenance of the Target Materials for Tektites and Possible Impactites as Inferred from Sm-Nd and Rb-Sr Systematics. Earth Planet Sci Lett 60:155-177.

672. Shaw, H.R., et al. (1971) Sierra Nevada Plutonic Cycle, Part II: Tidal Energy and a Hypothesis for Orogenic-Epeirogenic Periodicities. Geol Soc Amer Bull 82:869-896.

673. Shell Oil Co. Exploration Dept. (1975) Stratigraphic Atlas of North and Central America. NJ, Princeton Univ.

674. Shelley, D. (1975) Metamorphic Belt and Volcanic Arc Migration in New Zealand. Nature 258:668-672.

675. Shevchenko, V.A., Shilenko, V.B. (1969) Genetic Consequences of the Action of Radiation on Populations. [in Russian] in Current Problems of Radiation Genetics. Moscow.

676. Shklovskiy, I.S. (1974) Universe, Life and Reason. [in Russian] 3rd ed, Moscow.

677. Shoemaker, E.M., Wolfe, R.F. (1984) Crater Ages, Comet Showers and Putative "Death Star." Meteoritics 19:313.

678. Short, N.M. (1968) Nuclear-Explosion-Induced Micro-deformation of Rocks: An Aid to the Recognition of Meteorite Impact Structures. pp.185-210 in (ed) N.M. Short, Shock Metamophism of Natural Materials. Baltimore, Md, Mono Book.

679. Showen, R.L., et al. (1986) Ionospheric Detection of the July 1986 Bishop Earthquake. Eos 67:1122.

680. Sigurdsson, H., et al. (1991) Glass from the Cretaceous/ Tertiary Boundary in Haiti. Nature 349:482-499.

681. Silverman, S., et al. (1989) Satellite-Based Digital Data System for Low Frequency Geophysical Data. Bull Seismol Soc Amer 79:189-198..

682. Simberloff, D. (1986) Are We on the Verge of a Mass Extinction in Tropical Rain Forests? pp.165-180. in (ed) D.K. Elliott, Dynamics of Extinction. NY, Wiley.

683. Simpson, G.G. (1960) History of Life. pp.117-180 in (ed) S. Tax, Evolution of Life. Univ Chicago.

684. Sloan, R.E., et al. (1986) Gradual Dinosaur Extinction and Simultaneous Ungulate Radiation in the Hell Creek Formation. Science 232:629-633.

685. Smiley, C.J. (1976) Pre-Tertiary Phytogeography and Continental Drift—Some Apparent Discrepancies. pp.311-320 in (eds) J. Gray, A.J. Boucot, Historical Biogeography, Plate Tectonics and the Changing Environment. Proc 37th Ann Biol Colloq Oregon State.

686. Smirnov, A.M. (1968) Role of the Precambrian Basement in Strucural Evolution of the Pacific Mobil Belt (Particularly Its Northwestern Section). Pacific Geol 1:145-165.

687. Smit, J. (1984) Evidence for a Worldwide Microtektite Strewn Field at the Cretaceous-Tertiary Boundary. Geol Soc Amer Abstr Prog 16:659.

688. Smit, J. (1985) Catastrophe Events at the Terrestrial Cretaceous-Tertiary (K/T) Boundary. Geol Soc Amer Abstr Prog 17:720.

689. Smit, J., et al. (1988) Impact and Extinction Signatures in Complete Cretaceous-Tertiary (K/T) Boundary Sections. pp.182-183 in Global Catastrophes in Earth History: An Interdisciplinary Confernece on Impacts, Volcanism and Mass Mortality. Houston, Lunar & Planetary Instit.

690. Smit, J., Hertogen, J. (1980) An Extraterrestrial Event at the Cretaceous-Tertiary Boundary. Nature 285:198-200.

691. Smit, J., Klaver, G. (1981) Sanidine Spherules at the Cretaceous-Tertiary Boundary Indicate a Large Impact Event. Nature 292:47-49.

692. Smit, J., Kyte, F.T. (1984) Siderophile-Rich Magnetic Spheriods from the Cretaceous-Tertiary Boundary in Umbia, Italy. Nature 310:403-405.

693. Smit, J., Romein, A.J.T. (1985) Sequence of Events Across the Cretaceous-Tertiary Boundary. Earth Planet Sci Lett 74:155-170.

694. Smit, J., ten Kate, W.G.H. (1982) Trace-Element Patterns at the Cretaceous-Tertiary Boundary: Consequences of a Large Impact. Cretaceous Res 3:307-332.

695. Smith, A.G., Briden, J.C. (1977) Mesozoic and Cenozoic Palaeocontinental Maps. NY, Cambridge Univ.

696. Smith, B.E., Johnston, M. (1976) Tectonomagnetic Effect Observed Before a Magnitude 5.2 Earthquake Near Hollister, California. J Geophys Res 81:3556.

697. Sparrow, A.H., et al. (1965) Use of Nuclear and Chromosomal Variables in Determining and Predicting Radiosensitivities. pp.101-132 in Technical Meeting on the Use of Induced Mutations in Plant Breeding. Rome, 1964. The Use of Induced Mutations in Plant Breeding. NY, Pergamon.

698. Sparrow, A.H., Woodwell, G.M. (1962) Prediction of the Sensitivity of Plants to Chronic Gamma Irradiation. Radia Bot 2:9-26.

699. Spicer, R.A. (1989) Plants at the Cretaceous-Tertiary Boundary. Phil Trans Roy Soc London 325B:291-305.

700. Spicer, R.A., Parrish, J.D. (1986) Paleobotanical Evidence for Cool North Polar Climates in Middle Cretaceous (Albian-Cenomanian) Time. Geology 14:703-706.

701. Sporli, K.B. (1978) Mesozoic Tectonics, North Island, New Zealand. Geol Soc Amer Bull 89:415-425.

702. Srivastava, S.K. (1981) Evolution of Upper Cretaceous Phytogeoprovinces and Their Pollen Flora. Rev Palaeobot Playnol 35:155-173.

703. Stanley, S.M. (1979) Macroevolution: Pattern and Process. SF, Freeman.

704. Stearn, C.W., et al. (1979) Geological Evolution of North America. NY, Wiley.

705. Stebbins, G.L., Ayalal, F.J. (1981) Is a New Evolutionary Synthesis Necessary? Science 213:967-971.

706. Stevens, C.H. (1977) Was Development of Brackish Oceans a Factor in Permian Extinctions? Geol Soc Amer Bull 88:133-8.

707. Stevens, G.R., Clayton, R.N. (1971) Oxygen Istopic Studieson Jurassic and Cretaceous Belemnites from New Zealand and Their Biogeographic Significance. N Zealand J Geol Geophys 14:829-897.

708. Stoffler, D. (1972) Deformation and Transformation of Rock-Forming Minerals by Natural and Experimental Processes, 1: Behavior of Minerals Under High Pressure. Fortschritte der Mineralogie 49:50-113.

709. Stokes, W.L. (1982) Essentials of Earth History; An Introduction to Historical Geology. NJ, Prentice-Hall.

710. Stone, D.B., Packer, D.R. (1977) Tectonic Implications of Alaska Peninsula: Paleomagnetic Data. Tectonophysics 37:183-201.

711. Stothers, R.B. (1986) Periodicity of the Earth's Magnetic Reversals. Nature 322:444-446.

712. Stothers, R.B., et al. (1986) Basaltic Fissure Eruptions, Plume Heights and Atmospheric Aersols. Geophys Res Lett 13:725-728.

713. Stott, L.D., Kennett, J.P. (1989) Paleoceanographic and Plaeoclimatic Signature of the Cretaceous/Tertiary Boundary in the Antarctic: Stable Isotopic Results from ODP Leg 113 in Proc Ocean Drilling Prog, Intial Repts, Leg 113. Wash, DC, USGPO.

714. Strahler, A.N. (1987) Science and Earth History: The Evolution/Creation Controversy. Buffalo, NY, Prometheus.

715. Strangeway, D.W. (1970) History of the Earth's Magnetic Field. NY, McGraw Hill.

716. Stuiver, M., et al. (1981) History of the Marine Ice Sheet in West Antarctica During the Last Glaciations: A Working Hypothesis. pp.319-436 in (eds) G.H. Denton, T.J. Hughes. Last Great Ice Sheets. NY, Wiley.

717. Suarez, M. Pettigrew, T.H. (1976) An Upper Mesozoic Island-Arc-Back-Arc System in the Southern Andes and South Georgia. Geol Mag 113:305-328.

718. Sullivan, W. (1985) Geologists Add More Pieces to a Global Jigsaw Puzzle. Smithsonian 16(1):66-75.

719. Sun, Y.-Y., et al. (1984) Discovewry of Iridium Anomaly in the Permian-Triassic Boundary Clay in Changxing, Zhejiang, China, and Its Significance. pp.235-245 in Developments in Geoscience. Beijing, Science Press.

720. Swain, T. (1976) Angiosperm-Reptile Coevolution. Linn Soc Symp Series 3:107-122.

721. Sweet, A.R., Braman, D.R. (1988) Floral Changes within the Interval Containing the Cretaceous-Tertiary Boundary in Western Canada: A Stratigraphic, Palaeogeographic and Palaeoenvironmental Perspective. p.161 in Abstracts of the 7th International Palynological Congress, Brisbane.

722. Symposium on the Nature, Induction and Utilization of Mutations in Plants (1969) Induced Mutations in Plants. Proc Symp Joint IEAE/FAO Pullman, Wash, 14-18 July 1969. Vienna, Intl Atomic Energy Agency.

723. Symposium on the Effects of Neutron Irradiation Upon Cell Function (1974) Biological Effects of Neutron Irradiation; Proc Symp Effects Neutron Irradiation Upon Cell Function, Neuherberg, Munich, 22-26 Oct 1973. Vienna, Intl Atomic Energy Agency.

724. Symposium on the Late Biological Effects of Ionizing Radiation (1978) Late Biological Effects of Ionizing Radiation (Proc) Veinna, 13-17 March 1978. Vienna, Intl Atomic Energy Agency.

725. Tappan, H., Loeblich, A.R. (1971) Geobiologic Implications of Fossil Phytoplankton Evolution and Time-Space Distribution. Geol Soc Amer Spec Paper 127:247-340.

726. Tate, J., Daily, W. (1986) Evidence of Electro-seismic Phenomena. Eos 67:1086.

727. Taylor, S.R. (1973) Tektites: A Post-Apollo View. Earth Sci   Rev 9:101-123.

728. Teichert, C. (ed) (1970) Treatise on Invertebrate Paleontology. Vol.1. Boulder, Colo, Geol Soc Amer.

729. Terry, K.D., Tucker, W.H. (1968) Biologic Effects of Supernova. Science 159:421-423.

730. Thaler, L. (1965) [Dinosaur Eggs from Midi, France Yield the Secrets of Their Extinction]. Les Oeufs des Dinosaures du Midi de la France Livrent le Secret de leur Extinction. Sci Prog Nat (Feb):41-48.

731. Thierstein, H.R. (1982) Terminal Cretaceous Plankton Extinctions: A Critical Assessment. Geol Soc Amer Spec Paper 190:385-399.

732. Thierstein, H.R., Berger, W.H. (1978) Injection Events in Ocean History. Nature 276:461-6.

733. Thierstein, H.R., Okada, H. (1979) Cretaceous/Tertiary Boundary Event in the North Atlantic. Initial Rpts DSDP. Wash, DC, USGPO.

734. Thomas, D.K., Olsen, E.C. (eds) (1980) A Cold Look at the Warm-Blooded Dinosaurs. Boulder, Colo, Westview.

735. Thomas, W., et al. (1973) Structure and Origin of the Japan Sea. pp.415-434 in (ed) P.J. Coleman, Western Pacific: Island Arcs, Marginal Seas, Geochemistry. NY, Crane, Russak.

736. Tredoux, M., et al. (1987) Chemostratigraphy Across the Creatceous-Tertiary Boundary at Localities in Denmark and New Zealand: "A Case for Terrestrial Origin of the Plantinum-Group Element Anomaly." Proc Intl Workshop Cryptoexplosive Catastrophes in the Geological Record, 6-10 July 1987, Parys, South Africa.

737. Tremaine, S. (1986) Is There Evidence for a Solar Companion Star? pp.409-416. in (eds) R. Smoluchowski, et al., The Galaxy and the Solar System. Tucson, Univ Az.

738. Tschopp, H.J. (1956) Ecuador: Upper Amazon Basin Geological Province. Geol Soc Amer Memoirs 65:253-267.

739. Tschudy, R.H. (1971) Palynology of the Cretaceous-Tertiary Boundary in the Northern Rocky Mountain and Mississippi Embayment Regions. Geol Soc Amer Spec Paper 127:65-111

740. Tschudy, R.H., et al. (1984) Disruption of the Terrestrial Plant Ecosystem at the Cretaceous-Tertiary Boundary, Western Interior. Science 225:1030-1032.

741. Tschudy, R.H., Tschudy, B.D. (1986) Extinction and Survival of Plant Life Following the Cretaceous/Tertiary Boundary Event, Western Interior, North America. Geology 14:667-70

742. Tucker, W.H. (1977) Effect of a Nearby Supernova Explosion on the Cretaceous-Tertiary Environments. pp.111-124 in (eds) K-Tec Group, Creatceosu-Tertiary Extinctions and Possible Terrestrial and Extraterrestrial Causes. Proc Workshop Ottawa, Canada, 16-17 Nov 1976. Ottawa, Natl Museums.

743. Tudge, C. (1989) The Rise and Fall of *Homo sapiens sapiens*. Phil Trans Roy Soc London 325B:479-488.

744. Turco, R.P., et al. (1983) Nuclear Winter: Gobal Consequences of Multiple Nuclear Explosions. Science 222:1283-1292.

745. Turpin, L., et al. (1988) Isotopic (Sr, Nd) and Chemical Variations Across the K-T Boundary. Chem Geol 70:121.

746. United Nations, Scientific Commitee on the Effects of Atomic Radiation (1958) Report of the Scientific Committee on the Effects of Atomic Radiation. U N Official Record 13, Session Suppl 17. Document A/3838. NY, United Nations.

747. United Nations, Scientific Commitee on the Effects of Atomic Radiation (1972) Ionizing Radiation: Levels and Effects; A Report of the United Nations Scientific Committee on the Effects of Atomic Radiation to the General Assembly. NY, United Nations.

748. Unklesbay, A.G. (1987) Midwest Earthquakes: The Mississippi River Flowed Backward and Churches Overflowed When Quakes Occurred 175 Years Ago. Earth Sci 40(4):11-13.

749. Urey, H.C. (1973) Cometary Collisions and Geological Periods. Nature 242:32-33.

750. Vail, P.R., et al. (1977) Seismic Stratigraphy and Global Changes of Sea Level. AAPG Memoirs 26:51-143.

751. Vail, P.R., et al. (1977) Seismic Stratigraphy and Global Changes of Sea Level. Part Four: Global Cycles of Relative Changes of Sea Level. AAPG Memoirs 26:83-97.

752. Vail, P.R., Hardenbohl, J. (1979) Sea-Level Changes During the Tertiary. Oceanus 22:71-9.

753. Valentine, J.W., Moores, E.M. (1970) Plate-Tectonic Regulation of Faunal Diversity and Sea Level. Nature 228:657-659.

754. Van Andel, T.H. (1968) Structure and Development of Rifted Mid-Ocean Ridges. J Marine Res 26(2):144-161.

755. Van Hinte, J.E., et al. (1985) DSDP Site 603: First Deep (>1000 M) Core of the Continental Rise Along the Passive Margin of Eastern North America. Geology 13:392-397.

756. Van Hinte, J.E., et al. (1987) Sites 604 and 605. pp.277-413 in (eds) J.E. Van Hinte, et al., Initial Rpts DSDP. Wash, DC, USGPO.

757. Vannucci, R., et al. (1981) [Considerations About the Geochemistry of a Few Argillaceous and Marly-Calcareous Deposits in Comprehending the Cretaceous-Tertiary Boundary of the 'Red Cinders' of Umbro-Marchigiana]. Considerazioni Geochimiche su Alcuni Livelli Argillosi e Marnoso-Calceri in un Intervallo Comprendenta il Limite Cretaceo-Terziario Della Scaglia Rossa Umbro-Marchigiana. Rend Soc Ital Mineral Petrogr 38:413.

758. Van Valen, L.M. (1984) Catastrophes, Expectations and Evidence. Paleobiology 10:121-37.

759. Van Valen, L., Sloan, R.E. (1977) Ecology and the Extinction of the Dinosaurs. Evol Theory 2:37-64.

760. Varekamp, J.C., Thomas, E. (1982) Chalcophile Elements in Cretaceous/Tertiary Boundary Sediments: Terrestrial or Extraterrestrial? Geol Soc Amer Spec Paper 190:461-467.

761. Venkatesan, I.I., Dahl, J. (1989) Organic Geochemical Evidence for Global Fires at the Cretaceous/Tertiary Boundary. Nature 338:57-62.

762. Vogt, P.R. (1972) Evidence for Global Synchronism in Mantle Plume Convection and Possible Significance for Geology. Nature 240:338-342.

763. Vogt, P.R. (1975) Changes in Geomagnetic Reversal Frequency at Times of Tectonic Change: Evidence for Coupling Between Core and Upper Mantle. Earth Planet Sci Lett 25:313-321.

764. Vogt, P.R. (1979) Global Magmatic Episodes: New Evidence and Implications for the Steady-State Mid-Oceanic Ridge. Geology 7:93-98.

765. Vogt, P.R., et al. (1969) Discontinuities in Sea-Floor Spreading. Tectonophysics 8:285-317.

766. Wagner, W.S., Visvanathan, T.R. (1978) Earthquake Lights—A Potential Aid inn Earthquake Forecasting. Eos 59:329.

767. Wakita, H., et al. (1978) "Helium Spots": Caused by a Diapiric Magma from the Upper Mantle. Science 200:430-432.

768. Wakita, H., et al. (1989) Coseismic Radon Changes in Groundwater. Geophys Res Lett 16:417-420

769. Walker, D.A. (1987) Apparent Correlations Between Ninos and Tectonic Activity Along the East Pacific Rise. Eos 68:1336.

770. Ward, P.D., Signor, P.W. III (1983) Evolutionary Tempo in Jurassic and Cretaceous Ammonites. Paleobiology 9:183-198.

771. Warren, C.S. (1988) Theories of the Earth and Universe: A History of Dogma in the Earth Sciences. Ca, Stanford Univ.

772. Warwick, J., et al. (1982) Radio Emission Associated with Rock Fracture: Possible Application to the Great Chilean Earthquake of May 22, 1960. J Geophys Res 87B:2851.

773. Watkins, N.D. (1969) Crustal Spreading: A Critical Comparision of Hypothesis Requirements and Present Knowledge of Relevent Magnetic Properties and Polarity History (Abst). Tectonophysics 7:543.

774. Watts, A.B. (1978) An Analysis of Isostacy in the World's Oceans, I. Hawaiian-Emperor Seamount Chain. J Geophys Res 83:5989-6004.

775. Webb, P.N. (1973) Upper Cretaceous-Paleocene Foraminifera from Site 208 (Lord Howe Rise. Tasman Sea) DSDP, Leg 21. DSDP Intial Reports 21:541-573.

776. Webb, S.D. (1977) History of Savana Vetebrates in the New World. Part I. North America. Ann Rev Ecol Syst 8:355-380.

777. Wegener, A. (1924) Origin of the Continents and Oceans. Trans from German by J. Biram. NY, Dover, c1966.

778. Weisburd, S. (1986) Plunging Plates Cause a Stir. Sci News 130:106-109.

779. West, S. (1982) Patchwork Earth. Sciences 82(6):46-52.

780. Wesson, P.S. (1978) Cosmology and Geophysics. NY, Oxford.

781. Wezel, F.C., et al. (1981) [Discovery of a Widespread Iridium-Rich Layer of the "White Cinders" in the Ombrie-Marches of the Apennines (Italy)] Decouverte de Divers Niveaux Riches en Iridium dans le "Scaglia Bianc" de L'Apennine d'Ombrie-Marches (Italy). C Roy Acad Sci Paris 293:837.

782. Whalley, P. (1987) Insects and Cretaceous Mass Extinction. Nature 327:562.

783. Whitmire, D.P., Jackson, A.A. (1984) Are Periodic Mass Extinctions Driven by a Distant Solar Companion? Nature 308:713-715.

784. Whitmire, D.P., Matese, J.J. (1985) Periodic Comet Showers and Planet X. Nature 313:36-38, 744.

785. Whitmore, T.C. (1981) Introduction. pp.1-2 in (ed) T.C. Whitmore, Wallace's Line and Plate Tectonics. Oxford, Clarendon.

786. Whitmore, T.C. (1981) Wallace's Line and Some Other Plants. pp.70-80 in (ed) T.C. Whitmore, Wallace's Line and Plate Tectonics. Oxford, Clarendon.

787. Wiedmann, J. (1986) Macro-Invertebrates and the Cretaceous-Tertiary Boundary. pp.397-410 in (eds) O.H. Walliser, et al., Global Bio-Events: A Critical Approach. Lecture Notes in Earth Sciences, No 8. Berlin, Springer-Verlag.

788. Wilde, P. et al. (1986) Iridium Abundances Across the Ordovician-Silurian Stratotype. Science 233:339-341.

789. Winn, R.D., Jr (1978) Upper Mesozoic Flysch of Tierra del Fuego and South Georgia Island, A Sedimentologic Approach to Lithosphere Plate Restoration. Geol Soc Amer Bull 89:533-47.

790. Winterer, E.L. (1976) Anomalies in the Tectonic Evolution of the Pacific. pp.267-278 in (eds) G.H. Sutton, et al., Geophysics of the Pacific Ocean Basin and Its Margin. Monograph 19. Wash, Amer Geophys Union.

791. Wolbach, W.S., et al. (1985) Cretaceous Extinctions: Evidence for Wildfires and Search for Meteoritic Material. Science 230:167-170.

792. Wolbach, W.S., et al. (1988) Global Fire at the Cretaceous-Tertiary Boundary. Nature 334:665.

793. Wolfe, J.A. (1978) Paleobotanical Interpretation of Tertiary Climates in the Northern Hemisphere. Amer Sci 66:694-703.

794. Wolfe, J.A. (1985) Distribution of Major Vegetational types During the Tertiary. Mono Amer Geophys U Geophys 32:357-75.

795. Wolfe, J.A., Upchurch, G.R. (1986) Vegetation, Climatic and Floral Changes at the Cretaceous-Tertiary Bounadayr. Nature 324:148-152.

796. Wolfe, J.A., Upchurch, G.R. (1987) Leaf Assemblages Across the Cretaceous-Tertiary Boundary in the Raton Basin, New Mexico and Colorado. Proc Natl Acad Sci (USA) 84:5096-5100.

797. Woodwell, G.M. (1963) Ecological Effects of Radiation. Sci Amer 208(6):40-49.

798. Woodwell, G.M. (1967) Radiation and the Paterns of Nature. Science 156:461-470.

799. Woodwell, G.M., et al. (1978) Biota and the World Carbon Budget. Science 199:141-146.

800. Worsley, T. (1974) Cretacceous-Terrtiary Boundary Event in the Ocean. pp.94-125 in (ed) W.W. Hay, Studies in Paleoceanography. Spec Pub. Vol.20. Tulsa Oklahoma, SEPM.

801. Xu Dao-Yi, et al. (1985) Abundance Variation of Iridium and Trace Elements at the Permian-Triassic Boundary at Shangsi in China. Nature 314:154-156.

802. Yamada, I., et al. (1989) Electromagnetic and Acoustic Emission Associated with Rock Fracture. Phys Earth Planet Int 57:157-168.

803. Yen, T.F., Chilingarian, G.V. (1976) Oil Shale. NY, Elsevier.

804. Zachos, J.C., Arthur, M.A. (1986) Paleoceanography of the Cretaceous/Tertiary Boundary Event: Inferences from Stable Isotopic and Other Data. Paleoocenography 1:5-26.

805. Zachos, J.C., et al. (1989) Geochemical Evidence for Suppression of Marine Productivity at the Cretaceous/Tertiary Boundary. Nature 337:61-64.

806. Zahnle, K., Grinspoon, D. (1990) Comet Dust as a Source of Amino Acids at the Cretaceous/Tertiary Boundary. Nature 348:157-160.

807. Zeil, W. (1979) The Andes: A Geological Review. Berlin, Gebruder Borntrager.

808. Zhao, M., Bada, J.L. (1989) Extraterrestrial Amino Acids in the Cretaceous/Tertiary Boundary Sediments at Stevns Klint, Denmark. Nature 339:411-421.

809. Zimmerman, R.L. (1981) Relative Effects of Alpha and Beta Radiation. Radiation Effects 14:81-92.

810. Zinsmeister, W.J., Feldmann, R.M. (1984) Cenozoic High Latitude Heterochroneity of the Southern Marine Fauna. Science 224:281-283.

811. Zoback, M.D. et al. (1987) New Evidence on the State of Stress of the San Andreas Fault System. Science 238:1105-1111.

812. Zoback, M.L., Zoback, M. (1980) Stress in the Conterminous United States. J Geophys Res 85:6113-6156.

813. Zlotnicki, J., Le Mouel, J.L. (1990) Possible Electrokinetic Origin of Large Magnetic Variations at La Fournaise Volcano. Nature 343:633-635.

814. Zonenshain, L.P. (1973) Evolution of Central Asiatic Geosynclines Through Sea-Floor Spreading. Tectonophysics 19:213-232.